MW00990083

Grid Integration of Wind Energy Conversion Systems

Second Edition

Grid Integration of Wind Energy Conversion Systems

Second Edition

Siegfried Heier

Kassel University, Germany

Translated by

Rachel Waddington

Member of the Institute of Translation and Interpreting, UK

Originally published in the German language by B.G. Teubner Verlag as "Siegfried Heier: Windkraftanlagen. 3. Auflage".

Telephone (+44) 1243 779777

Email (for orders and customer service enquiries): cs-books@wiley.co.uk
Visit our Home Page on www.wiley.com

Reprinted December 2006, November 2007, January 2009

Other Wiley Editorial Offices

John Wiley & Sons Inc., 111 River Street, Hoboken, NJ 07030, USA

Jossey-Bass, 989 Market Street, San Francisco, CA 94103-1741, USA

Wiley-VCH Verlag GmbH, Boschstr. 12, D-69469 Weinheim, Germany

John Wiley & Sons Australia Ltd, 42 McDougall Street, Milton, Queensland 4064, Australia

John Wiley & Sons (Asia) Pte Ltd, 2 Clementi Loop #02-01, Jin Xing Distripark, Singapore 129809

John Wiley & Sons Canada Ltd, 22 Worcester Road, Etobicoke, Ontario, Canada M9W 1L1

Wiley also publishes its books in a variety of electronic formats. Some content that appears in print may not be available in electronic books.

Library of Congress Cataloging in Publication Data

Heier, Siegfried.
[Windkraftanlagen im Netzbetrieb, z. uberarbeitete und erweiterte Auflage. English]
Grid integration of wind energy conversion systems / Siegfried Heier;
translated by Rachel Waddington. — 2nd ed.
p. cm.
Includes bibliographical references and index.
ISBN-13: 978-0-470-86899-7 (cloth : alk. paper)
ISBN-10: 0-470-86899-6 (cloth : alk. paper)
1. Wind power plants. 2. Wind energy conversion systems. 3. Electric power systems. I. Title.
TK1541.H3513 2006
621.31′2136—dc22

2005023374

British Library Cataloguing in Publication Data

A catalogue record for this book is available from the British Library

ISBN-13 978-0-470-86899-7 (H/B)

Typeset in 10/12pt Times by Integra Software Services Pvt. Ltd, Pondicherry, India
Printed and bound in Great Britain by CPI Antony Rowe, Chippenham, Wiltshire

Contents

Preface

In the long run, an ecologically sustainable energy supply can only be guaranteed by the integration of renewable resources. Besides water power, which is already well established, wind energy is by far the most technically advanced of all renewable power sources, and its economic breakthrough the closest. With a few exceptions, wind power will be used mostly for generating electricity.

Just two decades on from the 50 kW class machines of the mid-1980s, the development of wind turbines has led to production converters with outputs in the 500 to 2500 kW range. Three to five megawatt turbines are currently being launched on to the market. In the development of these machines, successful techniques and innovations originating from small- and medium-sized turbines were carried over to larger ones, which has led to a considerable improvement in the reliability of wind turbines. The technical availability currently achieves average values of approximately 98%. Furthermore, economic viability has been increased enormously. As a result, wind energy has experienced an almost unbelievable upsurge and has exceeded the contributions of water power.

The rapid development of wind power has awakened strong public, political and scientific interest, and has triggered widespread discussion over the past few years, much of it concerning the degree to which nature, the environment and the electricity grid can withstand the impact of wind power.

If political requirements regarding the reduction of environmental pollution are to be met, long-term growth in the use of wind power must be the target. Since obtaining electricity from the wind currently offers the most favourable technical and economic prospects of all the sources of renewable energy, it must be assigned the highest priority. Due to the fact that turbine sizes are still increasing, a high degree of grid penetration must be expected (regionally at any rate), meaning that the connection of wind turbines could come up against its technical limits. This is already the case today in some instances.

The objective of a forward-looking energy supply policy must therefore be to utilize the existing grid as well as possible for the supply of wind power. This is made possible by the use of turbines with good grid compatibility in connection with measures for grid reinforcement. In assessing grid effects, control operations and the electrical engineering design of wind turbines play a significant role. The themes developed in the present work are therefore particularly concerned with this topic.

This edition of the book has been updated to cover important innovations in this rapidly changing technology in terms of energy converters, development and grid integration. The layout of the book has also been updated to achieve a consistent format and a number of new illustrations have been included. A great deal of new material has also been added to cover changes in legislation.

This book grew from my many years of work in research and development, especially as head of the wind energy research projects at the Universität Gesamthochschule Kassel, in the Electrical Energy Supply Department of the Institut für Elektrische Energietechnik. Close cooperation with the Institut für Solare Energieversorgungstechnik (ISET) e.V. brought with it a considerable broadening of the horizon of experience. My special thanks go to the founder, Professor Dr Werner Kleinkauf. His suggestions, and our technical discussions together, have had a considerable influence on the current work.

The help and support of Herrn Dr-Ing. Gunter Arnold, Herrn Dr Boris Valov, Herrn Dipl.-Ing. Michael Durstewitz, Herrn Dipl.-Ing. Martin Hoppe-Kilpper, Herrn Dipl.-Ing. Berthold Hahn, Herrn Dipl.-Ing. Martin Kraft, Herrn Dipl.-Ing. Volker König, Herrn Dipl.-Ing. Werner Döring, Herrn Dipl.-Ing. Bernd Gruß, Herrn Dr-Ing. Oliver Haas, M.Sc. Rajesh Saiju, Herrn Thomas Dörrbecker, Herrn Bernhard Siano, Herrn Martin Nagelmüller, Frau Dipl.-Des. Renate Rothkegel, Frau Melanie Schmieder, Frau Anja Clark-Carina and Frau Judith Keuch have contributed greatly to the success of this book.

My grateful thanks also go to ENERCON GmbH for kindly granting permission to use the image of the wind turbine in the design of the front cover.

This book is not only intended for students in technical faculties. Many procedural notes and experimental results will also be of great help to engineers in both theory and in practice.

For their readiness to publish this book and for the painstaking preparation involved, my special thanks must go to the publisher, John Wiley & Sons, Ltd. I must also thank my wife Hannelore for her patience and understanding, without which this task could not have been brought to fruition.

Kassel, October 2004 *Siegfried Heier*

Notation

a	Constant factor related to the pivot of the profile
a_a	Distance between point of application of lifting force and blade axis of rotation
a_p	Distance along blade axis between the points of application of torque and gravity
a_s	Blade deflection and slewing
A_1	Far-upstream cross-section of flow
A_2	Cross-section of flow at turbine
A_3	Broadening downstream cross-section of flow
A_{lt}	Long-term flicker factor
A_R	Rotor swept area
A_{st}	Short-term flicker factor
b	Acceleration of the rotor blade centre of gravity
b'	Acceleration of the rotor blade centre of gravity in the rotating coordinate system
b_c	Coriolis acceleration of rotor blade centre of gravity
b_{defl}	Blade bending in direction of deflection
b_o	Centripetal acceleration in the rotor head
b_R	Centripetal acceleration arising from ω_R
b_s	Bending acceleration of the rotor blade in the direction of deflection and slew
b_{slew}	Blade bending in direction of slew
$b\omega_A$	Centripetal acceleration arising from ω_A
c	Magnification factor for the initial short-circuit alternating current power or the maximum possible short-circuit current
c_a	Lift coefficient of blade profile
c_k	Capacitor bank capacitance
c_m	Torque coefficient of the turbine
c_p	Performance coefficient of the turbine
c_t	Torsional moment coefficient of blade profile (t_B/4-related)
c_w	Drag coefficient of blade profile
$\cos\varphi$	Power factor

Grid Integration of Wind Energy Conversion Systems, Second Edition S. Heier

$\cos\varphi_K$	Power factor in case of short-circuit
C	$= C(k)$, Theodorsen function
d	Half-profile depth
d_m	Average bearing diameter
dA_B	Blade element area
dF_A	Lift force on blade element
dF_{AW}	Resultant force on blade segment from lift and drag components
dF_{ax}	Axial force at blade element
dF_t	Tangential force at blade element
dF_W	Drag force on blade element
$d\dot{J}_{ax}$	Axial momentum losses by blade streaming
$d\dot{J}_t$	Change of tangential momentum of angular streaming
dM_L	Moment per unit of width during blade pitch adjustment due to acceleration of air mass and due to air damping
dM_{lift}	Torsional moment at blade element due to lifting forces
dM_T	Righting moment in direction of air flow on the blade element
dU	Voltage deviation, voltage drop
f	Frequency
f_1	Grid frequency
$f_{2\nu}$	Rotor current frequency (the νth harmonic) in asynchronous machines
f_G	Generator frequency
f_{L1}	Bearing coefficient dependent upon bearing type and loading
f_μ	Frequency of the μth subharmonic
f_v	Frequency of the vth harmonic
F_a	Axial force on bearing
F_N	Normal force
F_{Pr}	Force creating propeller moment
F_Q	Transverse force component
F_{St}	Actuating force on blade
F_Z	Centrifugal force
g_{L1}	Bearing load direction factor
i_{ABl}	Transmission ratio between adjustment mechanism and blade pitch adjustment
i_G	Total current (rotating pointer)
i_{G1}	Total current in phase 1
i_{G2}	Total current in phase 2
i_{G3}	Total current in phase 3
i_{Gd}	Total current in longitudinal direction of field coordinates
i_{Gq}	Total current in transverse direction of field coordinates
i_{MBl}	Transmission ratio between adjustment motor and blade rotation
$i_{MBl,rot}$	Transmission ratio between servomotor and rotor blade adjustment
$i_{MBl,lin-rot}$	Transmission ratio between servomotor and blade pitch adjustment in the case of direct motor drive
i_{ms}	Magnetizing current in the stator
i_R	Machine-side rotor current (rotating pointer)

i_{R1}	Machine-side rotor current in phase 1
i_{R2}	Machine-side rotor current in phase 2
i_{R3}	Machine-side rotor current in phase 3
i_{Rd}	Machine-side rotor current in longitudinal direction of field coordinates
$i_{Rd\ act}$	Actual value of i_{Rd}
$i_{Rd\ des}$	Desired value of i_{Rd}
i_{RN}	Grid-side rotor current (rotating pointer)
i_{Rq}	Machine-side rotor current in phase 1
$i_{Rq\ act}$	Actual value of i_{Rq}
$i_{Rq\ des}$	Desired value of i_{Rq}
I_0	No-load current in one machine phase
I_1	Stator current
I_1	Effective value of fundamental component current
I_2'	Rotor phase current acting on stator side
I_{an}	Starting current of asynchronous machines
I_E	Exciter current
I_{Fe}	Iron loss current in one machine phase
I_{St}	Electric current or hydraulic flow for blade pitch positioning
I_Z	Current of reactive power compensation system
I_μ	Magnetizing current in one machine phase
I_ν	Effective value of the νth harmonic current
J_B	Moment of inertia of blade during rotation around the hub
J_{Bl}	Moment of inertia of rotor blade when turned about its longitudinal axis
$J_{Bl(A)}$	Moment of inertia of rotor blade taken at the drive motor side
J_G	Moment of inertia of generator rotor
J_{LB}	Equivalent moment of inertia due to accelerated air mass
J_{Mot}	Moment of inertia of the drive motor
$J_{Mot(Bl)}$	Moment of inertia of the drive motor acting on the rotor blades
J_R	Moment of inertia of all rotating masses
J_{tot}	Total moment of inertia of blade pitch adjustment system
$J_{tot(A)}$	Total moment of inertia of the entire blade pitch adjustment system taken from the drive side
$J_{tot(Bl)}$	Total moment of inertia of the entire blade pitch adjustment system taken from the blade side
J_{trans}	Moment of inertia of transmission elements such as gears, couplings, etc., between drive motor and blade turning mechanism
k_A	Rate of change factor of the rotor displacement angle after falling out of step
k_d	Characteristic damping
k_{DB}	Coefficient of structural and aerodynamic damping
k_{DK}	Coefficient of damping for the drive train
k_{RL}	Coefficient of friction for bearing friction at rotor blade during blade pitch adjustment
k_t	Ratio of the acceleration moments of the drive-train component to the entire rotor system (M_{BT}/M_{BR})
k_{THD}	Total harmonic distortion

k_{THD0}	Grid-state-dependent and grid short-circuit power-dependent output value of total harmonic distortion
k_{THD1}	Gradient of relative harmonic content
k_{THD2}	Elongation factor of relative harmonic content
k_{TS}	Torsional stiffness of the drive train
k_u	Harmonic distortion of voltage
k_U	Factor for the maximum magnification of generator moment
m	Number of phases of three-phase current windings
m_B	Mass of a rotor blade
m_{dyn}	$= M_{KD}/M_{KS\,max}$, dynamic increase in moment
M	Moment
M	Torque
M_A	Driving torque
M_{act}	Actual value of moment
M_{AG}	Driving torque at generator
M_{AM}	Motor start-up torque
M_{An}	External torque of blade pitch adjustment drive taking into account spring and damping characteristics
M_{AT}	Driving torque of drive train including losses
M_{AV}	Internal torque of blade pitch adjustment drive
M_{AW}	Wind turbine driving moment
M_{bend}	Torsional moment at rotor blade due to bending
M_{BG}	Acceleration moment at the generator
M_{Bl}	Rotor blade torsional moment during turning about blade longitudinal axis
$M_{Bl\,max}$	Maximum blade torsional moment in extreme situations
M_{Bln}	Blade torsional moment in normal operation
M_{BR}	Acceleration moment in rotor system
M_{BT}	Acceleration moment in drive train
M_{BW}	Acceleration moment on wind turbine
M_{Cz}	Coriolis moment in relation to the z axis
M_D	Damping moment of the synchronous machine
M_{des}	Desired value of torque
M_{frict}	Moment of friction of all blade bearings during blade pitch adjustment
M_K	Breakdown torque of asynchronous machine
M_K	Pull-out torque of synchronous machine
M_{KD}	Dynamic breakdown or pull-out torque
M_{KG}	Generator breakdown or pull-out torque
M_{KM}	Motor breakdown or pull-out torque
$M_{K\,max}$	Maximum breakdown or pull-out torque
$M''_{K\,max}$	Maximum moment of synchronous machines due to subtransient short-circuit currents in the damping winding
$M'_{K\,max}$	Maximum value of pulsating short-circuit moment of synchronous machines due to transient currents
M_{KS}	Static breakdown or pull-out torque
$M_{KS\,max}$	Static breakdown or pull-out torque at maximum excitation
$M_{KS\,min}$	Static breakdown or pull-out torque with no-load excitation

M_{Ku}	Coupling torque at generator
M_L	Moment with blade pitch adjustment by acceleration of air masses and air damping
M_{LB}	Moment with blade pitch adjustment due to acceleration of air masses
M_{LD}	Moment with blade pitch adjustment due to air damping
M_{lift}	Torsional moment at rotor blade due to lift forces
M_{max}	Maximum moment
M_N	Nominal moment
M_{NG}	Generator nominal moment
M_{NM}	Motor nominal moment
M_{Pr}	Propeller moment
M_{res}	Reserve moment during acceleration of the blade pitch adjustment mechanism
M_{RL}	Load-dependent moment of friction of a bearing
M_{RLk}	Load-dependent moment of friction of bearing k
M_s	Steady-state torque
M_S	Pull-up torque
M_{SM}	Pull-up torque of a motor (asynchronous machine)
M_{St}	Moment exerted upon blade by actuator
M_T	Righting moment in direction of air flow on blade profile
M_{TD}	Damping component of drive train moment
M_{teeter}	Torsional moment at the blade due to teetering of the rotor
M_{TT}	Torsionally elastic component of drive-train moment
M_W	Load torque
M_{WG}	(Electrical) load torque of the generator
n	Rotational speed
n_1	Speed of rotating field or synchronous speed
n_A	Number of turbines
n_{act}	Actual value of rotational speed
n_{AV}	Rotational speed of blade pitch adjustment drive
n_{des}	Desired value for speed
n_{KG}	Generator breakdown-torque speed (asynchronous machine)
n_{KM}	Motor breakdown-torque speed (asynchronous machine)
n_{NG}	Generator nominal speed (asynchronous machine)
n_{NM}	Motor nominal speed (asynchronous machine)
n_ν	Speed of the harmonic field of ordinal number ν
p_1	Number of pairs of poles in the stator
p_2	Number of pairs of poles in the rotor
P	Average value of power
P_E	Power of producer in the grid
P_{el}	Electrical generator power
P_G	Total active power in rotor and stator
P_L	Power of the load in the grid
P_{L0}	Equivalent static bearing loading
P_{mech}	Mechanical input power of generator
P_N	Nominal power
P_O	Moving air mass power

P_{Stn}	Power for normal positioning procedures
P_{Sts}	Power for fast positioning procedures
$P_{T\ act}$	Actual value of total active power
$P_{T\ des}$	Desired value of total power
P_{W}	Wind turbine power
$P_{W\max}$	Maximum wind turbine power
P_{δ}	Air gap power of an electrical machine
P_{σ}	Standard deviation of power
Q_{C}	Compensation reactive power
Q_{G}	Total reactive power in rotor and stator
$Q_{T\ act}$	Actual value of total reactive power
$Q_{T\ des}$	Desired value of total reactive power
r	Radius of a blade element
r'	Radius of the rotor blade centre of mass
r_{o}	Distance between yaw and rotor blade fulcrums
R_{1}	Stator resistance of one machine phase
R'_{2}	Rotor resistance of one phase of an asynchronous machine transformed on the stator side
R_{a}	Outer radius of rotor blade
R_{grid}	Resistance of the connection elements between higher grid and point of common coupling
R_{i}	Inner radius of rotor blade
R_{L+T}	Ohmic resistance of lines and transformers
R_{L+T}	Resistance between wind turbine and point of common coupling
s	Slip (of an asynchronous machine)
s_{K}	Breakdown slip (asynchronous machine)
s_{N}	Nominal slip (asynchronous machine)
s_{ν}	Slip of the νth harmonic (asynchronous machine)
S_{grid}	Grid apparent power
S_{k}	Grid short-circuit power
S''_{k}	Initial value of alternating current short-circuit power
S_{load}	Load apparent power
S_{p}	Centre of gravity
S_{rG}	Generator rated apparent power
S_{rT}	Transformer rated apparent power
S_{supply}	Supply apparent power
t	Time
t_{0}	Time for rotor blade adjustment into a safe operating state
t_{0b}	Time for rotor blade adjustment into a safe operating state with pure acceleration processes
t_{Af}	Acceleration time of blade positioning drive in the case of fast positioning procedures
t_{APD}	Acceleration time of blade positioning drive system
t_{APDd}	Acceleration time of direct-driven blades
t_{APDz}	Acceleration time of z rotor blades adjusted by positioning drive system

t_B	Blade thickness
t_f	Duration of secondary effect of flicker
t_v	Delay time
t_v	Rotor blade adjustment time at constant speed
T_D	Time constant of damping of torque oscillation
T_E	Time constant of the exciter circuit
T_G	Generator acceleration time constant
T_n	$= 1/\omega_0 = 1/A_o$, time constant of the rotation speed integrator
T_R	Rotor system acceleration time constant
T_V	Time constant for the decaying dynamic pull-out torque to its steady-state value
T_W	Wind turbine acceleration time constant
T_ε	$= p/\omega_0$, time constant of integrator for the determination of the angle of torsion (generator side)
u_G	Total voltage (rotating pointer) corresponds to stator voltage ($u_T = u_S$)
u_{Gq}	Total voltage in quadrature-axis direction of the field coordinates ($u_{Tq} = u_{Sq}$)
u_{kASM}	Magnification factor of the short-circuit power of asynchronous machines
u_{kSM}	Magnification factor of the short-circuit power of synchronous machines
u_N	Nominal voltage
u_{R1}	Machine-side rotor voltage in phase 1
u_{R2}	Machine-side rotor voltage in phase 2
u_{R3}	Machine-side rotor voltage in phase 3
u_{Rd}	Machine-side rotor voltage in direct-axis direction of the field coordinates
u_{Rq}	Machine-side rotor voltage in quadrature direction of the field coordinates
u_{S1}	Stator voltage in phase 1
u_{S2}	Stator voltage in phase 2
u_{S3}	Stator voltage in phase 3
u_{Sq}	Stator voltage in quadrature direction of the field coordinates
$u_{\mu VT}$	Compatibility level of the μth subharmonic specific to the fundamental component
u_ν	Harmonic voltage of the νth-order specific to the fundamental component
$u_{\nu VT}$	Compatibility level of the νth harmonic specific to the fundamental component
U_0	Direct voltage component
U_1	Effective value of fundamental component voltage, grid voltage
U_1	Grid voltage
U_C	Reference conductor voltage of capacitor banks
U_{di}	Ideal direct voltage
U_g	Direct current link voltage
U_{Gen}	Generator voltage
U_{Grid}	Grid voltage
U_i	Induced machine voltage
U_{Mot}	Motor voltage
U_p	Rotary-field (internal) voltage of a synchronous machine
$U_{R\,max}$	Maximum rotor voltage
U_ν	Effective value of the νth harmonic voltage
U_Z	Ignition impulse voltage

v	Wind speed
$\bar{v}$	Average wind speed
v_0	Rotational speed of the rotor head
v_1	Undisturbed far-upstream wind speed
v_2	Wind speed at the turbine
v_{2ax}	Axial component of the decelerated wind speed at the rotor blade
v_{2t}	Tangential component of the decelerated wind speed at the rotor blade
v_3	Decelerated wind speed far downstream of the turbine
v_r	Resultant wind speed
v_u	Peripheral speed
v_w	Local wind speed
v_σ	Standard deviation of wind speed
V_a	Airstream volume element
W_W	Energy drawn by wind turbine
X''_d	Subtransient reactance of a synchronous machine
$X_{1\sigma}$	Leakage reactance of one stator phase
$X'_{2\sigma}$	Leakage reactance of one rotor phase specific to the stator
X_d	Synchronous direct-axis reactance of a synchronous machine
X'_d	Transient reactance of a synchronous machine
X_G	Leakage reactance from stator and rotor of an asynchronous machine
X_{grid}	Reactance of the connection elements between grid and point of common coupling
X_h	Main reactance of one machine phase
X_{L+T}	Reactance between wind turbine and point of common coupling
X_{L+T}	Reactance of lines and transformers
X_q	Synchronous quadrature reactance of a synchronous machine
X_u	Reactance of the frequency converter valves
z	Number of rotor blades
z_a	Number of driven rotor blades
Z_C	Impedance of the capacitor bank
Z_k	Grid impedance (in short-circuit)
Z_{load}	Load impedance
α	Local profile flow angle at the rotor blade
α	Trigger or firing angle of thyristors
α_0	Ignition angle output value
α_{max}	Maximum thyristor trigger angle/thyristor inverter stability limit
$\alpha_{max\,15}$	Maximum thyristor trigger angle/thyristor inverter stability limit at 15% voltage drop
β	Rotor blade pitch
$\dot{\beta}$	Adjustment speed of the rotor blade
$\dot{\beta}_n$	Normal adjustment speed of the rotor blade
$\dot{\beta}_s$	Fast adjustment speed of the rotor blade
$\ddot{\beta}$	Adjustment acceleration of the rotor blade
γ	Cone angle
ν	Ordinal number of one harmonic (integer)

δ	Angle between plane of rotation and resultant air flow velocity
Δ_n	Adjustment range of rotation speed
ΔP	Power fluctuation range
ΔU	Voltage drop
ΔU_r	Voltage rise
$\Delta U_{r\ perm}$	Permissible voltage rise
Δv	Fluctuation range of wind velocity
$\Delta \varepsilon$	Angle of torsion
ε_G	Generator rotor angle of rotation
$\dot{\varepsilon}_G$	Angular velocity of the generator rotor
ε_W	Wind turbine angle of rotation
$\dot{\varepsilon}_W$	Angular velocity of the turbine
η_{ABl}	Transfer efficiency between positioning system and rotor blade
θ	Rotor displacement angle (electrical) of a synchronous machine
θ	Angle (mechanical) between plane of rotation and chord
$\theta_{0.7}$	Angle between plane of rotation and chord at the rotor blade given at 0.7 times the radius
θ_B	Angle between $\theta_{0.7}$ and the chord of the rotor blade
λ	Load angle (electrical) in asynchronous machines between fixed grid slip voltage and load-dependent slip voltage
λ	Line angle
λ	Tip speed ratio (mechanical) of the rotor blade tip speed to wind velocity
λ_N	Tip speed ratio in nominal operating state
μ	Noninteger factor of subharmonics
ν	Ordinal number of harmonics
ρ	Air density
φ_{Gen}	Phase angle of generator current (input angle)
φ_{Mot}	Phase angle of motor current
φ_ν	Phase displacement angle of the νth harmonic
ψ	Position of rotor in relation to tower
ψ_{Bl}	Rotor blade position in relation to tower
ω	Resultant angular velocity from azimuth yaw and turbine rotation
ω_0	Steady-state grid (angular) frequency
ω_1	Angular velocity of the stator rotating field in two-pole winding design ($p = 1$)
ω_2	Angular velocity of the rotor rotating field in two-pole winding design ($p = 1$)
ω_A	Yaw control angular velocity
ω_{Bl}	Angular velocity of blade rotation about its longitudinal axis
ω_{BV}	Design value for angular velocity of blade pitch adjustment system
ω_G	Angular velocity of the generator rotor
ω_{mech}	Angular velocity of the mechanical rotation of the generator rotor
ω_{Mot}	Angular velocity of the servomotor
ω_N	Nominal angular velocity
ω_R	Angular velocity of the turbine rotor (vector quantity)
ω_{st}	Angular velocity of actuator
ω_W	Angular velocity of wind turbine
ω_ν	Angular frequency of the νth harmonic

1

Wind Energy Power Plants

Rising pollution levels and worrying changes in climate, arising in great part from energy-producing processes, demand the reduction of ever-increasing environmentally damaging emissions. The generation of electricity – particularly by the use of renewable resources – offers considerable scope for the reduction of such emissions. In this context, the immense potentials of solar and wind energy, in addition to the worldwide use of hydropower, are of great importance. Their potential is, however, subject to transient processes of nature. Following intensive development work and introductory steps, the conversion systems needed to exploit these power sources are still in the primary phase of large-scale technical application. For example, in Germany around 5 % of electricity is already being provided by wind turbines, while in the German province of Schleswig–Holstein and in Denmark this figure is more than 30 %. However, in Germany, more power is supplied by wind energy than by hydroelectric plants.

These environmentally friendly technologies in particular require a suitable development period to establish themselves in a marketplace of high technical standards.

The worldwide potential of wind power means that its contribution to electricity production can be of significant proportions. In many countries, the technical potential and – once established – the economically usable potential of wind power far exceeds electricity consumption. Good prospects and economically attractive expectations for the use of wind power are, however, inextricably linked to the incorporation of this weather-dependent power source into existing power supply structures, or the modification of such structures to take account of changed supply conditions.

1.1 Wind Turbine Structures

In the case of hydro, gas or steam, and diesel power stations (among others) the delivery of energy can be regulated and adjusted to match demand by end users (Figure 1.1(a)). In contrast, the conversion system of a wind turbine is subject to external forces (Figure 1.1(b)). The delivery of energy can be affected by changes in wind speed, by machine-dependent

Grid Integration of Wind Energy Conversion Systems, Second Edition S. Heier

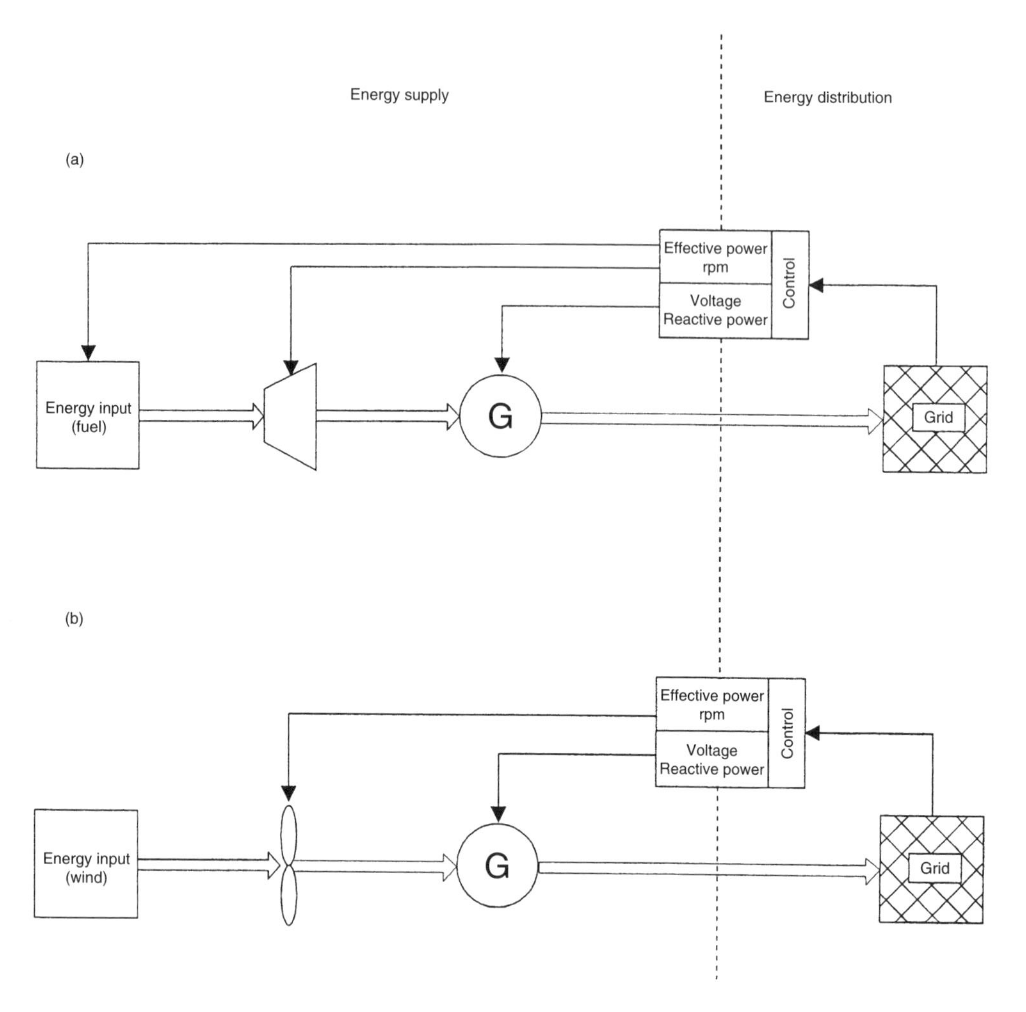

Figure 1.1 Energy delivery and control in electrical supply systems: (a) diesel generators, etc., and (b) wind turbines

factors such as disruption of the airstream around the tower or by load variations on the consumer side in weak grids.

The principal components of a modern wind turbine are the tower, the rotor, the nacelle (which accommodates the transmission mechanisms and the generator) and – for horizontal-axis devices – the yaw systems for steering in response to changes in wind direction. Switchgear and protection systems, lines, and maybe also transformers and grids, are required for supplying end users or power storage systems. In response to external influences, a unit for operational control and regulation must adapt the flow of energy in the system to the demands placed upon it. The next two figures show the arrangement of the components in the nacelle and the differences between mechanical–electrical converters in the modern form of wind turbines. Figure 1.2 shows the conventional drive train design in the form of a geared transmission with a high-speed generator. Figure 1.3, by contrast, shows the gearless

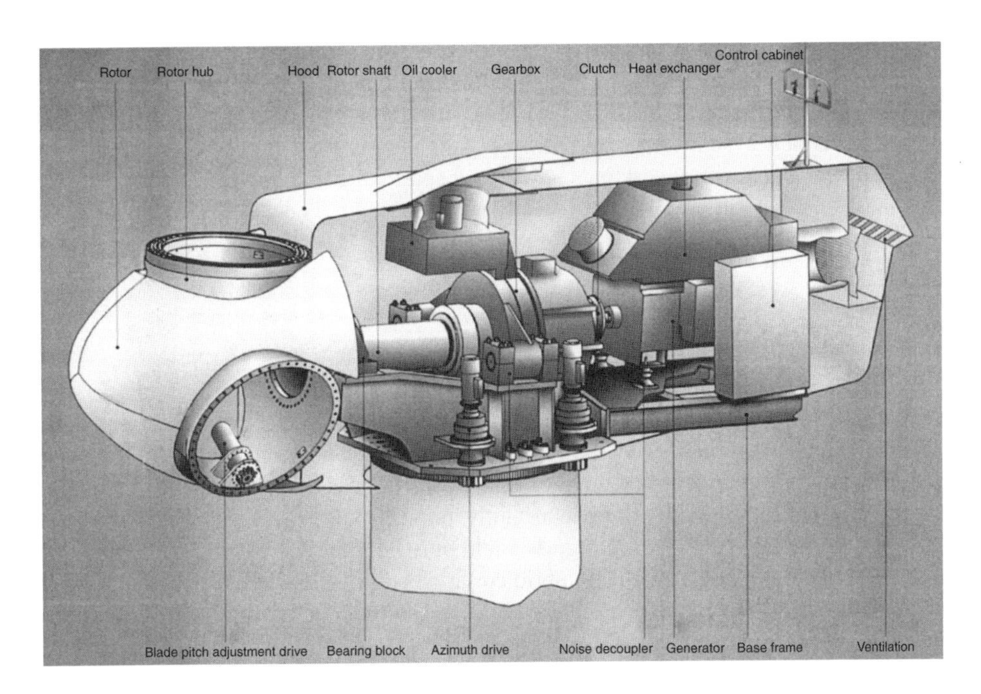

Figure 1.2 Nacelle of a wind turbine with a gearbox and high-speed 1.5 MW generator (TW 1.5 GE/Tacke). Reproduced by permission of Tacke Windenergie

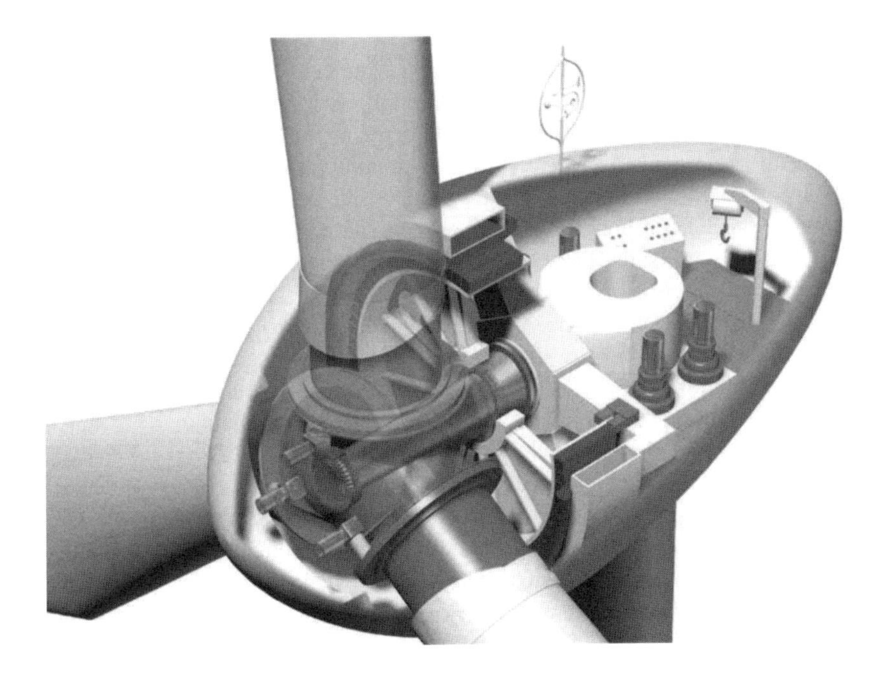

Figure 1.3 Schematic structure of a gearless wind turbine (Enercon E66, 70 m rotor diameter, 1.8/2 MW nominal output). Reproduced by permission of Enercon

variant with the generator being driven directly from the turbine. These pictures represent the basis for the functional relationships and considerations of the system.

Following a brief glance back into history, developmental stages and different wind turbine designs and systems will be briefly highlighted and the processes of mechanical–electrical power conversion explained. Moreover, particular importance is assigned to the interconnection of wind turbines to form wind farms and their combined effect in grid connection.

1.2 A Brief History

For thousands of years, mankind has been fascinated by the challenge of mastering the wind. The dream of defying Aeolos[1] and taming the might of the storm held generations of inventors under its spell. To attain limitless mobility by using the forces of Nature, thereby expanding the horizons of the then known world, was a challenge even in antiquity. Thus, sailing and shipbuilding were constantly pursued and developed despite doldrum, hurricane, tornado and shipwreck. Progress could only be achieved by employing innovative technologies. These, together with an unbridled lust for voyages of discovery, built up in the minds of sovereigns and scholars a mosaic of the world, the contours of which became ever more enclosed as time went by.

With wind-harnessing technology on land and on the sea, potentials could be realized and works undertaken that far outpaced any previously imagined bounds. For example, using only the power of animals and of the human arm, it would never have been possible for the Netherlands to achieve the drainage that it has through wind-powered pumping and land reclamation.

Archaeological discoveries relating to the use of wind energy predate the beginning of the modern era. Their origins lay in the Near and Middle East. Definite indications of windmills and their use, however, date only from the tenth century, in Persia [1.1]. The constructional techniques of the time made use of vertical axes to apply the drag principle of wind energy capture (Figure 1.4). Such mills were mostly found in the Arab countries. Presumably, news of these machines reached Europe as a result of the Crusades. Here, however, horizontal-axis mills with tilted wings or sails (Figure 1.5) made their appearance in the early Middle Ages.

The use of wind energy in Western Europe on a large scale began predominantly in England and Holland in the Middle Ages. Technically mature post mills (Figure 1.6) and Dutch windmills (Figure 1.7) were used mostly for pumping water and for grinding. More than 200 000 (two hundred thousand) of these wooden machines were built throughout North-West Europe, representing by far the greatest proportion of energy capture by technical means in this region. At the beginning of the twentieth century, some 20 000 (twenty thousand) windmills were still in use in Germany.

From the nineteenth century onwards, mostly in the USA, the so-called 'western wheel' type of turbine became widespread (Figure 1.8). These multibladed fans were built of sheet steel, with around 20 blades, and were used mostly for irrigation. By the end of the 1930s, some 8 million units had been built and installed, representing an enormous economic potential.

[1] Aeolos: Greek god of the winds.

Figure 1.4 Persian windmill (model)

Figure 1.5 Sail windmill

Figure 1.6 Post mill

1.3 Milestones of Development

The first attempt to use a wind turbine with aerodynamically formed rotor blades to generate electricity was made over half a century ago. Since then, besides the design and construction of large projects in the 1940s by the German engineers Kleinhenz [1.2] and Honnef [1.3], the pilot projects of the American Smith-Putnam (1250 kW nominal output, 53 m rotor diameter, 1941), the Gedser wind turbine in Denmark (200 kW nominal output, 24 m rotor diameter, 1957) and the technically trail-blazing Hütter W34 turbine (100 kW nominal output, 34 m rotor diameter, 1958) are worthy of mention (Figure 1.9).

The German constructor Allgaier started the first mass production of wind power plants in the early 1950s. They were designed to supply electricity to farmsteads lying far from the public grid. In coastal areas these turbines drove 10 kW generators; inland they were fitted

Figure 1.7 Dutch windmill

with 6 kW units. Their aerodynamically formed blades of 10 m diameter could be pitched about the longitudinal axis so as to regulate the power taken from the wind. Even today, some of these turbines (see Figure 1.10) are in operation with full functionality, after more than 50 years of service.

After the 1960s, cheaper fossil fuels made wind energy technology economically uninteresting, and it was only in the 1970s that it returned to the spotlight due to rising fuel prices. Some states then developed experimental plants in various output classes.

In particular in the USA, Sweden and the Federal Republic of Germany, turbines with outputs in the megawatt class have attracted most attention. Here, with the exception of the American MOD-2 (Figure 1.11) with five units and the Swedish–American WTS-4 (Figure 1.12) with five or two units, large converters such as the German GROWIAN (Figure 1.13), the Swedish WTS-75 AEOLUS model, the Danish Tvind turbine and the US MOD-5B variants in Hawaii were all one-offs. Despite many and varied teething troubles with the pilot installations, it was clear even then that technical solutions could be expected

Figure 1.8 American wind turbine

in the foreseeable future that would permit the reliable operation of large-scale wind turbines. Second-generation megawatt-class systems such as the WKA 60 (Figure 1.14) and the Aeolus II (Figure 1.15) have confirmed this expectation.

Mainly in the US state of California, but also in Denmark, Holland and the Federal Republic of Germany, considerable efforts were being made, independently of the development of large turbines, to use wind power to supply energy to the grid on a large scale. In the 1980s, wind turbines with total capacity of around 1500 MW were installed in California alone. In the initial phases, turbines of the 50 kW categories were used (Figure 1.16). Scaling-up the systems that were successful through the 100, 150 and 250 kW classes (Figures 1.17 and 1.19) and the 500/600 kW order of magnitude (Figures 1.18 and 1.20) has led to wind farms with turbines in the megawatt range (Figure 1.21).

Figure 1.9 Hütter W 34 turbine

This development has made the mass production of wind turbines possible. A considerable improvement of performance can thus be achieved. Progressively increasing turbine size (see Figures 1.22 to 1.25) using designs of widely differing types and costs has led to the development of machines in the 500 kW and megawatt classes that are remarkable for their high availability and good return-on-investment potential.

The individual manufacturers have chosen very different routes to market success in relation to this trend. NEG Micon has retained the classic Danish stall-regulated turbines with an asynchronous generator rigidly coupled to the grid in the power classes up to 1.5 MW (Figure 1.22). Bonus (Figure 1.23), Nordex (Figure 1.24) and Vestas (Figure 1.25) as well as GE/Tacke (Figure 1.26) have altered their turbine configuration in the different size classes, particularly with regard to the turbine regulation (stall or pitch) and generator systems (fixed-speed or variable-speed with a thyristor/ IGBT frequency converter). Currently 3 to 5 MW systems from all well-known manufacturers are being operated as prototypes or are available on the market.

One new development has been the trend towards gearless wind turbines. Several attempts have been made to introduce and establish in the market small, high-speed, horizontal-axis

Figure 1.10 Allgaier turbine

turbines with direct-drive generators. Up until now these attempts have met with limited success. Microturbines (Figure 1.27) with a permanent-magnet synchronous generator driven directly from the turbine are usually used as battery chargers. The success of such systems is rooted in their attractive design and low price as well as in the modern worldwide sales concept and the simple installation of the plants.

To some degree, companies that have entered into the production of wind generators at a later stage have been able to draw upon existing developments and techniques, thus allowing their first efforts to overtake the systems of established manufacturers. DeWind started its development (Figure 1.28) with a pitch-regulated 600 kW turbine and a variable-speed generator system (double-fed asynchronous machine), which could not have been produced at an economical cost a few years previously and which is currently favoured by most manufacturers. Then 1 and 2 MW systems of the same design followed.

The development of wind power systems has largely been carried out by medium-sized companies. Smaller manufacturers, however, face financial limits in the development of MW systems. The 1.5 MW turbine MD 70/MD 77 (Figure 1.29), again with the double-fed asynchronous generator design, which was developed by pro + pro for the manufacturers

Figure 1.11 MOD 2 in the Goodnoe Hills (USA): 2.5 MW nominal output, 91 m rotor diameter, 61 m hub height

Figure 1.12 WTS-4 turbine in Medicine Bow, USA.: 4 MW nominal output, 78 m rotor diameter, 80 m tower height

Figure 1.13 GROWIAN near Brunsbüttel, North Friesland: 3 MW nominal output, 100 m rotor diameter, 100 m hub height

BWU, Fuhrländer, Jacobs Energie (now REpower Systems) and Südwind / Nordex is opening up new developmental and market opportunities for smaller companies in the field of large-scale plants.

Vertical-axis rotors, so-called Darrieus turbines, are enchantingly simple in structure. In their basic form they have up until now mostly been built with gearing and generators at base level (Figure 1.30). Variants in the form of so-called H-Darrieus gearless turbines in the 300 kW class were first designed with rotating towers and large multiple generators at ground level (Figure 1.31(a)). Further development led to machines with fixed tripods and annular generators in the head (Figure 1.31(b)). These variants have not, however, been successful in establishing themselves widely in the wind power market.

Figure 1.14 WKA 60 in Kaiser-Wilhelm-Koog: 1.2 MW nominal output, 60 m rotor diameter, 50 m tower height

The Enercon E 40 horizontal-axis turbine was the first system in the 500 kW class with a direct-drive generator to establish itself in the market with great success in a very short time. Figure 1.32 shows the schematic construction of the nacelle. The generator, specially developed for this model, connects directly to the turbine and needs no independent bearings. In this way, wear on mechanical components running at high speed is reduced to a minimum. Operational run times of 180 000 hours have been quoted for many years.

The gearless E 30, E 40, E 58, E 66 and E 112 models from Enercon were produced as a development of the stall-regulated geared models E15/E16 and E17/E18, by way of the E 32/E 33 variable-pitch turbines (Figure 1.33). In parallel, but with a slight delay, the conversion from thyristors to pulse inverters was accomplished. This configuration thus unites the advantages of variable speeds (and the associated reduction in drive-train loading) with those of a grid supply having substantially lower harmonic feedback.

Figure 1.15 AEOLUS II near Wilhelmshaven: 3 MW nominal output, 80 m rotor diameter, 88 m tower height

Figure 1.16 Wind farm in California with turbines in the 50/100 kW class

Figure 1.17 Wind farm in California with turbines in the 250 kW class

In comparison to the gearless designs with electrically excited synchronous generators, as shown in Figure 1.33(d) to (h), permanent-magnet machines permit the arrangement of higher numbers of poles around the rotor or stator. By using high-quality permanently magnetic materials, relatively favourable construction sizes can thus be achieved (Figure 1.34) and very high efficiencies attained, particularly in the partial load range. Such a plant configuration of the 600 kW class (Figure 1.34(a)) has been able to achieve excellent returns over several

Figure 1.18 Wind farm in Wyoming with turbines in the 600 kW class

Figure 1.19 Wind farm in North Friesland with turbines of the 250 kW class

Figure 1.20 Wind farm on Fehmarn Island with turbines of the 500 kW class

(a) On land

(b) At sea. Reproduced by permission of GE Wind Energy

Figure 1.21 Wind farm with 1.5 MW turbines

years of fault-free operation. A 2 MW unit with such a generator design (Figure 1.34(b)) was designed with a medium-voltage generator of 4 kV system voltage.

A further possibility, which has been considered for large, slow-running turbines in particular, is the combination of a low-speed generator and a turbine-side gearbox, as shown in Figure 1.35. The single-stage gearbox turns the generator shaft at around eight times the turbine speed of approximately 100 revolutions per minute. Thus, even for units in the 5 MW

Figure 1.22 Size progression of stall-regulated turbines of the same design (fixed-speed, fixed-pitch machines) from NEG Micon / Nordtank. Reproduced by kind permission of NEG Micon

Figure 1.23 Size progression of Bonus turbines: (a,b) fixed-speed, stall-controlled turbines; (c,d) active (combi-)stall turbines with a slight blade pitch adjustment

(a) N 27/29 (150/250 kW) (b) N 43 (600 kW) (c) N 62 (1300 kW) (d) N 80/90 (2500 kW). Reproduced by permission of Nordex

Figure 1.24 Size progression of Nordex turbines: (a,b,c) fixed-speed, fixed-pitch machines; (d) a large-scale, variable-speed, variable-pitch unit

range, generators in compact and technically favourable construction sizes of approximately 3 m diameter can be used.

A further large-scale turbine in the 5 MW class with a rotor diameter of over 125 m is expected from REpower (Fig. 1.36) in the near future. A double-fed asynchronous generator with medium-voltage isolation in the low-voltage range (950 V stator-side or 690 V rotor-side) will be used in this system.

In the following we consider various real operational situations, the essential differences between the systems involved and the resulting effects on supply to the grid, taking as a basis the functional structure of wind power machines and their influences.

1.4 Functional Structures of Wind Turbines

For the following consideration, which is mainly concerned with the mechanical interaction of electrical components and with interventions to modify output, we will draw upon the nacelle layout shown in Figure 1.2. With the correct design, the influences of the tower and of steering in response to changes in wind direction can be handled separately (Section 2.2.1) or treated as changes in wind velocity. The block diagram shown in Figure 1.37 (see page 28), which illustrates the links between the most important components and the associated energy conversion stages, may serve as the basis for later detailed observations. This diagram also gives an idea of how operation can be influenced by control and supervisory

Figure 1.25 Size progression of Vestas turbines: (a) small, fixed-speed, fixed-pitch machine; (b,c,d) larger variable-pitch units; (d,e) machines with speed elasticity; (f) machines with double-fed asynchronous generators

processes. Furthermore, the central position occupied by the generator is made particularly clear.

The following pages therefore explain the physical behaviour of a wind energy extraction system and the conversion of this mechanical energy to electrical energy by means of generators. We examine how mechanical moments are handled in the drive unit when the generator is connected to the grid, the design of generators suitable for wind turbines and the

(a) TW 80 (80 kW, 21 m rotor diameter)

(b) TW 250 (250 kW, 26 m rotor diameter)

(c) TW 300 (300 kW)

(d) TW 600 (600 kW, 43 m rotor diameter)

(e) TW 1.5/1.5S (1500 kW, 65/70 m rotor diameter)

(f) GE 3.6 (3.6 MW, 100 m rotor diameter). Reproduced by permission of GE Wind Energy

Figure 1.26 Size progression of turbines from GE / Tacke: first (a,b) and second (c,d,e,f) generation machines, from fixed-speed, fixed-pitch turbines (a to d) to large-scale, pitch-controlled, variable-speed turbines (e,f)

Figure 1.27 Small system-compatible turbine from aerosmart. Reproduced by permission of Aerodyn Energiesystems GmbH

(a) DeWind 4 (600 kW, 46/48 m rotor diameter)

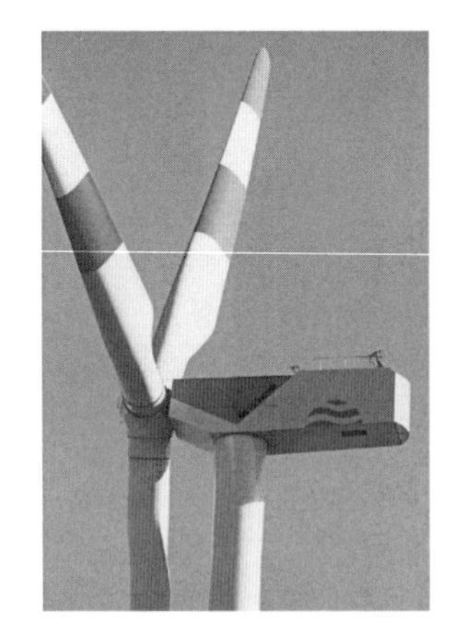

(b) DeWind 6 (1000/ 1250kW, 60/62/64 m rotor diameter)

Figure 1.28 DeWind 4 (600 kW, 46/48 m rotor diameter). Reproduced by permission of DeWind

Figure 1.29 Joint development of the 1.5 MW MD 70/MD 77 turbine (70/77 m rotor diameter)

Figure 1.30 Fixed-speed 300 kW Darrieus unit with gearing and a conventional generator

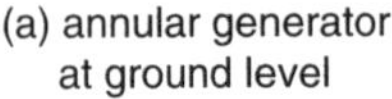

(a) annular generator at ground level

(b) annular generator in head

Figure 1.31 Variable-speed 300 kW gearless H-Darrieus unit

combined effects of turbines and power supply grids, as well as the regulation of turbines in isolation and in grid operation, bearing in mind the conditions imposed by the grid and the consumer.

From Figure 1.37 (see page 28), the functional structures for entire wind energy conversion systems, or for particular types of wind energy converter as shown in Figure 1.38(a) and (b) (see page 29), can be further developed. Such simplified block diagrams can help us to understand how the principal components of pitch- or stall-regulated horizontal-axis wind energy converters work and interact.

Wind energy converters with variable-blade pitch (Figure 1.38(a)) allow direct control of the turbine. Figure 2.6 shows that by varying the blade pitch it is possible, firstly, to influence the power input or torque of the rotor, with a smaller blade pitch angle β (or greater ϑ) leading to a lower turbine output and a greater β leading to a higher turbine output (pitch regulation). Secondly, by a few degrees adjustment of the rotor blades, the profile can be brought more fully into stall when β is greater (active stall regulation) and the turbine power falls. A slight reduction to the blade pitch angle, on the other hand, guides the rotor out of stall and power increases until laminar flow is achieved on the blade profiles. In this way, the speed of rotation, determined by integration of the difference between turbine torque and the generator's load torque, taking rotating masses (or mechanical time constants) into account, can be influenced at all performance levels – insofar as sufficient energy is available. The pitch control of a wind turbine therefore makes it possible to regulate energy extraction. In this way, adaptation to user needs (e.g. in standalone operation) can be achieved, as well as a measure of protection in storm conditions.

In (passive) stall-controlled converters (Figure 1.38(b)), the rotor speed is kept at an almost constant speed by the load torque of a rigidly coupled asynchronous (mains) generator,

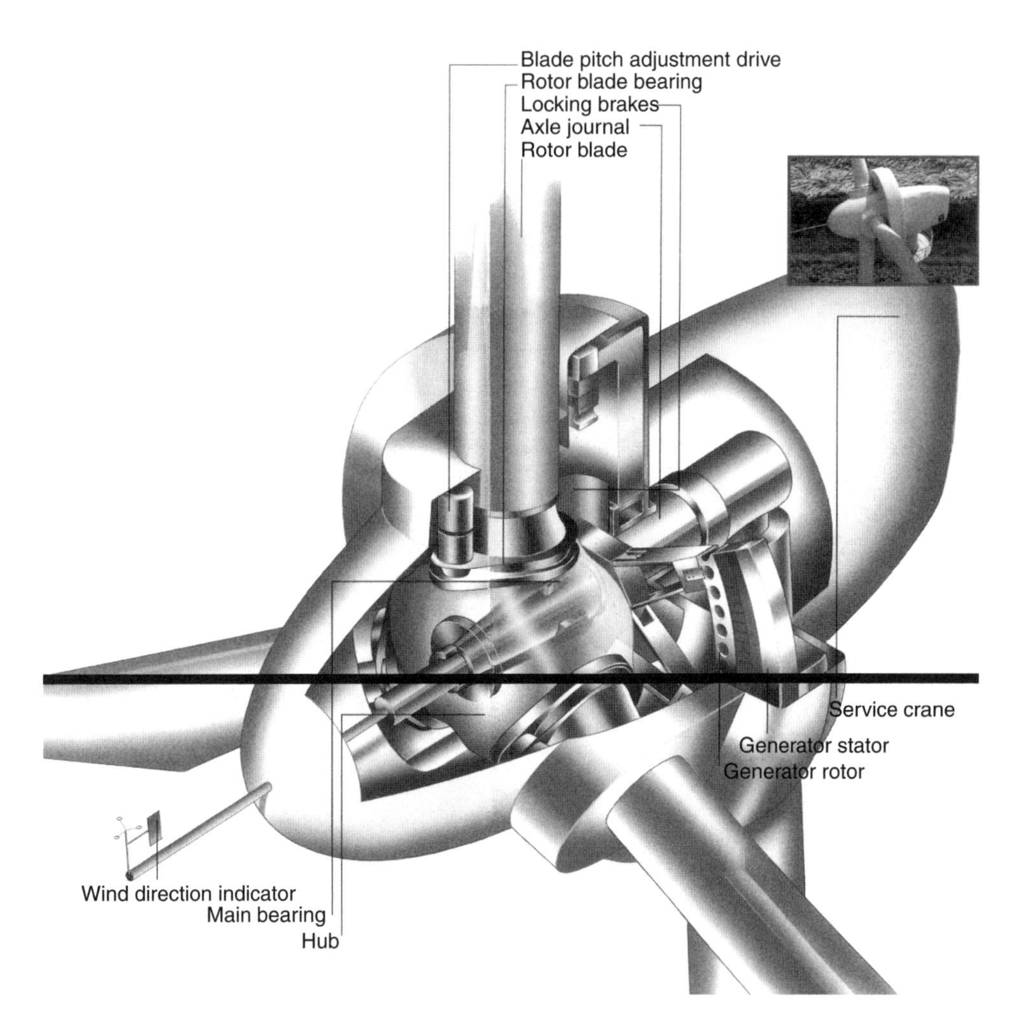

Figure 1.32 Schematic layout of the Enercon E 40 gearless turbine. Reproduced by kind permission of Enercon

usually of large dimensions. When wind strength rises above nominal levels, the flow over the rotor blades achieves partial or even total stall – whence the so-called 'stall regulation'. The power take-up of the turbine is thereby passively (i.e. design-dependently) limited under full loading to values such that under operational wind speed conditions the nominal output of the generator is not significantly exceeded.

The use of variable-speed generators in both regulation systems allows the reduction of sudden load surges, and considerably extends the range of operation. The optimal power can be produced by adjusting the speed of the rotor to the desired speed. For example, it is also possible, in cases where partially increased transitional loads must be handled, to influence the drive torque of stall-regulated turbines by varying the rotational speed of the generator.

(a) E 15/16 (55 kW) (b) E 17/18 (80 kW) (c) E 32/33 (300 kW) (d) E 30 (200 kW)

(e) E 40 (500 kW) (f) E 58 (850 kW/ 1000 kW) (g) E 66 (1500 kW / 1800 kW) (h) E 112 (4500 kW)

Figure 1.33 Enercon turbines from variable-speed geared models with thyristor inverters (a,b,c) to gearless configurations with pulse inverters (d,e,f,g,h); (a,b) with fixed and (c,d,e,f,g,h) with variable pitch. Reproduced by kind permission of Enercon

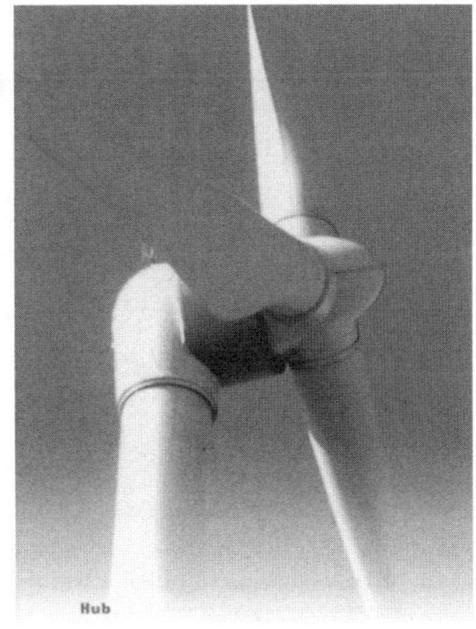

(a) Genesys 600

(b) Harakosan Z72 (70.65 m rotor diameter and 2 MW nominal output) with medium-voltage generator. Reproduced by permission of Harakosan

Figure 1.34 Gearless wind turbines with permanent-magnet synchronous generator (46 m rotor diameter, 600 kW nominal output)

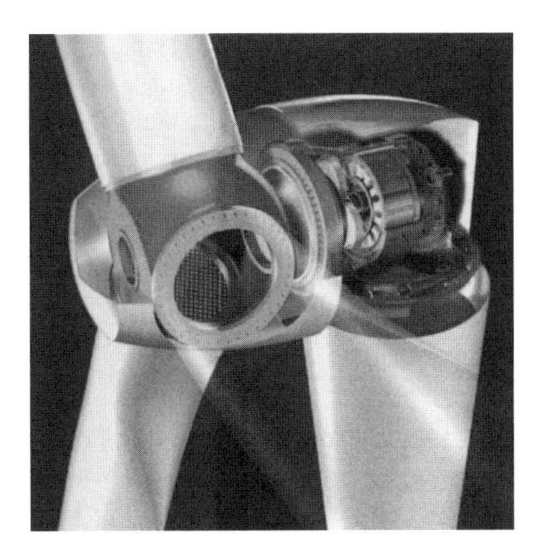

Figure 1.35 Nacelle of the large-scale Multibrid N 5000 (5 MW, 116 m rotor diameter) with single-stage gearing, integral hub and low-speed synchronous generator. Reproduced by permission of Multibrid Entwicklungsgesellschaft GmbH

Figure 1.36 5 MW offshore turbine from REpower (5 MW nominal output and 126.5 m rotor diameter)

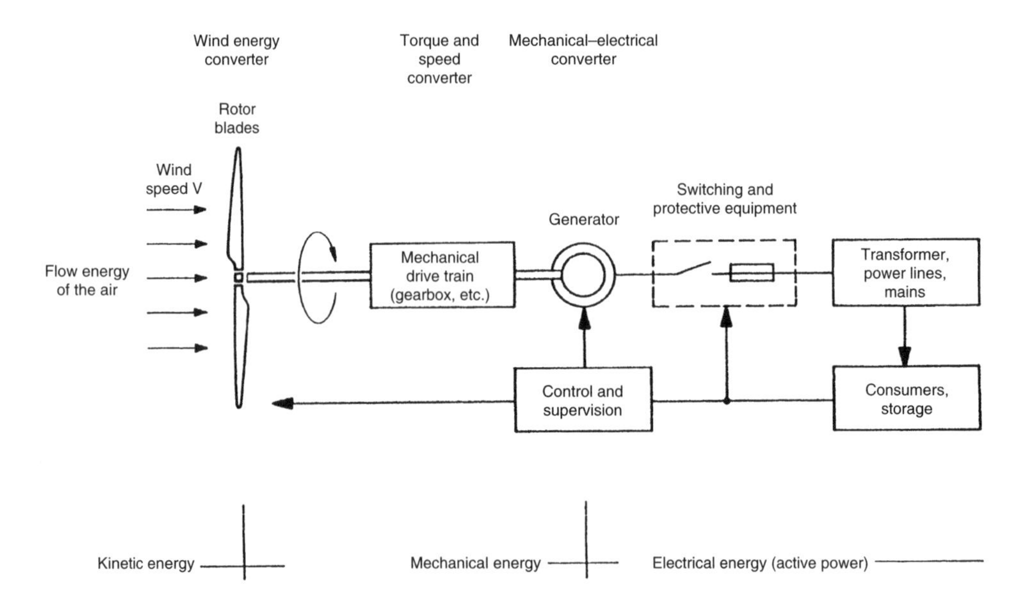

Figure 1.37 Functional chain and conversion stages of a wind energy converter

A detailed treatment of the generator and associated discussions on the theme of turbine regulation require knowledge of the physical processes and a review of the mathematical laws governing the entire converter system. The following text should encompass this, insofar as is necessary. More detailed studies are also necessary regarding the combined effects of wind turbines working together with existing grid systems and the measures that must be applied to control these effects throughout the entire system.

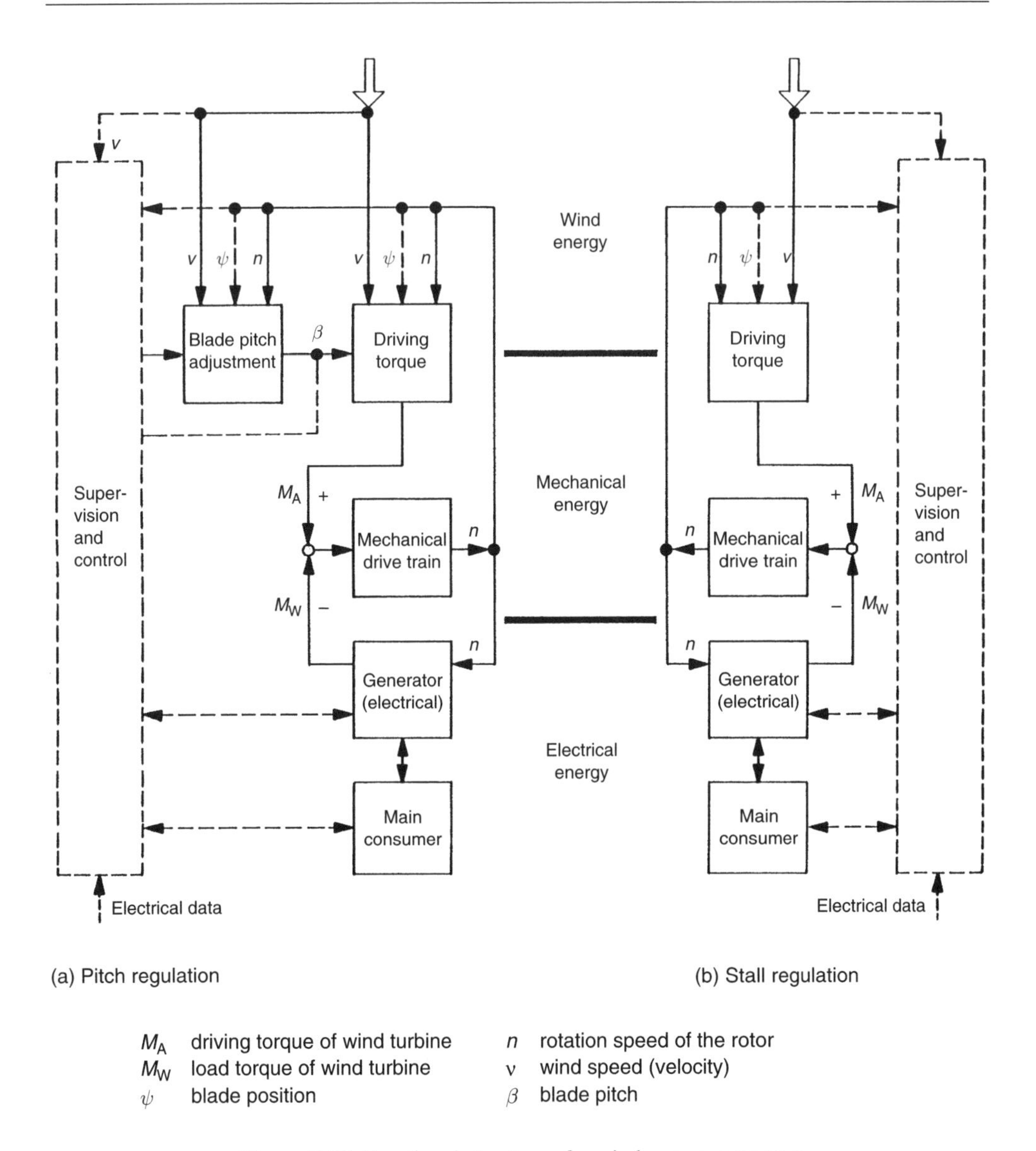

Figure 1.38 Functional structure of a wind energy converter

The following chapters summarize the results of years of research and development. Through practical references achieved from completed projects, particular weight has been given to the usefulness of the results for plant conception and design.

2

Wind Energy Conversion Systems

As shown in Figure 1.37, the blades of a wind turbine rotor extract some of the flow energy from moving air, convert it into rotational energy and then deliver it via a mechanical drive unit (shafts, clutches and gears), as shown in Figures 1.2 and 1.3 respectively, to the rotor of a generator and thence to the stator of the same by mechanical–electrical conversion. The electrical energy from the generator is fed via a system of switching and protection devices, lines and if necessary transformers to the grid, consumers or an energy storage device. In the further discussion of the relevant system components, special attention must be given to the drive torque or performance properties, the structures based thereon and the actions necessary to limit turbine speed, together with reaction effects of the transmission on the turbine.

2.1 Drive Torque and Rotor Power

In contrast to the windmills of yesteryear, modern wind turbines used for generating electricity have relatively fast-running rotors. A few blades with high lift-to-drag ratio profiles that utilize lift attain much higher levels of efficiency than drag-type rotors. They also make it possible to alter the power of the turbine more quickly. Such turbines can, given sufficient wind, attain operating conditions akin to those of conventional power stations.

In examining the operational characteristics of a wind energy converter we must determine the variables influencing the turbine, beginning with the forces on the rotor blades or on small areas thereof, and thence derive the resulting drive torque and corresponding output power.

Grid Integration of Wind Energy Conversion Systems, Second Edition S. Heier

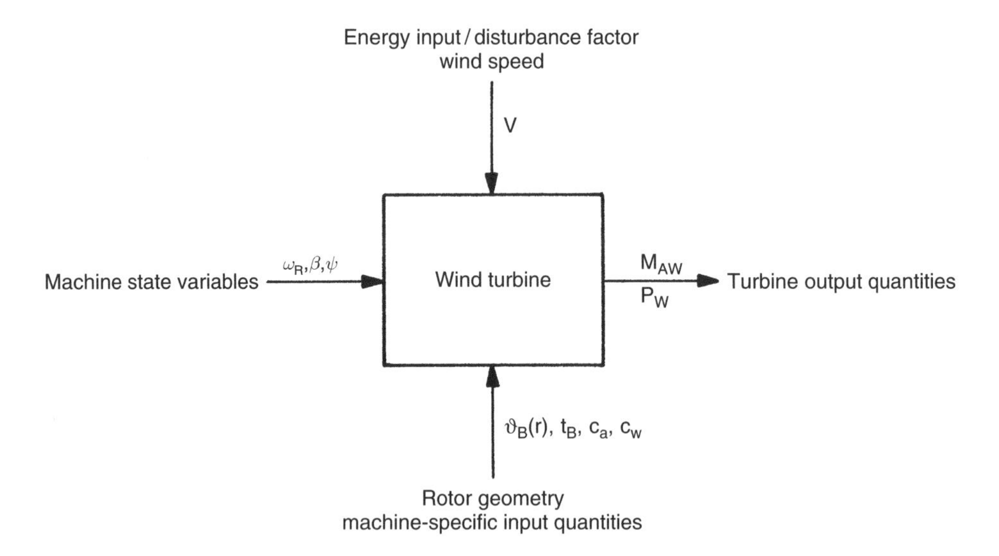

Figure 2.1 Wind turbine inputs and outputs

2.1.1 Inputs and outputs of a wind turbine

To model the power or torque properties of a wind turbine, we can consider the block labelled 'driving torque' in Figure 1.38(a) as a single structure having inputs and outputs as shown in Figure 2.1. These can be subdivided as follows:

- the independent input quantity 'wind speed', which determines the energy input but which can act at the same time as an interference quantity;
- machine-specific input quantities, arising particularly from rotor geometry and arrangement and
- the state variables 'turbine speed', 'rotor blade position' and 'rotor blade angle' arising from the transmission system of the complete wind-power unit, and with the aid of which
- turbine output quantities, such as 'power' or 'drive torque', may be controlled.

The power or torque of a wind turbine may be determined by several means. The best results might be obtained by considering the axial and radial momentum of the airstream, or of small-diameter infinite flow-tubes impinging on the rotor blades, or of elements thereof, thus allowing us to determine local flow conditions and the resulting forces or rotational action on the turbine blades. The following considers this briefly. Calculations based on circulation and vortex distribution over aerofoils [2.1], which allow the derivation of solutions by the Biot–Savart theorem, are not taken into consideration here.

2.1.2 Power extraction from the airstream

Within its effective region, the rotor of a wind turbine absorbs energy from the airstream, and can therefore influence its velocity. Figure 2.2 represents the flow that develops around

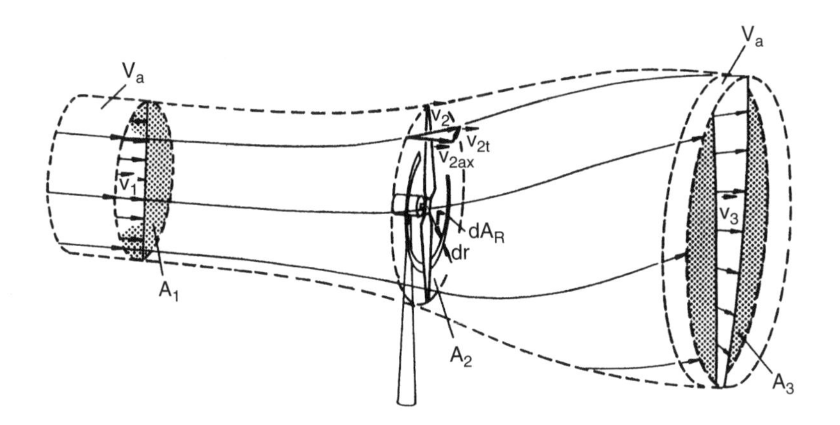

Figure 2.2 Airstream around the turbine

a converter in an unrestricted airstream in response to prevailing transmission conditions, whereby the airstream is decelerated axially and deviated tangentially in the opposite direction to the rotation of the rotor.

From references [2.2], [2.3] and [2.4], the energy absorbed from an air volume V_a of cross-section A_1 and swirl-free speed of flow v_1 far upstream of the turbine, which results in a downstream reduction of flow speed to v_3 with a corresponding broadening of the cross-sectional area (wake decay) to A_3, can be expressed as

$$W_w = V_a \frac{\rho}{2} \left(v_1^2 - v_3^2 \right) \tag{2.1}$$

The wind turbine power may therefore be expressed as

$$P_w = \frac{dW_w}{dt} = d\frac{\left(V_a \frac{\rho}{2} \left(v_1^2 - v_3^2 \right) \right)}{dt}. \tag{2.2}$$

An air volume flow in the rotor area $(A_2 = A_R)$ of

$$\frac{dV_a}{dt} = A_R v_2 \tag{2.3}$$

yields, in the quasi-steady state,

$$P_W = A_R \frac{\rho}{2} \left(v_1^2 - v_3^2 \right) v_2. \tag{2.4}$$

The power absorption and operating conditions of a turbine are therefore determined by the effective area A_R, by the wind speed and by the changes occurring to these quantities in the field of flow of the rotor. The power of the turbine can thus be influenced by varying the flow cross-sectional area and by changing flow conditions at the rotor system.

According to Betz [2.2], the maximum wind turbine power output

$$P_{W_{max}} = \frac{16}{27} A_R \frac{\rho}{2} v_1^3 \tag{2.5}$$

is obtained when

$$v_2 = \frac{2}{3}v_1 \qquad \text{and} \qquad v_3 = \frac{1}{3}v_1. \tag{2.6}$$

Under normal operating conditions up to the nominal output capacity, this reduction of wind speed is approached. When the rotor is idling, or running under light load, the value of v_2 approaches that of v_1.

The ratio of the power P_W absorbed by the turbine to that of the moving air mass

$$P_0 = A_R \frac{\rho}{2} v_1^3 \tag{2.7}$$

under smooth flow conditions at the turbine defines the dimensionless performance coefficient

$$c_p = \frac{P_W}{P_0}. \tag{2.8}$$

The above expression is based upon the assumption that tubular axial air mass transport only occurs from the leading side of the entry area A_1 to the exit area A_3. A more detailed examination of the turbine or rotor blades can be carried out using the modified blade element theory, by introducing a radial wind speed gradient and by taking into account any angular movement of the airstream.

2.1.3 Determining power or driving torque by the blade element method

If, instead of a circular section, we consider an annular section of radius r, width $\mathrm{d}r$ and area at the turbine of

$$\mathrm{d}A_R = 2\pi r \mathrm{d}r, \tag{2.9}$$

then the following is valid for the mass flow rate $\mathrm{d}\dot{m}$ in front of, at and behind the rotor in a quasi-steady state:

$$\mathrm{d}\dot{m}_1 = \mathrm{d}\dot{m}_2 = \mathrm{d}\dot{m}_3 \tag{2.10}$$

or

$$\rho \mathrm{d}A_1 v_{1ax} = \rho \mathrm{d}A_{2ax} = \rho \mathrm{d}A_3 v_{3ax}. \tag{2.11}$$

The force that brakes the air axially from v_{1ax} to v_{3ax} may be derived from the loss of momentum from entry to exit by

$$\mathrm{d}\dot{J}_{ax} = \mathrm{d}F_{ax} = \mathrm{d}\dot{m}\,(v_{1ax} - v_{3ax}). \tag{2.12}$$

In the rotor area, by application of Froude's theorem,

$$v_{2ax} = \frac{v_{1ax} + v_{3ax}}{2} \qquad \text{or} \qquad v_{1ax} - v_{3ax} = 2\,(v_{1ax} - v_{2ax}) \tag{2.13}$$

and the thrust of the air tubes

$$\mathrm{d}F_{\mathrm{ax}} = 4\pi\rho\,\mathrm{d}r\; v_{2\mathrm{ax}}\,(v_{1\mathrm{ax}} - v_{2\mathrm{ax}}) \tag{2.14}$$

can be obtained as a function of the axial wind speed on the rotor (to be determined). The tangential change of momentum may be determined in the same fashion:

$$\mathrm{d}\dot{J}_{\mathrm{t}} = v_{1\mathrm{t}}\,\mathrm{d}\dot{m}_1 - v_{3\mathrm{t}}\,\mathrm{d}\dot{m}_3. \tag{2.15}$$

When the air entering the flow tube is swirl-free, no other moment is applied to it. The tangential force, which brings the air into rotational flow, is derived as

$$\mathrm{d}F_{\mathrm{t}} = \mathrm{d}\dot{J}_{\mathrm{t}} = -v_{3\mathrm{t}}\,\mathrm{d}\dot{m}_3 = -v_{2\mathrm{t}}\,\mathrm{d}\dot{m}_2 \tag{2.16}$$

or, in applications,

$$\mathrm{d}F_{\mathrm{t}} = -2\pi\rho r\,\mathrm{d}r\; v_{2\mathrm{t}}v_{2\mathrm{ax}}. \tag{2.17}$$

The force thus depends on the radius and the axial and tangential airstream at the turbine. The air, according to equations (2.14) and (2.17), exerts identical forces on the rotor blades.

For the sake of clarity, the physical processes will be shown for a single rotor blade. Multiblade arrangements for fast-running turbines (e.g. with $z = 2$, 3 or 4 lift-type blades) can be handled by extension of this system, considering conditions at a single blade of z-fold depth. Depending on blade radius, Figure 2.2 shows that there is different flow behaviour at the profile for different blade angles (Figure 2.3).

The combined effect of velocity components and the resultant forces are shown for a single blade element in Figure 2.4. Total values (forces, moments, power) are obtained by the integration of the corresponding values over the blade radius, or by summation of the components of individual blade sections.

A segment at radius r of a blade rotating with angular velocity ω_{R} experiences two airflows: that due to the wind deceleration across the swept area,

$$\boldsymbol{v}_2 = \boldsymbol{v}_{2\mathrm{ax}} + \boldsymbol{v}_{2\mathrm{t}} \tag{2.18}$$

and that due to the speed of the rotating element at the given radius,

$$\boldsymbol{v} = -\boldsymbol{\omega}_{\mathrm{R}} \times \boldsymbol{r}. \tag{2.19}$$

If we disregard the cone angle then, in the direction of the resultant velocity component,

$$v_{\mathrm{r}} = \sqrt{v_{2\mathrm{ax}}^2\,(\omega_{\mathrm{R}} r + v_{2\mathrm{t}})^2}, \tag{2.20}$$

a drag of

$$\mathrm{d}F_{\mathrm{W}} = \frac{\rho}{2} t_{\mathrm{B}} v_{\mathrm{r}}^2 c_{\mathrm{w}}(\alpha)\,\mathrm{d}r \tag{2.21}$$

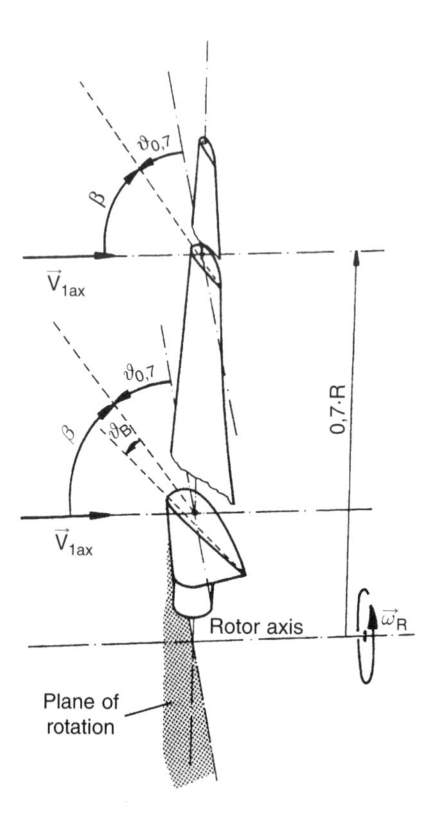

Figure 2.3 Definition of blade pitch angle

is exerted, acting against the movement of the blade, while an orthogonally directed lift of

$$dF_A = \frac{\rho}{2} t_B v_r^2 c_a(\alpha)\, dr \tag{2.22}$$

exerts a propulsive component. The force on the blade segment resulting from the lift and drag components is

$$d\boldsymbol{F}_{AW} = d\boldsymbol{F}_A + d\boldsymbol{F}_W. \tag{2.23}$$

Separating this into axial and tangential components leads, for z blades, to a drive torque-generating value of

$$dF_t = z\frac{\rho}{2} t_B v_r^2 (c_a \sin\delta - c_w \cos\delta)\, dr \tag{2.24}$$

and an axial thrust on a rotor blade and on the hub of

$$dF_{ax} = z\frac{\rho}{2} t_B v_r^2 (c_a \cos\delta - c_w \sin\delta)\, dr. \tag{2.25}$$

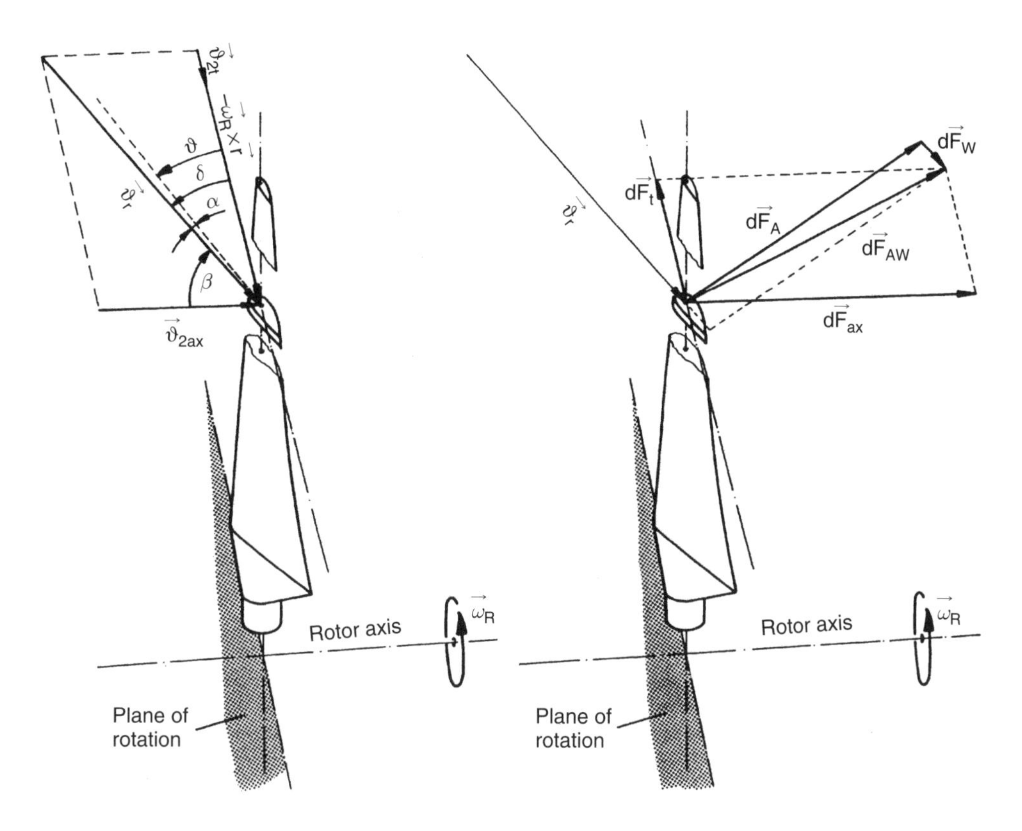

Figure 2.4 Airflow and forces on a rotor blade segment

According to this, and for the respective wind and blade tip velocities, these forces are essentially dependent on:

- the pitch angle ϑ of the blade element with respect to the plane of rotation or β with respect to the rotor axis;
- the local angle of attack α between the resultant wind velocity and chord of the aerofoil;
- the lift and drag coefficients (c_a, c_w) for the blade profile (Figure 2.5), the Reynolds number and the roughness of the blade surface, which should be ignored here, as should skewed airflow impingement due to the cone angle, etc.

The drive torque of the wind turbine is therefore

$$M_{AW} = \int_{R_i}^{R_a} r \, dF_t \int_{R_i}^{R_a} rz \frac{\rho}{2} t_B v_r^2 \left(c_a \sin\delta - c_w \cos\delta\right) dr. \tag{2.26}$$

To determine the forces, it is necessary to determine the local wind velocity v_2 at the rotor or the deceleration relative to v_1. By equating the axial and tangential forces to the corresponding losses in momentum, the components v_{2ax} and v_{2t} or the vector sum v_2, and thus the air flow acting on the rotor, may be derived.

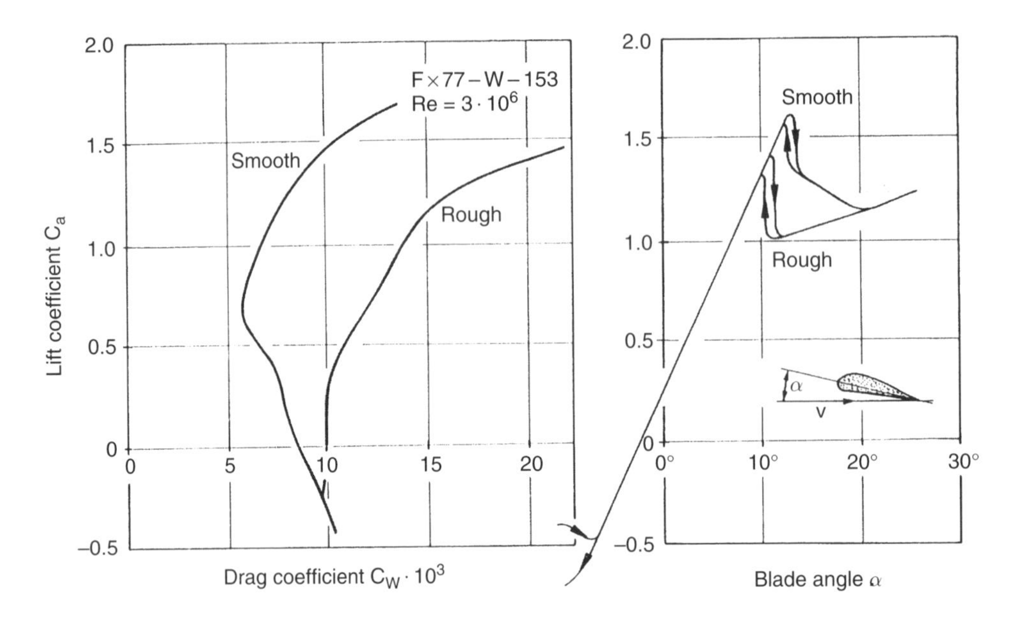

Figure 2.5 Lift and drag coefficients of a Wortmann aerofoil section [2.4]

Marginal vortex losses, due mainly to flow around the blade tips and roots, which cause a vortex field downstream of the rotor, can be allowed for by applying a correction factor of

$$F_S = \frac{2}{\pi}\arccos\left\{\exp\left[-\frac{z}{2}\frac{1}{\sin\delta}\left(1-\frac{r}{R_a}\right)\right]\right\} \tag{2.27}$$

for the blade tips [2.5] and

$$F_N = \frac{2}{\pi}\arccos\left\{\exp\left[-\frac{z}{2}\frac{1}{\sin\delta}\left(\frac{r}{R_i}-1\right)\right]\right\} \tag{2.28}$$

for near the hub. The product of these two factors

$$F = F_S F_N \tag{2.29}$$

allows the influence of the free peripheral eddies to be described with sufficient exactitude.

When the wind deceleration due to the turbine is very high, a turbulent vortex field develops in the downstream of the machine (turbulent wake state) [2.6]. However, we only need to allow for this state in a few machines, when the outer blade areas are running close to the cut-in speed. If necessary, various empirical approximations (see references [2.6] and [2.7]) may be used.

2.1.4 Simplifying the computation method

The method detailed above yields very good results that are in agreement with practice. The effort needed to determine operational conditions of the turbine or to examine the behaviour of wind energy plants is, however, enormous. Very reliable estimates of operational performance can be arrived at by computational methods that give errors in the order of a few percent. The objective of further research was to considerably simplify the method, while keeping errors within predetermined limits. Starting with the method detailed above, various possibilities were examined with the objective of reducing computational effort while holding the errors within acceptable bounds (e.g. 2.5 %) [2.8]. To this end, the complete method is compared with the chosen simplification for two wind turbines of different sizes (12.5 m and 60 m rotor diameter), divided into 50 blade elements.

This examination showed the following:

- To calculate the drop in wind velocity v_2 at the turbine, swirl effects can be ignored. Under operational conditions, the errors relative to the long method are less than 2.5 %.
- To determine thrust for variable-pitch machines, it is possible to remain within the same margin of error while ignoring the drag component. For stall-regulated machines, values that are dependent upon the operating state must also be incorporated.
- Blade tip and hub losses can be approximated to the same level of exactitude using the given wind speed v_1 far upstream of the turbine instead of the iteratively determined wind speed v_2 at the rotor in determining $\sin\delta$, or

$$\frac{1}{\sin\delta}=\sqrt{1+\left(\frac{v_{\mathrm{u}}}{v_2}\right)}\approx\sqrt{1+\left(\frac{v_{\mathrm{u}}}{v_1}\right)^2}\approx\sqrt{1+\left(\frac{\omega_{\mathrm{R}}r}{v_{1\mathrm{ax}}}\right)^2}. \tag{2.30}$$

The reduction factor is then determined just once for each blade element according to its own operating conditions at the beginning of the iteration.

- The torque, instead of being calculated by summing partial moments over 50 blade elements by the step-by-step method

$$M_{\mathrm{AW}}=\int_{R_{\mathrm{i}}}^{R_{\mathrm{a}}}\mathrm{d}M_{\mathrm{AW}}\approx\sum_{k=0}^{n}\Delta M_{\mathrm{k}}, \tag{2.31}$$

may be determined using Simpson's rule

$$M_{\mathrm{AW}}=\int_{R_{\mathrm{i}}}^{R_{\mathrm{a}}}\mathrm{d}M_{\mathrm{AW}}\approx\frac{R_{\mathrm{i}}-R_{\mathrm{a}}}{n}\left(\frac{\mathrm{d}M_0}{\mathrm{d}r}+4\frac{\mathrm{d}M_1}{\mathrm{d}r}2\frac{\mathrm{d}M_2}{\mathrm{d}r}+2\frac{\mathrm{d}M_{n-2}}{\mathrm{d}r}+4\frac{\mathrm{d}M_{n-1}}{\mathrm{d}r}+\frac{\mathrm{d}M_n}{\mathrm{d}r}\right) \tag{2.32}$$

to roughly the same precision and may be made even faster by taking

$$n=20 \tag{2.33}$$

elements, i.e. 21 points. Then, because of the flow around the blade tips and roots, the boundary values

$$\mathrm{d}M_0=\mathrm{d}M_{\mathrm{n}}=0. \tag{2.34}$$

Using this simplification, a considerable reduction of computation time – to around 1 % – may be achieved by:

- series expansion of the transcendent arctan and root functions;
- approximation of the aerofoil section with polar equations $c_A = f(\alpha)$ in the form of a polynomial to the third power;
- solving the resulting cubic equations by the cardanic method to determine the wind velocity v_{2ax} at the turbine. Time-consuming iterations thereby become superfluous.

The results can then be plotted, on the basis of the performance coefficient c_p in relation to the tip speed ratio λ, to give the usual c_p–λ characteristics (see Section 2.2), as shown in Figure 2.6 (a) and (b). The tip speed ratio is obtained here from the quotient of the peripheral to the undisturbed wind velocity

$$\lambda = \frac{v_u}{v_1}. \tag{2.35}$$

This two-dimensional representation of turbine power calculations shows the dependencies both upon blade pitch angle ϑ (or β) and also upon the speed n, which is proportional to the peripheral speed v_u and the wind speed v_1.

Further studies have shown that the result of the cardanic method under normal operating conditions is essentially limited to one of the three possible partial solutions. These conditions can thus be described in the form of a closed solution path for wind turbine power or torque calculation. The errors arising from this, using narrower element widths in comparison with blade element theory and including swirl, etc., are shown in Figure 2.7.

Because of the cubic approximation, the simplified method described here has its limits. Just below cut-in and just above cut-out wind speed, it compares unfavourably with the blade element method. However, blade element theory also yields results that differ considerably from measurements taken in these conditions, so such conditions can only be reconstituted with limited success.

The results of the calculation were shown as examples for two model turbines. Despite differences in size, number of blades, tip speed ratio, geometry, etc., hardly any difference can be seen in the calculated results within the operating ranges of the machines, indicated by continuous lines (see Figure 2.7(a) and (b)). Particularly for smaller machines, errors of more than 5 % are few and occur mostly at very low and very high wind speeds. However, the conventional computation-intensive methods do not yield any exact error-margin data under these conditions either. It should therefore be possible to apply this method to high-speed wind turbines with variable pitch. For stall-regulated machines, however, some limitations should be expected, since working out the lift using a third-degree polynomial can yield significant errors with the high angles of attack used. Furthermore, the drag component should then be included in determining thrust.

The correct conditions for using the method described, and the resulting limitations to its validity, can thus only be defined after appropriate investigation. On the whole, however, with this considerably simplified method, a wide range of machines and operational areas can be handled. In examining complete systems, the physical characteristics of turbines in quasi-steady states can be taken into consideration. Thereafter the dynamic effects and the resulting deformations of system components can be included. These will be determined

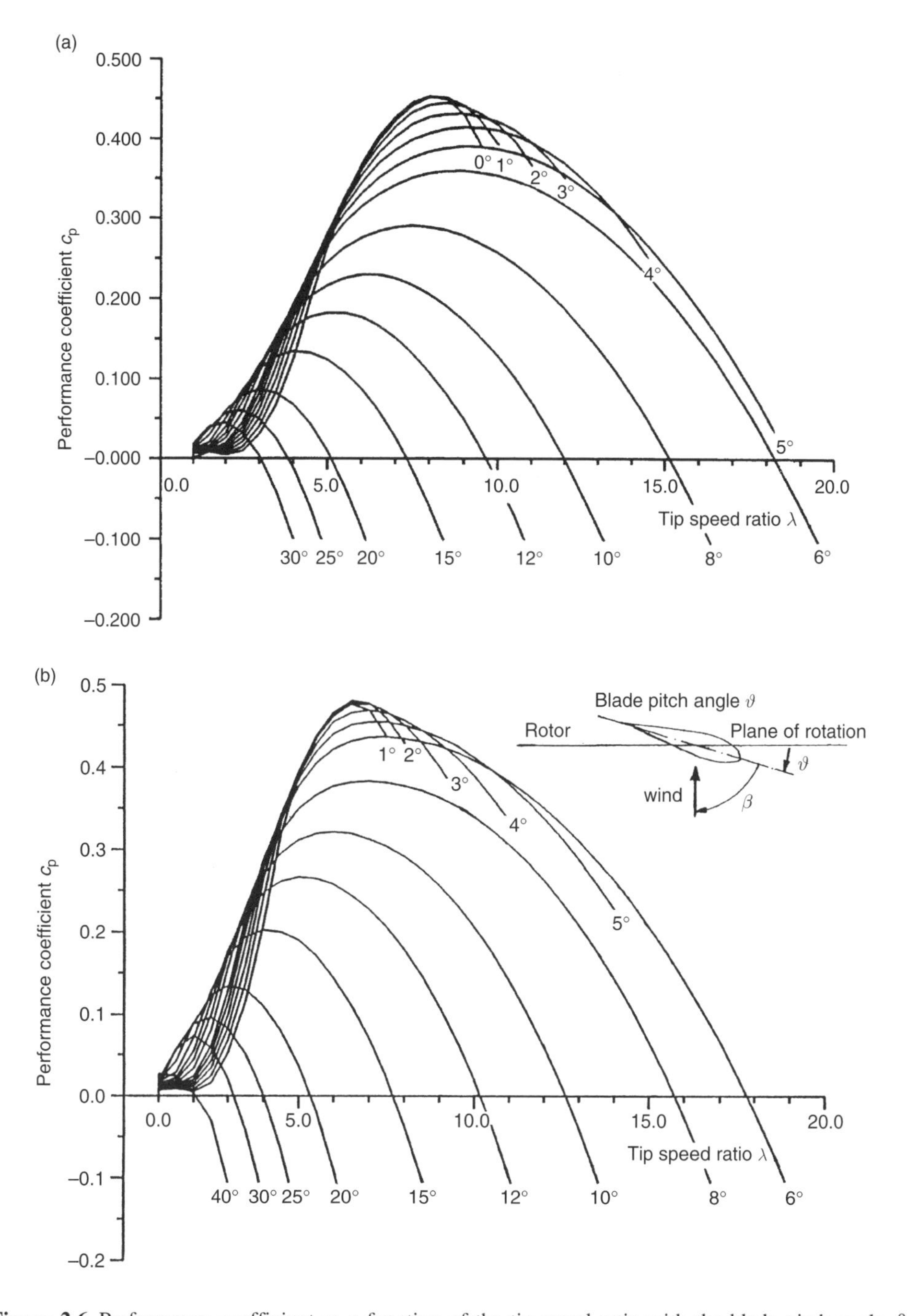

Figure 2.6 Performance coefficient as a function of the tip speed ratio with the blade pitch angle ϑ as a parameter, so-called c_p–λ characteristics; v_{2ax} analytically with cubic equation, torque calculated over 21 points and integrated using Simpson's rule: (a) two-bladed turbine with 12.5 m rotor diameter, 20 kW nominal output (Aeroman 12.5); (b) three-bladed turbine with 60 m rotor diameter, 1.2 MW nominal output (WKA-60)

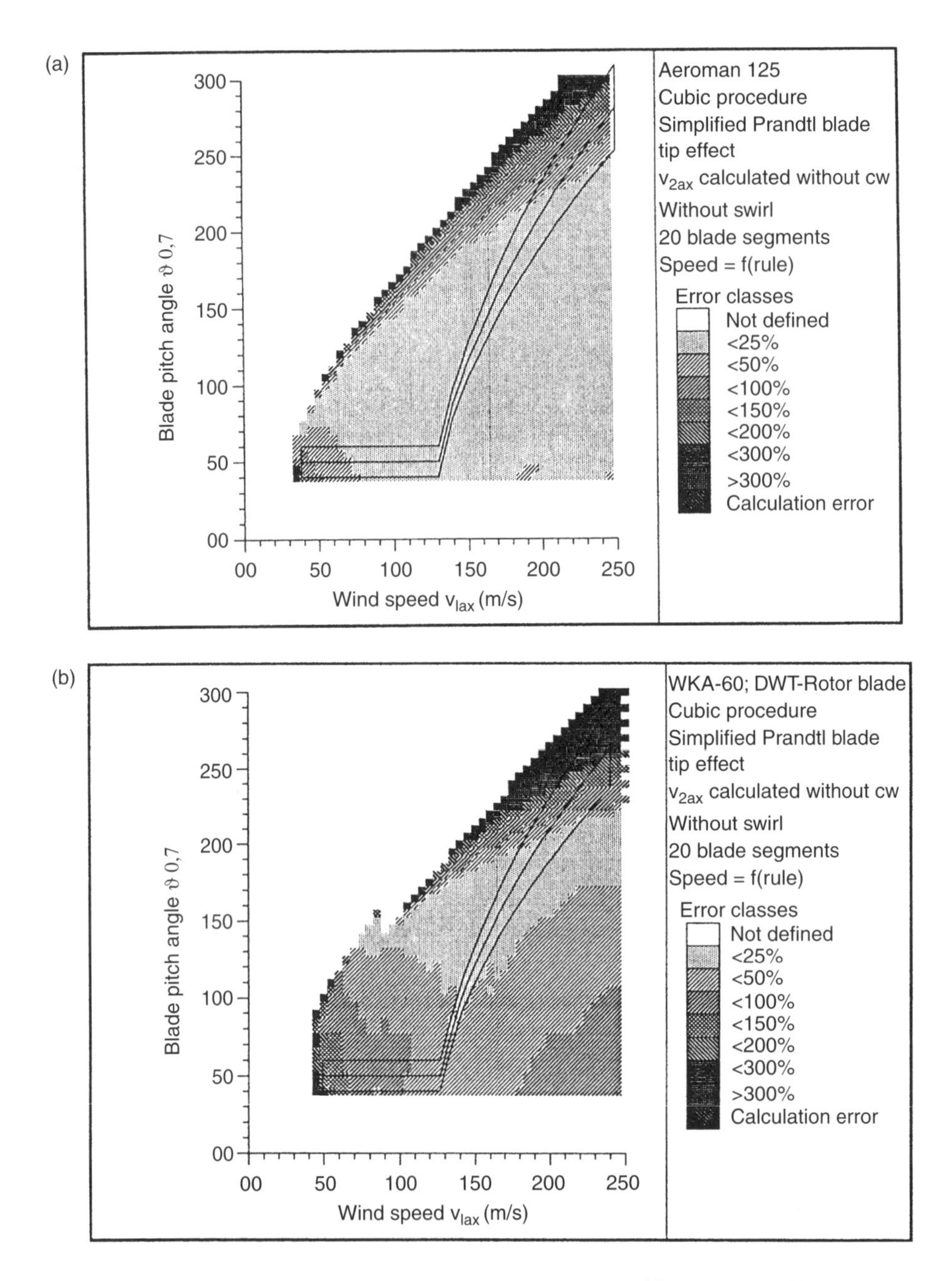

Figure 2.7 Relative errors in simplified rotor speed calculation for (a) a small two-bladed Aeromann 12.5 turbine and (b) a large three-bladed WKA-60 rotor

essentially by external effects and by whatever direct manipulations are performed on the turbine to limit and regulate its energy extraction performance. This forms the main theme of Section 2.2. In the text that follows we now examine simple methods of depicting turbine characteristics.

2.1.5 Modelling turbine characteristics

For studies centred on transient mechanical or electrical effects in wind energy machines, the frequency ranges of the components to be examined take on special significance. For simulations that target power fluctuations in wind energy units and transient effects in wind farms, the time constants between the pitch setting range (Section 2.3) and the rotor range (Section 2.5) are decisive. The resulting frequencies vary from 10 to 0.01 Hz. In modelling electrical effects, frequencies of around a thousand times this must be considered.

As the basis for the determination of rotor performance, Equations (2.4), (2.7) and (2.8) yield the relation

$$P_W = \frac{\rho}{2} c_p (\lambda, \vartheta) A_R v_W^3. \tag{2.36}$$

From this, the performance coefficient c_p can be determined according to Sections 2.1.3 or 2.1.4, etc. The most interesting torque levels of the rotor are thus obtained by the simple mechanical relation

$$M_W = \frac{P_W}{\omega_W}. \tag{2.37}$$

If the c_p–λ performance characteristics (see Figure 2.6), derived from calculations or from direct measurement, are available, the turbine parameters can also be modelled from:

- data fields containing the group of curves derived from measurement or from calculation or
- analytical functions.

Both methods are explained briefly below to provide the user with simple simulation and design tools.

2.1.5.1 Determining the performance coefficient from data fields

If turbine characteristics as in Figure 2.6, or the data for plotting them, are available, data fields can be created by reading off the various values or entering them directly. These then form the basis for power computation in system simulations. The validity of the results thereof will then be dictated by that of the data. When enough data are available, linear interpolation can be used to arrive at intermediate values, although a quadratic derivation is usually better for ongoing processes.

To arrive at a complete data set for the operation of a turbine, however, it may be necessary to extend the characteristics plot. By extending the characteristic curves for small, or even negative, or large angles and by supplementing incomplete characteristics, undefined operating states can be avoided. Also, when the characteristics of a turbine in such areas cannot be forecast exactly, useful decisions and courses of action in extreme situations can be indicated or deduced. Furthermore, many interruptions to the program that would otherwise occur are largely avoided.

2.1.5.2 Approximating the performance coefficient by analytic functions

Groups of c_p–λ curves obtained by measurement or by computation can also be approximated in closed form by nonlinear functions. Following reference [2.9], a model can be derived in the form

$$c_p = c_1 \left(c_2 - c_3 \vartheta - c_4 \vartheta^x - c_5\right) e^{-c_6(\lambda, \vartheta)}. \tag{2.38}$$

According to reference [2.10],

$$c_1 = 0.5, \quad c_2 = v_w/\omega_w, \quad c_3 = 0, \quad x = 2$$
$$c_4 = 0.022, \quad c_5 = 5.6, \quad c_6 = 0.17 v_w/\omega_w$$

can be used for the MOD 2 turbine, where v_w represents wind velocity and ω_w is the angular velocity of the turbine. On the other hand, procedures in reference [2.11] yield

$$c_1 = 0.5, \quad c_2 = 116/\lambda_i, \quad c_3 = 0.4$$
$$c_4 = 0, \quad c_5 = 5 \quad c_6 = 21/\lambda_i$$

and

$$\frac{1}{\lambda_i} = \frac{1}{\lambda + 0.08\vartheta} - \frac{0.035}{\vartheta^3 + 1}.$$

Figure 2.8 was obtained using slightly modified factors with an exponent $x = 1.5$.

According to the characteristics chosen, the above figures must be modified to obtain a close simulation of the machine in question. To manage this, however, demands a non-negligible investment of time and effort, even for those with a long experience of performing such approximations.

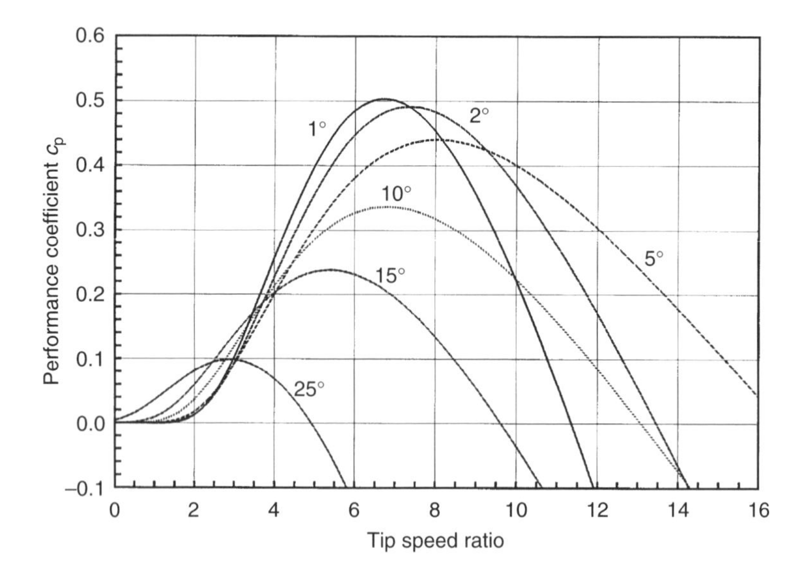

Figure 2.8 Analytical approximation of c_p–λ characteristics (blade pitch angle ϑ as the parameter)

2.1.5.3 Wind-speed profiles, tower effects and transmission losses

The above sections assume a constant distribution of wind speed over the area of the rotor. However, wind-speed profiles result in different wind speeds at the blades nearest to the ground level compared to those at the top of blade travel, which in turn produce corresponding flow and power effects on the turbine (Figure 2.9). Furthermore, gusts and changes in wind speed do not affect the entire rotor at the same instant.

For wind speeds that lie in the operational range of the turbine and exceed about 4 m/s, the wind speed at a given height can be found from the relation

$$v_{\mathrm{w}}(h) = v_{10}\left(\frac{h}{h_{10}}\right)^{a},$$

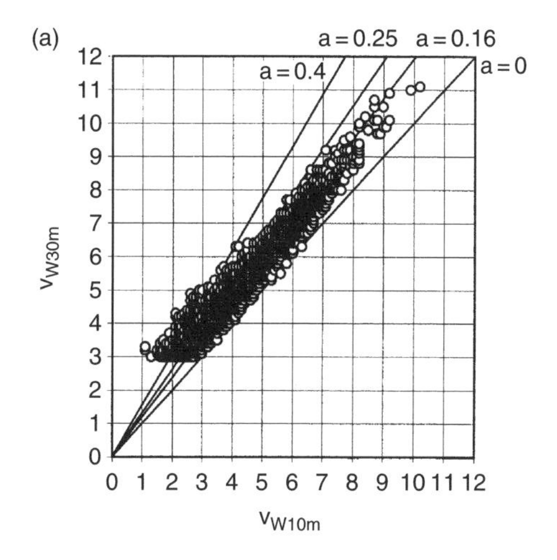

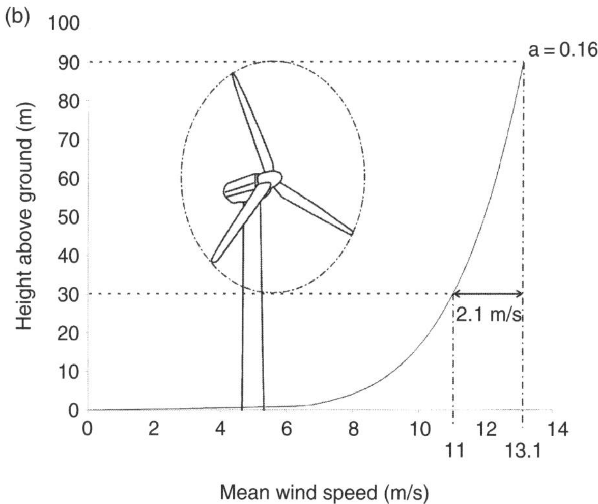

Figure 2.9 Wind-speed gradient: (a) Hellmann exponent (measured values); (b) vertical profile effects on turbines

where $0.14 \leq a \leq 0.17$ and where v_{10} is the determined or computed wind speed for $h_{10} = 10\,\mathrm{m}$, $h = h(\psi)$ corresponding to blade position and a is the so-called Hellman exponent.

In addition to the vertical speed gradient, influences due to tower wind-shadow or windbreak effects cause fluctuations in power and hence torque during blade rotation. Depending upon the number of blades (usually two or three), the height of the turbine, the positioning of the rotor upwind or downwind of the tower, the turbulence depending on tower diameter and the influence of the general surroundings, the results can be very different. The changes in power or torque can only be determined reliably if these influences can be well defined. Once the variations in wind speed over the rotation of a blade have been determined or estimated, the variation in power or torque can be read off the performance characteristic. As a first approximation for research purposes, during a blade rotation the wind conditions at $r = 0.7R$, i.e. at 70 % of the rotor radius, may be used.

In single-blade turbines – as yet the exception and so not studied very closely – power/torque levels at the transmission oscillate widely. In symmetrically arranged multibladed rotors (given identical wind conditions at all blades) the torque produced at the transmission shaft is more or less constant. In these systems, however, flow disturbances also lead to power/torque fluctuations. These results are mostly generated from tower-created turbulence. Nonlinearities in the transmission system (c_p–λ characteristic) lead to further oscillations in the drive train. These can be represented in simplified form using the pitch-dependent torque

$$M(\psi) = M_u - \frac{1}{z}\left[M_o\left(z\omega_w t\right)\right] \tag{2.39}$$

where M_u represents torque under undisturbed wind distribution at the rotor and M_o is the oscillating component. This exerts an effect on the transmission shaft that is inversely proportional to the number of blades and has a recurrence frequency of $z\omega_m$. This can be represented in ramp form or – closer to what actually happens – in cosine form (Figure 2.10(a) and (b)).

Mechanical and electrical losses depend mostly on performance and rotor speed. Effects due to thermal changes, etc., cannot easily be handled mathematically. For simple modelling, machine-dependent losses $P_v = f(P)$ can be handled using constants (P_{vo}) and a power-dependent component (P_{vN}), as shown in Figure 2.11.

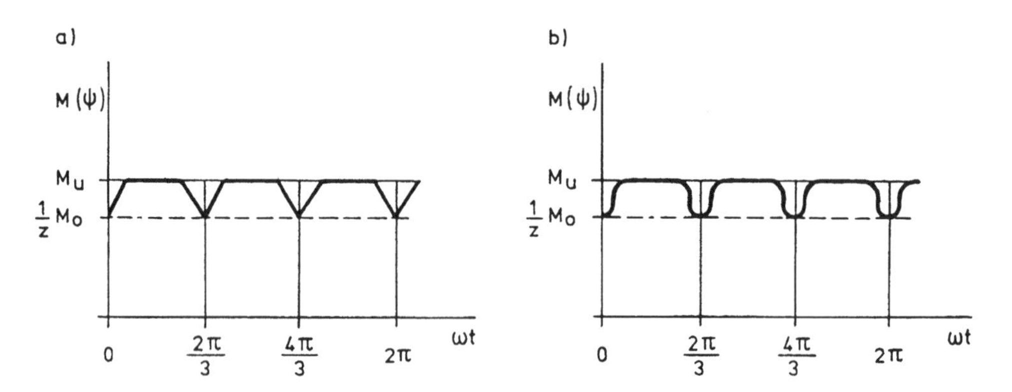

Figure 2.10 Pitch-dependent torque ($z = 3$ rotor blades): (a) ramp representation; (b) cosine representation

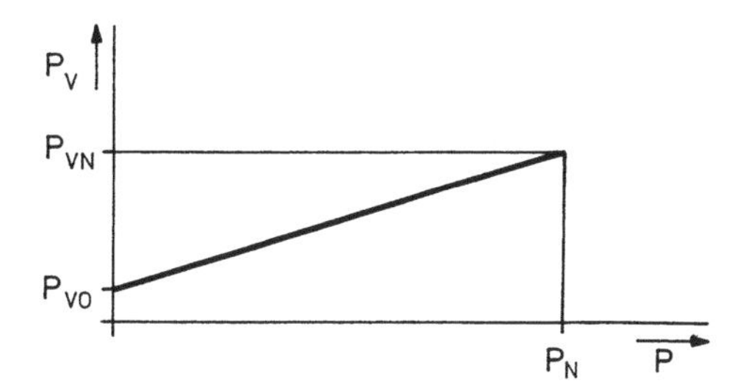

Figure 2.11 Power-dependent loss modelling

2.2 Turbines

The conversion of kinetic energy from the wind for technical applications is effected by means of a variety of turbine types [2.4] (Figure 2.12). Machine designs are divided into horizontal and vertical axis types. Depending on the way in which energy is extracted from the wind, wide variations are discernible between converters, depending on whether they use the drag developed at the surface of the moving parts or the lift exerted on the blades.

In turbines that use drag only, e.g. cup types, board constructions and other surfaces set against the wind, the energy derived is lower than that developed by lift types. Because of lower speeds of rotation, the use of such machines is limited essentially to driving mechanical devices. Their construction is usually simple and very heavy; they will not be considered here.

Turbines using concentrating or suction effects (Figure 2.13), which have, for example, been developed and tested in windy New Zealand, plus thermal turbines (Figure 2.14), a prototype of which has been constructed in sunny southern Spain, are the exceptions in wind turbine technology.

Wind power machines for generating electricity are produced in horizontal and vertical axis format. The turbines are so constructed that they can utilize the force of lift. Lift originates in the flow of air past the rotor blade, which causes an overpressure on the underside of the blade and an underpressure on the top. The tangential component of the lifting forces causes the rotor blade to rotate.

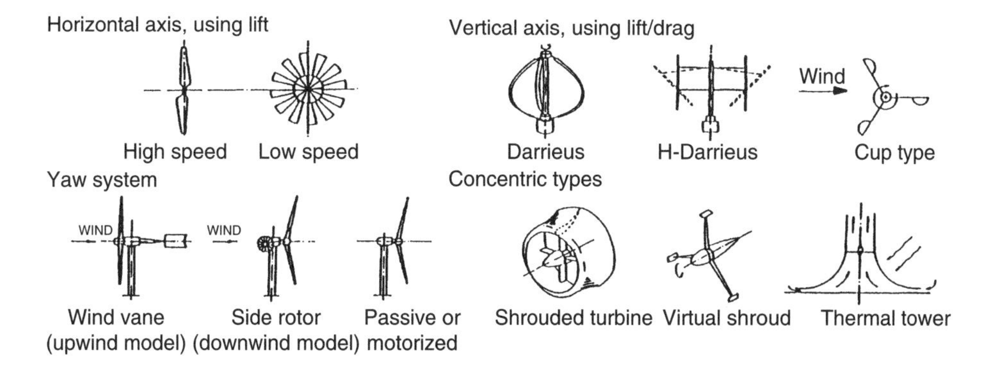

Figure 2.12 Main turbine types

Figure 2.13 Vortex turbine

According to Betz [2.2], a lifting rotor can only extract 60 % of the energy/power from an airstream. The remaining 40 % of power must remain in the air flowing past (see Figure 2.2). Complete braking of the moving air mass to $v_3 = 0$ would result in air backing up at the turbine; the inflow of air would be halted and energy extraction would no longer be possible. In practice, after conversion losses, lower levels in the order of 45 % are achieved. Therefore, in contrast to water-driven turbines, for example, no efficiency levels are quoted for wind turbines – rather the performance coefficient c_p is used. For operating machines, this figure gives the ratio of the power taken from the wind to that contained by it (Equation 2.8).

A deceleration of the air as in Figure 2.2 can occur just as well with many blades moving slowly as with a few blades rotating at high speed. Simple wood and sheet metal

Figure 2.14 Thermal turbine

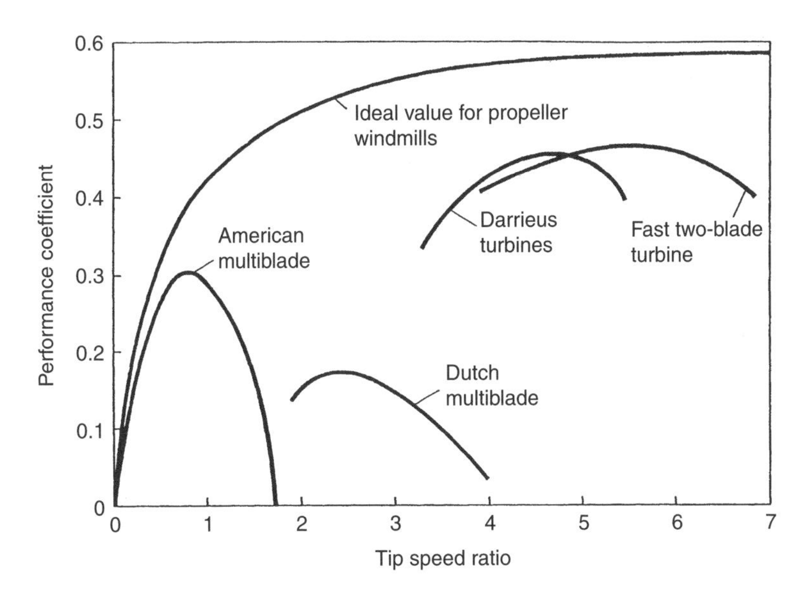

Figure 2.15 Performance coefficient as a function of tip speed ratio for various turbine types, compared with the ideal value [2.12]

constructions only allow slow rotation with high blade counts (e.g. more than six). The torques to be transmitted are correspondingly high, and the constructions must, in consequence, be massive. A few, swiftly spinning blades (e.g. one to three) attain higher power extraction levels and thus better performance coefficients (Figure 2.15). Such figures are, however, only possible using correctly formed aerofoil sections having a low structural area and low vortex creation and which generate little resistance to rotational movement. In such machines, the torque transmitted at high speed is lower, allowing the transmission components to be correspondingly lighter. The rpm, or so-called tip speed ratio, therefore has a critical influence on torque and power figures, and on the coefficients that apply to them.

The tip speed ratio $\lambda = v_u/v_1$ gives the ratio between blade tip speed v_u and the wind speed v_1 upstream of the rotor. The blade tip speed, or rather its maximum value, determines the rotor blade loading. Tip speed usually lies between 50 and 150 m/s. However, due to the problem of noise, tip speeds greater than 75 m/s are not usually possible. The tip speed is critical for establishing blade dimensions in both small and large machines. Since it is defined as the product of the blade radius and the rotational speed, low rotational speeds result for large machines, and vice versa. Hence kilowatt machines reach speeds of around 3 rps, i.e. 180 rpm, while in megawatt machines we observe about one revolution every 3 to 6 seconds, or 10 to 20 rpm.

As shown in Figure 2.13, for high-speed rotors (Darrieus, or two-blade rotors) having aerodynamically formed blades, at tip speed ratios of approximately $\lambda = 4$ to 7, performance coefficients from 0.4 to 0.5 are attained. Slow-running machines (American multiblade or Dutch four-blade rotors) with nonaerodynamic wood and sheet metal blades ($\lambda = 1$ to 2.5) have significantly lower performance coefficients, i.e. $c_p = 0.15$ to 0.3.

For operation over a wide range of wind speeds, horizontal-axis machines with high tip speed ratios (Figures 1.9 to 1.27, etc.) have advantages over vertical-axis machines (Figures 1.30 and 1.31). They also offer advantages in start-up behaviour and control options.

In addition to high aerodynamic quality, wind turbines must also exhibit sufficient rigidity and strength for the lowest capital outlay. Here the number of blades, the rotor speed or tip speed ratio, the constructional technique, the geometry of the blades and the pitch control arrangements, plus the hub construction, play a decisive role.

2.2.1 Hub and turbine design

In stall-regulated machines the rotor blades are fixed directly to the hub (Figure 2.16(a)). In variable pitch turbines, on the other hand, the blades must be flange-mounted such that they can be rotated about their longitudinal axes (see Figure 2.54). In both systems, all forces, moments and vibrations resulting from gusts, tower shadow, etc., are transmitted via the hub to the tower. Improving smooth running and minimizing blade loading can be achieved using a teetering hub design (Figure 2.16(b)). Independently moving blades with cone hinges allow a freely self-adjusting cone angle (Figure 2.16(c)). This corresponds with the direction of the resultant of wind thrust (due to lift on the blade) and centrifugal forces as a result of the rotating blade masses, such that the articulation is free of bending moment in the direction of the wind. However, the two last-mentioned systems require relatively complicated constructions and are thus susceptible to repairs. They have therefore been unable to establish themselves on the market.

As shown in Figure 2.12, the rotor (seen from upwind) can be run in front of the tower (the upwind model, as in historical windmills) or behind the tower (downwind model). Three-bladed rotors (Figure 2.17) are by far the most widely used horizontal-axis types in all power ranges. Two-bladed rotors – widespread in the 1980s at the beginning of the modern

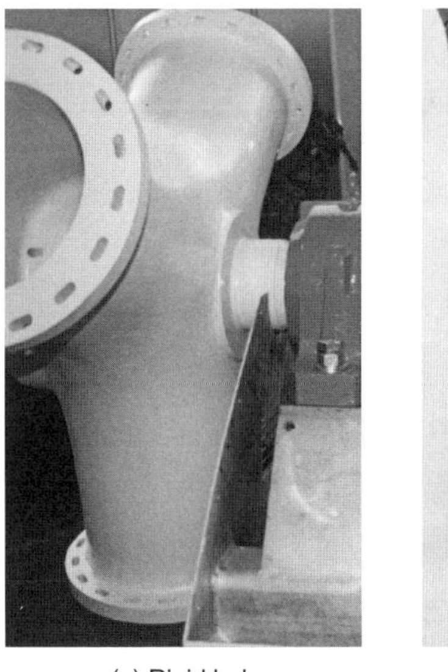

(a) Rigid hub

(b) Teetering hub

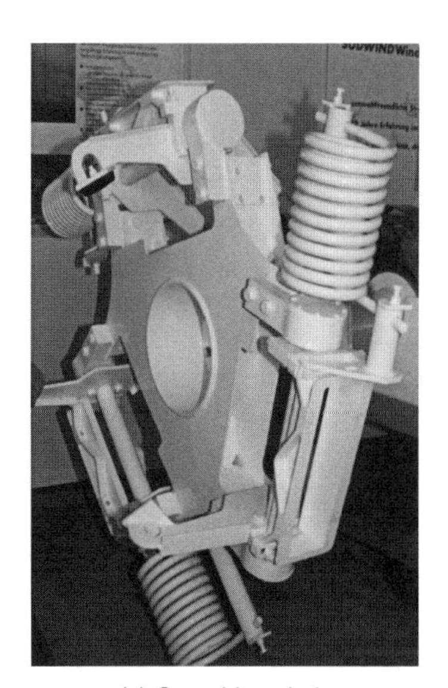

(c) Cone-hinge hub

Figure 2.16 Hub types

Figure 2.17 Three-bladed turbine: Nordex N54, 1000 kW nominal output, 54 m rotor diameter

Figure 2.18 Two-bladed type: Ned Wind, 1 MW nominal output, 52.6 m rotor diameter. Reproduced by kind permission of Nedwind

development of wind turbines – today represent the exception (Figure 2.18). Turbines with a single blade (Figure 2.19), which also appeared in this phase of development, currently occupy an absolutely exceptional position. They have not been able to establish themselves on the market. In the machine shown in Figure 2.17 the blades are rigidly connected to the hub, while in Figure 2.18 the blades are attached to the hub such that they can rotate about their longitudinal axes. The single-blade rotor (Figure 2.19), on the other hand, is fitted with a cone-hinge hub. Multibladed arrangements are commonly found in small machines in the kilowatt range and below.

2.2.2 Rotor blade geometry

When designing a machine to extract energy from the wind and to develop the resulting torque in the turbine, the airstream pattern shown in Figure 2.2 is the goal. It is then possible to achieve energy extraction using many blades at low speed or a few blades at high speed. Further, the optimal wind deceleration for the same speed of rotation can be attained using

Figure 2.19 Single-bladed type, Monopteros (MBB), 640 kW nominal output, 56 m rotor diameter

one very broad blade, or two or three blades of correspondingly smaller breadth. The optimal blade chord can be derived from the blade radius by the formula

$$t_{\mathrm{B}}(r)=\frac{2\pi r}{z}\frac{8}{9}\frac{1}{c_{\mathrm{a}}}\frac{v_1^2}{v_{\mathrm{u}}(r)\,v_{\mathrm{r}}(r)}. \tag{2.40}$$

A rotor tip speed of

$$v_{\mathrm{u}}(r)=\omega r=\lambda\frac{r}{R}v_1, \tag{2.41}$$

a target wind deceleration at the turbine of

$$v_2=\frac{2}{3}v_1 \tag{2.42}$$

and the resultant relative wind speed of

$$v_{\mathrm{r}}(r)=\sqrt{v_{\mathrm{u}}^2+v_2^2}=v_1\sqrt{\left(\frac{\lambda}{R}r\right)^2+\frac{4}{9}} \tag{2.43}$$

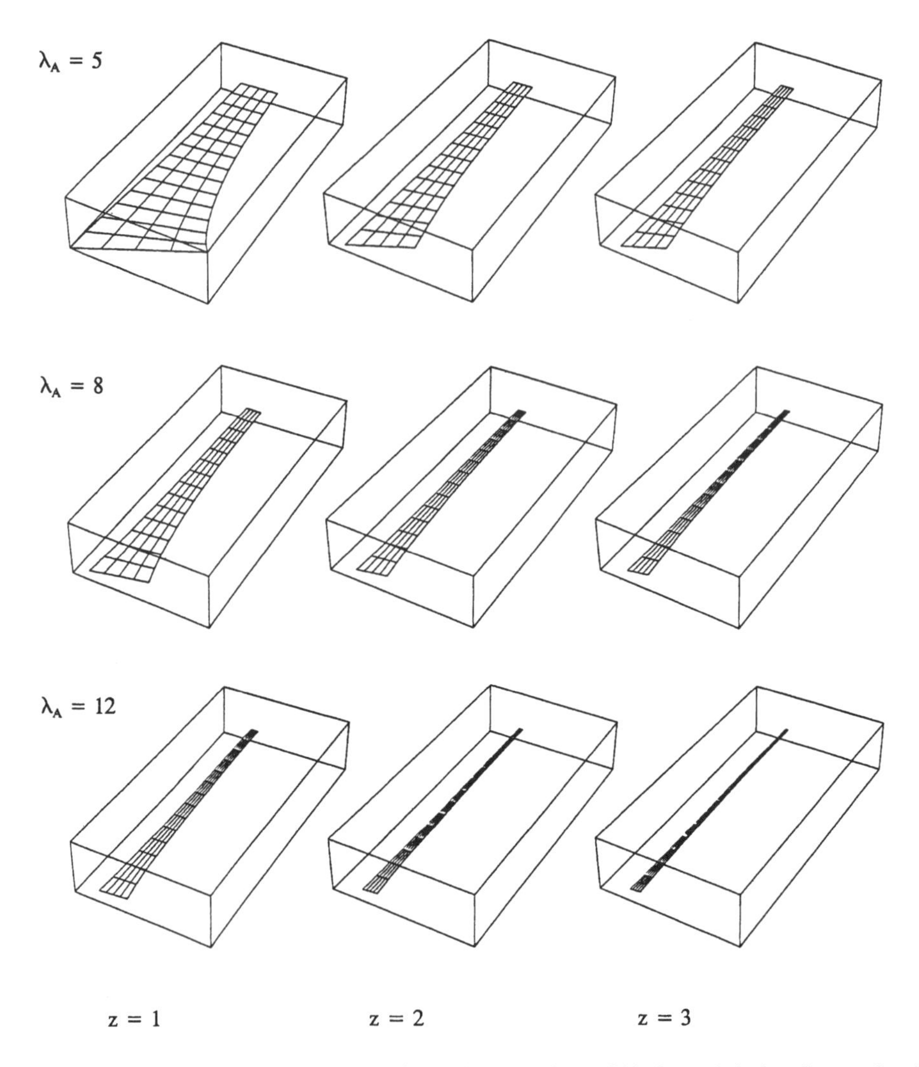

Figure 2.20 Optimal rotor blade geometry for various numbers of blades and design tip speed ratios

yield a blade chord of

$$t_{\mathrm{B}}(r) = \frac{16\pi R}{9c_{\mathrm{a}}\lambda z}\frac{1}{\sqrt{\left(\frac{\lambda}{R}r\right)^2 + \frac{4}{9}}}. \tag{2.44}$$

According to this, hyperbolic blade contours are to be expected. These, however, differ widely for different (design) tip speed ratios and numbers of rotor blades [2.13]. Figure 2.20 shows the relative blade contours for one-, two- and three-bladed machines having design tip speed ratios λ_{A} of 5, 8 and 12. This presentation makes it absolutely clear that single-bladed turbines are best for high-speed and multibladed turbines for low-speed machines.

(a) Optimal shape (Südwind). Reproduced by kind permission of Südwind GmbH

(b) Trapezoidal blade (Windmaster). Reproduced by kind permission of Windmaster GmbH

(c) Rectangular blade (Krogmann). Reproduced by kind permission of H. J. Krogmann

Figure 2.21 Produced rotor blade shapes

When producing rotor blades, the optimal blade contour is approximated, mostly in trapezoidal or even in rectangular form (Figure 2.21). Trapezoidal blades, by far the most widely used, reach performance coefficients approaching those of optimally shaped blades. Rectangular outlines, in contrast, yield a markedly lower maximum performance coefficient at the design tip speed ratio. Outside the design area, however, extensive ranges of operation exist, some with even better performance ratios. In addition to blade outline, the position of the rotor blades relative to the wind direction or to the rotor plane is decisive in extracting energy from the wind.

The peripheral speed of the turbine is high at the blade tip and relatively low at the hub. This results, for the same airstream effect, in a low blade chord at the tip and a larger blade area near the hub. To obtain similar flow conditions over the entire length of the blade when running, similar flow directions relative to the aerofoil section must be achieved at all points between the tip and the hub. This is attained when the relative airspeed at design performance (i.e. nominal operation conditions) is the same at all radii (Figure 2.20).

Following Figures 2.4 and 2.22, the angle of attack α is the difference between the pitch angle ϑ and the resultant relative airflow direction. This is the vector sum of the wind deceleration ($v_2 = \frac{2}{3}v_{\mathrm{w}} = \frac{2}{3}v_1$) and the peripheral speed resulting from blade rotation. Hence the blade pitch angle

$$\vartheta(r) = \arctan\frac{2v_1}{3v_{\mathrm{u}}} - \alpha \tag{2.45}$$

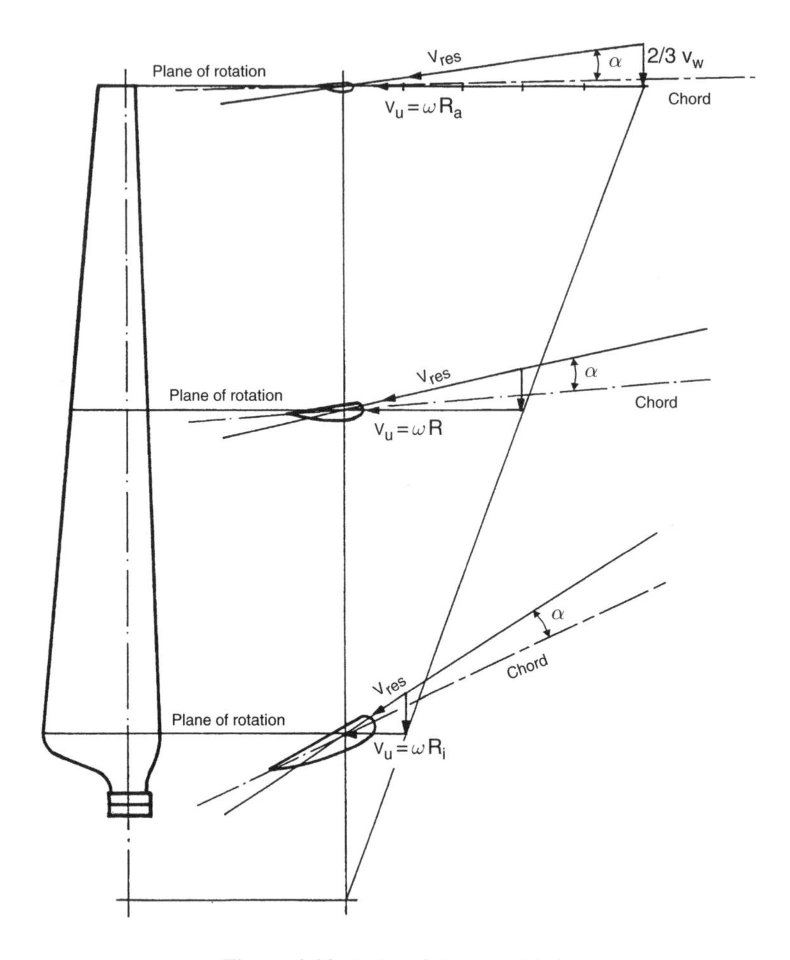

Figure 2.22 Twist of the rotor blade

and its equivalent power complement

$$\beta(r) = \frac{\pi}{2} - \vartheta(r). \tag{2.46}$$

For electricity-generating machines, high-speed turbines are usually used. The rotor blades must exhibit the lowest possible moving mass and must be very strong. Further, glass-fibre-reinforced polyester and epoxy resin composite materials are mostly used to yield aerodynamically correct blade profiles and high strength. Even greater strength may be achieved using carbon fibre materials and the use of such materials is on the increase, in spite of the higher costs involved.

Figure 2.23 shows the manufacture of rotor blades in the megawatt range. In large turbines, in particular, the necessary distance from the tower can be achieved even in operation at high wind speeds with bent blades (Figure 2.24). Figure 2.25 shows the complete rotor being lifted for fitting to the rotor shaft.

Figure 2.23 Manufacturing a rotor blade. Reproduced by permission of Nordex

Figure 2.24 Bent rotor blade (NOI 37.5 / 77 m rotor diameter)

Figure 2.25 Installing the rotor. Reproduced by permission of DeWind

Figure 2.26 Transport of rotor blades (Nordex N80). Reproduced by permission of Nordex

Rotor blades are currently manufactured in two shells and put together into a single unit in the production halls for transporting and fitting. Figure 2.26 illustrates the dimensions of the rotor blades and the demands this places upon manufacture, logistics and fitting. Furthermore, it becomes evident that the limits of transportability can easily be exceeded in many regions. The importance of factories in harbour areas will therefore increase as the size of turbines – particularly those for offshore use – increases.

The construction and the fitting of rotor blades and turbines, which are discussed in more detail in references [2.3], [2.4] and [2.13], will not be considered further here.

2.3 Power Control by Turbine Manipulation

In earlier centuries, wind energy machines were protected from over-revving by manual intervention, e.g. by reefing the sails or by turning the plane of rotation to lie parallel to the wind. In American (slow-running) wind turbines it was even possible to control power take-up by yawing or tilting the rotor, depending on the structure. Since the middle of the twentieth century, principles founded on aerodynamics have come into use to limit power using blade stall or to modify it by varying blade pitch. In special designs, even in the 10 kW class, variable-speed generators allow an adjustable adaptation of turbine output to the prevailing wind conditions.

2.3.1 Turbine yawing

Power extraction in slow-running turbines can be limited by yawing or tilting the plane of rotation in the direction of wind pressure. In this way the effective flow cross-section of the rotor is reduced and the flow incident on each blade considerably modified. Figure 2.27 shows the drastic drop in performance coefficients resulting from turning wind turbines out of the wind, with consequent blade stall.

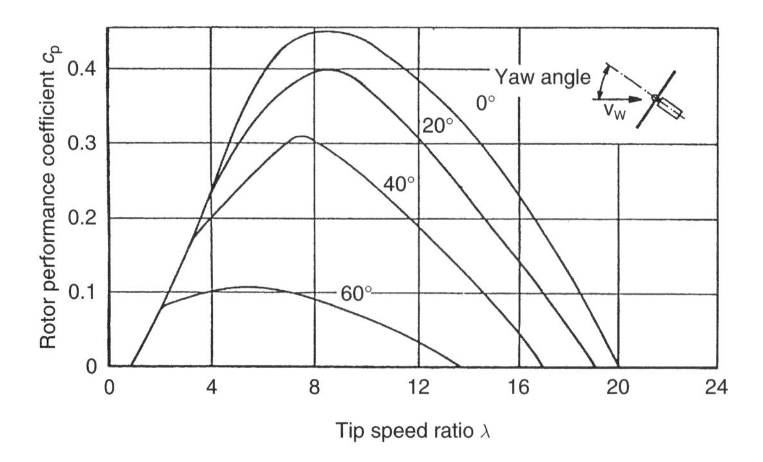

Figure 2.27 Rotor performance coefficients under skewed airflow [2.14]

For multibladed symmetrical models that run at low speeds, this operation is particularly effective. The following shows why.

2.3.1.1 Control procedures and yaw accelerations

For high-speed wind turbines, turning the plane of rotation out of the wind by a predetermined amount provides a slow-acting means of protecting the machine. This procedure can be carried out by an active yaw-control system. Here, skewing the rotational plane with respect to the wind exerts moments on the rotating blades, the behaviour of which must be formalized if their influence on the control system is to be determined. Further, the following considers the accelerations and moments occurring in a two-bladed model such as that shown in Figure 2.28.

The frame of reference x', y', z' is fixed with respect to the rotor. Then, for the centre of mass of a stiff blade, for the speed in this coordinate system

$$\frac{\mathrm{d}r'}{\mathrm{d}t} = v' = 0$$

holds, with the equivalent acceleration

$$\frac{\mathrm{d}v'}{\mathrm{d}t} = b' = 0.$$

For a yaw velocity of ω_A at the head pivot, a turbine rotating with a velocity of ω_R gives a resultant angular velocity of

$$\boldsymbol{\omega} = \boldsymbol{\omega}_A + \boldsymbol{\omega}_R \tag{2.47}$$

and a rotor head yaw velocity of

$$\boldsymbol{v}_0 = \boldsymbol{\omega} \times \boldsymbol{r}_0 = \boldsymbol{\omega}_A \times \boldsymbol{r}_0. \tag{2.48}$$

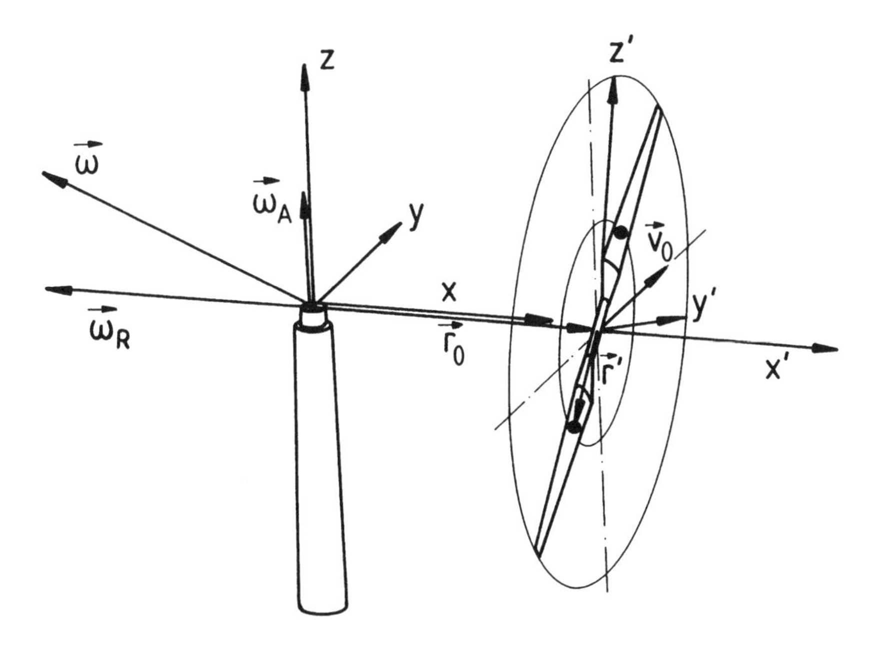

Figure 2.28 Relative motion of rotating blades for a wind turbine under azimuth angle yawing

The total velocity of the centre of mass [2.15] is then

$$\boldsymbol{v} = \boldsymbol{v}_0 + \boldsymbol{v}' + \boldsymbol{\omega} \times \boldsymbol{r}' = \boldsymbol{\omega}_\mathrm{A} \times \boldsymbol{r}_0 + \boldsymbol{\omega} \times \boldsymbol{r}' \tag{2.49}$$

and

$$\boldsymbol{b} = \boldsymbol{b}' + \boldsymbol{b}_0 + 2\boldsymbol{\omega} \times \boldsymbol{v}' + \frac{\mathrm{d}\boldsymbol{\omega}}{\mathrm{d}t} \times \boldsymbol{r}' + \boldsymbol{\omega} \times (\boldsymbol{\omega} \times \boldsymbol{r}'), \tag{2.50}$$

where

$$\boldsymbol{b}' = 0 \qquad \text{and} \qquad 2\boldsymbol{\omega} \times \boldsymbol{v}' = 0.$$

The centripetal acceleration of the head

$$\boldsymbol{b}_0 = \omega_\mathrm{A} \times (\omega_\mathrm{A} \times \boldsymbol{r}_0)$$

and

$$\frac{\mathrm{d}\boldsymbol{\omega}}{\mathrm{d}\boldsymbol{t}} = \boldsymbol{\omega}_\mathrm{A} \times \boldsymbol{\omega} = \boldsymbol{\omega}_\mathrm{A} \times (\boldsymbol{\omega}_\mathrm{A} + \boldsymbol{\omega}_\mathrm{R}) = \boldsymbol{\omega}_\mathrm{A} \times \boldsymbol{\omega}_\mathrm{R}$$

gives the acceleration of the centre of gravity

$$\boldsymbol{b} = \boldsymbol{\omega}_\mathrm{A} \times (\boldsymbol{\omega}_\mathrm{A} \times \boldsymbol{r}_0) + (\boldsymbol{\omega}_\mathrm{A} \times \boldsymbol{\omega}_\mathrm{R}) \times \boldsymbol{r}' + (\boldsymbol{\omega}_\mathrm{A} + \boldsymbol{\omega}_\mathrm{R}) \times ((\boldsymbol{\omega}_\mathrm{A} + \boldsymbol{\omega}_\mathrm{R}) \times \boldsymbol{r}')$$

or with components

$$b = \underbrace{\boldsymbol{\omega}_{\mathrm{A}} \times (\boldsymbol{\omega}_{\mathrm{A}} \times (\boldsymbol{r}_0 + \boldsymbol{r}'))}_{\substack{\text{centripetal acceleration} \\ \text{from}\omega_{\mathrm{A}}}} + \underbrace{2\boldsymbol{\omega}_{\mathrm{A}} \times (\boldsymbol{\omega}_{\mathrm{A}} \times \boldsymbol{r}')}_{\text{Coriolis acceleration}} + \underbrace{\boldsymbol{\omega}_{\mathrm{R}} \times (\boldsymbol{\omega}_{\mathrm{R}} \times \boldsymbol{r}')}_{\substack{\text{centripetal acceleration} \\ \text{from}\omega_{\mathrm{R}}}} \tag{2.51}$$

which we will now examine in detail, so as to be able to interpret the individual accelerations clearly.

Centripetal acceleration due to yawing
When the offset of the rotor head is

$$\boldsymbol{r}_0 = |\boldsymbol{r}_0|\, \boldsymbol{e}_x = r_0 \boldsymbol{e}_x$$

or that of the rotor blade centre of mass from the tower axis is

$$\boldsymbol{r}' = y' \boldsymbol{e}_y + z' \boldsymbol{e}_z$$

then from Figure 2.29 the resultant is

$$\boldsymbol{r}_0 + \boldsymbol{r}' = r_0 \boldsymbol{e}_x + y' \boldsymbol{e}_y + z' \boldsymbol{e}_z.$$

Allowing for a yaw control angular velocity of

$$\boldsymbol{\omega}_{\mathrm{A}} = \boldsymbol{\omega}_{\mathrm{A}} \boldsymbol{e}_z$$

yields the determinant for yaw speed

$$\boldsymbol{\omega}_{\mathrm{A}} \times (\boldsymbol{r}_0 + \boldsymbol{r}') = \begin{vmatrix} \boldsymbol{e}_x & \boldsymbol{e}_y & \boldsymbol{e}_z \\ 0 & 0 & \omega_{\mathrm{A}} \\ r_0 & y' & z' \end{vmatrix} = -\omega_{\mathrm{A}} y' \boldsymbol{e}_x + \omega_{\mathrm{A}} r_0 \boldsymbol{e}_y$$

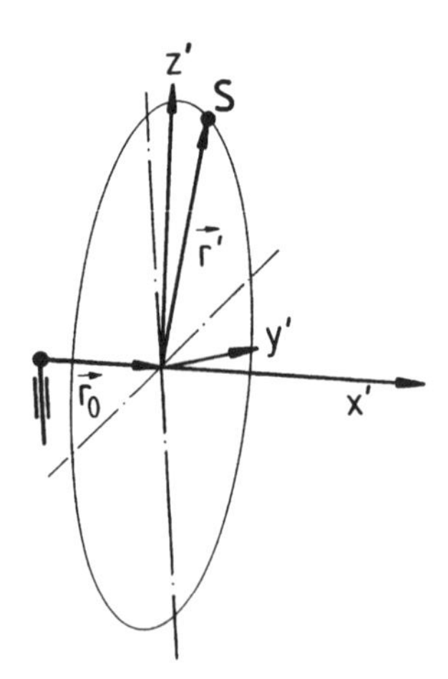

Figure 2.29 Centre of gravity of rotating blade

and thus the centripetal acceleration from ω_A

$$\boldsymbol{b}_{\omega_A} = \boldsymbol{\omega}_A \times (\boldsymbol{\omega}_A \times (\boldsymbol{r}_0 + \boldsymbol{r}')) = \begin{vmatrix} \boldsymbol{e}_x & \boldsymbol{e}_y & \boldsymbol{e}_z \\ 0 & 0 & \omega_A \\ -\omega_A y' & \omega_A r_0 & 0 \end{vmatrix}$$

or

$$\boldsymbol{b}_{\omega_A} = -\omega_A^2 r_0 \boldsymbol{e}_x - \omega_A^2 y' \boldsymbol{e}_y. \tag{2.52}$$

For rotors with two symmetrical blades, the component becomes

$$\omega_A^2 y' \boldsymbol{e}_y = 0$$

and thus engenders an intrinsic moment that loads the blade structure and, in pitch-regulated machines, the blade bearings.

The residual acceleration

$$\boldsymbol{b}_{\omega_A} = -\omega_A^2 r_0 \boldsymbol{e}_x \tag{2.53}$$

only results in a moment along the x axis. Yaw control can generally attain only low angular velocities ω_A, so that b_{ω_A} and the resulting stresses play only a secondary role.

Centripetal acceleration due to blade rotation

From Figure 2.29, the radius of rotation of the centre of gravity is

$$\boldsymbol{r}' = y' \boldsymbol{e}_y + z' \boldsymbol{e}_z$$

and its angular velocity is

$$\boldsymbol{\omega}_R = -\omega_R \boldsymbol{e}_x.$$

Therefore the velocity at S_p is

$$\boldsymbol{\omega}_R \times \boldsymbol{r}' = \begin{vmatrix} \boldsymbol{e}_x & \boldsymbol{e}_y & \boldsymbol{e}_z \\ -\omega_R & 0 & 0 \\ 0 & y' & z' \end{vmatrix} = \omega_R z' \boldsymbol{e}_y - \omega_R y' \boldsymbol{e}_z$$

and the centripetal acceleration is

$$\boldsymbol{b}_R = \boldsymbol{\omega}_R \times (\boldsymbol{\omega}_R \times \boldsymbol{r}') = \begin{vmatrix} \boldsymbol{e}_x & \boldsymbol{e}_y & \boldsymbol{e}_z \\ -\omega_R & 0 & 0 \\ 0 & \omega_R z' & -\omega_R y' \end{vmatrix}$$

Therefore

$$\boldsymbol{b}_R = -\omega_R^2 y' \boldsymbol{e}_y - \omega_R^2 z' \boldsymbol{e}_z. \tag{2.54}$$

Although the forces or moments resulting from centripetal acceleration place loads on both blades and bearings, these components have no effect on yaw control in symmetrically arranged rotors.

Coriolis acceleration

From the velocity of the centre of mass

$$\boldsymbol{\omega}_\mathrm{R} \times \boldsymbol{r}' = \omega_\mathrm{R} z' \boldsymbol{e}_y - \omega_\mathrm{R} y' \boldsymbol{e}_z$$

and the corresponding yaw angular velocity

$$\boldsymbol{\omega}_\mathrm{A} \times (\boldsymbol{\omega}_\mathrm{R} \times \boldsymbol{r}') = \begin{vmatrix} \boldsymbol{e}_x & \boldsymbol{e}_y & \boldsymbol{e}_z \\ 0 & 0 & \omega_\mathrm{A} \\ 0 & \omega_\mathrm{R} z' & -\omega_\mathrm{R} y' \end{vmatrix} = -\omega_\mathrm{A} \omega_\mathrm{R} z' \boldsymbol{e}_x$$

the Coriolis acceleration may be derived:

$$\boldsymbol{b}_\mathrm{c} = 2\boldsymbol{\omega}_\mathrm{A} \times (\boldsymbol{\omega}_\mathrm{R} \times \boldsymbol{r}') = -2\omega_\mathrm{A} \omega_\mathrm{R} z' \boldsymbol{e}_x. \tag{2.55}$$

From this, we obtain, for an observer in a coordinate system x'', y'', z'' having a yaw velocity of ω_A but not a rotation of ω_R, a Coriolis acceleration b_C in the negative x'' direction when $z'' > 0$. For $z'' < 0$, b_C acts in the opposite direction (see Figure 2.30).

2.3.1.2 Yaw moments

Coriolis acceleration engenders moments that act on the rotor blades and on the yaw system in the tower head. When the nacelle is yawed the moment in the x direction acts not with a continuously damping effect, but either to accelerate or decelerate the movement: for a rotor blade of mass m_B the moment resulting from Coriolis acceleration with respect to the z'' or z axis is

$$M_{\mathrm{C}z} = -m_\mathrm{B} b_\mathrm{C} y''. \tag{2.56}$$

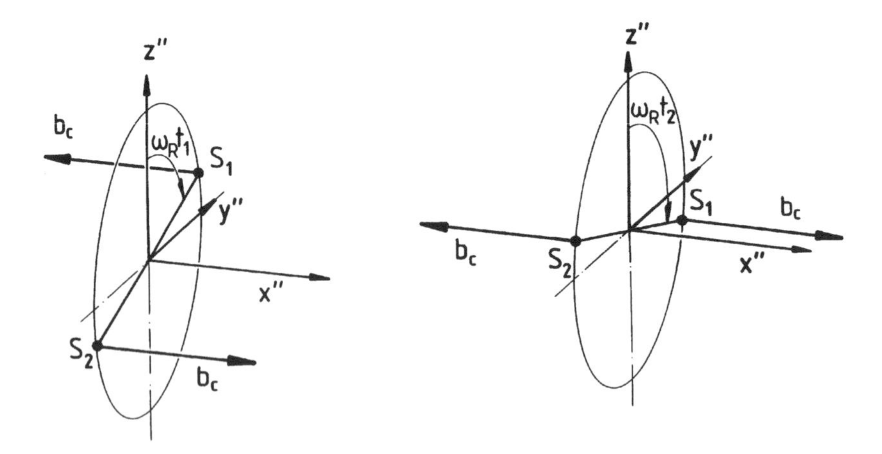

Figure 2.30 Coriolis acceleration of a rotor blade

Where

$$z'' = r' \cos \omega_R t,$$

$$y'' = r' \sin \omega_R t$$

and

$$\sin \omega_R t \cos \omega_R t = \frac{1}{2} \sin 2\omega_R t$$

the moment in the x direction can be written as

$$M_{Cz} = -m_B r'^2 \omega_A \omega_R \sin 2\omega_R t \quad (2.57)$$

With a rotor blade moment of inertia of

$$J_B = m_B r'^2$$

this becomes

$$M_{Cz} = -J_B \omega_A \omega_R \sin 2\omega_R t. \quad (2.58)$$

In the angular ranges

$$0 \leq \omega_R t \leq \frac{\pi}{2} \quad \text{and} \quad \pi \leq \omega_R t \leq \frac{3}{2}\pi$$

the Coriolis acceleration for both blades, $2M_{Cz}$, works against the yaw movement.

In the ranges

$$\frac{\pi}{2} \leq \omega_R t \leq \pi \quad \text{and} \quad \frac{3}{2}\pi \leq \omega_R t \leq 2\pi$$

the $2M_{Cz}$ works with the yaw movement and tries to accelerate it. Thus damping or accelerating torques result from this overlaid rotation, depending upon the position of the rotor. This effect can be observed in two-bladed machines as jerkiness in the yaw movement. In three- and four-bladed machines, among others, these accelerating torques add up to zero and so have no effect on yawing.

Because of the moment along the y axis, which engenders a yawing moment on the rotor, swinging the turbine too fast can subject all arrangements of rotor blades to severe and even dangerous stresses. Such steering operations should therefore only be carried out very slowly.

2.3.1.3 Yaw control mechanisms

Very slow and controlled yawing can be achieved using active positioning mechanisms. Electric azimuth drives are most commonly used. Figure 2.31 shows all the main components of a typical control mechanism, with its motor–gearbox unit, ring gear and yaw control.

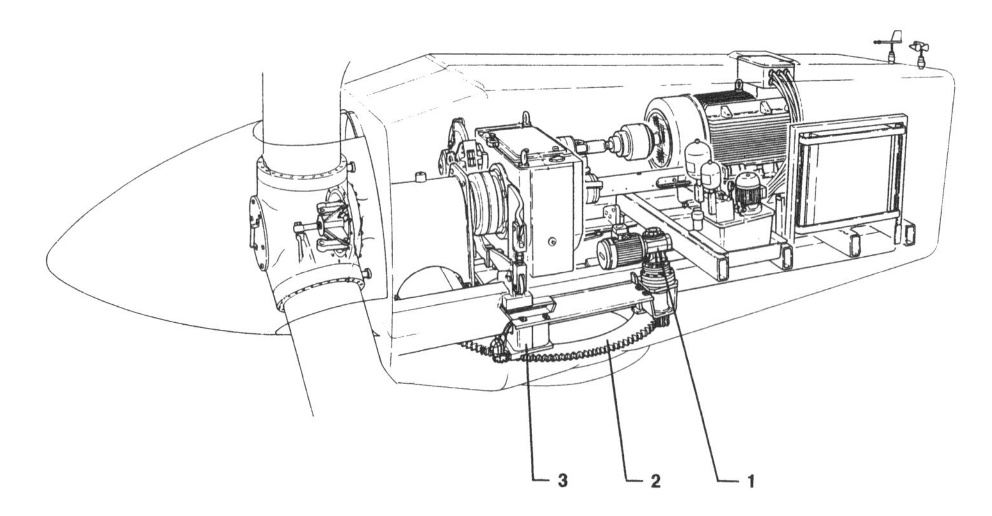

Figure 2.31 Electrically driven turbine yaw system (Vestas V27) with an external ring gear: 1, geared motor; 2, ring gear; 3, yaw control

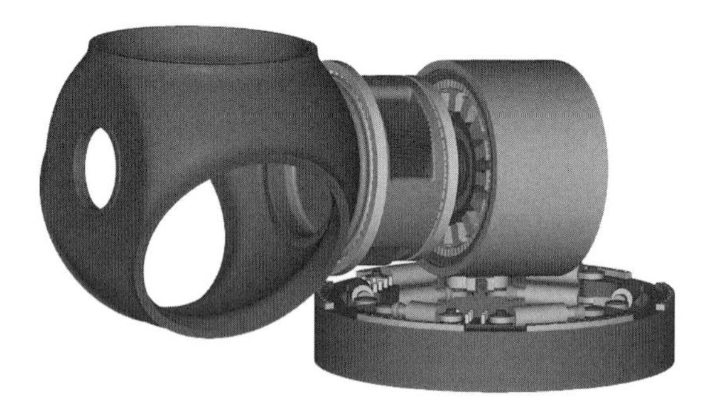

Figure 2.32 Hydraulic turbine yaw system (Multibrid N5000, 5 MW). Reproduced by permission of Aerodyn Energiesystems GmbH

Hydraulic systems are usually of similar construction. They are, however, used only in large turbines (Figure 2.32).

In large turbines, internal ring gears (Figure 2.33(a)) are generally used instead of the external ring gear shown in Figure 2.31. Furthermore, several centrally mounted geared motors (Figure 2.33(b)) are established in the market. Their greater axial length, among other things, makes it easier to climb into the nacelle from the tower. Brake systems (Figure 2.33(c)) stop the yaw movement and thus protect the yaw gears and ring gears in particular.

Side-rotor yaw systems were widespread in earlier times, also being found in Dutch windmills (Figure 1.7), but for cost reasons they are hardly used today. They are mostly limited to wind energy machines that are not connected to the public grid supply (Figure 2.34).

Figure 2.33 Electrically driven turbine yaw system for large turbines: (a) internal ring gear, (b) geared–motor pair and (c) yaw–brake pair

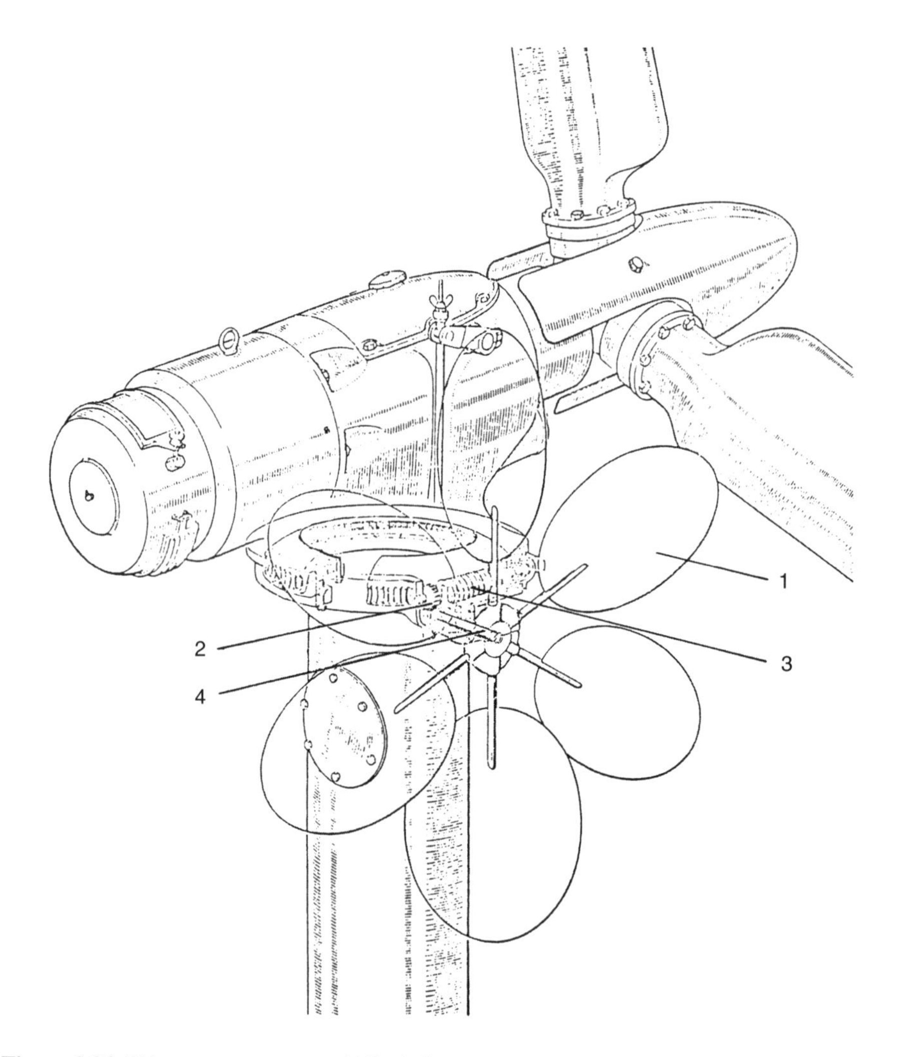

Figure 2.34 Side rotor yaw system (Allgaier): 1, side rotor; 2, pinion; 3, worm; 4, worm shaft

Passive yaw systems are only used on small machines. The wind vane system (Figure 2.35(a)) was very popular, being used by the million on the American wind pumps known as the 'Western wheel' and similar machines (Figure 1.8). On the other hand, downwind turbines (Figure 2.35(b)), steered by wind pressure of the rotor blades, are not used so often. In these systems an azimuth drive is unnecessary, but Coriolis accelerations caused by changes in wind direction can exert considerable forces on the blades unless some damping or stabilizing system is provided.

Because of continual changes in wind speed, wind turbines are subjected to ongoing and occasionally enormous available-energy gradients. To extract energy from the wind under such conditions, it must be possible to act swiftly on the turbine system. Rotor blade pitch control is particularly suitable for this purpose.

(a) Upwind turbine with wind vane (Enercon)

(b) Downwind turbine with wind-pressure yawing

Figure 2.35 Passive yaw systems

2.3.2 Rotor blade pitch variation

The rules developed in Section 2.1 have shown that the performance coefficient and therefore power production from the turbine is strongly influenced by variation of the blade pitch relative to the direction of the wind or the plane of rotation. As the example in Figure 2.6(b) clearly shows, at a tip speed ratio of $\lambda = 7$ the performance coefficient, and thus the output of the turbine, can be roughly halved by varying the blade pitch angle from 5 to 12°.

Rotating the blades about their longitudinal axes (see Figure 2.36) thus facilitates swift and active modification of the drive power on the rotor. To do this, a pitch variation mechanism must apply the necessary moment to the blade. In addition to the moment that the variation mechanism is designed and constructed to exert, moments arising from the following must also be taken into account:

- the inertia of the entire adjustment mechanism (motor, clutches, shafts, brake discs and lever mechanism or hydraulic cylinder, etc.);
- the springing and damping characteristics of the mechanism;
- adjustment springs (if present).

In lever-operated pitch adjustment mechanisms, these moments are also dependent upon the pitch angle. As a result, nonlinear relationships arise between drive power and blade pitch, which have an influence on transmission behaviour and can, for example, cause variations in the amplification ratio.

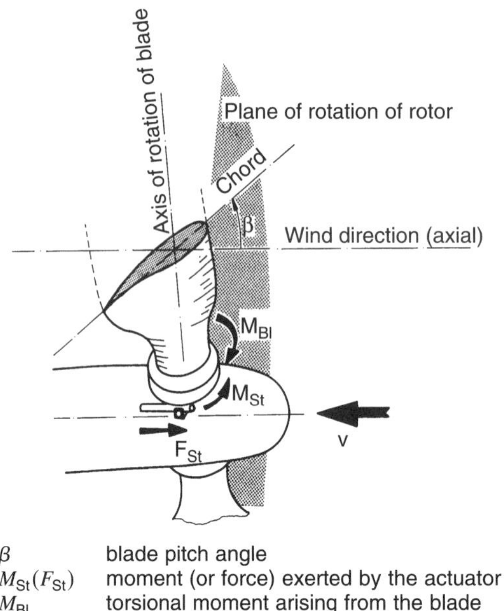

β	blade pitch angle
$M_{St}(F_{St})$	moment (or force) exerted by the actuator
M_{Bl}	torsional moment arising from the blade

Figure 2.36 Blade pitch variation mechanism

In the following, the individual moments are explained, illustrated and examined in detail, with a view to establishing methods for handling them mathematically.

2.3.2.1 Moments on the rotor blade

The following design characteristics of a rotor blade:

- lateral and longitudinal geometry,
- stiffness,
- mass distribution, dictated by the choice of constructional materials,
- possible degrees of freedom in the directions of teeter and wind thrust and in the rotational direction of the rotor and
- bearings,

can, depending on the operating conditions of a machine, give rise to the torsional moments described in what follows.

Propeller moments

Propeller moments arise as a result of the unequal mass distribution of the rotor blade (shown in Figure 2.37 as an aerofoil element) with respect to the axis of rotation of the blade due to the centrifugal force F_z acting on every partial centre of mass.

Breaking this down into its normal and transverse components F_N and F_Q yields the quantity F_{Pr}, which is largely dependent on the speed of rotation and blade pitch angle. Multiplying F_{Pr} by a_P, which is the offset from the blade axis at which F_{Pr} applies and

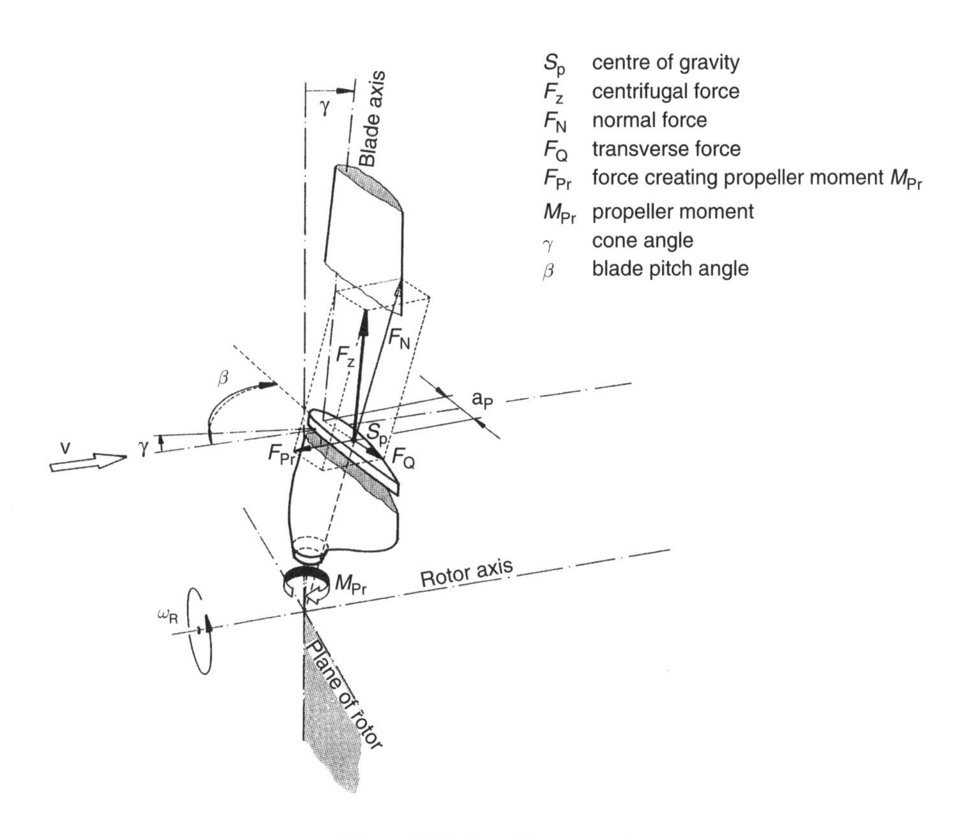

Figure 2.37 Propeller moment

integrating the product over the length of the blade yields the propeller moment M_{Pr}. For the usual operating blade angle of approximately 90° and the always-small blade angle with respect to the vertical axis ($\gamma < 10°$), a_P can be approximated for the moment-building offset, so that the propeller moment can be expressed as

$$dM_{Pr} \approx dF_{Pr}a_p. \tag{2.59}$$

The corresponding worked examples will be further detailed in Section 5.5. Equation (5.11) shows, for example, the entire propeller moment of a blade.

The lifting force F_A, acting to one side of the axis of the blade, develops moments M_{lift} as in Figure 2.38, which are largely dependent on the resultant wind velocity, the blade pitch angle, the blade profile and the offset at $t_B/4$ between the point of action of the lifting force and the axis of the blade. For a single profile element the following is approximately true:

$$dM_{lift} = c_a(\alpha)\, v_r^2 \frac{\rho}{2} a_a \cos\alpha dA_B. \tag{2.60}$$

The total moment resulting from lift forces on the blade is found in accordance with Equation (5.12). The restoring torques M_T cause, among other effects, a twisting of the profile into the direction of flow and are dependent on the resultant wind velocity v_r, the

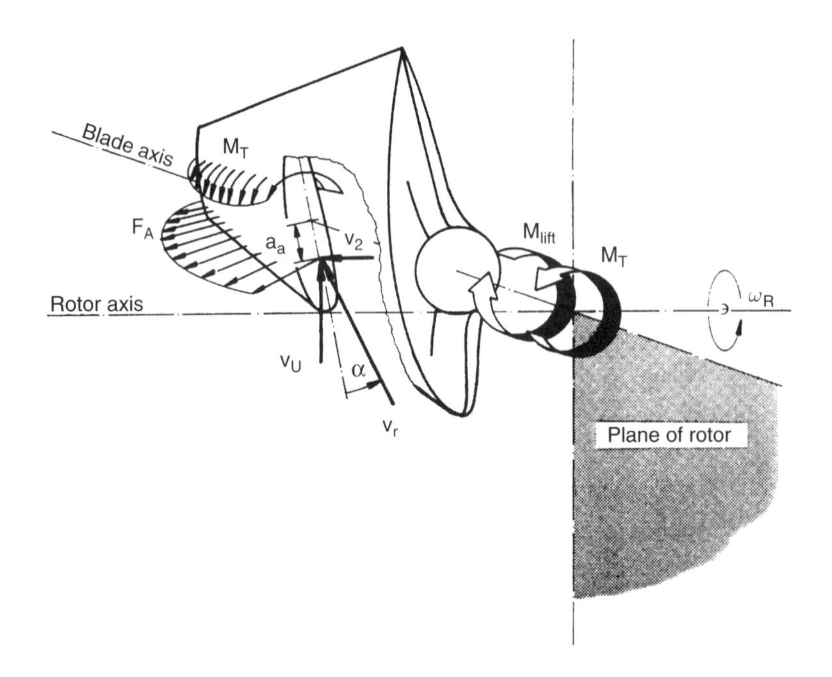

Figure 2.38 Moments arising from lift forces and aerodynamic restoration

blade area A, the chord of the blade t_B and the airstream angle and blade profile, with the associated moment coefficient c_t.

For a single element, relative to $t_B/4$, the following statement is generally true:

$$dM_T = c_t\,(\alpha)\,v_r^2 \frac{\rho}{2} t_B\,dA_B. \tag{2.61}$$

The total restoring torque on the blade can be determined using Equation (5.13).

Blade bending causes displacement of the lift and centre of mass. This generally results in an increase to the angular moment working against blade restoration, i.e. the moment M_{lift}, developed as a result of lifting forces, usually becomes larger due to aerofoil deflection. Due to mass shifts, extra propeller moments are generally developed. Furthermore, the moment of inertia of the deformed blade becomes significantly greater than it is in its undeformed state (see Figure 2.39).

Moments resulting from rotor teetering and the associated changes in pitch, which arise mainly in teetered hub models, are mainly dependent on blade angle and the amplitude of teeter during blade rotation (see Figure 2.40). In symmetrical blade arrangements, moments engendered (e.g. due to the acceleration of inert masses) cancel each other out with respect to the external drive. Propeller moments, on the other hand, can change considerably.

Because of the significant influences and the continually changing conditions during the rotation of a blade, these effects cannot be handled as they stand without unacceptable computing effort. The determination of extreme conditions is often sufficient, however, for dimensioning purposes.

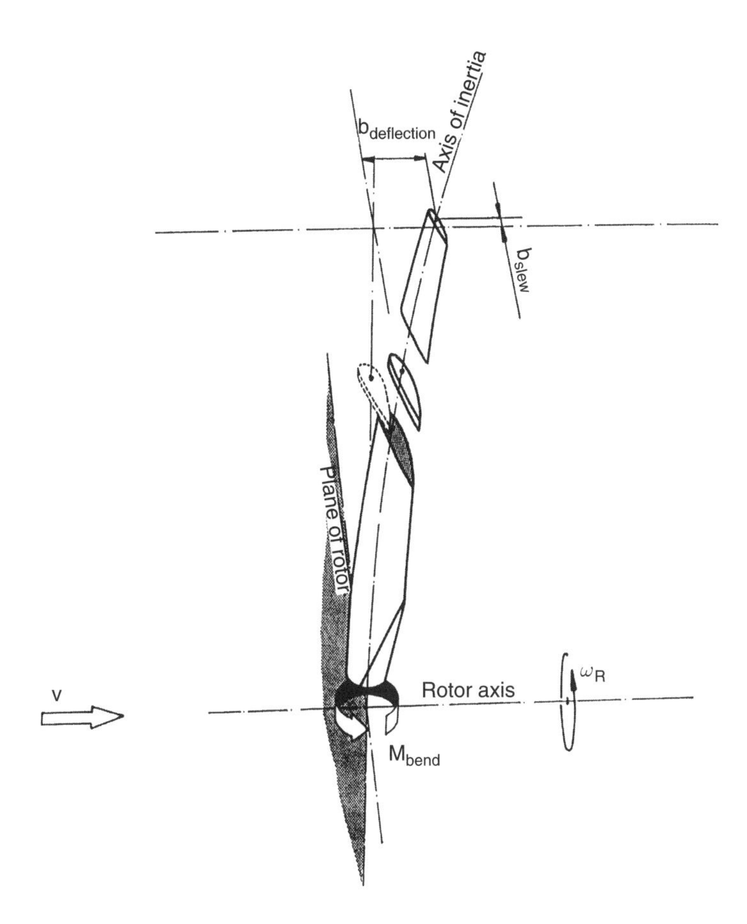

Figure 2.39 Moments arising from blade bending

Frictional moments in the blade bearings always work against blade movement (Figure 2.41). They depend upon the rotor position and speed of revolution, the speed of pitch variation and the wind speed, and have a quasi-damping character.

The frictional moment of a bearing can be expressed as the sum of a load-independent component and a load-dependent component. The load-independent component depends on hydrodynamic losses in the lubricant. This depends on lubricant viscosity and quantity, and also on rolling speed, and is dominant in swift-running lightly loaded bearings. The pitch angle varies only very slowly (maximum $\pi/6$ per second) and the load on the bearing is high, so this component can be neglected.

The load-dependent frictional moment M_{RL} arises as a result of elastic deformation and partial local slippage of contact surfaces. This component predominates in bearings rotating slowly under load. It can be determined by the relation

$$M_{RL} = f_{L1} g_{L1} P_{L0} d_{m}. \tag{2.62}$$

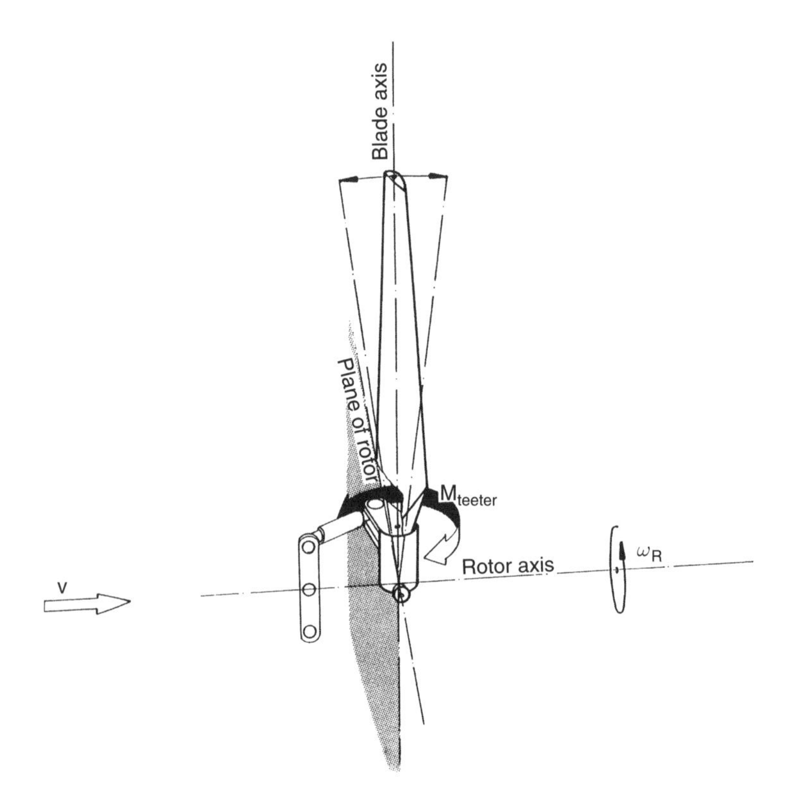

Figure 2.40 Moments arising as a result of teetering

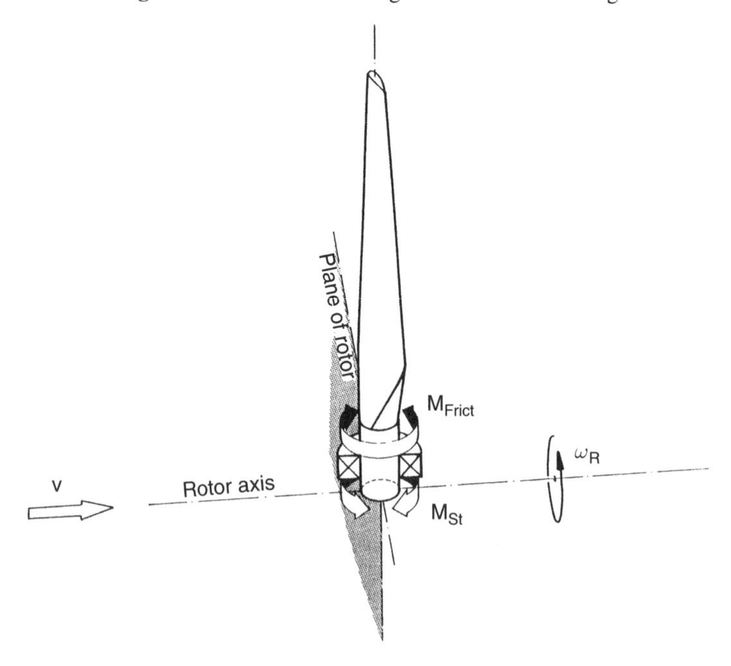

Figure 2.41 Frictional moments in blade bearings

The quantities

f_{L1} coefficient depending on bearing type and loading
g_{L1} factor determined by load direction
P_{L0} equivalent static loading
d_m average bearing diameter

used here can be found in the manufacturer's catalogue (e.g. the SKF Main Catalogue).

The equivalent static loading is characterized by axial and radial bearing forces. These vary, however, in rotating rotors, depending on the blade angle of rotation ψ_{Bl}, so that

$$P_{L0} = f(\psi_{Bl}), \tag{2.63}$$

e.g. for axial bearings

$$g_{L1} P_{L0} = F_a(\psi_{Bl})$$

and is therefore equal to the axial force. The simplified assumption for small deviations from the mean values

$$P_{L0} \approx \text{constant} \tag{2.64}$$

must, however, be investigated for each individual case.

Frictional moments are also dependent on the speed of rotation. As an approximation, when starting a movement the load-dependent component can usually be doubled.

The sum of frictional moments over all bearings active during pitch variation yields the total frictional moment

$$M_{Frict} = \sum_{k=1}^{n} M_{RLk}, \tag{2.65}$$

where n is the number of bearings involved. When the pitch of a rotating blade is altered, further moments act on the system, originating from the acceleration of air masses around the profile and from air damping (see Figure 2.42). A description of these quantities, which hamper movement, is only possible under predefined conditions.

For the limited case of

- blades that are stiff under bending and twisting,
- with an axis of rotation at a quarter of the chord ($t_B/4$),
- when the blade is turned about the pitch variation axis only (i.e. no blade deflection),

these moments can be approximated per span-width element by the relation [2.16]

$$\frac{dM_L}{dr} = \pi \rho d^4 \Bigg\{ \underbrace{-\left(\frac{1}{8} + a^2\right)\ddot{\beta}}_{\text{inertia of accelerating air masses}} + \underbrace{\left[a - \frac{1}{2} + 2\left(\frac{1}{4} - a^2\right) C(k)\right] \frac{v_r}{d} \dot{\beta}}_{\text{damping of the moved air}} \Bigg\}$$

where

β	is the angle of rotation (pitch variation angle),
d	is half the profile chord,
ad	is the offset from the centre of the profile to the point of rotation,
$C(k)$	is the Theordorsen function (quasi-stationary $C(k) = 1$) and
v_r	is the flow speed.

The geometrical relations that underlie further considerations are shown in Figure 2.43 for the most common aerofoil sections.

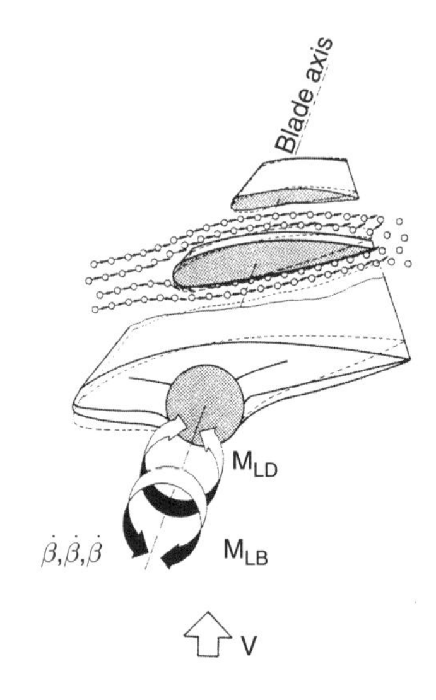

Figure 2.42 Moments arising from the acceleration of air masses and from air damping during blade positioning

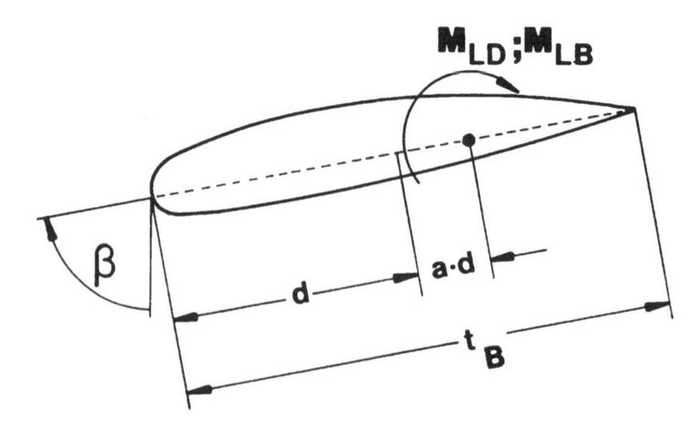

Figure 2.43 Definition of geometrical profile quantities for calculating moments arising from the acceleration of air masses and from air damping [2.16]

In a quasi-stationary state the total moment resulting from air forces is found by integrating over the blade span from the inner to the outer radii (R_i to R_a) where $a = -\frac{1}{2} = \text{constant}$ and $d = t_B/2 = f(r)$ substituted. Therefore

$$M_L = \int_{R_i}^{R_a} dM_L = -\frac{3}{128}\pi\rho\ddot{\beta}\int_{R_i}^{R_a} t_B^4\,dr - \frac{1}{8}\pi\rho\dot{\beta}\int_{R_i}^{R_a} v_r t_B^3\,dr \tag{2.66}$$

or

$$M_L = M_{LB} + M_{LD} \tag{2.67}$$

where

$$M_{LB} = J_{LB}\ddot{\beta} \qquad \text{and} \qquad M_{LD} = k_{LD}\dot{\beta}.$$

For approximations of this kind, it is usually sufficient to divide the blade into five to ten sections, to derive the partial moments using the average chord and flow speed values, and to sum these to yield the total moment.

The results of such rough estimates show that components arising from accelerated air masses play a completely subordinate role in comparison with the large inert masses of the blades and pitch-setting mechanism, and can thus be disregarded. Moments arising from air damping are in general smaller than the associated frictional values, but achieve similar orders of magnitude and should therefore be taken into account.

The following section uses the equations of motion of the system to develop a model for rotor blade positioning.

2.3.2.2 Blade positioning model

In addition to the dependencies mentioned, the moments depicted above are also influenced by the momentary state of the rotor, i.e. by the rotor blades' vibration characteristics. If the wind-speed profile and tower-shadow effects are also taken into account, the position of the rotor during one rotation plays a decisive role. The block diagram in Figure 2.44 shows the quantities acting and the combined effects of all moments acting on the rotor blades.

In the following pages the quantities influencing the rotor blade will be more closely detailed. Starting from the general equation of motion of the blade turning about its longitudinal axis, the following differential equation may be obtained:

$$\frac{d\left((J_{LB} + J_{Bl})\frac{d\beta}{dt}\right)}{dt} + \frac{d\left((k_{DB} + k_{RL})\beta\right)}{dt} = M_{St} - M_{Bl}. \tag{2.68}$$

After differentiating and separating out the values we can write

$$(J_{LB} + J_{Bl})\frac{d^2\beta}{dt^2} + \left(\frac{dJ_{Bl}}{dt} + \frac{dJ_{LB}}{dt} + k_{DB} + k_{RL}\right)\frac{d\beta}{dt}$$
$$+\left(\frac{dk_{DB}}{dt} + \frac{dk_{RL}}{dt}\right)\beta + M_{Pr} + M_{lift} + M_T + M_{bend} + M_{teeter} = M_{St} \tag{2.69}$$

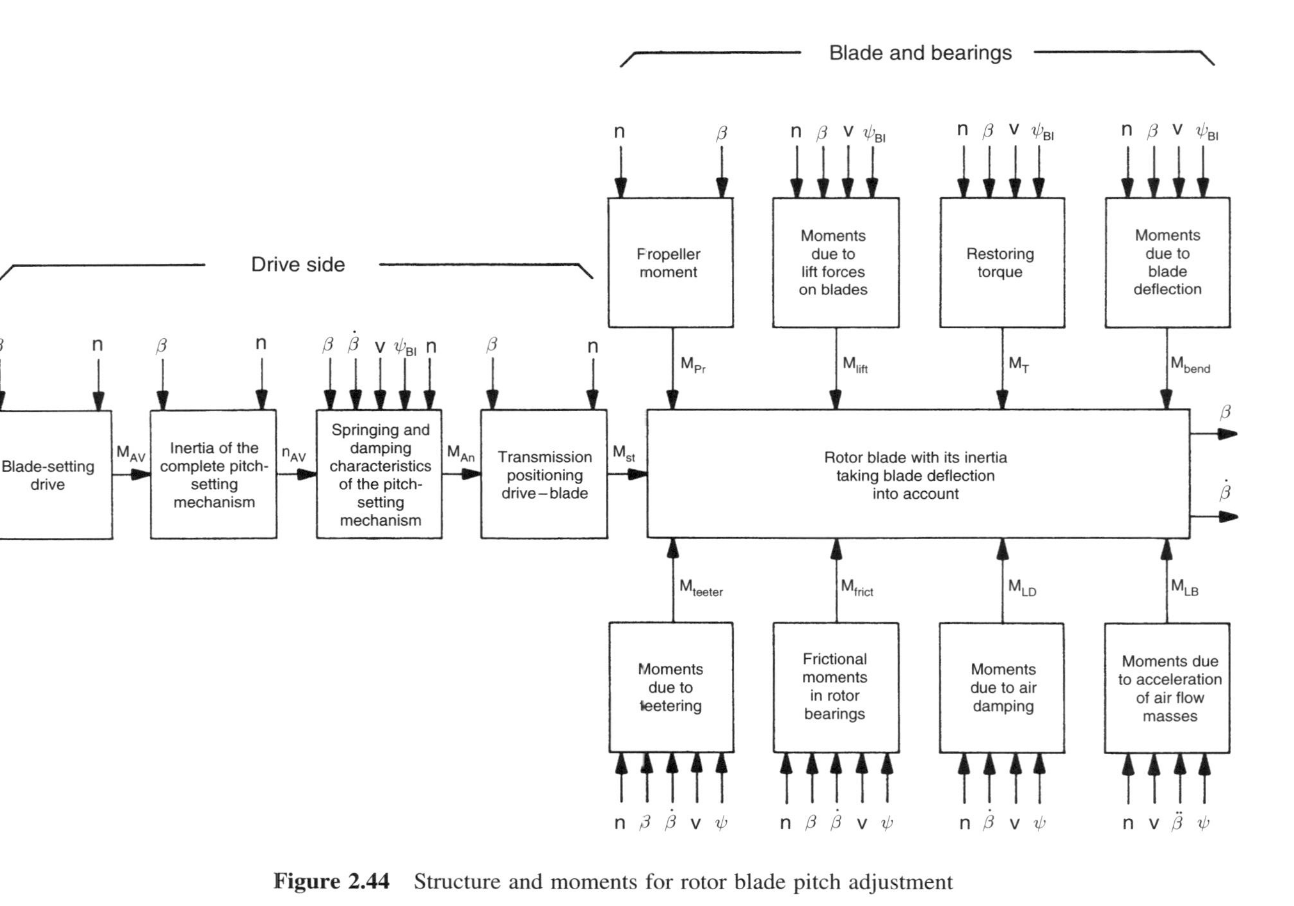

Figure 2.44 Structure and moments for rotor blade pitch adjustment

where

J_{LB} is the value equivalent to the inertia of masses due to accelerated air ($M_{LB} = J_{LB}\,d^2\beta/dt^2$ – forming),
J_{Bl} is the mass moment of inertia of the rotor blade about its longitudinal axis, where J_{Bl} is subject to the influence of the centre of mass displacement due to blade bending,
k_{DB} is the damping coefficient for the mostly negligible structural component (k_{DS}) and the generally dominating aerodynamic damping component (k_{LD}),
k_{RL} is the coefficient of friction for bearings,
M_{St} is the drive moment of pitch regulation taking into account inertia, springing and damping characteristics of the regulation mechanism and the gearing ratio to the blade,
M_{Bl} is the load torque during blade positioning about the longitudinal axis, including propeller moments, aerodynamic components and moments due to blade deflection and teetering of the rotor.

Essential dependencies can be given for:

Moment of inertia of air mass	$J_{LB} = f(v, \psi_{Bl}, n, b_s, t)$
Moment of inertia of mass	$J_{Bl} = f(\beta, v, \psi_{Bl}, a_s, t)$
Damping coefficient	$k_{DB} = f(v, \psi_{Bl}, n, t)$
Coefficient of friction	$k_{RL} = f(\beta, v, \psi_{Bl}, n, t)$
Propeller moment	$M_{Pr} = f(\beta, n, t)$
Moment due to lift on blade	$M_{lift} = f(\beta, v, \psi_{Bl}, n, t)$
Moment due to blade bending	$M_{bend} = f(\beta, v, \psi_{Bl}, n, a_s, t)$
Moment due to teetering	$M_{teeter} = f(\beta, \dot{\beta}, v, \psi_{Bl}, n, t)$
Torsional effect due to aerodynamic restoring torques in the direction of flow	$M_T = f(\beta, v, \psi_{Bl}, n, t)$

Here the factors

- blade pitch angle β,
- wind speed v,
- rotor blade position relative to the tower, ψ_{Bl},
- rotor speed n,
- deflection of the blade a_s in the thrust and slew directions,
- acceleration of the blade b_s in the thrust and slew directions,
- observation time t during which the current status has remained unchanged,
- derivation of the named quantities

must be taken into account.

The great number of components and the many effects preclude (without unreasonable computing effort) a complete description of the movement processes of rotor blades with a treatment of the corresponding moments. A simplified mathematical representation of events therefore makes good sense, even if the processes being examined are subject to certain limitations. This will be shown in the following pages.

2.3.2.3 Simplified rotor blade positioning model

In order to dimension a rotor blade adjustment drive and determine its governing parameters, the size and effect of the individual moments must be recorded as precisely as possible. Even calculations based on quasi-steady states, with the derivation of corresponding characteristic curves, allow the estimation of extreme situations, and are sufficient for dimensioning purposes. To this end, all the moments already mentioned must be taken into account under various operating conditions, e.g. even in the event of changes in surface roughness and icing up of the rotor blades, which particularly affect lift forces, propeller effects and blade inertia. Such variable operating conditions can best be mastered by self-optimizing regulation procedures based on parameters adapted to the locality. It is, however, possible to obtain good results by simply adjusting the most important parameters (see Section 5.5.4).

To examine the operational characteristics of wind power units, it was sometimes possible to use characteristic curves (see references [2.17], [2.18] and [2.19]) to determine load torque during blade positioning. In this manner, restoring torques, moments due to lifting forces, blade bending and propeller effects dependent on wind speed, speed of rotor rotation and blade pitch angle could be taken into account. Extensive calculations and simulations (e.g. in reference [2.20]) have shown that in a reliable design for a blade pitch drive and regulator mechanism different air damping and bearing frictional processes have little influence on general behaviour. Considerable simplifications can thus be achieved.

In slim plastic blades, torsional moments resulting from blade bending predominate. The behaviour of very stiff blades of heavy construction is, on the other hand, mostly determined by propeller moments. An estimate of individual moments based on constructional parameters allows recognition of the most important contributions and points the way towards further simplifications.

When the less relevant components are left out, the dominant moments that remain can be catered for by adding a safety margin to arrive at a total torsional moment M_{Bl}. When the transmission of moments between the pitch regulator drive and rotor blade is sufficiently rigid, springing and damping effects can also be left out of the reckoning and the inertia of the regulation mechanism can be attributed to the blades.

This simplified model of pitch variation moments, whether the pitch control system be electrical or hydraulic, is shown in Figure 2.45. Here I_{St} indicates the magnitude of electric current or hydraulic flow from the regulator required to position the blades. Air flow changes due to the tower as a result of compression effects in upwind turbines or shadow effects in downwind turbines can also be included in the calculations. Changes in transmission, e.g. between linear drive and radial blade adjustment, are also taken into account. Working under these last assumptions thus brings the effort involved in designing turbine pitch regulation and power control mechanisms within reasonable bounds.

2.3.2.4 Blade pitch control mechanisms and safety systems

The multiplicity of effects detailed in Section 2.3.2.1 demands a very good match between the rotor blade and the pitch regulation mechanism. Here the mechanical construction and the arrangement play an important role. Therefore, it is not possible to simply swap blades between different models and makes.

Adjusting the pitch of the rotor blades, as shown in Figure 2.6, provides an effective means of regulating or limiting turbine performance in high wind speeds or storm conditions,

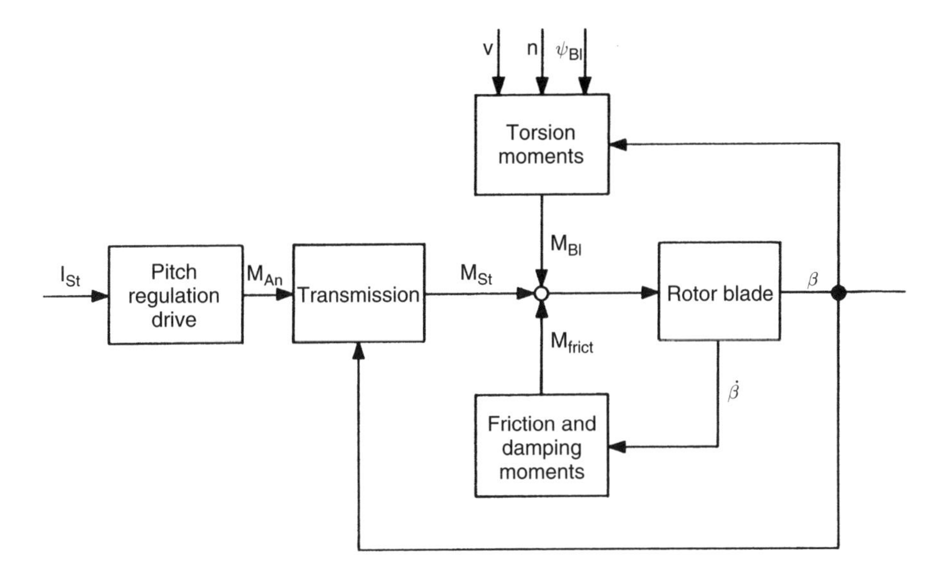

Figure 2.45 Simplified model of blade pitch regulation

or changing the speed of rotation. To put the blades into the necessary position, various mechanisms are employed. These may allow repositioning of the entire blade or just its tip. The necessary moment at the blade root can be applied by means of rotating masses, hydraulic lines or electric motors. A few concrete examples of pitch regulation mechanisms are presented in the following.

Mechanical systems of rotational speed control that exploit the principles of forces and moments in rotating masses have long been part of electrical engineering. Such mechanisms are usually simple and their solid construction usually makes for reliable operation. The energy they deploy for control processes is derived from the movement of rotating masses. To use such an arrangement for pitch regulation, however, the system must be accurately dimensioned, which in turn requires that the precise moments engendered in the blades during operation must be known. This is necessary, since modifying and adapting the parameters of mechanical devices usually involves considerable expense.

Figure 2.46 shows a pitch-regulation mechanism in which rotation of the turbine sets up a centrifugal force in a set of flyweights. Should the centre of mass lie outside the plane of rotation, the moments acting tend to bring it back into the plane. Load and restoring torques are applied by means of springs. By varying their length of travel the springs can be adjusted to suit machine and wind characteristics.

In the model shown in Figure 2.46, the pitch of each blade is regulated separately. In this way, every blade can be acted upon to protect the turbine from over-revving. Such in-built redundancy promotes reliability.

The three separate regulation mechanisms can, however, lead to different regulation procedures. These cause aerodynamic 'imbalance', which can place the rotor, hub and yaw gear under considerable loads.

Figure 2.47 shows a centrifugal-force-based arrangement, used in the 1950s on Allgaier machines (see Figure 1.10). This mechanical regulation and positioning system was supposed

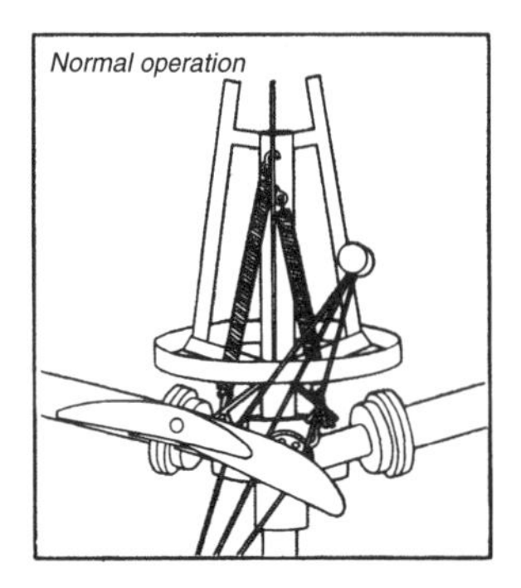

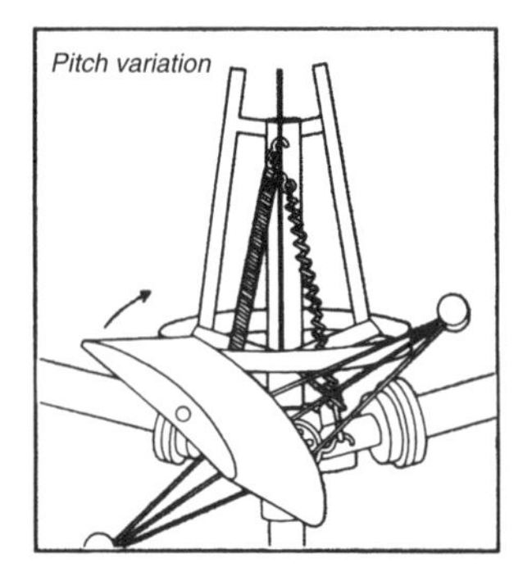

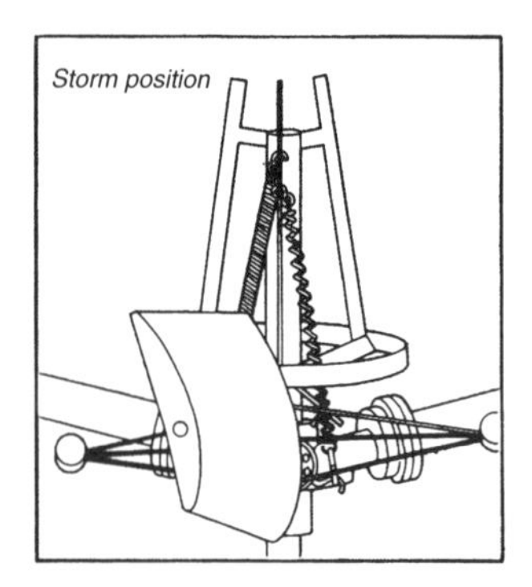

Figure 2.46 Pitch-regulation mechanism using rotating masses, with the possibility of varying the return spring moment (Brümmer). Reproduced by kind permission of Hermann Brümmer

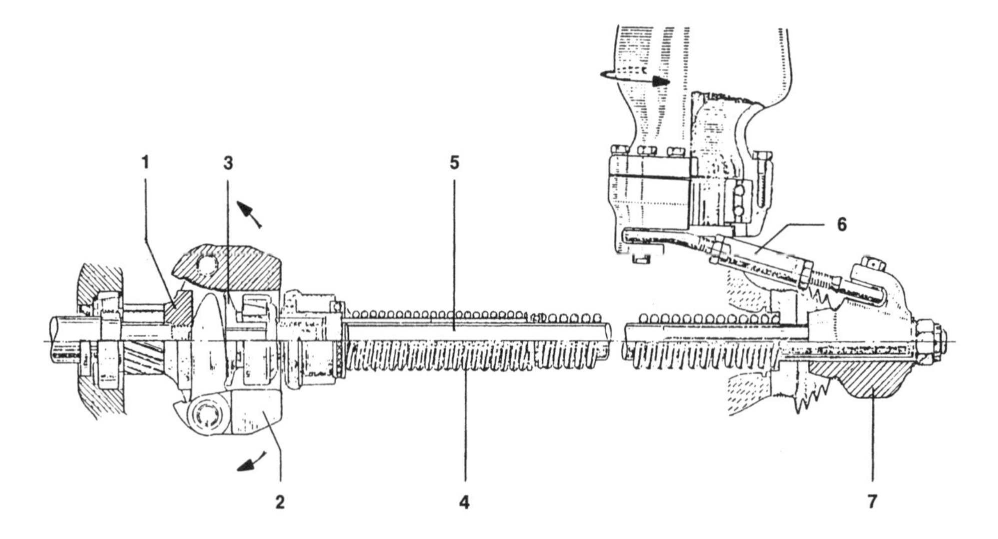

Figure 2.47 Centrifugal force regulator and control linkages (Allgaier): 1, regulator carrier; 2, centrifugal weight; 3, pressure plate; 4, return spring; 5, control shaft; 6, control rod; 7, spider

to limit the speed of the turbine to its nominal speed. The regulator is built into the gearing and rotates at the same speed as the generator. The 2.5 kg centrifugal weights generate a centrifugal force of about 5000 N. They are therefore capable of working via the thrust rod and positioning crank, against the force of the return springs, to bring the rotor blades into a position appropriate for the current operating conditions. This regulation and positioning system thus makes it possible to control or limit the speed and power of the turbine.

In contrast to regulator systems that depend on rotating masses, hydraulic and electrical systems require an external energy source. The feeding of this energy to the rotating hub and blade positioning device considerably increases construction costs. In spite of this, hydraulic drives are in widespread use, principally in smaller and medium-sized machines, but are also found in large machines. Electrical adjusters, on the other hand, are generally found only in machines of 200 kW nominal output or above.

The power for hydraulic pitch-regulation mechanisms is usually supplied from units in the machine house. For turbines in the 10 to 100 kW class, the drive is usually fed directly via the rotor or generator shaft.

Figure 2.48(a) shows the pitch-setting and power supply systems for a turbine of the 30 or 40 kW class. The hydraulic pump is driven from the generator shaft. From the generator standpoint, the speed of rotation and power of the turbine are handled as electrical quantities

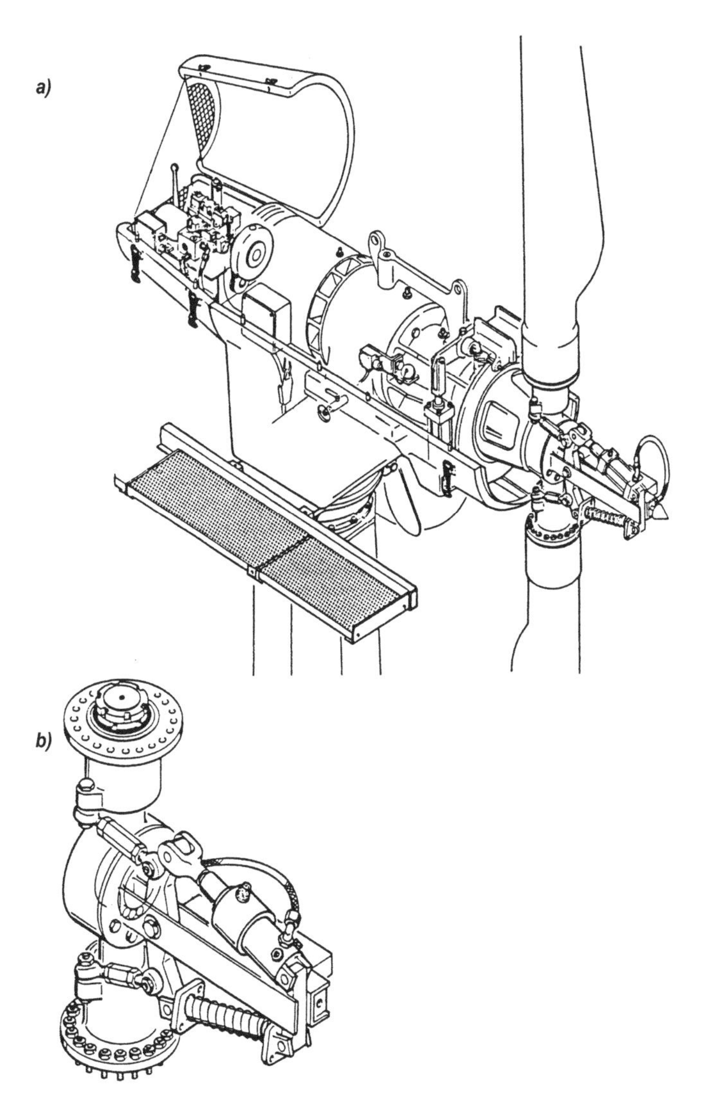

Figure 2.48 Hydraulic blade pitch regulators (Aeroman, MAN): (a) pitch adjustment and supply system in the nacelle and (b) setting cylinder and crank system

and are compared with given control values. Should any deviation occur, electrohydraulic valves are opened or closed according to control characteristics and the positioning cylinder (see Figure 2.48(b)) is activated. When hydraulic power is applied to the cylinder, it turns the blades towards the plane of rotation. In the absence of hydraulic power, springs acting in the opposite direction return the blades to their original positions. This means that, if the hydraulic system fails or another emergency situation arises, the turbine is returned to a safe operating mode. Seen from the turbine, the control valves and the hydraulic power unit are located behind the generator.

The regulation and adjustment system shown here consists primarily of standard hydraulic drive products. In this way, in addition to the high reliability needed for wind turbines, a high level of solidity and cost-effectiveness can be attained.

Figure 2.49 shows the pitch-regulation system layout in the nacelle of a 4 MW wind power unit. Four positioning cylinders acting directly upon the blade positioning rods bring the 80 metre-diameter blades of the teetered-hub rotor into the positions dictated by the control system. This positioning system thus makes it possible to control power in both directions.

The hydraulic pressure supply with its pumps, fluid cooler and control computer are located at the rear of the nacelle. To provide a redundant backup supply for use during power outages, a hydraulic reservoir is located in front of the rotor hub.

A similar idea for pitch setting is to be found in the WKA 60 1.2 MW turbine shown in Figure 2.50. Here, too, the hydraulics unit and the cabinet with the electronics for control and regulation are located at the rear of the nacelle. Each of the three blades has its own direct-acting positioning cylinder. Double redundancy is thus afforded.

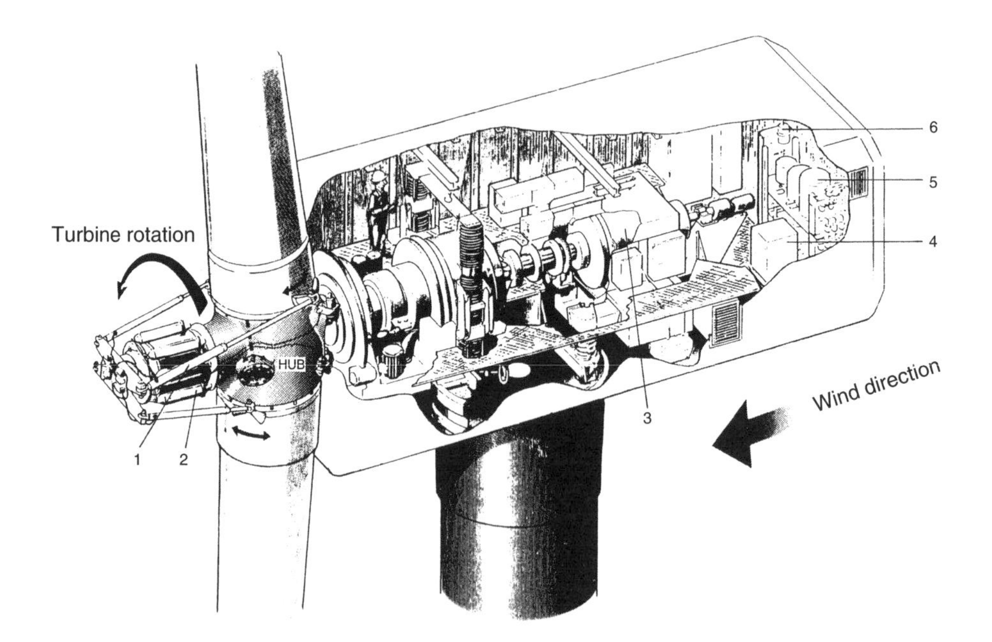

Figure 2.49 Hydraulic blade pitch-regulation system with direct cylinder transmission (WTS 4, Hamilton Standard): 1, hydraulic cylinder; 2, hydraulic reservoir; 3, generator; 4, control computer; 5, hydraulic pumps; 6, hydraulic fluid cooler

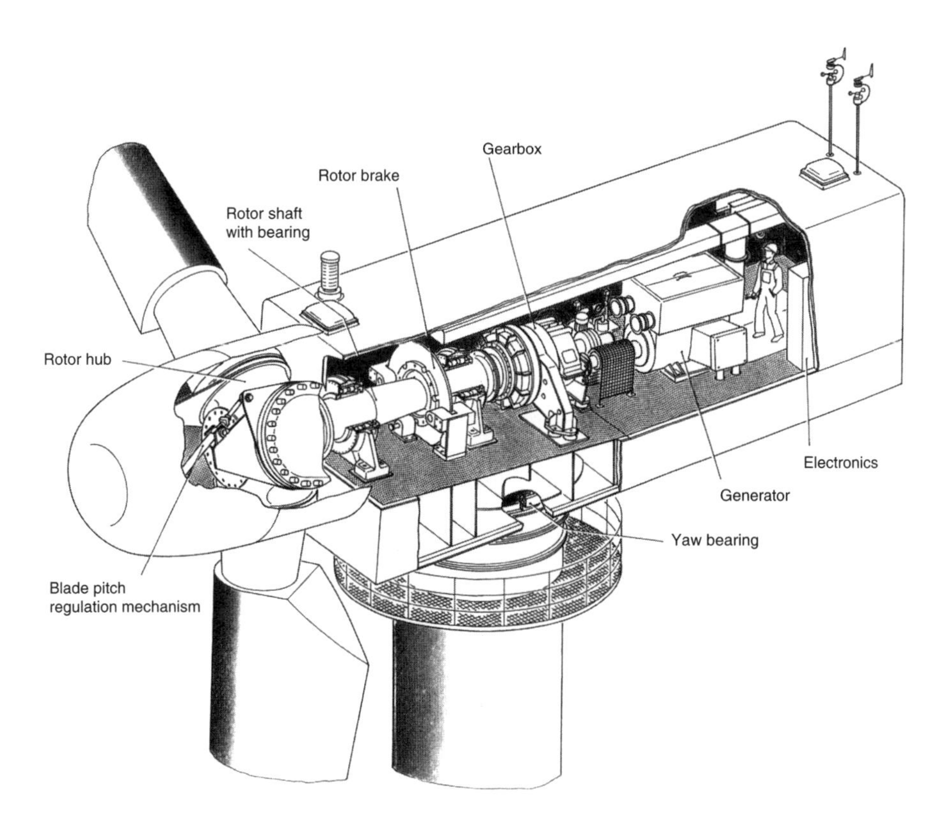

Figure 2.50 Hydraulic blade pitch-regulation system with positioning cylinder and lever transmission (WKA 60, MAN)

Manufacturers of wind turbines prefer mass produced, reliable components for the hydraulic pitch adjustment systems, which are currently in use. The buckling load on the positioning cylinder can be kept low if the mechanism is designed such that positioning routes are kept short and direct (Figure 2.51).

The pitch-setting systems shown above have the pressure supply in the nacelle and the pitch adjustment mechanism arranged around the rotor. This means that the connection between the supply and mechanism must be by rotary lead-through. A rotating hydraulic system, as used in the turbine shown in Figure 1, makes this unnecessary. Such a system is described briefly below (Figure 2.52).

In the pitch adjustment systems shown above, the blade is rotated over its entire length, e.g. into the wind, to reduce rotor power. Similar power changes can be attained by moving the blade tips only (Figure 2.52(a). This requires much smaller positioning forces. This means that the positioning cylinders (Figure 2.52(b)) and the rotating hydraulic power supply, with its pumps, motor, fluid reservoir and pressure reservoir (Figure 2.52(c)), can be correspondingly smaller. However, high turbulence levels must be expected in the area of transition between fixed and movable parts of the rotor blade.

As mentioned earlier, electrically driven pitch-regulation systems are used only in large machines. Figure 2.53(a) shows the entire blade pitch-regulation mechanism of the largest

Figure 2.51 Positioning system arrangement with cardanic bearing and short positioning routes

turbine built in the twentieth century (100 metres in diameter), which operated in the mid-1980s (Figure 1.13). The pitch-adjustment mechanism, with its single motor (Figure 2.53(b)), is mounted to the fore of the rotor head and acts on its two blades via axial gearing and connecting levers. The drive used for positioning is an inverter-powered asynchronous motor. A hydraulic emergency shut-down system can, when necessary, force the entire electromechanical drive through a fast positioning procedure and bring the turbine into a safer operation mode.

The Enercon system shown in Figure 2.54 dispenses with the need for a redundant safety system. Three mutually independent electric servomotors drive one blade each. A precise pitch angle measurement system takes care of synchronous blade positioning. Figure 2.54(a) shows the electrically driven single-blade positioning system of a 200 kW Enercon E30. Figure 2.54(b) shows two positioning drives for a single blade of the 500 kW Enercon E40 turbine. Figure 2.54(c) shows the arrangement in the 1.5 or 1.8 MW E66 turbine. The motors, rotating with the turbine, always act with the same gear ratio during positioning.

Electric blade pitch adjustment drives predominate for turbines in the megawatt class. However, in contrast to the systems shown in Figure 2.54(a) and (b), these are usually fitted within the hub (Figure 1.2). The blade pitch adjustment system shown in Figure 2.55 is based upon a fundamentally different functional principle.

In the 1.2 MW VENSYS turbine with a permanently excited synchronous generator a safety and blade pitch adjustment system is used that requires no batteries, pressure reservoir or other energy storage devices for redundant operation and makes slip ring transmission superfluous. This is possible because the electromechanical safety system uses the rotor energy to bring the rotor blades into feathered pitch, e.g. in the event of an emergency stop [2.20]. In stationary normal operation the turbine rotor and servomotor run synchronously

(a) Blade tip positioning

(b) Positioning cylinder

(c) Rotating hydraulic power unit

Figure 2.52 Hydraulic blade tip positioning mechanism with a rotary hydraulic supply (MOD2, Boeing)

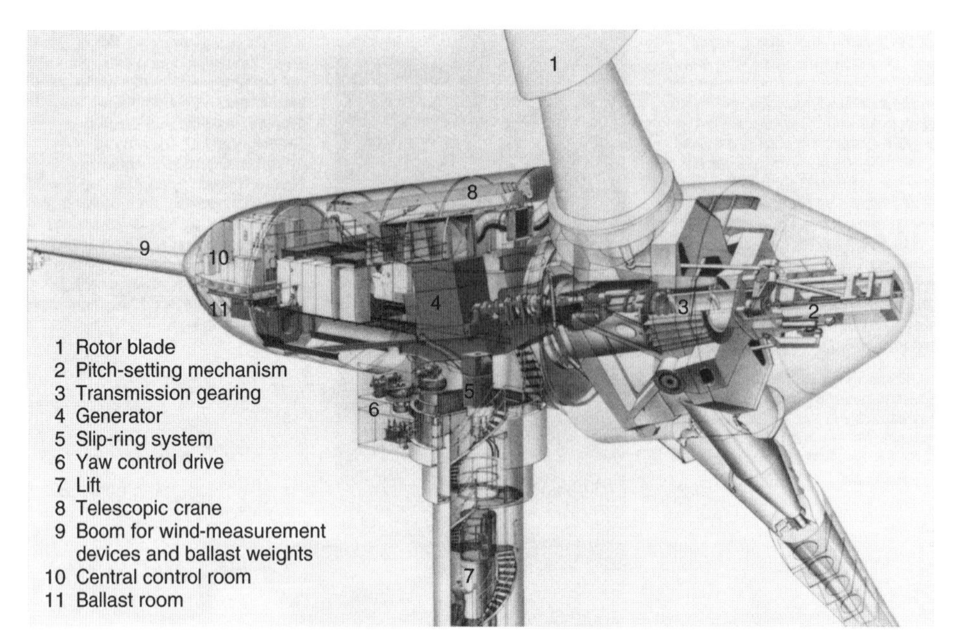

(a) Complete system. Reproduced by kind permission of MAN Technologie/MDE

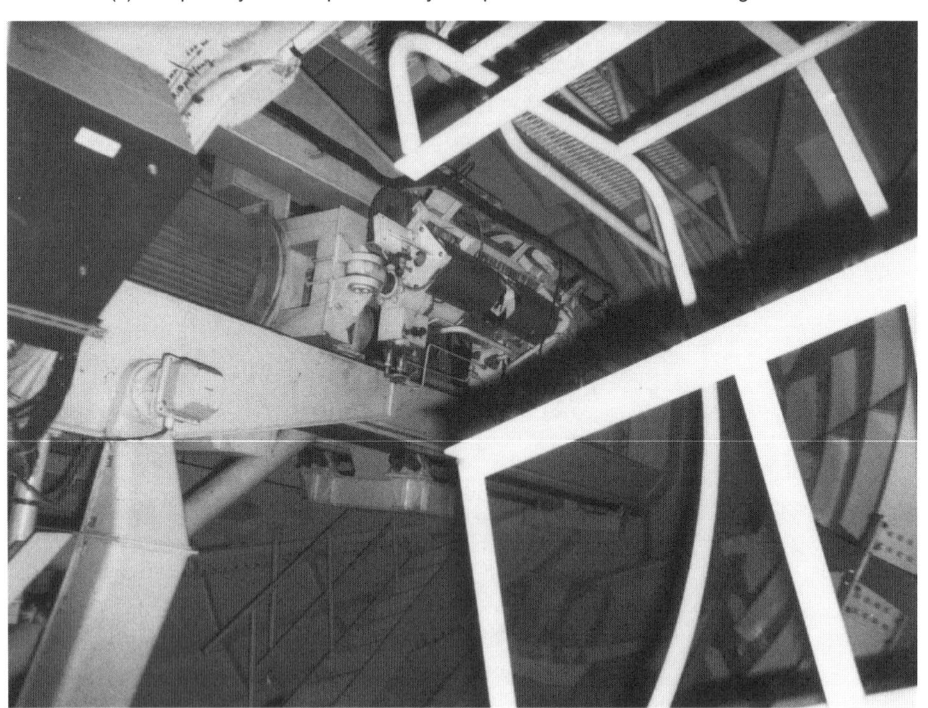

(b) Pitch adjustment mechanism. Reproduced by kind permission of LM Glasfiber A/S

Figure 2.53 Electrically driven blade positioning system with a hydraulic emergency positioning system (GROWIAN, MAN)

(a) Enercon E 30. Reproduced by kind permission of Enercon

(b) Enercon E 40. Reproduced by kind permission of Enercon

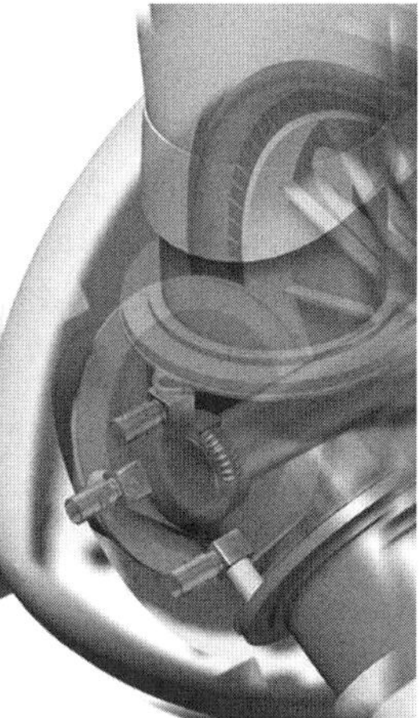

(c) Enercon E 66/E70.

Figure 2.54 Electrically driven blade positioning systems

so there are no differences in speed and angle of rotation in the distributor gearbox and the blade pitch angle thus remains the same. If the adjustment shaft is driven more quickly than the turbine, the rotor blades move in the direction of their working position. A slower adjustment shaft, on the other hand, adjusts the blade pitch angle in the direction of the blade's feathered pitch. The brakes, which in normal operation are actively released, drive the blades in the direction of their feathered pitch in the event of partial or total failure of the power supply and safety systems.

In contrast to direct motor drives, nonlinear transmission ratio differences always occur in the above-mentioned systems, due to translation of the linear positioning cylinder or

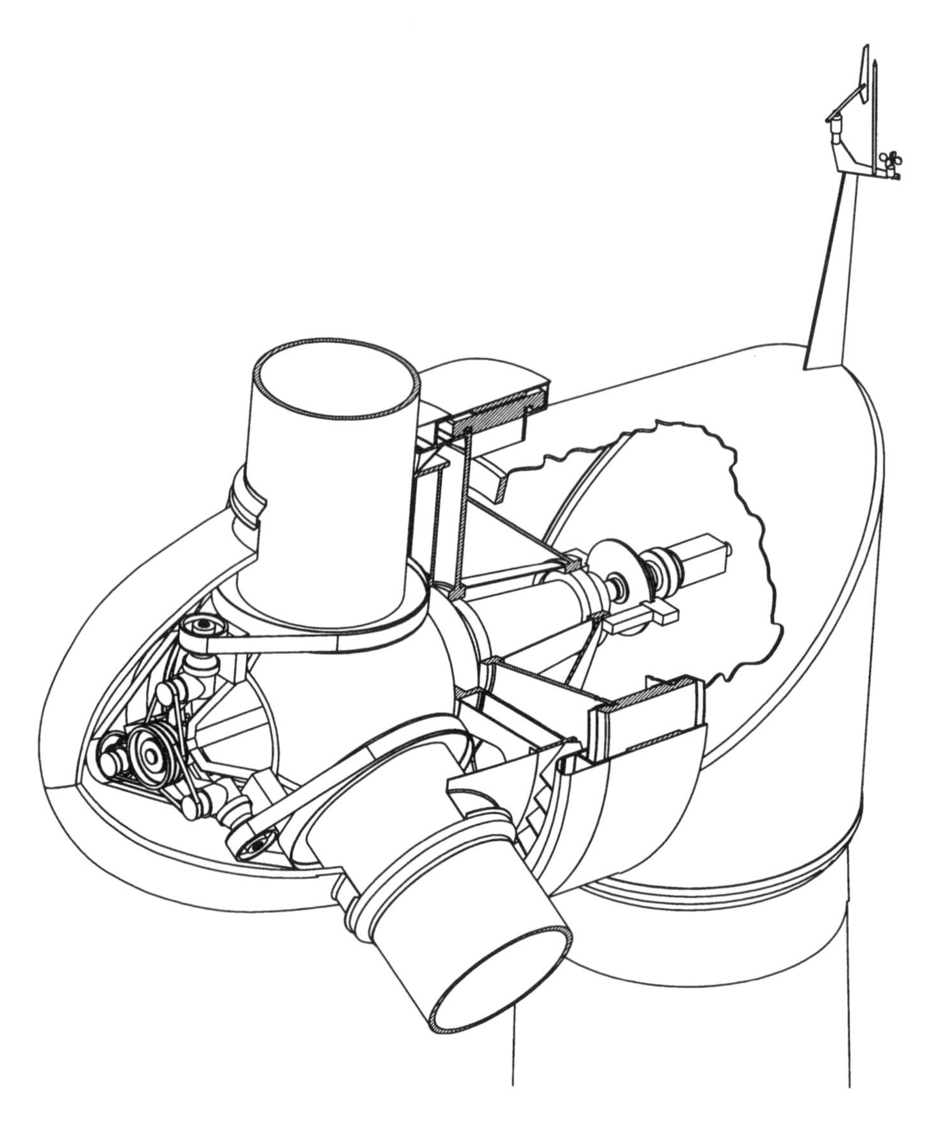

Figure 2.55 Electromechanical blade pitch adjustment with power transmission by means of toothed belts

drive-rod movements into changes of the blade angle. These must be taken into account when dimensioning the system. The above paragraphs have looked at a few of the most important blade positioning systems out of the wide spectrum available, some of which also bring the blades into active stall (see the systems shown in Figures 1.23(c) and (d) and Figure 1.26(g)).

Fixed-speed machines without pitch variation, which will be described in Section 2.3.3, do not allow the manipulations described here. Such machines must be protected against overspeeding as a result of power outages (during which the generator develops no load torque and so cannot dictate the speed of rotation) by a so-called 'passive control' system. This may take the form of braking flaps in the blade profile (see Figure 2.56(a)) or the

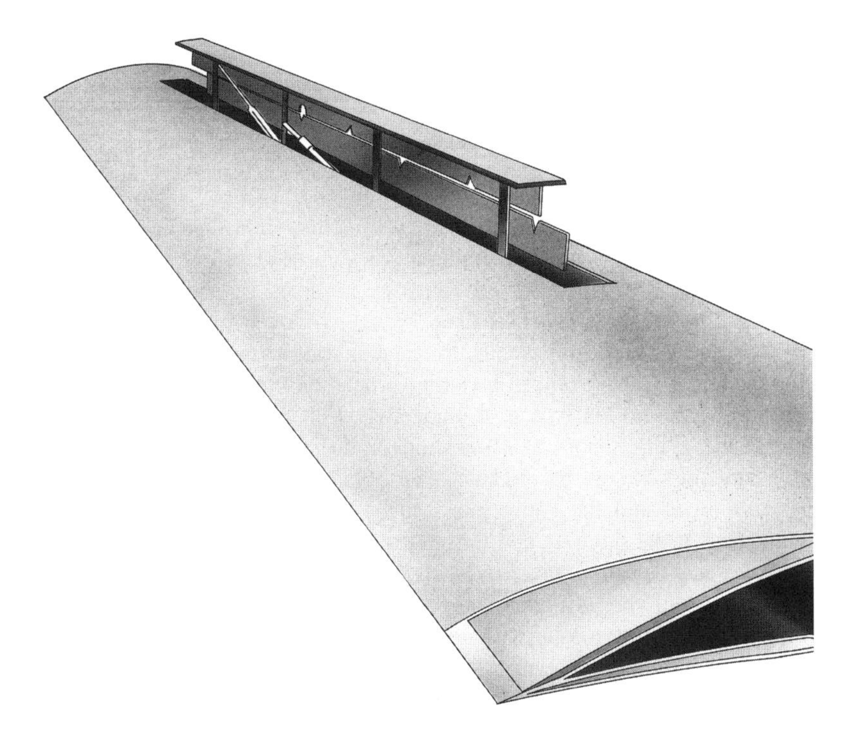

(a) Brake flaps in the blade profile

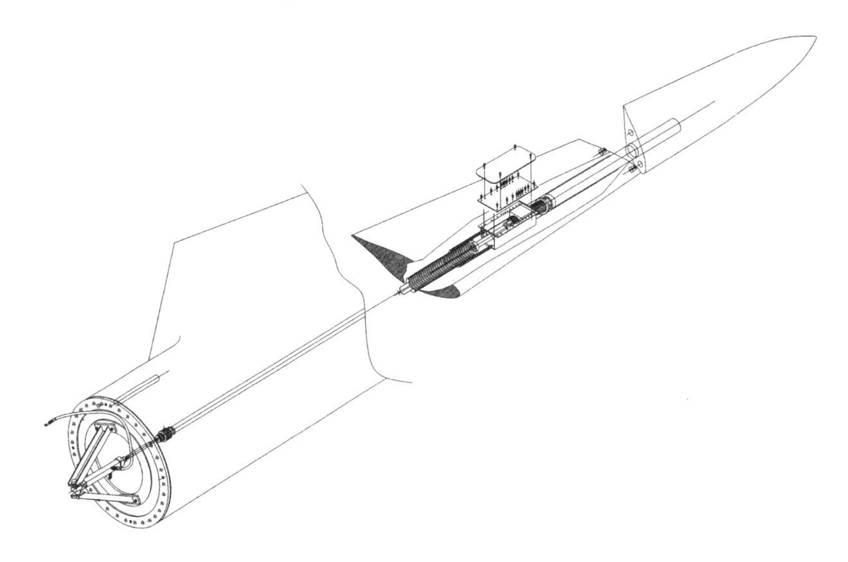

(b) Brake flaps at the blade tips. Reproduced by kind permission of LM Glasfiber A/S

Figure 2.56 Aerodynamic brakes

blade tips (Figure 2.56(b)). In addition, rotor brakes, usually in the form of disc brakes (see Figures 1.2 and 2.48 to 2.50), afford the possibility of bringing the rotor to a standstill in most machines.

2.3.2.5 Blade positioning system design

Blade positioning drives – as control and safety systems – must guarantee the controlled operation of wind power units in all conditions. They must therefore ensure that in critical conditions the rotor blades can reduce the energy extracted by the turbine as quickly as possible to protect the machine from possible damage.

Stationary operating states can usually be used in estimating positioning forces or for dimensioning the required drive power. Dynamic processes, however, must also be taken into account, considering the normal speed of blade swivelling in the direction of the operating position against the greatest possible opposing torque and the maximum return speed in the feathering direction against the highest blade torsional moment. In addition, the rotor-accelerating component and the entire positioning mechanism cannot be neglected when determining the size of the drive. In addition to the positioning moments, the forces required to keep the blades at the desired angle must also be considered in choosing the drive.

Hydraulic blade positioning systems that remain under steady pressure in active systems are able to exert this holding torque without extra energy expenditure. In these systems the moved masses, being limited to relatively small components such as fluid and positioning pistons, can normally be neglected in comparison to the inertia of the rotor blades.

Electric drives, on the other hand, can usually exert the necessary holding torque only when there is a continuous current flow. This can be eliminated when self-limiting reduction gears (e.g. axial or worm drives) are used to transmit motor torque to the rotor blades. Such systems, however, demand an extra redundant system in order to guarantee secure operation in case the drive motor goes down. Furthermore, during the operation of electric motors, internal decelerations due to the acceleration of rotating masses must be considered. The moments of inertia of motor rotors, gears, etc., taken in comparison with those of the rotating blades, may well be small but they work through large reduction gears and thus play a significant role in the design.

Wind turbines, particularly in turbulent gusty wind conditions, are subject to tremendous dynamic loads. These arise due to high winds and the resulting power gradients. To allow the turbine to continue to function safely, even under extreme conditions, changes to the turbine's torque must follow as swiftly as possible after the initiation of the pitch-adjustment sequence. This can be attained if the pitch-adjustment drive has a very short start-up time. Under constant acceleration torque and moment of inertia, this is determined by the following relation. For a blade moved directly by the drive,

$$t_{\mathrm{APDd}} = \frac{\omega_{\mathrm{BV}} J_{\mathrm{tot}}}{M_{\mathrm{St}} - M_{\mathrm{Bl}} - M_{\mathrm{Frict}}}. \tag{2.70}$$

If the drive acts upon a number z_a of blades then

$$t_{\mathrm{APDz}} = \frac{\omega_{\mathrm{BV}} J_{\mathrm{tot}}}{M_{\mathrm{St}} - z_a \left(M_{\mathrm{Bl}} - M_{\mathrm{Frict}}\right)}. \tag{2.71}$$

An inverter-powered drive with a current regulator allows such constant-torque procedures to be achieved.

The variables in the above are as follows:

ω_{BV} design value for the angular velocity of the pitch-setting system
J_{tot} moment of inertia of the entire pitch-setting system
M_{St} positioning drive torque
z_a number of rotor blades driven
M_{Bl} load torque during blade pitch adjustment
M_{Frict} moments of friction for all blade bearings in the pitch-adjustment system

The quantities given can relate to either the drive or to the blade(s).

For electrically driven positioning systems, the total moment of inertia is the sum of the motor component, the transmission elements and the rotor blades. Thus

$$J_{tot(A)} = J_{Mot} + J_{trans} + z_a J_{Bl(A)} \tag{2.72}$$

where

J_{Mot} is the moment of inertia of the drive motor,
J_{trans} is the moment of inertia of transmission gearing, clutches, etc., and
$J_{Bl(A)}$ is the moment of inertia of a rotor blade relative to the drive.

The moment of inertia of a blade relative to the drive is derived as

$$J_{Bl(A)} = \frac{J_{Bl}}{i_{MBl}^2}, \tag{2.73}$$

where

J_{Bl} is the moment of inertia of a single blade and
i_{MBl} is the transmission ratio between the servomotor and blade swivelling, i.e.

$$i_{MBl} = \frac{\omega_{Mot}}{\omega_{Bl}}. \tag{2.74}$$

For direct motor drives, i.e. where only rotational motion is involved, when

$$i_{MBl,rot} = \text{constant} \tag{2.75}$$

the transmission ratio always remains the same. On the other hand, when the adjustment mechanism converts linear motion to rotary motion (e.g. in crank or lever transmission), the transmission ratio

$$i_{MBl,lin\text{-}rot} = f(\beta) \neq \text{constant} \tag{2.76}$$

and is thus dependent on the pitch angle of the blade.

The use of electric motor drives opens up the possibility of selecting the motor speed (e.g. by setting the number of poles in three-phase motors) according to the transmission ratio. The aim should be to achieve the fastest possible blade positioning. This is attained by a short run-up time. It must be kept in mind here that, seen from the blade side, according

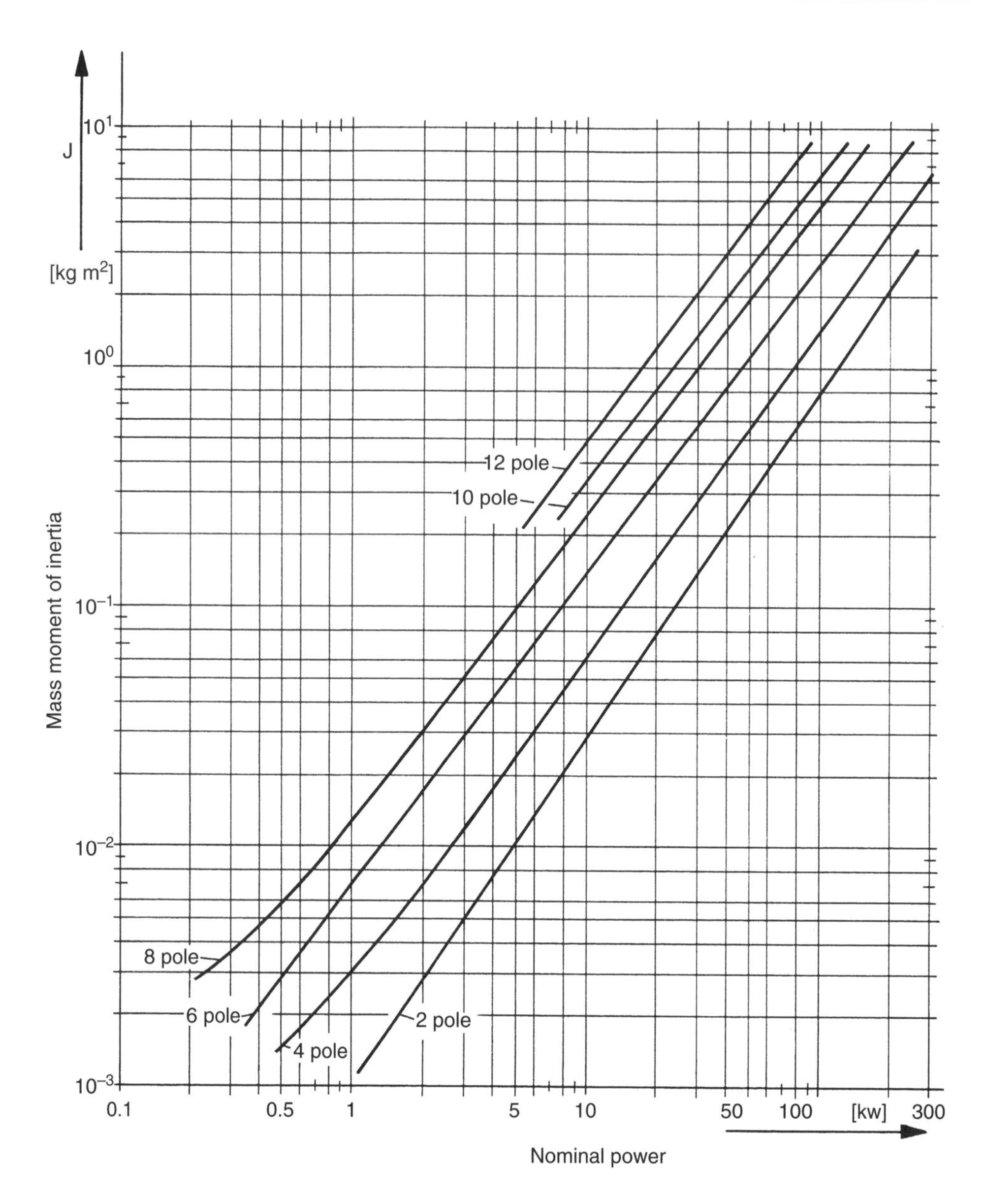

Figure 2.57 Moment of inertia of asynchronous three-phase motors as a function of nominal power with the number of poles $(2p)$ as a parameter

to Figure 2.57, the moment of inertia of three-phase motors with greater numbers of poles increases less quickly than the moment of inertia of the motor increases in relation to the value relative to the rotor blade

$$J_{\mathrm{Mot(Bl)}} = J_{\mathrm{Mot}} i_{\mathrm{M\,Bl}}^2. \tag{2.77}$$

Thus, in general, motors with higher pole counts, with correspondingly lower idling and maximum speeds,

$$n_{\mathrm{OM}} = \frac{n_1}{p}, \tag{2.78}$$

manage shorter run-up times than high-speed motors with low pole counts. Furthermore, with high-pole-count motors, lower transmission ratios may be attained and so extra gearing levels may possibly be avoided.

In dimensioning blade positioning systems, both normal and fast positioning procedures must be considered. Both areas of operation are examined in the following.

Normal positioning procedures

During normal operation, blade pitch adjustments with rotational speeds of approximately

$$\dot{\beta}_{\mathrm{n}} = 5 \text{ to } 10°/\mathrm{s} = 0.09 \text{ to } 0.17\,\mathrm{rad/s}$$

are expected. This ensures that during normal regulation procedures the turbine will be protected from braking conditions, which place considerable loads on the drive train.

For normal positioning procedures the nominal torsional moments on the rotor blade $M_{\mathrm{Bl\,n}}$ and the normal speed of displacement for blade positioning can be used. Under steady positioning conditions, ignoring the accelerating component, this yields a positioning torque of

$$M_{\mathrm{St}} = \frac{z_{\mathrm{a}}\,(M_{\mathrm{Bl\,n}} + M_{\mathrm{Frict}})}{i_{\mathrm{A\,Bl}}\,\eta_{\mathrm{A\,Bl}}} \tag{2.79}$$

for the drive side, where

$i_{\mathrm{A\,Bl}}$ is the transmission ratio between positioning drive and blade rotation and
$\eta_{\mathrm{A\,Bl}}$ is the efficiency of the transmission between the drive and blade.

An average torsional and frictional moment therefore acts against the positioning drive for the z_{a} blades driven. This figure is increased by the transmission efficiency and takes into account the transmission ratio between the drive and blade. The power required for positioning thus becomes

$$P_{\mathrm{Stn}} = \frac{z_{\mathrm{a}}\,(M_{\mathrm{Bl\,n}} + M_{\mathrm{Frict}})\,\dot{\beta}_{\mathrm{n}}}{\eta_{\mathrm{A\,Bl}}}. \tag{2.80}$$

For turbines in the 100 kW to MW class, positioning systems for normal operational ranges need between 1 and 10 kW. These figures should be understood as reference values for ongoing operation; as design values for a positioning system they are insufficient. In dimensioning the drive, the accelerating components and extreme situations must be allowed for.

Fast positioning procedures

In emergencies arising as a result of extreme wind conditions, sensor failure or similar states, fast positioning procedures with positioning speeds of

$$\dot{\beta}_s = 10 \text{ to } 20°/\text{s} = 0.17 \text{ to } 0.31\,\text{rad/s}$$

are used to bring the turbine as swiftly as possible into safe operational ranges.

When dimensioning the drive needed, the greatest possible moment working on the rotor blade against rotation into the feathering position – arising, for example, due to gusting or over-revving – must be allowed for. In addition, the accelerating components of all parts of the system must be taken into account in the design calculations.

The power necessary to accelerate the inert masses of the positioning system is derived from the angular velocity and the accelerating torque of the drive, transmission and rotor blade components. Thus the power necessary for the positioning drive is

$$P_{\text{Sts}} = \frac{z_a\,(M_{\text{Bl max}} + M_{\text{Frict}})\,\dot{\beta}_s}{\eta_{\text{A Bl}}} + \dot{\beta}_s J_{\text{tot(Bl)}} \ddot{\beta}_s. \tag{2.81}$$

This quantity is generally decisive for the dimensioning of the positioning system. For turbines in the 100 kW to MW class, the drive rating usually lies between 5 and 50 kW. For the design of a drive, however, torques and the dynamic characteristics of the entire system must additionally be included, as in the following study.

Positioning drive design

In designing the drive, it must be ensured that the positioning system retains sufficient power reserves to overcome the highest possible torsional moments on the blades. Uncertainties in working out torsional moments and production variations in components (the rotor blades in particular) must be allowed for with a healthy safety factor. However, changes in the dynamic behaviour of the entire system due to these power reserves must be taken into account. For example, a significant increase in the power of a two-pole motor will also considerably increase its inertia (see Figure 2.57), as a result of which blade positioning reaction time will be correspondingly longer. Such correlations show clearly that augmenting power reserves may be achieved at the expense of drive dynamics, which could possibly even compromise the safety of the entire system.

Therefore, during the design phase the transmission ratios and the drive power of technically feasible variants, such as, for example, those for electric motors, should be considered. In doing this, the time taken to attain the speed necessary for a fast positioning procedure

$$t_{\text{Af}} = \frac{J_{\text{tot(M)}} \dot{\beta}_s i_{\text{A Bl}}}{M_{\text{St}} - \dfrac{M_{\text{Bl max}} + M_{\text{R}}}{i_{\text{A Bl}} \eta_{\text{A Bl}}} z_a} \tag{2.82}$$

should first be determined. Thereafter the displacement processes, or the phases thereof, can be better described.

For blade positioning, the time t_0 required to turn a rotor blade through an angle of $\Delta\beta_0$ from a critical to a safe position can be determined. The process by which the blade is brought to this angle

$$\Delta\beta_0 = \frac{\ddot{\beta}_s}{2} t_{0b}^2 \tag{2.83}$$

can, in high-inertia systems, be regarded as consisting only of acceleration over the acceleration time t_{0b}. In rapidly accelerating systems, on the other hand, after the run-up time t_{Af} the fast positioning speed is attained and movement progresses thereafter at the constant angular speed $\beta_{s,}$ and can therefore be characterized by the relation

$$\Delta\beta_0 = \frac{\ddot{\beta}_s}{2} t_{Af}^2 + \dot{\beta}_s t_v. \tag{2.84}$$

Similarly, for pure accelerated motion until reaching the desired angle, the time

$$t_0 = t_{0b} \tag{2.85}$$

or, for overlying phases of acceleration and constant-speed displacement,

$$t_0 = t_{Af} + t_v. \tag{2.86}$$

Observations of differently designed positioning systems have shown that for motorized drives the dynamics of the mechanism are strongly influenced by the number of poles in the motor.

Two-pole, three-phase motors may well be lighter than multipole machines (see Figure 2.58), but as a blade positioning drive they take significantly more time to run up or to reach a safe blade angle ($\Delta\beta_0$) than four- or six-pole motors. Furthermore, it is clear that increasing drive power above the level needed for fast positioning does not improve system dynamics. Generally, however, doubling the power of a two- or four-pole motor is sufficient for the run-up time to exceed the minimum. Further increasing power, particularly in two-pole motors, leads to a deterioration of drive dynamics. Six-pole motors, on the other hand, achieve even more favourable running-up behaviour if the power is increased further.

When designing connection, control and protection devices for asynchronous machines (Figure 2.59), it helps to know the nominal current. If no current limiters are provided, high start-up currents (see Figure 3.54) should be expected. The higher current consumption of multipole motors is due mainly to the larger air gaps, which are necessary because of the greater mass and diameter.

If the behaviour under acceleration of the positioning systems is given, then the safety margin during acceleration procedures can be derived in the form of a moment:

$$M_{res} = M_{St} - z_a \frac{M_{Bl} - M_{Frict}}{i_{ABl}\eta_{ABl}} - J_{tot(A)} \frac{d\omega_{St}}{dt}. \tag{2.87}$$

Calculations show that, here too, fast positioning procedures using two-pole motors can only be obtained with the help of blade torsional moments. Torque reserves are thus not available. Machines with greater numbers of poles, on the other hand, offer greater torque surpluses and thus higher safety margins because of more power and poles. Definite statements on the choice and behaviour of blade positioning systems and their effects on the system as a whole can only be arrived at with the aid of simulations. The basic contexts for this are reproduced here.

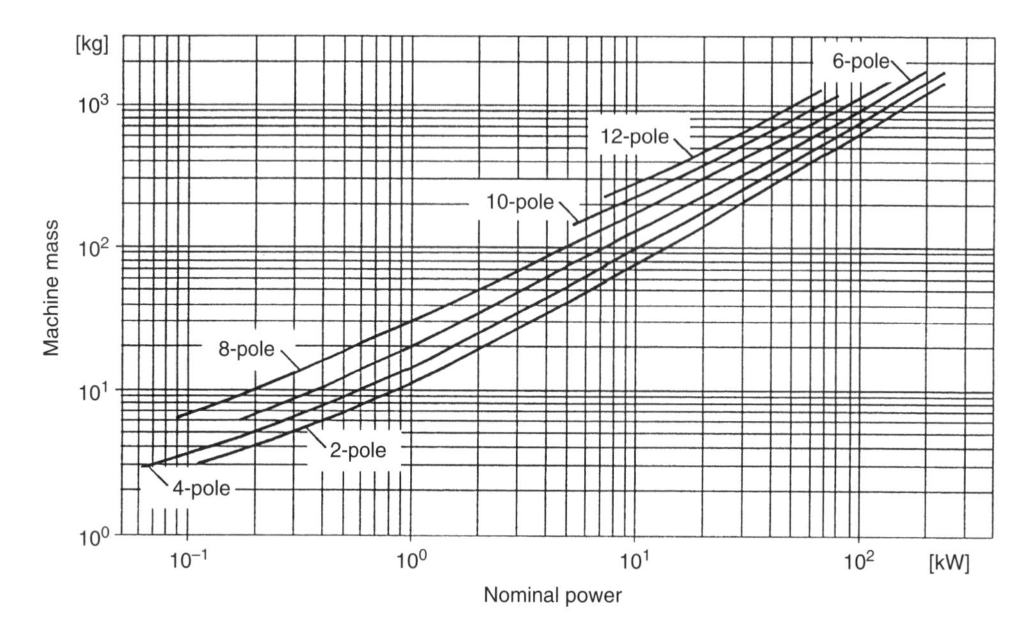

Figure 2.58 Mass of three-phase asynchronous motors as a function of the nominal power with the pole number $2p$ as a parameter

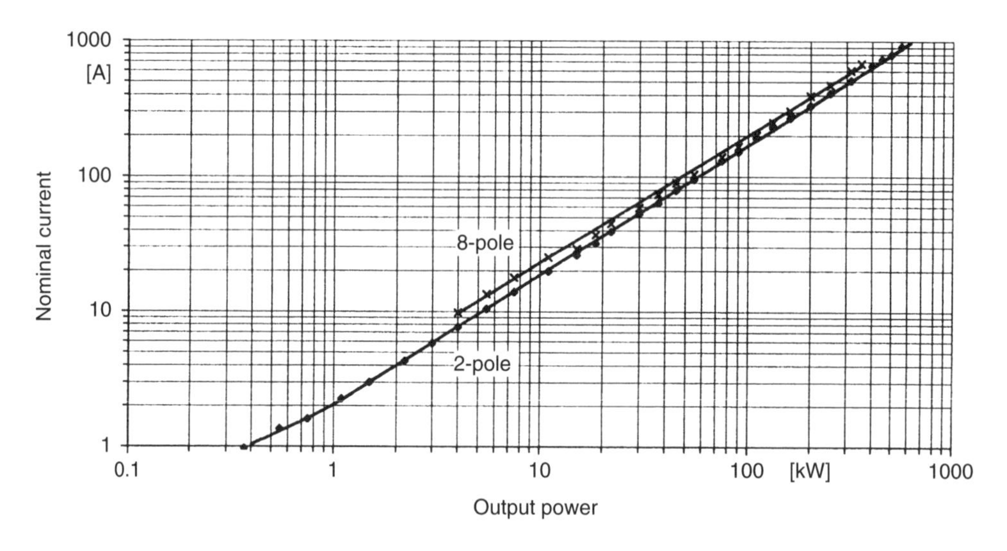

Figure 2.59 Nominal current of three-phase asynchronous motors as a function of nominal power with the pole count $2p$ as a parameter

2.3.3 Limiting power by stall control

In aviation, aerofoils, rotary wings and propellers are not generally run in stall for safety reasons. The power absorbed by wind turbines, on the other hand, can be limited by intentionally inducing stall conditions. In wind turbines with fixed rotor blades and asynchronous generators, coupled to the grid at a fixed speed, stall operation can be brought about passively by the altered profile flow that results from wind-speed increases. It can also be intentionally induced by varying the blade pitch angle or the turbine speed. In this active approach, however, a blade pitch adjustment system similar to that used in pitch-controlled machines must alter the blade position in the direction of stall or frequency converter systems (see Chapters 3 and 4) must alter the generator's (rotary) frequency, thereby changing the peripheral speed of the turbine and the flow over the blades. We will first discuss passive stall operation, which has predominated up until now.

2.3.3.1 Passive stall regulation

The vast majority of all installed wind power plants in the class up to approximately 1000 kW that supply power to the public grid are without blade positioning mechanisms, instead using so-called stall control to limit power extraction. Such plants are constructed up to an order of magnitude of 1.5 MW and are generally fitted with asynchronous generators (see Figure 1.22(d)).

In normal operation, laminar flow predominates at the rotor blades (Figure 2.60(a)). In these conditions, the lift values corresponding to the angle of attack are reached at low drag components (see Figure 2.60(c)). Thus in partial loading ranges a high degree of aerodynamic efficiency is attained. If, on the other hand, the wind speed approaches the value at which the generator reaches its maximum permanent load (usually the nominal power), further torque development at the rotor must be inhibited.

The largely rigid grid connection means that the generator (within the relatively narrow slip range of asynchronous machines) keeps the turbine at a near-constant speed; i.e. the peripheral speed v_{p} is approximately constant. Wind speeds exceeding nominal levels cause higher angles of attack and thus (in the appropriate design) stalling (Figure 2.60(b)), when the airflow 'unsticks' from all or part of the blade profile. Depending upon the angle of attack, therefore, as shown in Figure 2.5, the lift coefficient $c_{\mathrm{a}} = f(\alpha)$ and the lift forces $\mathrm{d}F_{\mathrm{A}}$ (see Figure 2.4) are reduced in certain ranges and the drag coefficients $c_{\mathrm{w}} = f(c_{\mathrm{a}}, \alpha)$ or the drag forces increase. As a result, the torque-creating tangential force F_{t} (the sum of all partial forces $\mathrm{d}F_{\mathrm{t}}$) does not significantly exceed its nominal values (Figure 2.60(d)). When the turbine is under full load and the wind speed climbs beyond the nominal range, this results – in spite of the greater levels of energy available – in lower rotor torque and lower performance coefficients.

The performance characteristics (Figure 2.61, top) of machines such as this are therefore largely dictated by their construction. In comparison with pitch-regulated turbines, stall-regulated machines are often designed with asynchronous generators of higher nominal output. Rigid grid coupling – a basic requirement for safe operation – is thereby obtained.

The complex aerodynamic and vibrational processes involved in stall regulation will not be examined in this study. For stall-regulated machines with nominal outputs of 30 kW and above, variable-speed turbine systems with synchronous generators and frequency converters are gaining in importance. Such configurations allow control manipulations similar to those of active-stall or pitch-regulated units. Limited performance control must, however, be balanced against the advantages of smoothed output power.

Manoeuvres directed at regulating the flow of energy into the supply system (highly desirable in wind turbines running under partial load in isolated operation) are not possible with fixed-speed units as depicted in Figure 1.38(b). These call for active interventions to the stall operation.

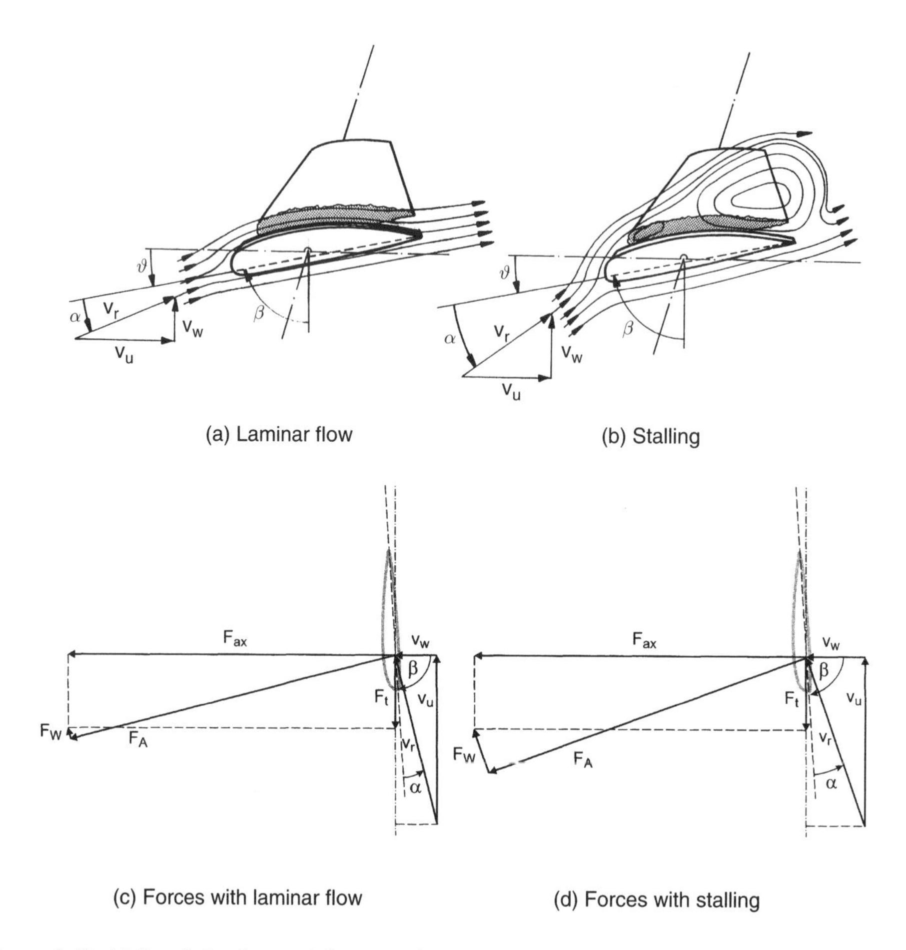

Figure 2.60 Airflow behaviour and forces at the rotor blade as wind speed increases from (a) to (b)

2.3.3.2 Active stall regulation

By adjusting the rotor blades (usually in the opposite direction to that for pitch control) as shown in Figure 2.36 or by changing the turbine speed with the aid of the generator (see Figure 2.68), stall operation and thus turbine output can be actively influenced in accordance with Figure 1.38(a) and (b) and matched to the desired grid or consumer requirements. The term 'active stall regulation' is generally used to mean the former process.

In machines with blade pitch-adjustment systems as described in Section 2.3.2, the rotor blades can be moved in the direction of the feathering pitch in order to reduce the power absorbed by a turbine. In contrast, active-stall-controlled machines require that the blades are rotated in the direction of the plane of rotation in order to move into the stall range and thus reduce the power drawn from the airstream. In general, a blade pitch-adjustment range of a few degrees is sufficient to protect the machine from overload, for example, or to adjust output to the desired levels.

Figure 2.62(a) and (b) illustrates the change in the torque-producing tangential force F_t given the same wind conditions and only slightly changed blade pitch angle or angle of

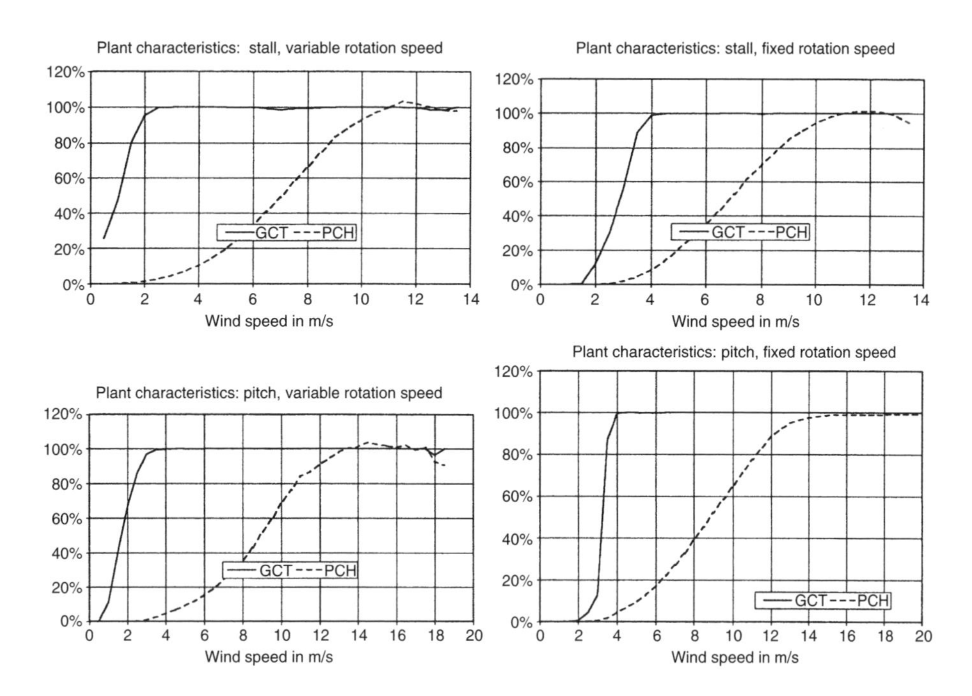

Figure 2.61 Grid connection time (GCT) and performance characteristics (PCH) of wind power plants (WMEP measurements ISET)

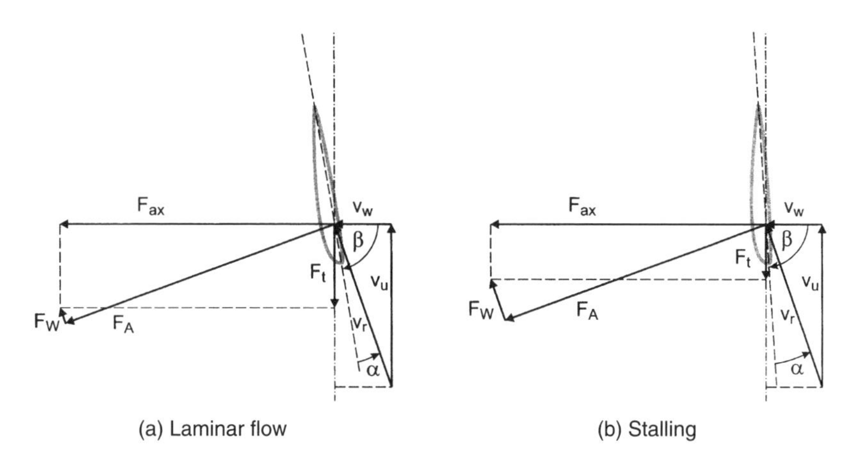

Figure 2.62 Flow and forces on the rotor blade given the same wind speed for active-stall adjustment (a) Laminar flow (b) Stalling

attack (β or α). On the other hand, stall-controlled machines with a synchronous generator and frequency controller or double-fed asynchronous generator permit variable-speed turbine operation. Such configurations mean that the peripheral speed of the blades and thus the angle of attack can be altered by adjusting the turbine speed (see Figure 2.63). Regulatory

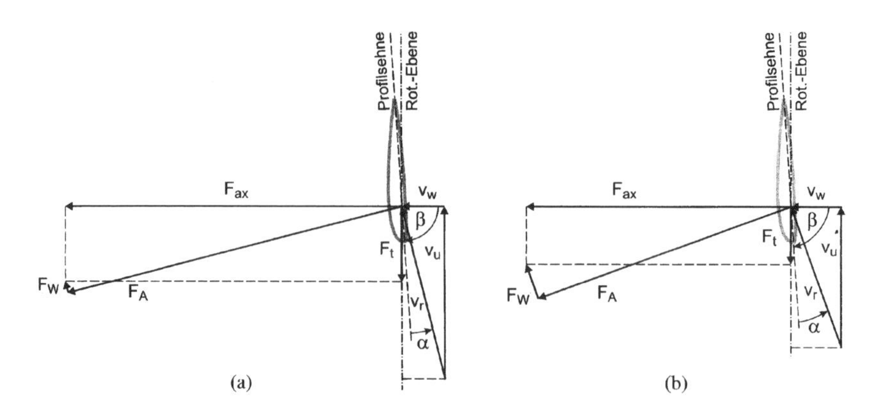

Figure 2.63 Flow conditions and forces at the rotor blade with (a) high and (b) low blade peripheral speed or rotor speed

interventions similar to those of blade pitch-adjustable (active-stall or pitch-controlled) turbines are possible, such that the machine output can be matched to grid or consumer requirements by the adjustment of the generator speed. As a result, optimal output ranges can be approached and, as described in what follows, only a proportion of the available power drawn for the protection of machine components and energy consumers.

2.3.4 Power control using speed variation

In spite of the constant frequency of the grid, the speed of a wind turbine can be influenced

- mechanically by varying the transmission ratio if the generator speed is constant (cf. Figure 2.64(a) and (b)) or
- electronically by frequency converter if the speed of rotation is variable over the entire drive train (turbine, gearing and generator).

The turbine speed can thus be adapted to meet grid and operational requirements (see Section 3.2). Varying the rotor speed in turn varies the power of the turbine, allowing it

- to be brought into the maximum range, depending on available wind, or
- to be reduced if required to meet grid or user requirements (see Figure 2.64(c) and (d)).

Such processes are currently used by machines in the 30 kW to MW range that employ power electronics (rectifiers, previously line-commutated and now self-commutated inverters, see Section 4.1) to convert the frequency to that required. Within their design limits, such models permit

- extensive running at optimal performance within the partial loading range without blade pitch adjustment, i.e. even in stall-regulated machines, and
- the reduction of power at turbine level under certain conditions.

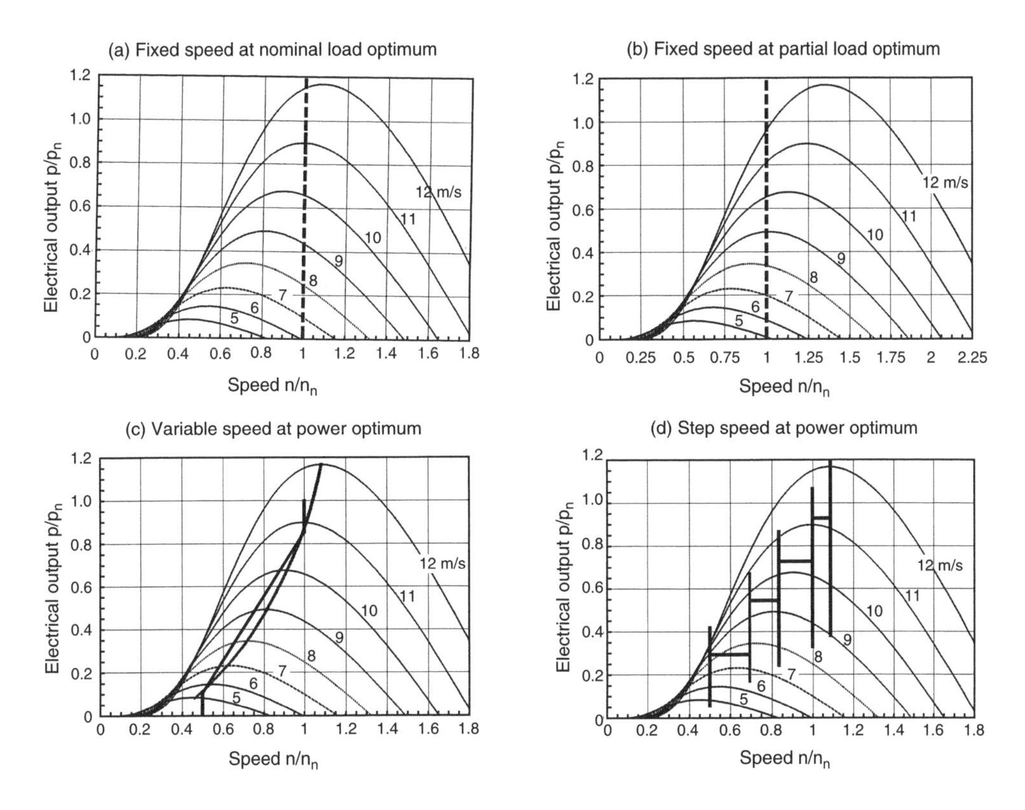

Figure 2.64 Output/rotational speed characteristics and dynamic operating ranges of wind turbines, with wind speed as a parameter

In addition, when flexible grid coupling is employed, the effect of fluctuations in available wind can be reduced by making use of the rotating mass of the drive train (rotor, gearing, generator rotor) to:

- smooth out variations in rotor speed and
- reduce the dynamic load on the entire system, particularly in mechanical drive trains.

In this way, the power converter allows a very short-time intervention at the generator, which allows the requirements and desires to be achieved at the turbine via the drive train. Thorough explanations of different technical variants of

- possible aspects of design,
- control procedures and
- dynamic processes and behaviour

in systems of this type have already been given in reference [2.21] and, for subareas such as double-fed asynchronous generators, in references [2.22] and [2.23], and more detailed explanations will be given in the course of this book, particularly in Chapter 5.

2.4 Mechanical Drive Trains

The drive power of a wind turbine engenders torques in its mechanical drive train or generator that are subject to fluctuation as a result of both periodic and aperiodic processes, such as

- changes in wind speed,
- tower-shadow or tower-occasioned upwind overpressure,
- blade asymmetry,
- blade bending and skewing, and
- tower oscillation.

In addition, load moments in the generator and converter due to

- static,
- dynamic and
- electromechanical

behaviour also act on the wind turbine via the drive train. The interaction of all torque effects works together with the flywheel-effect-dependent acceleration components to determine conditions in the mechanical drive train. A brief look at this is now given.

Figure 2.65 shows the structure of a typical drive train in a conventional geared turbine. From left to right, the rotor hub, the gearbox inclusive rotor bearing and the connecting flange of the generator may be discerned.

The considerably simpler construction of a gearless model is shown in Figure 2.66, which depicts the essential rotor–head components of a 200 kW Enercon E 30 turbine (see Figure 1.33(d)). The drive-train configuration is reduced to the rotor hub, the drive shaft and

Figure 2.65 Mechanical drive train of a geared wind turbine

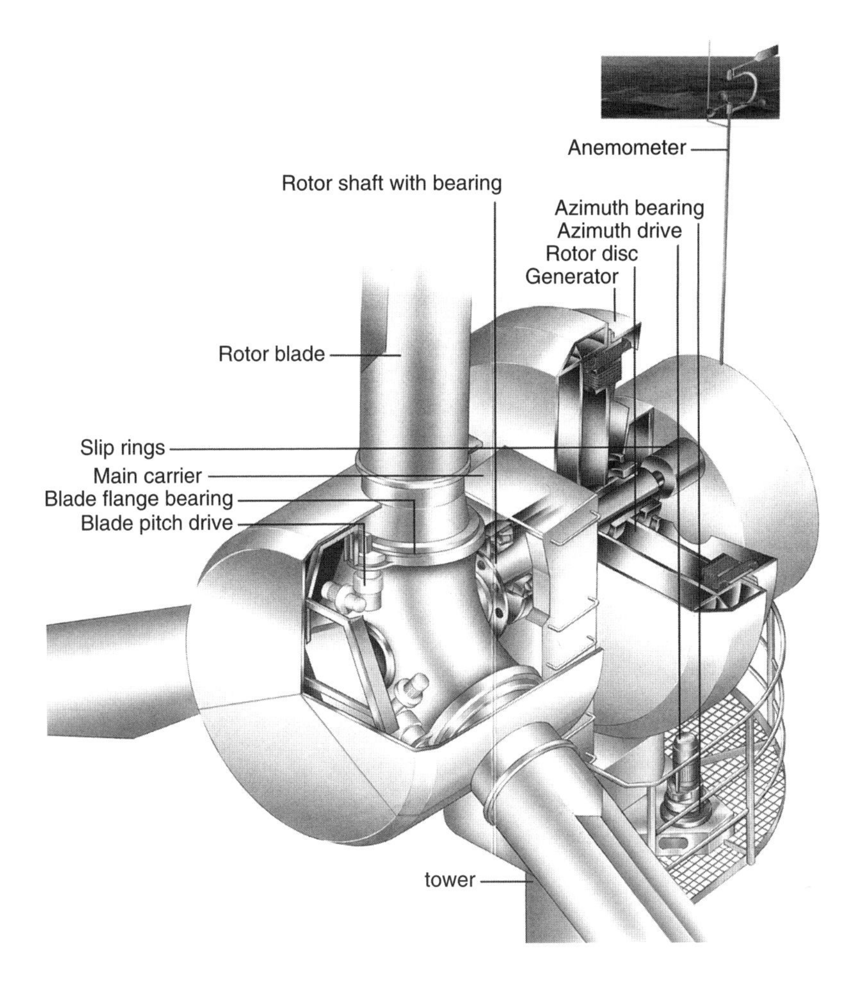

Figure 2.66 Schematic layout of an Enercon E 30 gearless wind turbine. Reproduced by kind permission of Enercon

the disc generator. This system represents a further step in the development of medium-sized turbines, which are distinguished in having variable-speed rotor or generator operation and grid-compatible coupling to power utilities via frequency converter systems.

Turbine effects, which are not further quantified here, depend upon the wind conditions at the site, i.e. wind speed and wind gradients (due to gusting and turbulence), as well as the constructional characteristics of the wind power unit as a whole. Tower-shadow and tower-occasioned upwind overpressure are especially important here as integer multiples of the rotor rotation frequency and blade count, occasioning torque oscillations in the hertz range that are generally more prominent in downwind than in upwind turbines. Load-alternation components arising due to

- rotor asymmetry (which acts according to the rotor frequency),
- blade bending and skewing, and
- tower oscillations,

which may be subject to influences caused by the respective resonant frequencies, are overlaid on the wind speed and thus show their effects additionally as wind gradients, in general play a subordinate role (from measurements, $\Delta M < 5\,\%\ M_N$) [2.24].

Torque oscillations at the meshing frequency of individual gear ratios are also possible, and measurements have shown that in unfavourable cases they can attain $\pm 10\,\%$ of nominal torque [2.4]. Particular influence is exerted by the design and type of the generator integrated with the grid and by the degree of elasticity of the entire system.

As shown in Figure 1.38(a) and (b), the drive torque M_A of the turbine acts on the mechanical drive train from one side and the load torque M_W of the generator acts on it from the other. Additionally, between these two main components there exists a speed-of-rotation or angle-of-rotation coupling via the mechanical elements connecting them. The rotating masses of the rotor blades, hub, gear train, brake disc, clutch, shafts and generator rotor are accelerated according to the difference in moments. The rigidity and damping characteristics of the individual components, and any play in the clutch and gears, exert a decisive influence on the behaviour of the transmission.

Studies of designed and constructed machines have shown that in general the inertia of gears, brake discs, clutches and shafts play a subordinate role in comparison with the dominant components exerted by the rotor blades, the hub and the generator rotor. If necessary, the relevant inertial components can be allowed for by simply adding them to the slow-revving side (hub, rotor blades) or to the high-speed side (generator rotor) according to their effect or, depending on the coupling, they can be taken into account as multiple-mass systems.

Because of the shock load that can be expected and the possible reversal of the energy flow, play in gearing and clutch is to be avoided, or at least kept as low as possible. It will therefore be ignored in the following.

Calculations based on existing machines show that to make an approximate determination of the behaviour of the transmission, the mechanical drive train, gears, clutches, etc., can usually be regarded as having zero mass according to the low moment of inertia within the system as a whole. This leaves only the two main components 'rotor' and 'generator', with an elastic, damped coupling existing between them (see Figure 2.67). Frictional components in the drive train can be ignored.

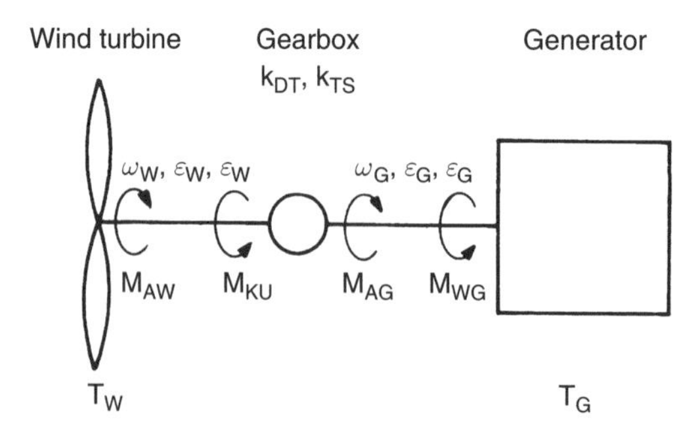

Figure 2.67 Simplified representation of a mechanical drive train

The torque at the generator coupling M_{Ku} can be represented by the simplified relation

$$M_{\mathrm{Ku}} = k_{\mathrm{TS}}\,(\varepsilon_{\mathrm{W}} - \varepsilon_{\mathrm{G}}) + k_{\mathrm{DK}} \frac{\mathrm{d}\,(\varepsilon_{\mathrm{W}} - \varepsilon_{\mathrm{G}})}{\mathrm{d}t} \tag{2.88}$$

or

$$M_{\mathrm{Ku}} = M_{\mathrm{TT}} + M_{\mathrm{TD}} \tag{2.89}$$

as the sum of moments of torsion–elastic (M_{TT}) and damping (M_{TD}) properties, where

M_{AW}	drive moment at the turbine rotor
M_{AG}	drive moment at the generator
M_{Ku}	coupling torque at the generator
M_{TD}	damping component of the drive train moment
M_{TT}	torsionally elastic component of the drive train moment
M_{WG}	electrical load torque in the generator
ε_{W}	angle of rotation of the rotor
ε_{G}	angle of rotation of the generator
$\Delta\varepsilon = \varepsilon_{\mathrm{W}} - \varepsilon_{\mathrm{G}}$	angle of torsion
k_{TS}	torsion resistance
k_{DK}	damping constant
T_{G}	run-up time constant of the generator
T_{W}	run-up time constant of the rotor

In stationary-state operation $\mathrm{d}(\varepsilon_{\mathrm{W}} - \varepsilon_{\mathrm{G}})/\mathrm{d}t = 0$, so that the coupling torque reduces to its torsionally elastic component

$$M_{\mathrm{Ku}} = M_{\mathrm{N}} = M_{\mathrm{TT}} = k_{\mathrm{TS}}\,(\varepsilon_{\mathrm{W}} - \varepsilon_{\mathrm{G}})\,. \tag{2.90}$$

Thus the torsional stiffness of the mechanical drive train under stationary component loading

$$k_{\mathrm{TS}} = \frac{M_{\mathrm{Ku}}}{\varepsilon_{\mathrm{W}} - \varepsilon_{\mathrm{G}}} = \frac{M_{\mathrm{Ku}}}{\Delta\varepsilon} \tag{2.91}$$

can be determined from the difference in angle between the transmission elements, which is relatively simple to measure.

Figure 2.68 shows a possible structure for the equation of moments as described above, including the rotating mass of the turbine rotor (J_{W} or T_{W}), on the one hand, and the inertia of the generator (J_{G} or T_{G}), on the other. In this figure,

M_{BG}	acceleration torque at the generator
M_{BW}	acceleration torque at the turbine rotor
$T_{\varepsilon} = p/\omega_0$	integrator time constants for determining the torsion angle (generator side)
ω_{G}	angular velocity of the generator
ω_{W}	angular velocity of the turbine rotor

In this way, torques and angular velocity values are generally related to the generator side, taking transmission ratios into account.

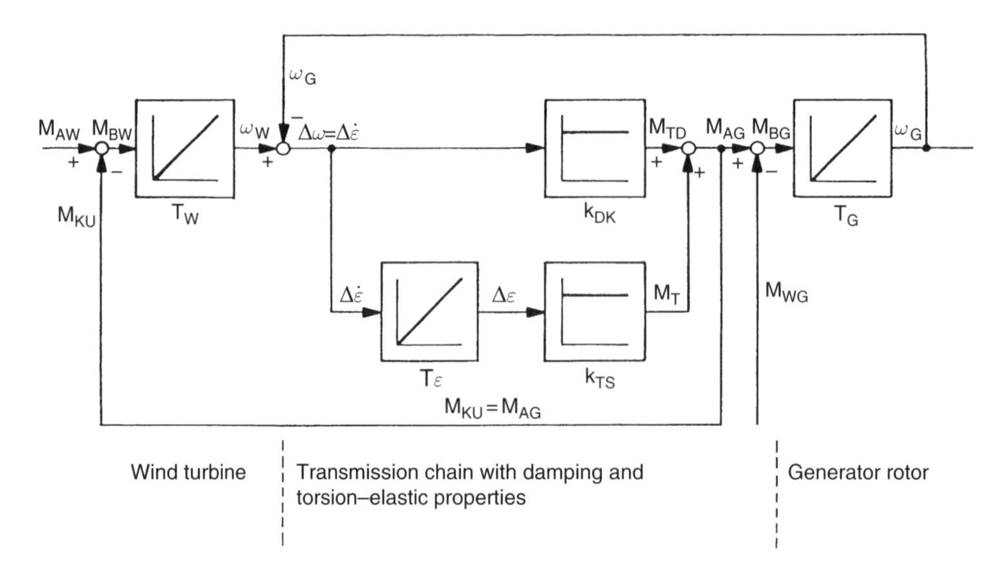

Figure 2.68 Structure of a mechanical drive train

As predicted by calculation, existing machines show nontrivial angles of torsion (e.g. $\Delta\varepsilon = \pi/4$ at nominal torque). These are certainly of considerable significance for the design of the drive train, but this is not the object of the current study, which seeks an approximate determination of transmission behaviour. In normal wind power plant designs a run-up time of around 10 seconds can be expected (see Section 2.5), while delays arising from the drive train are on the order of milliseconds.

Because of the large difference in time constants, in a good approximation a mechanical drive train can be regarded as constituting a perfectly rigid transmission, and can thus be described as having proportional behaviour. Accordingly, the moment of inertia of mass of all rotating parts can be taken as a whole, once any transmission ratios have been allowed for. In doing so, the generator side is the preferred frame of reference.

We thus obtain the simple equation of moments for the drive train

$$\frac{M_{\mathrm{AW}}}{M_{\mathrm{N}}} - \frac{M_{\mathrm{WG}}}{M_{\mathrm{N}}} = \frac{M_{\mathrm{BR}}}{M_{\mathrm{N}}} = T_{\mathrm{R}} \frac{\mathrm{d}\,(\omega/\omega_{\mathrm{N}})}{\mathrm{d}t}, \tag{2.92}$$

where the integration or run-up time constant of the rotor system

$$T_{\mathrm{R}} = \frac{J_{\mathrm{R}}\omega_{\mathrm{N}}}{M_{\mathrm{N}}} \approx T_{\mathrm{W}} + T_{\mathrm{G}}, \tag{2.93}$$

which corresponds approximately to the sum of the time constants of the turbine and generator, where

J_{R} is the moment of inertia of all rotating masses,
M_{BR} is the acceleration torque of rotating components,
M_{N} is the nominal moment and
ω_{N} is the nominal angular velocity.

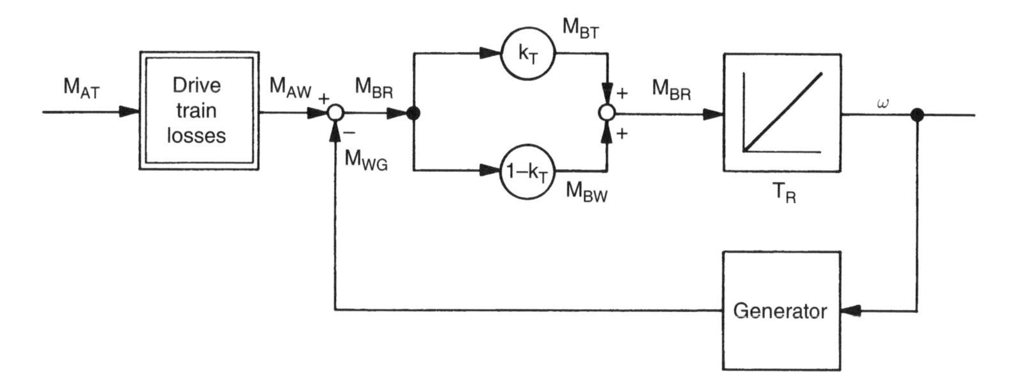

Figure 2.69 Highly simplified structure of a mechanical drive train

For cases of interest (e.g. dimensioning the drive train), the acceleration torque M_{BR} on the entire rotor system can be split into

$$M_{BW} = (1 - k_T)\, M_{BR} \tag{2.94}$$

on the turbine rotor side, and

$$M_{BT} = k_T M_{BR} \tag{2.95}$$

on the generator side, according to the inertia of their respective masses, as shown in Figure 2.16.

Transmission losses can be taken into account either by including them in the drive torque (Figure 2.11) or by introducing an extra block in the drive train structure, as in Figure 2.69. Here the input quantity M_{AT} stands for the drive torque of the drive train, which must be reduced by the losses therein.

Figure 2.69 illustrates the possibilities of controlling turbine speed by means of the generator, mentioned in Sections 2.3.3 and 2.3.4. This theme is complemented and continued in Chapter 3, which shows various ways of using generators to convert mechanical energy into electrical energy.

2.5 System Data of a Wind Power Plant

Standard system values are of great significance when planning machines, working out details and estimating everything from loads and expected behaviour to transport and assembly problems. The following sets out essential construction data and currently available reference values for machine costs as a function of capacity.

The installation of turbines, and in particular screw connections, between highly loaded construction elements (rotor blades, tower flange, etc.) that facilitate the attainment of the defined prestressing forces with reproducible accuracy where there is no torsion or side loading (see end of Chapter 2) will not be discussed further here.

2.5.1 Turbine and drive train data

Within the normal ranges of wind speed (5 to 15 m/s), airstreams do not carry much energy. In consequence, wind energy converters need to have turbines of large diameter rotating at low speed, which means that construction costs are high. The following reference values for essential physical quantities of mechanical converters (turbine, drive train) permit estimates and calculations of the dynamic behaviour of entire systems to be made. These values are given in Figure 2.70 in the form of a nomograph.

Reference values for nominal power, the necessary rotor radius, construction-dependent rotor speed and the resulting gear ratios, as well as run-up time constants for mechanical drive trains (see Section 2.4, Equation (2.93)) for use in system studies, are summarized in Figure 2.70. This figure is based on data from over 100 different turbines, allowing for various design-specific parameters and constructional boundary conditions, such as differences in power or diameter in systems designed for inland as opposed to coastal areas, and the associated considerations of stall-versus pitch-regulated turbines.

In addition to the blade profile and tip speed ratio, blade tip velocity (which is limited by aerodynamics, rigidity and noise development) is decisive in determining the very low speed of turbines, particularly for large machines. The mechanical transfer of energy to the generator is usually kept at a constant speed by the use of gearing. In wind turbines, as in other types of power plant, four- and six-pole generators are usually used. Such generators, together with the associated gear trains, represent a low total mass for the converter system. This is why the gear ratios of these common generator variants, for use in 50 Hz and 60 Hz grids or for variable-speed generator variants, were taken as a basis for Figure 2.70. The data for wind turbines that have directly driven generators with a high number of poles, designed for gearless units, cannot be directly read off. Their 'virtual' speed, which corresponds with the electrical rotational frequency, is found from the product of the pole pair number and mechanical speed of rotation according to the general relation $n_{el} = pn_{mech}$.

In conclusion, we give standard values for rotor system run-up time constants as a function of the flywheel effect of all rotating masses. Here, in addition to nominal torque, the speed of rotation and tip speed ratios in nominal operating conditions are given for two characteristic turbine configurations, where $\lambda_N = 4$ and 6. In accordance with Section 2.3, a friction-locked torque transfer from the turbine rotor to generator was taken as a prerequisite in determining run-up time constants. The explanations above contain wind energy converter data that must be allowed for in calculating and designing electromechanical energy converters.

In the following, essential data for common wind power machines on the market are given in the form of standard values for further studies. Manufacturer's data for around one hundred of the most common standard models in the 20 W to 2500 kW class have been evaluated and are presented in Figures 2.71 to 2.80.

2.5.2 Machine and tower masses

We limit our study of the masses in a wind power machine to the transportable and dynamically loaded components. These mainly comprise those items found in the head of the tower (the nacelle and the rotor blades) and the tower itself. Much heavier items such as the foundations, control room and connecting cables are ignored. In Figures 2.71 to 2.73, specific values for machine nacelle mass including rotor blades, are plotted as functions of

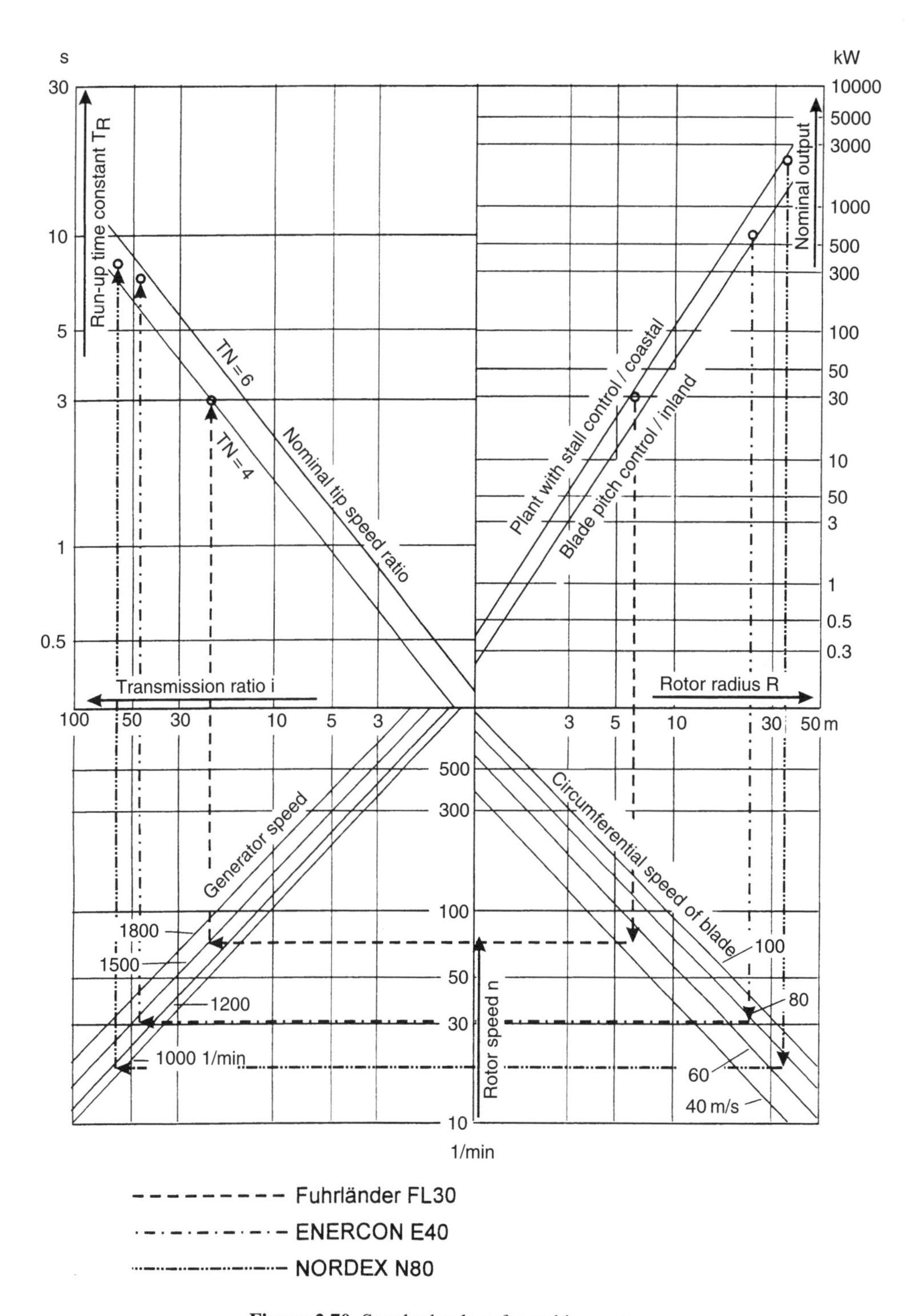

Figure 2.70 Standard values for turbine systems

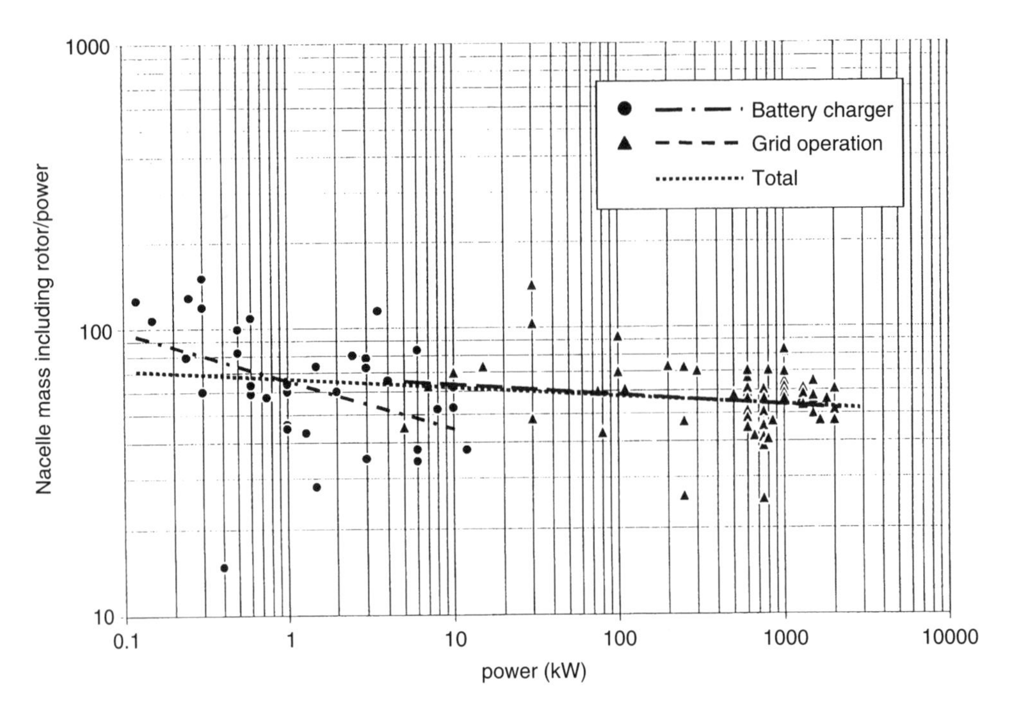

Figure 2.71 Specific mass per kW of nacelle plus rotor as a function of nominal power

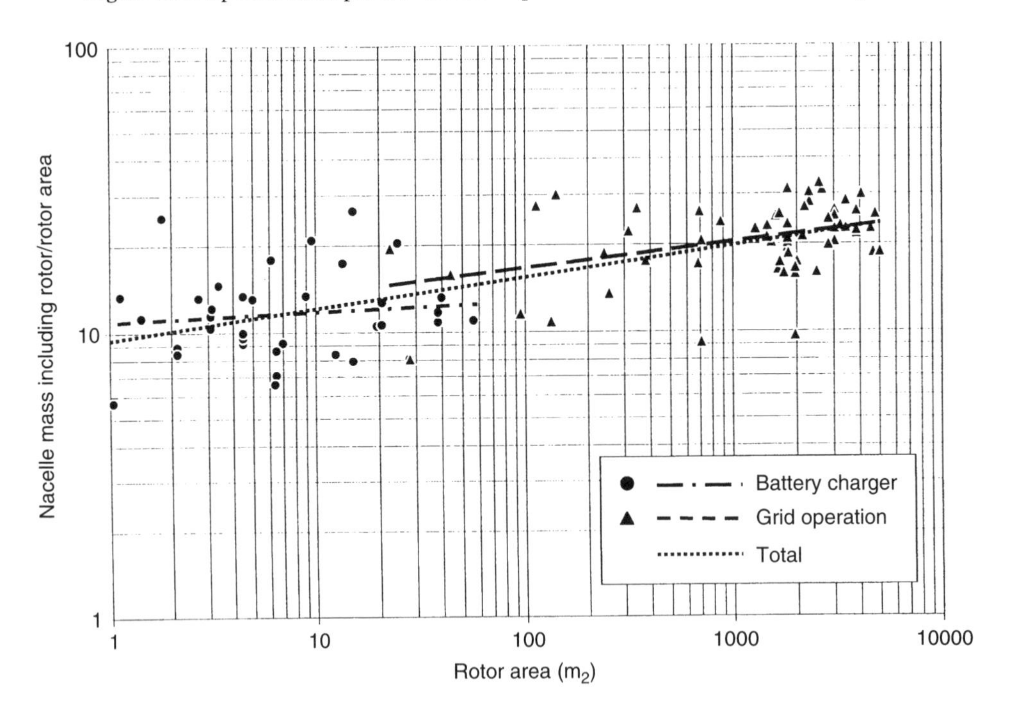

Figure 2.72 Specific mass per m^2 of nacelle and rotor blade as a function of swept rotor area

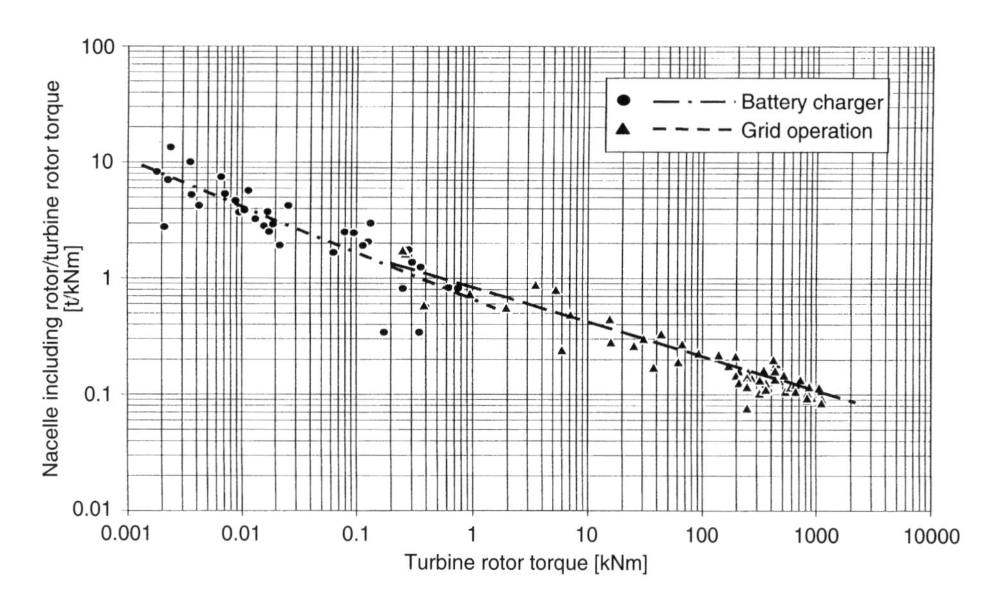

Figure 2.73 Specific mass per kNm of nacelle and rotor blade as a function of turbine rotor torque

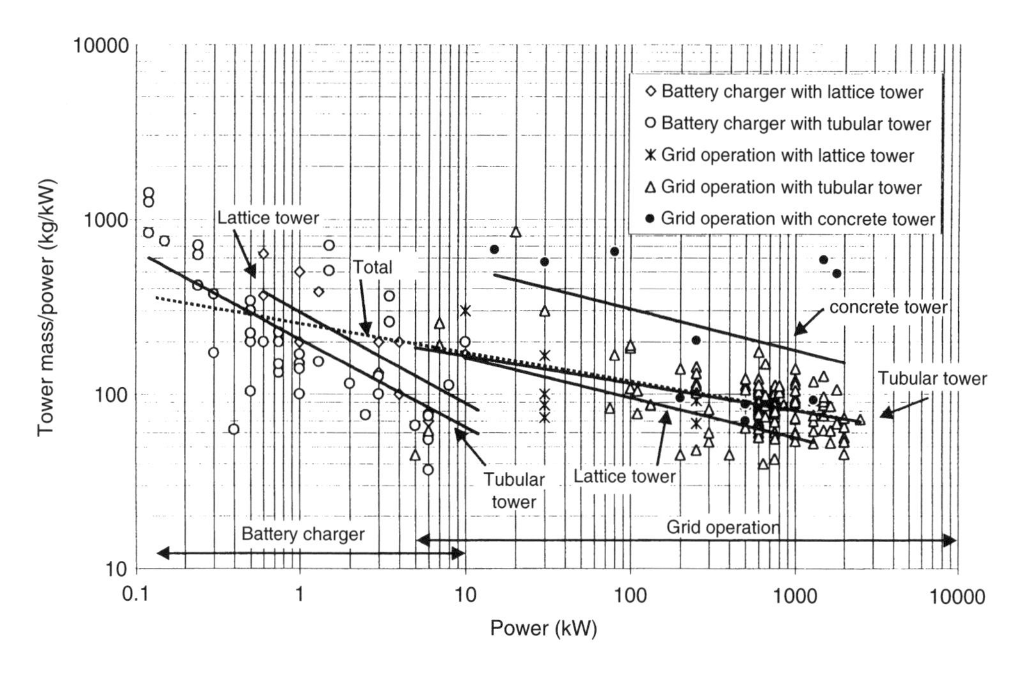

Figure 2.74 Specific mass per kW of tower as a function of turbine power

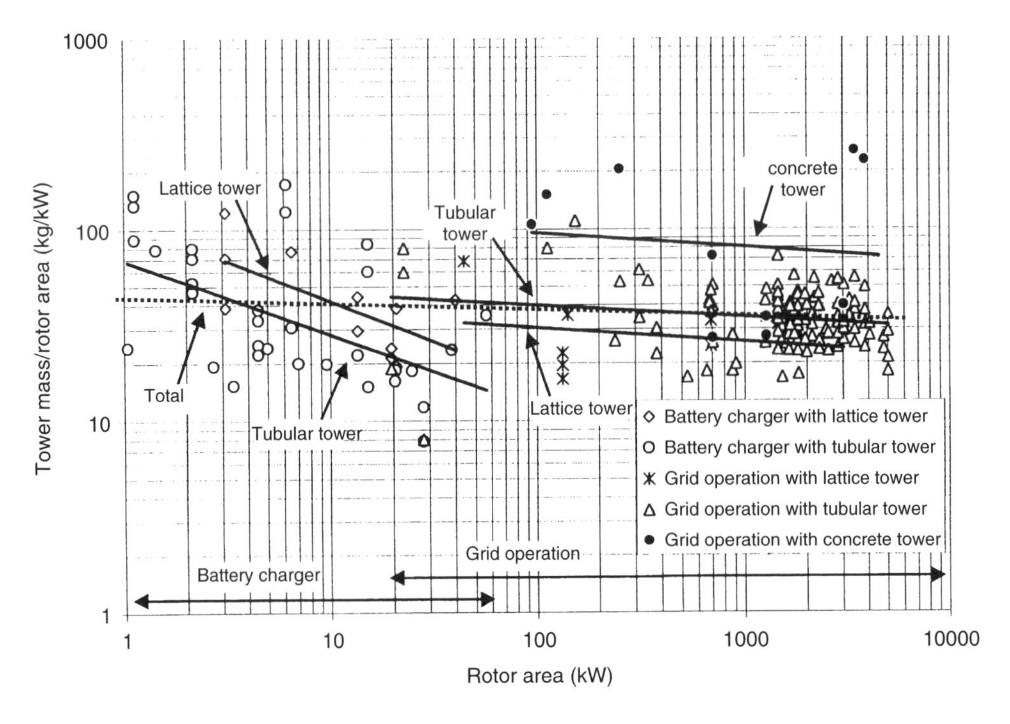

Figure 2.75 Specific mass per m^2 of a tower as a function of swept rotor area

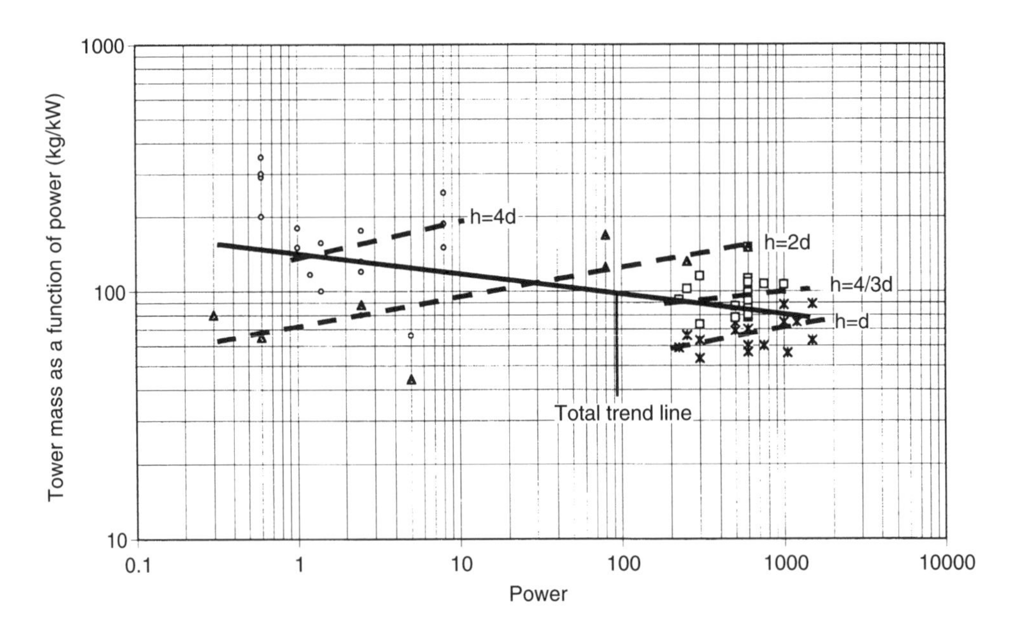

Figure 2.76 Specific mass per kW of a wind turbine tower as a function of the turbine output for various tower height to turbine diameter ratios

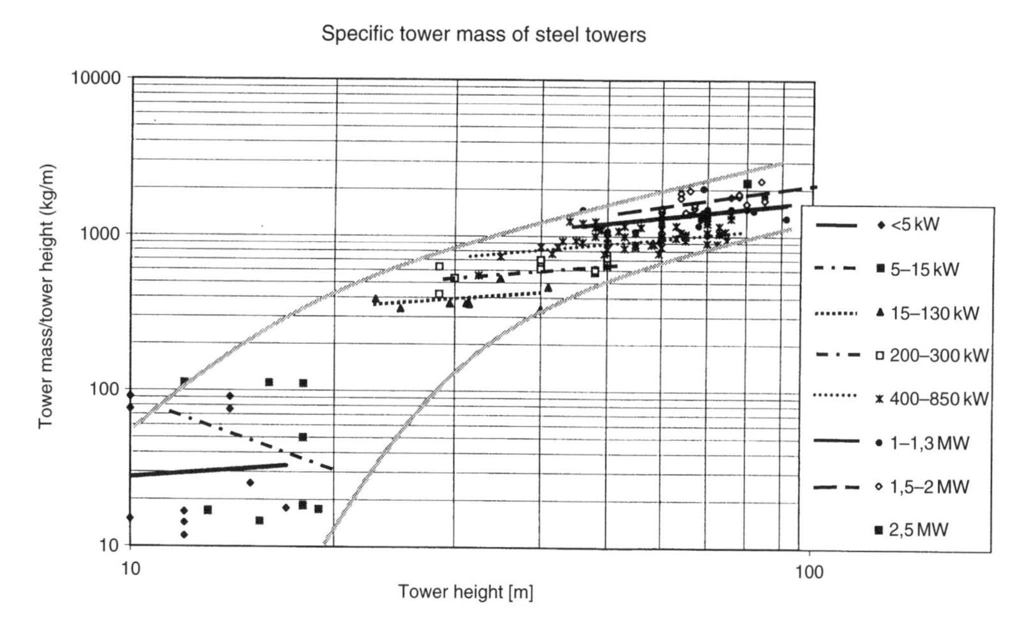

Figure 2.77 Specific mass per m of a wind turbine tower as a function of tower height with the turbine output as a parameter

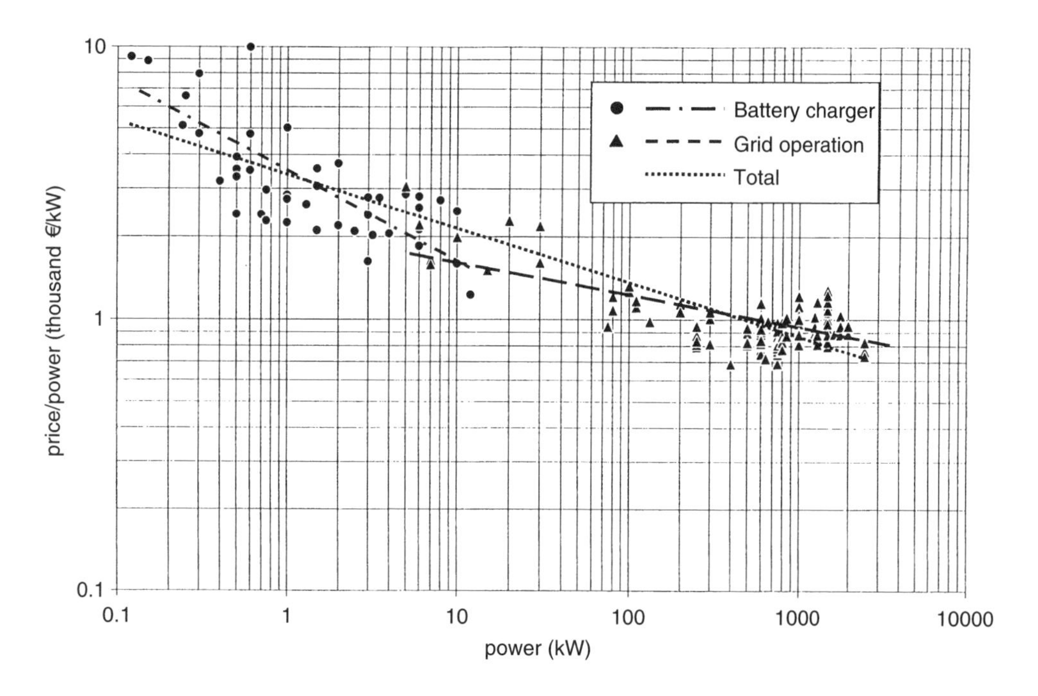

Figure 2.78 Specific machine cost per kW as a function of nominal output power

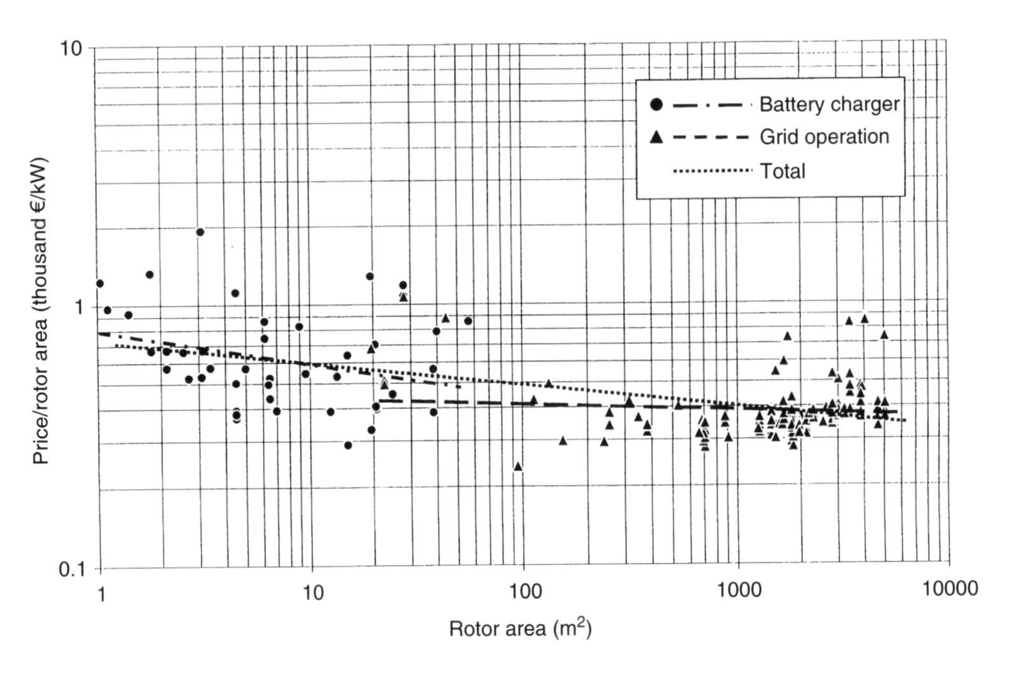

Figure 2.79 Specific machine cost per m^2 as a function of swept rotor area

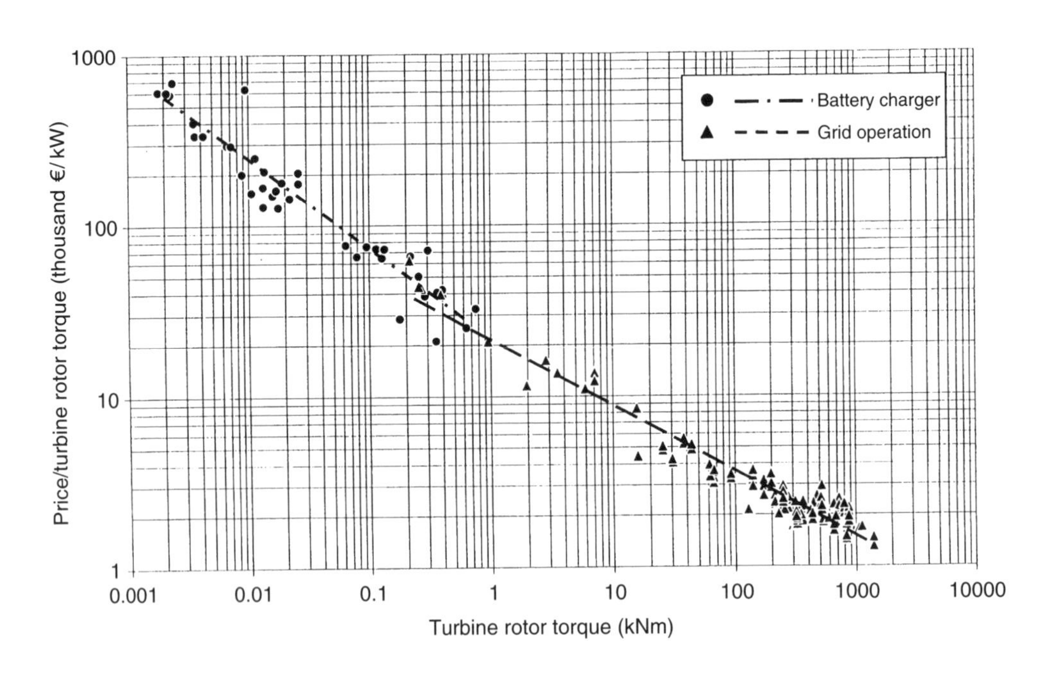

Figure 2.80 Specific machine cost per N m as a function of rotor torque

nominal power, swept rotor area and rotor torque. Comparing the figures shows that the mass per square metre of swept area climbs slowly with machine size, while the values related to machine power and turbine torque fall off distinctly. Furthermore, it can be clearly seen from the low degree of scattering of the turbine torque data that a reliable relation for nacelle mass exists.

Figures 2.74 and 2.75 show the difference in mass between steel tube, concrete and lattice towers as a function of nominal machine output and of the swept rotor area of the turbine. Because of the wide scattering of results, no standard rules can be immediately deduced. It can be seen, however, that in small plants steel tube masts have a lower mass than concrete or lattice towers. As swept area increases, however, they tend steadily to larger specific values. The masses of concrete and lattice towers, on the other hand, decrease as the nominal power and swept area increase.

Figure 2.76 shows that turbines with outputs below 100 kW mainly have specific tower masses greater than 100 kg per kW, while turbines with outputs above 100 kW generally exhibit figures less than 100 kg per kW. High ratios of tower height to turbine diameter (e.g. $h/d = 4$) are generally selected for small turbines, while turbines in the megawatt class tend to exhibit constant height to diameter ratios ($h = d$). Although the tower mass per kW tends towards smaller values as the turbine size increases, at the same height to diameter ratio there is a clear increase in mass.

A completely different trend is seen in the relationship between the tower mass and tower height. Figure 2.77 illustrates the sharp increase of tower mass per metre mast height as the turbine output increases. This figure rises from approximately 20 kg/m for 2.5 kW plants to around 2000 kg/m for those in the 1.5 MW class. Furthermore, it is clear that small plants exhibit the same mass per metre at different tower heights, while large plants tend to have a significantly greater mass per metre as the mast height increases.

2.5.3 Machine costs

Figure 2.78 shows machine costs as a function of nominal turbine power. In spite of differing design variants, which are allowed for in this presentation, costs deviate only slightly from the average values. However, clear differences can be seen (in particular in small machines) between battery chargers and grid-compatible machines. Furthermore, newly developed large-scale machines in the market introduction phase tend to have higher specific purchase costs. However, this trend could be discerned even in the previous developmental phases, meaning that cost digressions can be expected here too.

Better predictions, particularly as regards expected energy levels and economic questions, may be obtained with the help of a plot of cost against swept rotor area, as in Figure 2.79. A further variant is offered in the plot of machine costs against rotor torque as shown in Figure 2.80. In this, not only the nominal output power but also the turbine speed dependencies are included in the reckoning. The low scattering of the data shows that here too a clear relationship between cost and torque is to be observed.

In contrast to nacelle mass, it may be seen that with increasing machine size the area-related machine cost slowly decreases, while for nominal power and torque, cost falls off rapidly.

3

Generating Electrical Energy from Mechanical Energy

In examining the functionality and behaviour of wind power plants, particular significance must be accorded the generator as a source of load torque, depending on its situation and on how it is coupled to the system (see Figure 1.37). Seen as a whole, wind power stations, like other electricity supply sources of comparable output (e.g. hydropower, gas and diesel generators), should

- be simple to use,
- have a long useful lifetime,
- have a low maintenance outlay and
- have as low as possible initial cost.

To meet these requirements a suitable generator must be chosen. Because of their robust construction, the generators used in wind power plants for the conversion of mechanical energy to electrical energy should be almost exclusively synchronous or asynchronous. In making the choice, particular attention must be paid to the constraints imposed by the environment and the demands made of the electrical machines in consequence, as well as to the torques engendered in the generator.

3.1 Constraints and Demands on the Generator

On the one hand, drive torque characteristics are dictated by the properties of the rotor and the amount of wind energy delivered in the airstrip; on the other, the demands and, as far as possible, the wishes of the electricity supply company and the users must be catered for at the interface between mechanical and electrical interactions – the generator. Furthermore, the peculiarities of wind power machines must be particularly considered.

Grid Integration of Wind Energy Conversion Systems, Second Edition S. Heier

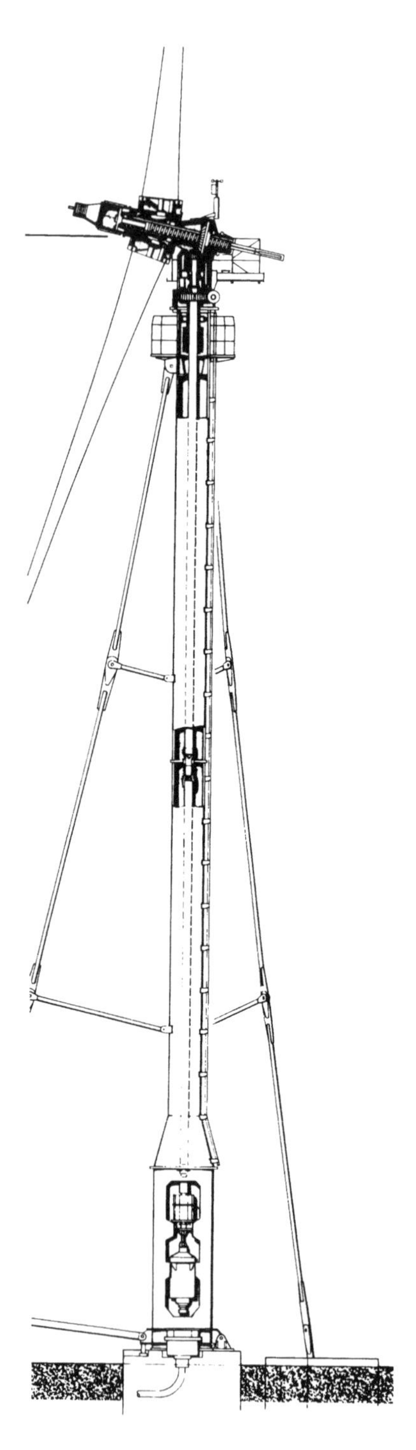

Figure 3.1 Turbine with a generator in the tower foot (Voith). Reproduced by kind permission of J.M. Voith AG

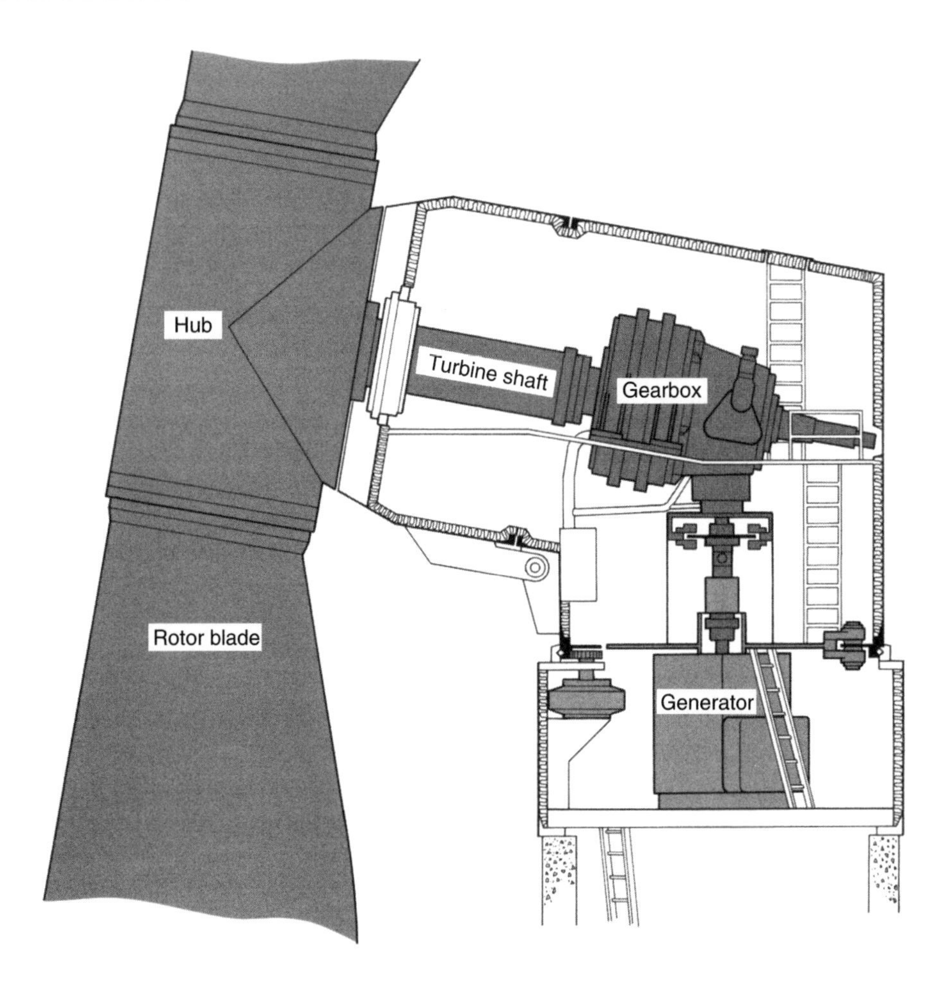

Figure 3.2 Turbine with a generator in the tower head (Aeolus). Reproduced by kind permission of Magnet Motor GmbH

In horizontal-axis machines, schemes attempting to accommodate the generator at the foot of the tower (see Figure 3.1) were beset with difficulties, e.g. those resulting from torsional oscillations in the drive shafts. Because of this, the generator is also put in the head of the tower (Figure 3.2) or more usually in the nacelle (Figure 3.3). Keeping costs and possible tower oscillation in mind, the lightest possible generator should be sought. As mentioned in Section 2.4, four- and six-pole generators used with gear trains yield good results. However, the gearbox mass associated with such designs is eliminated in gearless machines.

Periodic effects resulting from disturbed flow around the tower, asymmetries in the system, etc., and aperiodic effects due to wind-speed changes (e.g. gusts) make high demands on components. These should be eliminated as far as possible, since service life is essentially determined by component loading.

Shock moments to the drive train, the mechanical effects of which can propagate as far as the machine anchorings and the electrical effects of which can propagate across the windings

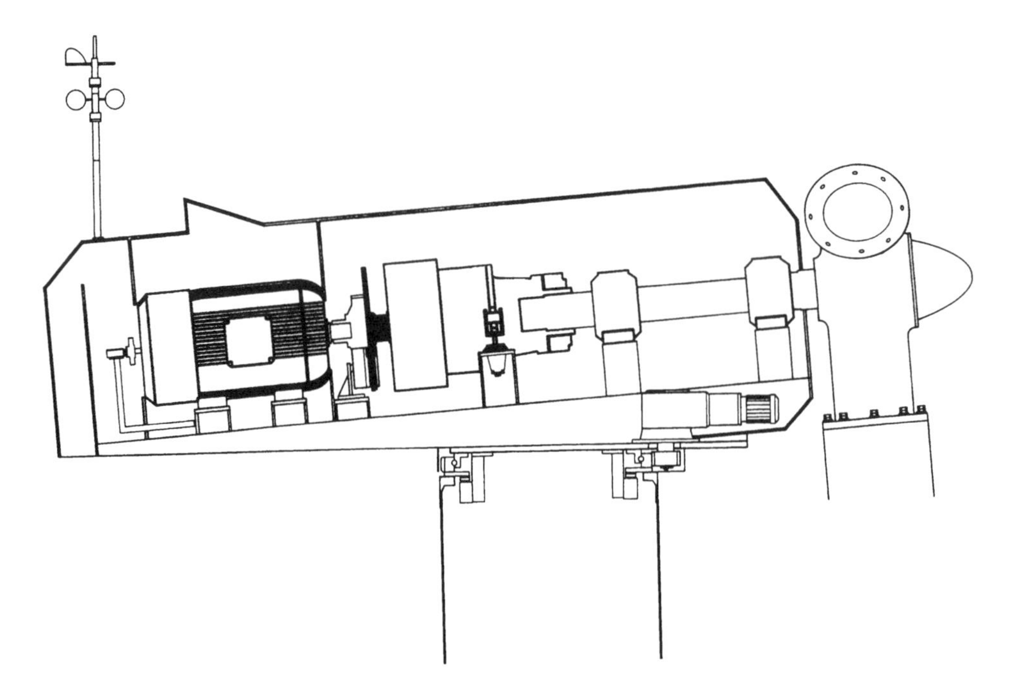

Figure 3.3 Turbine with a generator in the nacelle (Seewind). Reproduced by kind permission of Seewind

into the grid, are essentially dependent on the degree of elasticity, i.e. the torque/rotational speed gradient of the generator. Therefore, mechanical–electrical converters that are flexible both in their speed of rotation and in the way in which they are linked to the drive train have an advantage for use in wind turbines. By completely disconnecting the turbine or rotor speed from the grid, a greater degree of efficiency can be obtained under partial loading (see Figure 2.62) or a lesser amount of power can be drawn from the wind when the amount of energy required is lower.

Availability is of critical importance in guaranteeing security of supply and in the dimensioning and rating of storage and back-up plants. It has a great influence on capital investment and on the profitability of entire systems. Wind conditions at the turbine site play a significant role with regard to this aspect. Delivery of energy by wind power plants should start at low wind speeds and should be capable of maintaining small base loads. To accomplish this, standstill and run-up torques must be kept low. These can arise due to static friction in any bearings and brushes, etc. Also possible are dwell torques, arising mainly in permanent-magnet generators due to the reluctance effect. Further, the generator should show low no-load and exciter losses and high efficiency, even under low loading.

Systems for converting mechanical turbine power into electricity, which are suitable for incorporation into electricity supply schemes while fulfilling the above conditions and requirements, will be presented in the following. Based on personal studies and many years of experience in the design and optimization of such systems, guide values for dimensioning will be put forward, along with proposals for a forward-looking utilization of wind power.

3.2 Energy Converter Systems

Rotating field generators are used almost exclusively in the conversion of mechanical energy to electrical energy. Here, principally asynchronous and synchronous generators with direct grid coupling or with full or partial (rotor) inverter coupling are of significance. Three-phase generators employ a rotating magnetic field, known as a rotary field. This may be obtained by the use of rotating permanent magnets or by the rotation of excitation windings with the aid of current fed via brushes and slip rings. Such rotary fields excite an electric voltage in stationary conductors – the stator windings of the generator – the frequency of which is synchronized by mechanical rotation of the machine. In these synchronous generators, three (or an integer multiple thereof) coils are spatially offset by 120°. Such machines therefore produce three-phase voltages, which are displaced by a 120° vector in relation to one another – so-called three-phase alternating voltage. The voltage is dependent on the generator type, the speed of the rotor, the excitation and the load characteristics; in isolated and standalone operation this value can be regulated by varying the excitation. When connected to the public supply, both voltage and frequency are dictated by the grid.

If the three-phase alternating current stator of a generator is supplied with alternating current from the grid, this also sets up a rotary field. This field excites currents in the rotor windings of the generator, the frequency of which corresponds to the difference between the field rotation frequency and the mechanical speed of the rotor. These currents produce torques in the rotor, which, in synchronous machines, have a damping effect.

Asynchronous motors, on the other hand, cannot follow the rotary field. In their rotors a torque is set up that is proportional to the frequency difference (or the so-called slip) and acts in the direction of the rotary field. Asynchronous machines supplied at almost constant frequency and voltage from the grid go over to the generating mode when they are driven (e.g. by a wind turbine) past the synchronous rotation frequency. They then deliver power to the grid. Asynchronous motors and generators require inductive reactive power to build up their magnetic rotary fields, i.e. their electromagnetic excitation. This may be supplied by the grid or, for example, by capacitor banks. Rotary fields therefore set up torques in the rotors of grid-driven synchronous and asynchronous devices, which in turn enables them to extract energy from the driving turbine and convert it into electricity.

It is therefore important to differentiate between generator systems that draw their excitation or reactive power from the grid, and can thus serve only as grid support machines, and grid primary supply machines, which possess their own possibilities for voltage and reactive power control. All the most important configurations for converting mechanical energy into electrical energy are shown in Figure 3.4, which will serve as a reference for subsequent sections.

The energy user (e.g. the grid) usually sets low tolerance limits on frequency, voltage, harmonics, etc. According to the design, these can be met in various ways. Designs (a) and (g) in Figure 3.4 show extremely rigid grid coupling. In all the other systems shown, the design allows the mechanical speed of rotation to be dissociated in varying degrees from the electrical frequency or voltage in alternating or direct current systems by the incorporation of power electronics and the associated regulation procedures. In addition, these systems provide short-term energy smoothing during gusting, with the consequent reduction of loads on mechanical components. Variants (f) and (g), in addition to the requisite performance control, also allow a controlled delivery of reactive power. These systems can therefore be used as a primary grid supply or as grid support for alternating current (a.c.) grids, as

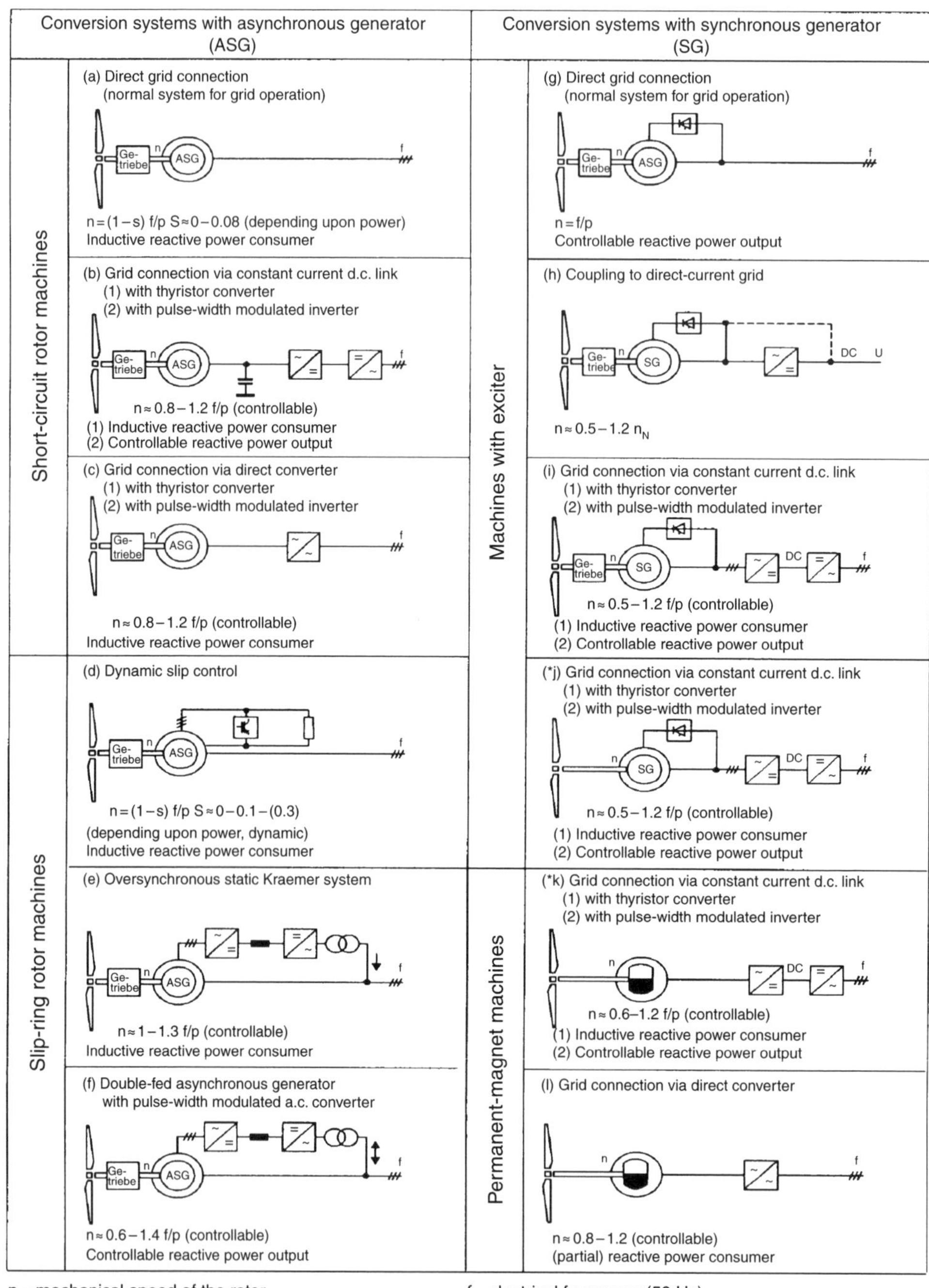

Figure 3.4 Systems for converting mechanical energy to electrical energy

can variant (h) for direct current supplies. Synchronous generators delivering power via grid-commutated frequency converters (models (i), (j) and (k)), which were used up until a decade ago, must draw their reactive power from the grid. By contrast, the self-commutated frequency converters in use today, such as designs (f) and (g), are themselves able to provide the reactive power necessary and to control the voltage in grid branches.

Direct converters in double-fed asynchronous generators, as shown in design (f), were used as far back as in the GROWIAN, for example. Further applications such as designs (c) and (l) are possible in both asynchronous and synchronous machines. Grid coupling by means of direct converters has not yet been able to establish itself on the market due to the high number of power semiconductor components that they incorporate. For the gearless systems (j), (k) and (l), only synchronous generators are used, excited either electrically (j) or by permanent magnets, (k) and (l).

The following section examines the main possible operational ranges for the rotary field machines mentioned above and the layout of the most important machines.

3.2.1 Asynchronous generator construction

With a view to the exclusion of dust and damp and to prevent accidental contact, asynchronous generators (Figure 3.5) for wind turbines are usually manufactured as closed units. The essential components are the copper stator windings: three windings spatially offset by 120° in two-pole designs and six windings offset by 60° in four-pole designs. These windings are insulated from one another. The so-called three-phase alternating current windings are laid into slots in the laminated core. Since the core must conduct the magnetic

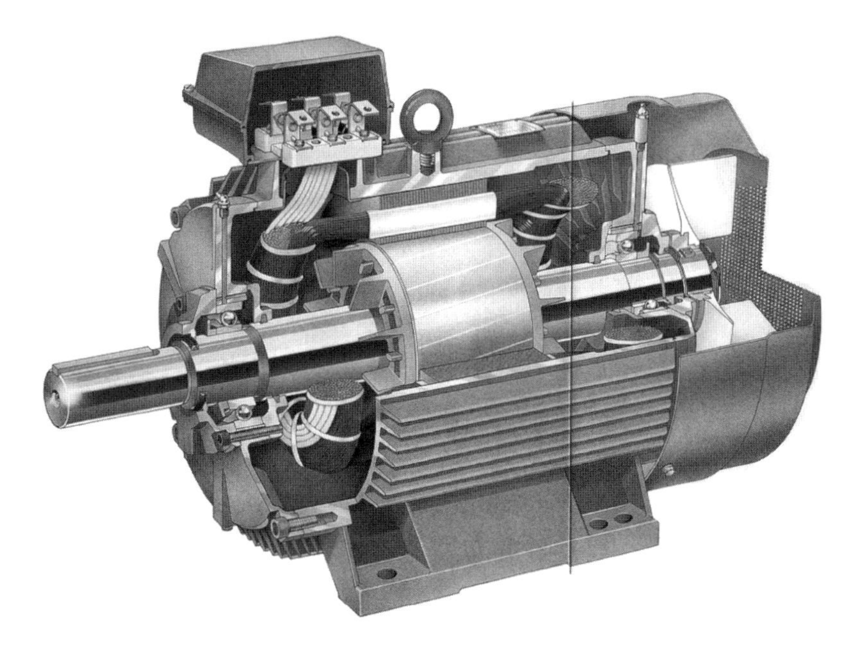

Figure 3.5 Cutaway of an asynchronous generator with a cage rotor (ABB). Reproduced by kind permission of ABB Motors

flux of the rotary field around it, it is built up of laminations, which are insulated from each other to reduce losses.

In slip-ring rotor machines, equivalent alternating current windings are also laid into slots in the rotor core. Current is fed to or collected from them via rotating slip rings running against carbon brushes mounted on the generator chassis.

By contrast, cage rotors, as shown in Figure 3.5, have an aluminium diecast or bar winding laid into the slots of the rotor core. These conductors are connected to one another on the end faces of the rotor; in other words, they are short-circuited. Cooling vanes on these end rings improve heat dissipation.

The rotor and the fan are carried on a common shaft. This runs on two roller bearings mounted concentrically with the stator on two end plates, one on the shaft side and one on the fan side. The stator core with its windings and connectors is enclosed by the stator frame, the outside of which is fitted with cooling fins. The base of the generator provides mounting lugs and also serves as a support against transmitted torque. The grid connection box provides a well-protected electrical connection to the grid.

Cage generators are remarkable for their extremely simple layout, which permits robust construction and high operational reliability, even in the event of rough handling. Above all, these machines can be turned out in large numbers, making them correspondingly cheaper.

3.2.2 Synchronous generator construction

The synchronous machine is much more complicated than its asynchronous counterpart. Figure 3.6(a) shows a cutaway of the relatively complex layout of a brushless self-regulating generator. Following the sequence of the symbolic circuit shown in Figure 3.6(b) from left to right, the cutaway shows the main a.c. generator, the exciter and the pilot exciter. The stator with its a.c. windings, the shaft carrying the fan, the bearings and the housing with its base and connection box are as on the asynchronous machine.

Large generators in the 100 MW and GW range are usually of synchronous nonsalient pole construction (turbogenerators). Their rotor slots carry exciter and damper windings. Generators suitable for wind turbines, however, lie in the range between a few kW and around 5 MW. As shown in Figure 3.6(a), conventionally constructed synchronous machines of this size use salient-pole rotors. The rotor comprises the pole shoes, the poles lying beneath and the exciter windings. The stator consists of the stator core and a.c. windings. The rotor and stator together make up the generator.

In the brushless model shown here, the rotor is supplied by the exciter, which consists of exciter poles in the stator, rotary field windings and a rectifier bridge, to be seen at the end of the shaft (far right). The exciter or the exciter coils in the stator are supplied from the pilot exciter. Rotating an outer permanent magnet (right) generates current in the pilot exciter coils, which is fed via a voltage controller to the main exciter coils. In this way, the necessary magnetic field is set up in the poles of the exciter and crosses the air gap to the a.c. windings of the exciter. This sort of construction is preferred for grid-independent power supplies.

If a grid is available to power the exciter windings, the pilot exciter becomes superfluous. In brushless machines the exciter current is fed from the stator winding, across the air gap of the exciter and via the rotating rectifier bridge to the rotor. Its rotating magnetic field

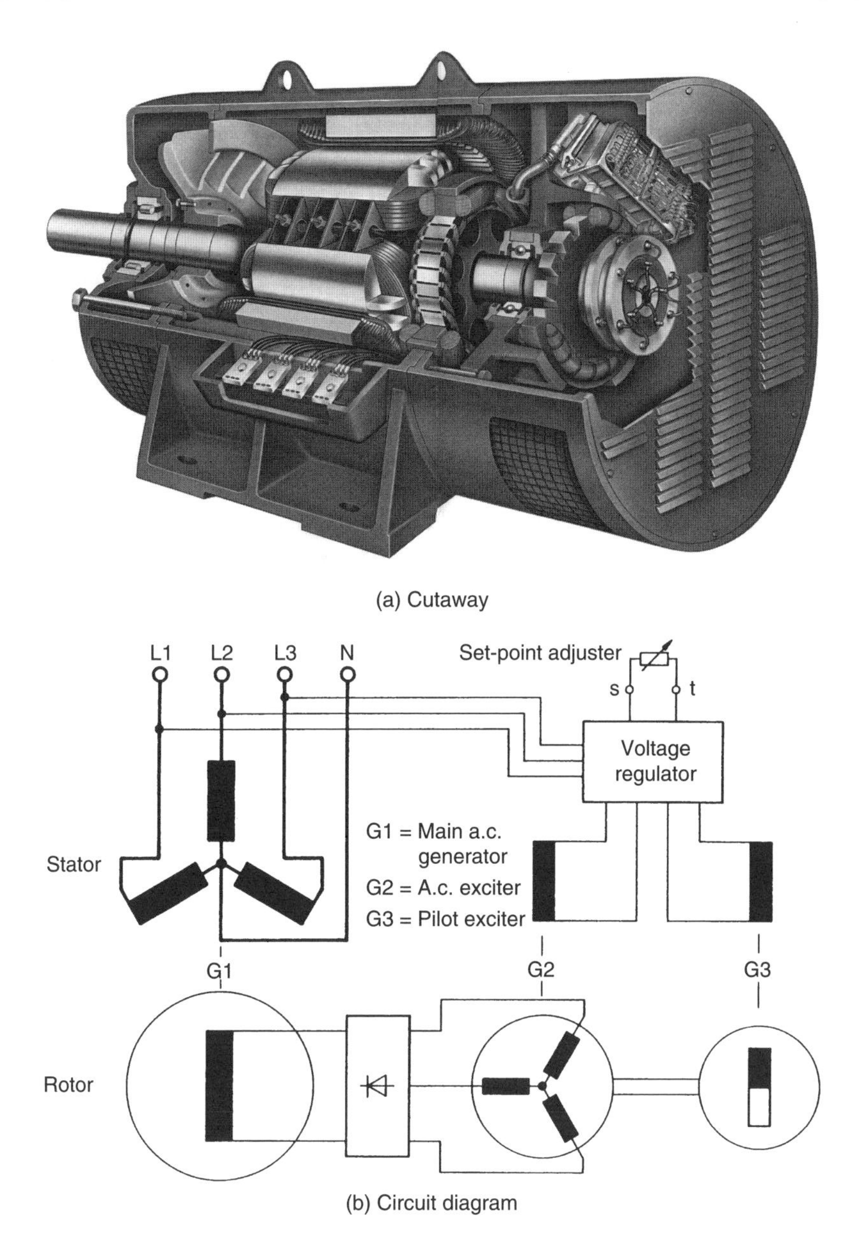

Figure 3.6 Synchronous generator structure (self-regulating brushless model, AvK). Reproduced by kind permission of AvK

(rotary field) is transferred to the stator. Voltage regulation procedures must therefore be carried out over all systems of the windings, and are thus subject to delay times of some 100 ms. Much faster regulation procedures can be achieved if the exciter current is fed directly to the rotor of the main generator via slip rings. In this arrangement, delays of about 20 ms can be expected. The much simpler construction of the machine and the gain in dynamic characteristics must, however, be set against the severe disadvantage of feeding electricity through brushes and slip rings. Higher frictional losses, brush and slip-ring erosion and higher maintenance costs are the consequences.

In the following discussions, the operating ranges and the static and dynamic characteristics of these machines will be more closely examined.

3.3 Operational Ranges of Asynchronous and Synchronous Machines

Working from the structure, function and voltage equations of synchronous and asynchronous machines, the largely familiar equivalent circuits shown in Figure 3.7(a) and (b) may be deduced for one machine phase. In the interests of clarity, a nonsalient pole synchronous machine with d.c.-fed exciter winding has been chosen by way of approximation. Permanent-magnet generators exhibit a similar equivalent circuit, but no interventions by means of exciter current are possible.

In the equivalent circuit diagrams, the variables with the subscript 1 characterize the stator values while the variables with the subscript 2 characterize the rotor values, i.e.:

R_1 represents the resistance of the stator winding,
R_2' represents the rotor resistance in relation to the stator side,
R_2'/s represents the slip-dependent value,
$X_{1\sigma}$ represents the leakage reactance of the stator winding,
$X_{2\sigma}'$ represents the leakage reactance of the rotor winding in relation to the stator side,
X_h represents the air-gap reactance and
R_{fe} represents the iron losses of the machine.

Furthermore,

U_1 represents the stator voltage,
I_1 represents the stator current,
I_2' represents the rotor current in relation to the stator side,
I_μ represents the excitation current,
I_{fe} represents the current equivalent to iron losses,
I_0 represents the open-circuit current (of the asynchronous machine),

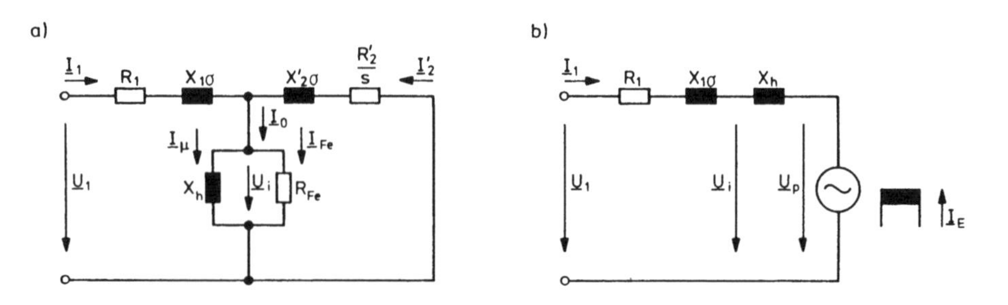

Figure 3.7 Equivalent circuits for one phase of (a) asynchronous and (b) synchronous machines

I_E represents the excitation current of the rotating rotor of the synchronous machine, transformed on the stator side,
U_i represents the induced voltage and
U_p represents the synchronous generated voltage of the machine.

In addition,

s represents the slip as a relative variable between the speed deviation from synchronism to the synchronous speed in accordance with

$$s = \frac{n_1 - n}{n_1} \tag{3.1}$$

In wind turbines there is a trend towards large systems in the kilowatt and megawatt range. For machines of this category, certain simplifications can be undertaken, while still obtaining a very good approximation for the following considerations. By ignoring

- iron ($R_{fe} = 0$) and frictional losses in asynchronous machines and
- resistance losses ($R_1 = 0$) in synchronous machines,

the simplified equivalent circuits shown in Figure 3.8 may be derived.

By grouping the reactances

$$X_1 = X_h + X_{1\sigma} \quad \text{and} \quad X_2' = X_h + X_{2\sigma}' \tag{3.2a}$$

or

$$X_d = X_h + X_{1\sigma} \tag{3.2b}$$

respectively, the stator current for asynchronous and synchronous machines

$$\underline{I}_1 = \frac{R_2'/s + jX_2'}{(R_1 + jX_1)(R_2'/s + jX_2') + X_h^2}\underline{U}_1 \quad \text{or} \quad \underline{I}_1 = -j\frac{\underline{U}_1}{X_d} + j\frac{\underline{U}_p}{X_d} \tag{3.3}$$

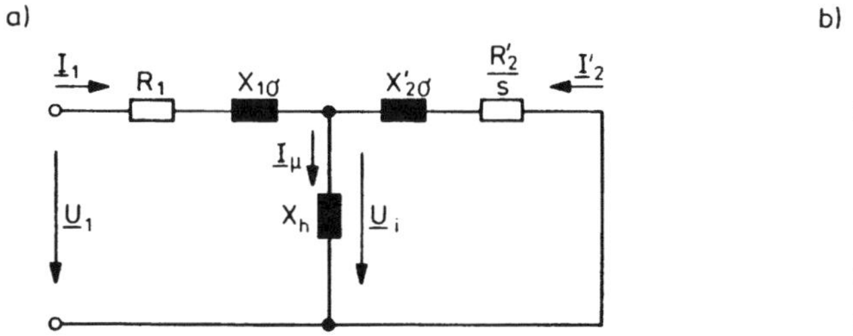

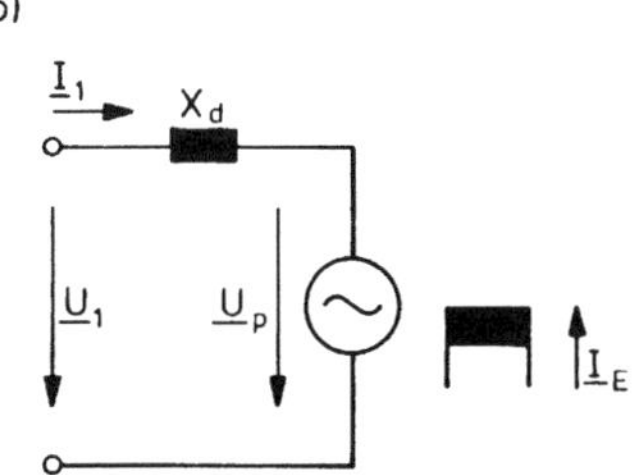

Figure 3.8 Simplified equivalent circuits for one phase of (a) asynchronous and (b) synchronous machines

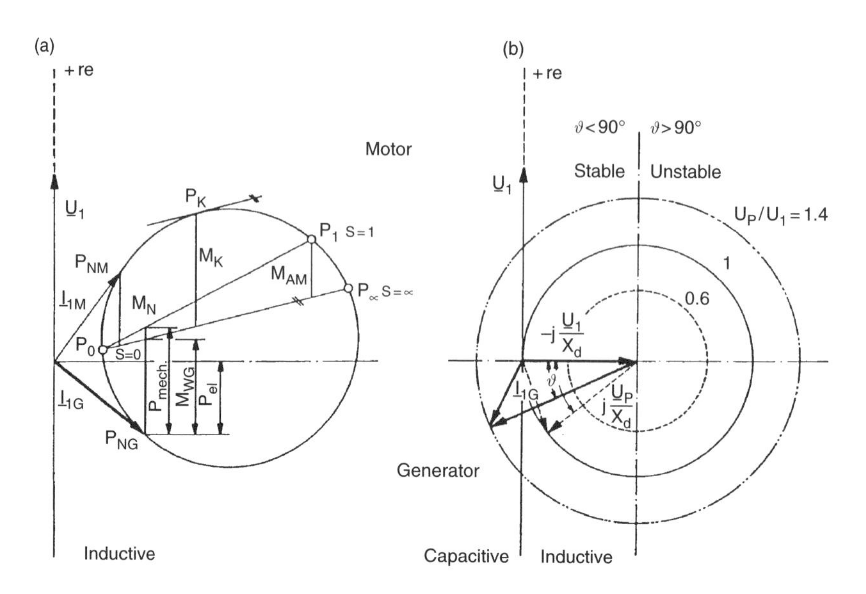

Figure 3.9 Current circle diagrams in simplified form for (a) asynchronous and (b) synchronous machines in grid operation (U_1, f = constant)

may be derived in the usual form. From this, the simplified current circle diagrams as in Figure 3.9 may be obtained. From these, the relationships between the mechanical input values for torque and speed of rotation (see Figure 3.13 and Equations (3.9), (3.11), (3.12) and (3.15)) and the output values for electric current, voltage and power output may be recognized.

In Figure 3.9 P_0 is the no-load point, P_N the nominal point, P_K the transition point and P_1 the short-circuit point, while M_N, M_K and M_{AM} are the corresponding torques when running as a motor. The quantities shown in the figure represent the following:

I_{1M} stator current in the motor mode
I_{1G} stator current in the generator mode
M_{WG} generator load torque
P_{el} electric output power of the generator
P_{mech} mechanical power absorbed by the generator

Particularly in view of the power station operation, regulation of the reactive power conditions is of great importance. From the current circle diagrams, it is clear that:

- For asynchronous machines, only motor and generator currents in the negative imaginary half-plane are possible, and these are always inductive.
- For synchronous machines, on the other hand, currents are possible in all four quadrants, i.e. in the motor and generator, and in inductive and capacitive ranges.

For power generation, therefore, synchronous generators offer distinct advantages. In addition to the positions of the circles, the radii also have differing effects on the operating state of these machines.

The full and simplified vector diagrams for the excitation status of asynchronous machines running under almost full load and synchronous machines under partial load, as shown in Figure 3.10, show different constellations. Here, the analogy between underexcited synchronous generator operation, which is highly variable, and design-determined fixed asynchronous generator operation becomes clear when, based upon the torque-producing rotor displacement angle ϑ for synchronous and asynchronous machines, a corresponding load angle λ (but with lower values) is introduced. Comparison of the vector diagrams in Figure 3.10 shows that the voltage U_1 induced at the magnetizing inductance

- is specified as a fixed value in asynchronous machines (examined more closely in Section 3.6), but
- is adjustable by means of the excitation state (over- or underexcited) in synchronous machines.

If the real axis is set in the direction of the stator voltage, it becomes clear that both types of machine allow current variations up to their stability limits in the real part of the plane. In generators, these arise as a result of changes in drive torque or active power. On the other hand, in asynchronous machines (in contrast to synchronous machines) the imaginary component of the currents, i.e. the reactive power alone, is not variable without resort to supplementary devices such as balancing capacitor banks.

If, however, reactive components can be pre-set by means of supplementary devices, e.g. capacitors, rotary phase shifters or static converters, then the same operational possibilities may be achieved with asynchronous as with synchronous machines. In asynchronous machines, the radius of the current circle is dictated essentially by the grid voltage. Deviations from circularity, which are ignored here, arise principally due to the saturation state and the construction of the machine (e.g. double-cage rotor). Fixed operating points can thus be explicitly assigned to predetermined load levels.

In synchronous machines, on the other hand, the radius of the current circle can be influenced by the excitation. This allows free selection of the operating conditions corresponding to a fixed load depending upon magnetization (over- or underexcited) or the reactive power demand (inductive or capacitive) as needed. In asynchronous machines, a supplementary device can deliver reactive current to the entire system. Looking at the asynchronous machine with adjustable reactive power delivery, shown in Figure 3.11, it is more or less possible to shift the axes to obtain, for an asynchronous machine, an operational range that covers all four quadrants, similar to that of a synchronous machine (see Figure 3.12).

Asynchronous generators working as shown here are usually coupled with cascade-switchable capacitors, and occasionally also static or rotary phase shifters, for standalone or isolated applications. Balancing capacitors in combination with machine and cable inductances can, however, set up grid resonance (see Section 4.3).

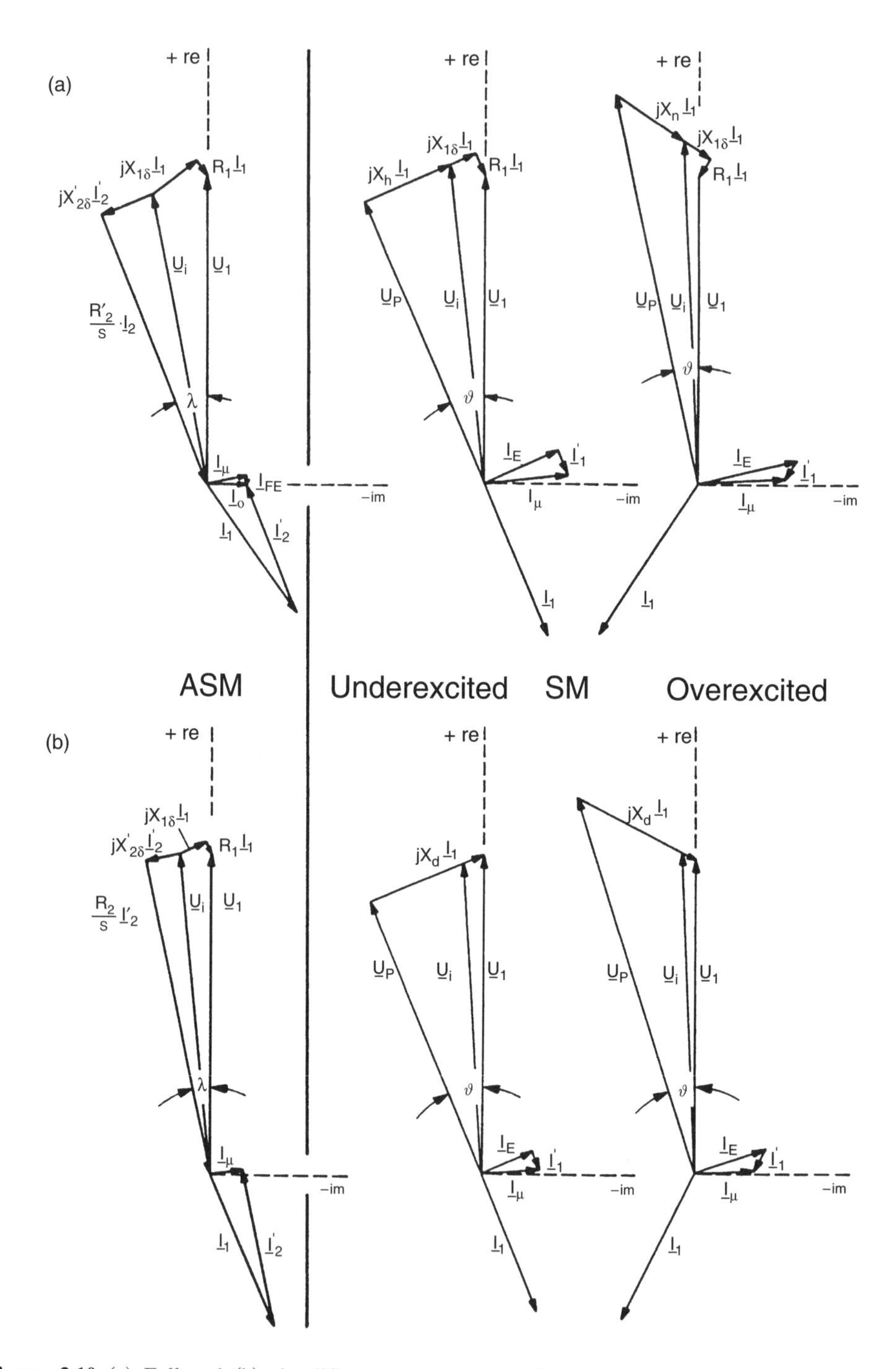

Figure 3.10 (a) Full and (b) simplified vector diagrams for asynchronous machines (ASM) and synchronous machines (SM) running as underexcited and overexcited generators

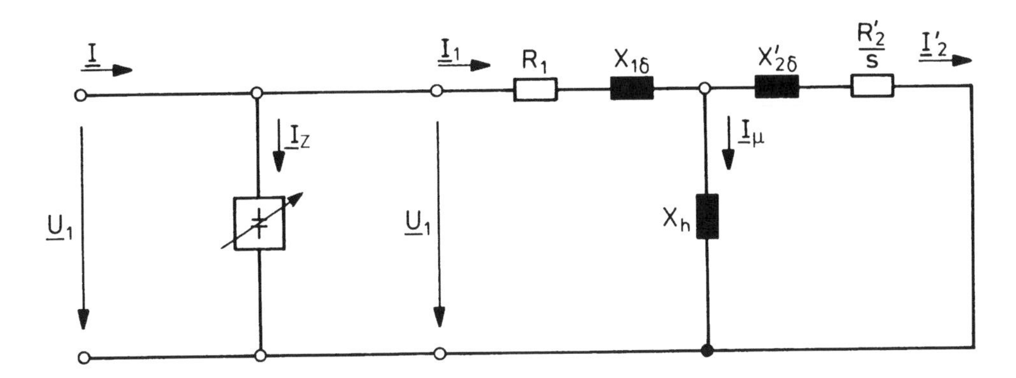

Figure 3.11 Equivalent circuit for an asynchronous machine with adjustable reactive power delivery

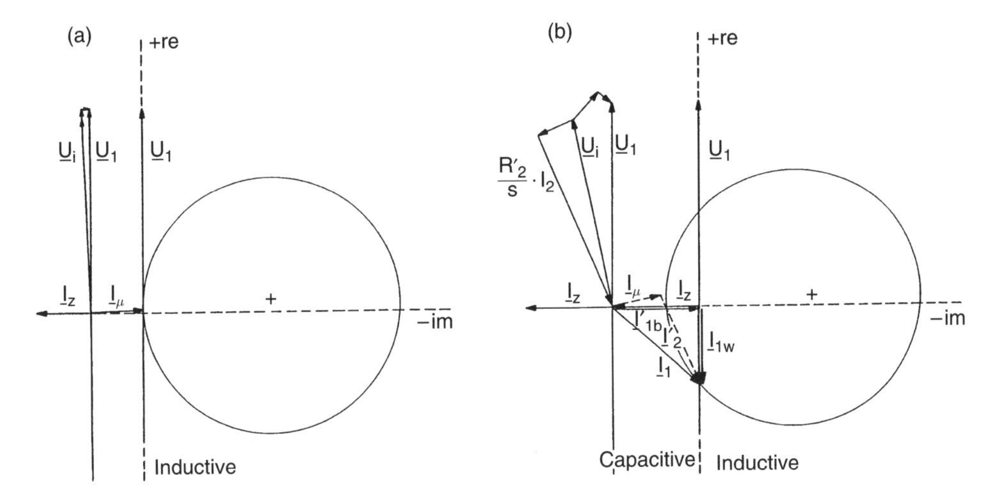

Figure 3.12 Operational ranges of asynchronous machines with reactive power delivery devices as entire systems with (a) no-load and (b) load compensation

3.4 Static and Dynamic Torque

In the discussion of generators for use with wind turbines, when allowing for torque development, the distinction must be drawn between the fast processes of electrical balancing in the windings (and in the magnetic circuit), which are also influenced by the design of the mechanical drive train, and the much slower processes that have a direct effect on the dynamic behaviour of the wind power unit. Knowledge of the torques involved in the complete system is important for deciding which measures are to be applied to protect the drive train.

Plots of steady-state torque against rotational speed are valid for speed or torque changes that take place much more slowly (quasi-steady-state) than dynamic processes (run-up and braking). Dynamic torques are decisive for fast processes, occurring within a short period immediately after any change of status. In wind turbines, their practical implications for dimensioning purposes are limited to special cases, such as switching-on procedures, line disturbances and generator short-circuits. The following examines parameters for estimating the order of magnitude and extreme conditions of static and dynamic torques.

3.4.1 Static torque

During quasi-static processes, the generator sets up a static load torque in opposition to the wind turbine. This load has a static and a dynamic component. The steady states of a rotary field machine are essentially determined by its fundamental-frequency torques. Synchronous and asynchronous harmonic components also exert their influence. In the generators mentioned, such harmonics can have considerable mechanical and electrical effects.

Due to the stepped magnetic waveform of successive phases, polyphase symmetrical stator windings produce rotary fields of order

$$\nu = \pm km + 1, \tag{3.4}$$

where $k = 0, 2, 4, 6, \ldots$ (even integer) and, usually, the number of phases $m = 3$. Apart from the fundamental frequency, harmonics arise, the ordinals of which are uneven integers indivisible by 3:

$$\nu = 1, -5, 7, -11, 13, -17, 19, \ldots.$$

Negative ordinals indicate rotation opposed to that of the fundamental field.

In assessing harmonic feedback in low-voltage networks, compatibility levels up to the 40th or 50th harmonic must be allowed for [3.1].

3.4.1.1 Asynchronous machines

The behaviour of asynchronous generators is primarily determined by the steady-state torque/rotational speed characteristic of the fundamental-frequency field (see Figure 3.13). This is largely dictated by the design of the machine. In addition to the nominal torque, characteristic values include starting torque, pull-up torque and breakdown torque, plus the corresponding values for speed of rotation and slip. Nominal and breakdown values are of particular importance in the discussion that follows.

The ratio of breakdown to nominal torque M_K/M_N normally lies in the range of 1.8 to 3.5 and the ratio of start-up to nominal torque M_{AM}/M_N is usually between 1 and 3. Pull-up torque M_S/M_N, which usually lies between 1 and 2, generally gives rise to no dominant loads and so, like start-up torque, plays an important role only during the motorized run-up of a wind turbine.

Here, M stands for torque at the general slip value s, M_K is the breakdown torque, s_K the breakdown slip and n_1 the speed of rotation of the rotary field. Individually, the quantities represent the following:

Motor mode		*Generator mode*	
M_{AM}	start-up torque		
M_{SM}	pull-up torque	M_{SG}	pull-up torque
M_{KM}	breakdown torque	M_{KG}	breakdown torque
M_{NM}	nominal torque	M_{NG}	nominal torque
n_{KM}	breakdown-torque speed	n_{KG}	breakdown-torque speed
n_{NM}	nominal speed of rotation	n_{NG}	nominal speed of rotation

Ignoring stator resistance ($R_1 = 0$), the following is valid for the internal torque, which is transmitted by the air gap of the stator:

$$M_i = \frac{mU_1^2}{2\pi n_1 \sigma X_1} \frac{1-\sigma}{s\frac{\sigma X_2'}{R_2'} + \frac{1}{s}\frac{R_2'}{\sigma X_2'}}. \tag{3.5}$$

The maximum torque is given by the derivative

$$\frac{\mathrm{d}M_i}{\mathrm{d}s} = 0$$

for the so-called breakdown slip

$$s_K = \frac{R_2'}{\sigma X_2'}. \tag{3.6}$$

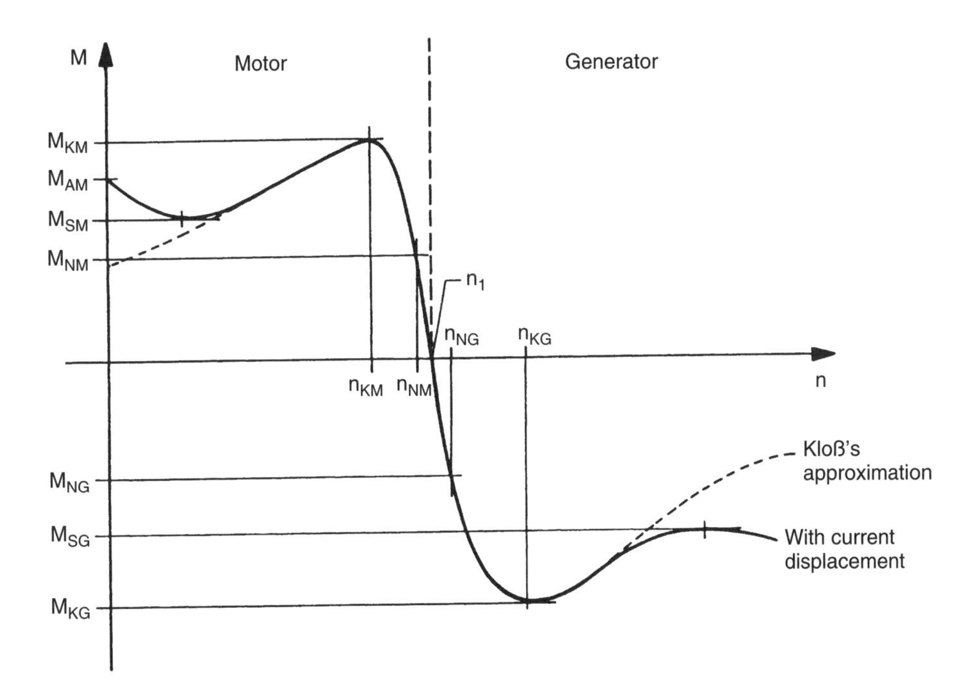

Figure 3.13 Torque/speed characteristic of an asynchronous machine

Defining the primary leakage $\sigma_1 = X_{1\sigma}/X_{\text{h}}$ and the secondary leakage $\sigma_2 = X'_{2\sigma}/X_{\text{h}}$, the total leakage can be calculated:

$$\sigma = 1 - \frac{X_{\text{h}}^2}{X_1 X_2'} = 1 - \frac{1}{(1+\sigma_1)(1+\sigma_2)}.$$

Further, Equations (3.2a) define X_1 and X_2', where

R_2' represents the rotor resistance transformed to the stator side,
X_σ represents the reactance leakage from the stator and rotor ($X_\sigma = X_{1\sigma} + X'_{2\sigma}$),
m represents the number of phases (in three-phase systems $m = 3$) and
U_1 represents the grid voltage.

For low leakage values σ_1 and σ_2, which have very small values or tend to σ as they approach zero,

$$\sigma X_1 \approx X_{1\sigma} + X'_{2\sigma} = X_\sigma$$

and

$$\sigma X_2' \approx X_{1\sigma} + X'_{2\sigma} = X_\sigma.$$

Thus

$$s_{\text{K}} \approx \frac{R_2'}{X_\sigma} \tag{3.7}$$

and

$$M_{\text{K}} = \frac{mU_1^2}{2\pi n_1} \frac{1}{2X_\sigma} \tag{3.8}$$

Ignoring friction, the torque $M_{\text{i}} = M$. When the slip $s = (n_1 - n)/n_1$, the torque characteristic can be approximately determined according to Kloß's equation, referring to Equation (3.5) from the breakdown values by the relation

$$\frac{M}{M_{\text{K}}} \approx \frac{2}{\frac{s}{s_{\text{K}}} + \frac{s_{\text{K}}}{s}} \tag{3.9}$$

where $s_{\text{K}} \approx$ (5 to 10) s_{N}.

In normal operation between no-load ($n = n_1$) and motor and generator nominal load or permitted overload, a simplified approximation for torque development is given by the equation

$$\frac{M}{M_{\text{K}}} \approx \frac{2s}{s_{\text{K}}} \tag{3.10}$$

This means that, within the ranges mentioned, the machine characteristic is determined essentially by the breakdown moment and the breakdown slip. Allowing for stator resistance by applying the machine-specific parameters

$$s_K = \frac{R_2'}{X_2'}\sqrt{\frac{R_1^2+X_1^2}{R_1^2+\sigma^2 X_1^2}} \tag{3.11}$$

and

$$M_K = \frac{3U_1^2}{4\pi n_0}\frac{1-\sigma}{R_1(1-\sigma)+\sqrt{(R_1^2+\sigma X_1^2)\left(1+\frac{R_1^2}{X_1^2}\right)}}, \tag{3.12}$$

both quantities can be determined.

According to this, torque development in asynchronous machines is determined by the relation of ohmic to leakage components, particularly in the rotor windings. With slip-ring rotors, these can be altered over a relatively wide spectrum using supplementary resistors or power converters in the rotor circuit.

The size of the machine physically dictates the limits of slip. Large machines of this kind, in particular, are highly limited in elasticity in the event of torque fluctuations. Figure 3.14 reproduces average,size-dependent nominal slip values for given nominal power levels of production machines.

Figure 3.14 shows that machines of conventional design in the 100 kW range exhibit nominal slip values of around 1 %. Six-pole generators show much greater elasticity than

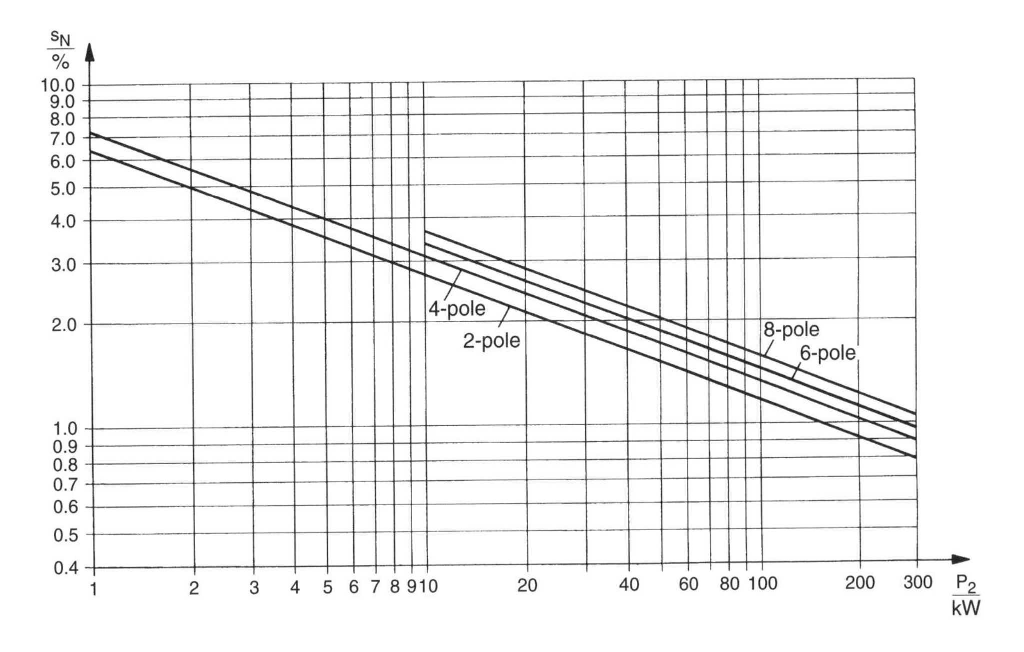

Figure 3.14 Nominal slip s_N for asynchronous machines as a function of nominal output power P_N for production machines (motors) with the pole count as a parameter

two-pole machines. In machines with such inflexible speeds of rotation, turbine torque fluctuations place heavy loads on the drive train because of the almost rigid coupling between the grid and generator. By raising design slip levels such systems may be made more flexible and thus offer greater protection to the drive train. However, this comes at the expense of greater volume, mass and losses.

This greater flexibility is desirable since the rotating blades of the turbine are subject to considerable airstrip turbulence, especially near to the tower. These result in fluctuations in turbine performance. Even when the blades react swiftly to limit turbine performance, the effects of tower-shadow fluctuations cannot be ironed out entirely. At low slip values, small variations in the speed of rotation therefore result in relatively large fluctuations in power.

Fluctuations in electrical output power can, as already mentioned, be considerably diminished by the use of asynchronous machines with higher nominal slip. Figure 3.15(a) and (b) shows comparative values for a 20 kW machine running under partial load, i.e. without application of blade pitch regulation. Considerable differences in output behaviour may be observed.

While fluctuations in power of more than $0.2P_N$ can be seen for generators with 2 % nominal slip (Figure 3.15(a)) due to tower-shadow effects, at 8 % nominal slip (Figure 3.15(b)) these are reduced to some 5 % nominal output ($0.05P_N$) [3.2]. The dominant variations in power shown here are measurable in seconds and are clearly due to variations in wind speed.

In accordance with Equations (3.6) and (3.11), high slip values can be achieved by designing the generator with

- high rotor resistance,
- low total leakage factor and
- lower rotor inductance,

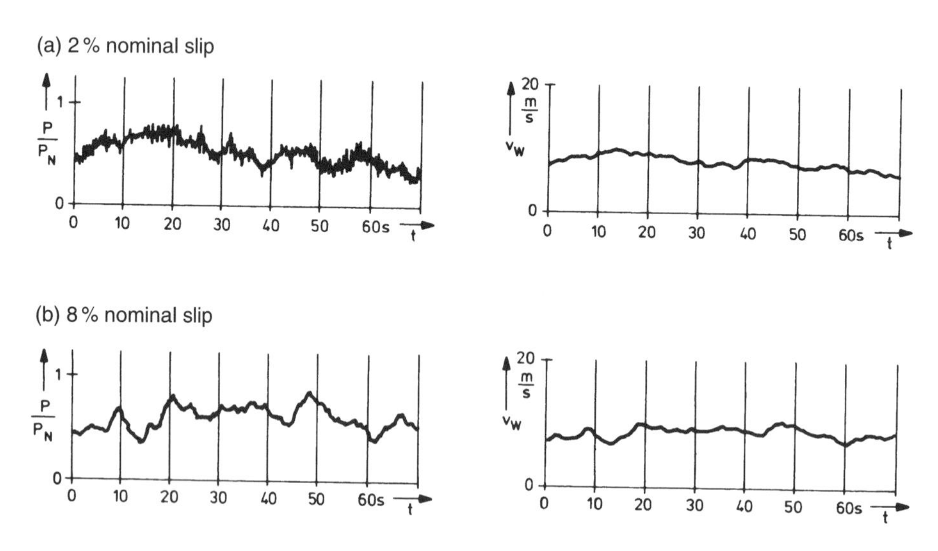

Figure 3.15 Power and associated wind speeds (measured values) for a wind power plant as a function of time for differently designed asynchronous generators

e.g. by attention to such constructional details as

- slot form and
- choice of conductor materials and conductor cross-sections in the rotor.

Applying power smoothing in this way can also considerably reduce demands on turbine components.

In addition to setting higher fixed slip values by using extra resistors in the rotor circuit, slip-ring machines offer the possibility of dynamically adjusting or controlling slip to adapt to output power variations (Figure 3.16). The slip value and thus the proportional loss in power are thereby kept low and system efficiency is increased by varying the resistance of the rotor. It is also possible to limit the torque on the drive train to values around the nominal value [3.3] and to smooth the power output. This principle is employed in the so-called 'opti-slip control' in the Vesta turbines V 44 and V 60/63/66 and the USA export variant V 80.

Asynchronous harmonic rotary field torque

Harmonic fields induce a current in the rotor winding of frequency

$$f_{2\nu} = s_\nu f_1 \tag{3.13}$$

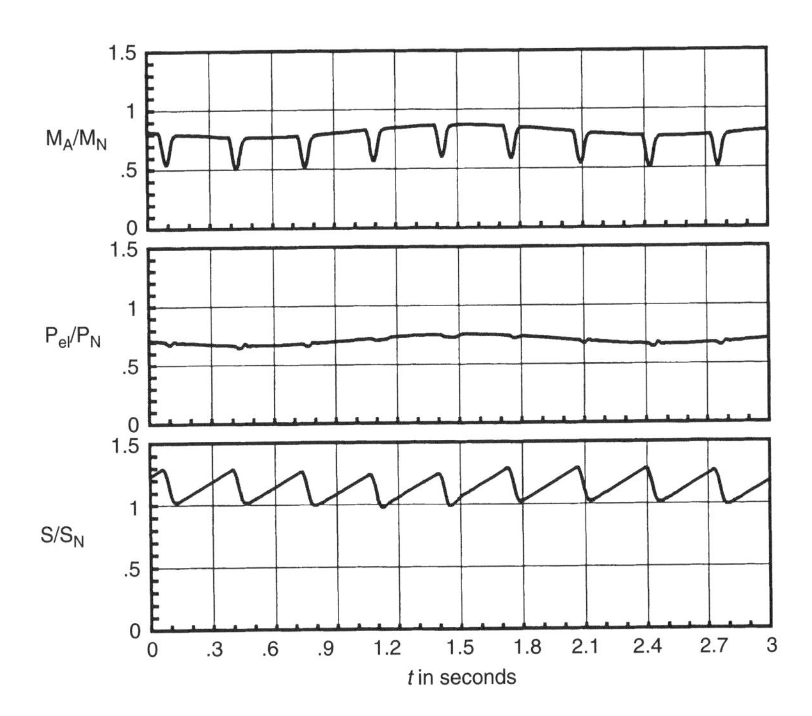

Figure 3.16 Dynamic slip control in slip-ring rotor asynchronous machines (simulation results)

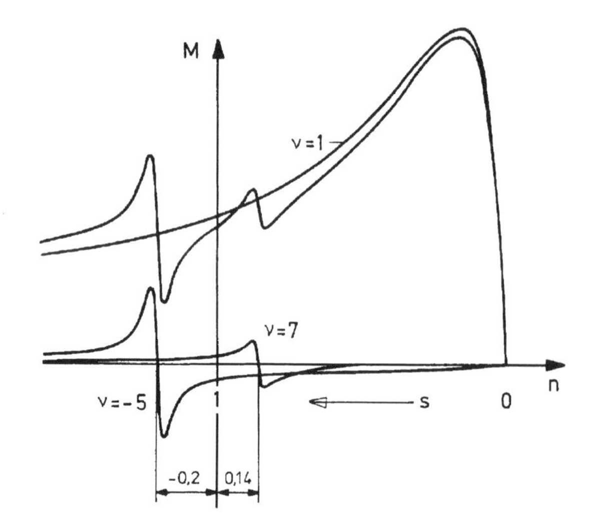

Figure 3.17 Fundamental (M_1) and harmonic torques (M_ν) and resultant torque (M) for an asynchronous machine

with slip

$$s_\nu = \frac{\Delta n_\nu}{n_\nu} = 1 - \nu(1 - s). \tag{3.14}$$

When the harmonics are synchronized the slip value $s_\nu = 0$ and the harmonic torque $M_\nu = 0$, i.e. the slip $s = (1 - 1/v)$. Outside these values, i.e. when running asynchronously ($s \neq 0$), these rotor currents set up a torque. All harmonic field torques are overlaid on that of the fundamental waveform (Figure 3.17).

Harmonic torques arise during run-up in the motor mode. They play a subordinate role if the windings are chosen appropriately and are not noticeable in normal generator operation.

Synchronous harmonic rotary field torque

Just as the stator current generates harmonic rotary fields, so too can the rotor current. If the stator harmonic and the rotor harmonic have the same frequency for a given operating point, they can generate torque together.

Such torques, however, are only significant during motor operation with high slip. During run-up in the motor mode, they can attain considerable levels relative to the start-up or nominal torque. In the generator mode, on the other hand, within the usual slip range of $s = 0$ to $10\,\%$ synchronous harmonic torques are of little importance, so they will not be examined further below.

Parasitic torque

Skewed slots in a cage rotor build cross-currents from bar to bar across the iron, thereby generating torque, as do further eddy current losses due to slot harmonics. These torques are only of interest during start-up and will not be considered further.

3.4.1.2 Synchronous machines

For a utility connection, synchronous generators are used almost exclusively. For wind power plants in particular, they are used with and without gearing in isolated operation and via frequency converters in grid operation. Direct connection of synchronous machines to the grid is used in a few isolated cases in the megawatt class only, and these are mostly beset with problems (see Figures 1.11 and 1.12). While in conventional power stations nonsalient pole generators (turbogenerators) are predominant, wind turbines more often use multipole salient pole machines.

We must further differentiate between slip-ring fed rotors with small excitation time constants and brushless machines, which require much longer times for excitation changes (see Section 3.2.2). The torque in nonsalient pole machines is given by the general relation

$$M = -\frac{mU_1}{2\pi n_1}\frac{U_\mathrm{p}}{X_\mathrm{d}}\sin\vartheta \qquad (3.15)$$

and for salient pole machines the following is valid:

$$M = -\frac{mU_1}{2\pi n_1}\left[\frac{U_\mathrm{p}}{X_\mathrm{d}}\sin\vartheta + \frac{U_1}{2}\left(\frac{1}{X_\mathrm{q}} - \frac{1}{X_\mathrm{d}}\right)\sin 2\vartheta\right], \qquad (3.16)$$

where

- m number of phases (in three-phase systems, $m = 3$)
- n_1 synchronous speed of rotation
- U_1 grid voltage
- U_p rotary field voltage
- X_d synchronous direct-axis reactance
- X_q synchronous quadrature-axis reactance
- ϑ rotor displacement angle

According to this, the torque and the output power depend on the rotor displacement angle ($\sin\vartheta$ and $\sin 2\vartheta$). Further, by altering the rotary field voltage by means of the excitation, the torque curve can be influenced.

Directly coupled synchronous generators form a rigid system with the grid, which propagates sudden changes in torque more or less immediately in the form of energy flux changes. In wind power plants this leads to heavy loads being placed on the drive train, with correspondingly large fluctuations in output power.

Depending on the design and level of excitation, running at nominal rating, the following are common:

- for nonsalient pole generators,
  - rotor displacement angle $\vartheta = 25$ to $30°$ and
  - overload capacity $M_\mathrm{K}/M_\mathrm{N} \approx 2$;

- for salient pole generators,
  - rotor displacement angle $\vartheta = 20$ to $25°$ and
  - overload capacity $M_\mathrm{K}/M_\mathrm{N} \approx 2$ to 2.5;

and, depending on the maximum exciter voltage,

— rotary field voltage $U_p/U_N = 0$ to 2.

Thus in quasi-steady operation a maximum (pull-out) torque of

$$\frac{M_{max}}{M_N} = 2 \text{ to } 5$$

is to be expected.

To make matters worse, periodic torques arise (particularly due to periodic tower-shadow or upwind tower effects and their influences on the rotor blades) that are only weakly damped, which place considerable loads on the rotor, bed plate and tower. As a result, extra measures (oscillation dampers, etc.) are usually necessary (see Figure 3.18). Therefore, the only synchronous generators to make it into widespread use in grid-coupled wind turbines are those that use frequency converters to provide elastic grid coupling.

3.4.1.3 Generators with power converters

Because of their extremely short reaction times (e.g. 2 to 20 ms), the use of power converters to couple generators to the grid allows very effective and dynamically efficient torque limiting values. This is the case in systems that feed their entire output power to the grid via rectifiers or inverters (see Figure 3.4, models (b), (c), (h), (i), (j) and (k)) and in configurations that feed only part of their output to the grid via power converters (see Figure 3.4, models (e) and (f)) or take off slip power via supplementary resistors (model (d)). If the regulation system is properly dimensioned, limiting values close to the nominal torque, such as, for example,

$$M_{max} = (1.2 \text{ to } 1.5)M_N,$$

can be attained. In doing so, the entire system must be designed such that fluctuations and instabilities resulting from internal and external disturbances are allowed for and eliminated by choosing component dimensions so as to protect the drive train.

3.4.1.4 Generator and turbine torque

The behaviour of a wind power unit is determined by the interaction between the turbine drive torque and the load torque of the generator. The real drive of a machine can be approximated from its quasi-steady-state behaviour. In this way, taking into account turbine and generator manipulations and using the torque–rotational speed curve, the operating and adaptive possibilities of using the characteristics of the energy converter system according to Sections 2.2 and 3.2 to control the power of the wind turbine can be determined. In doing so, the possibilities for the optimal exploitation of wind energy and for the adaptation of power to the grid will be considered [3.4].

From the performance coefficient (cf. equation (2.8))

$$c_p = \frac{P_W}{A_R v_1^3 \frac{\rho}{2}}$$

(a) Gear layout

(b) Elastic suspension

Figure 3.18 Torsionally elastic gearbox suspension with oscillation damper (Hamilton Standard WTS 4 turbine) and the characteristic given in Figure 2.6,

$$c_{\mathrm{p}} = f(\lambda)$$

or from the equivalent torque parameter

$$c_{\mathrm{m}} = \frac{M_{\mathrm{AW}}}{R_{\mathrm{a}} A_{\mathrm{R}} v_1^2 \frac{\varrho}{2}} \tag{3.17}$$

and the corresponding family of characteristics

$$c_{\mathrm{m}} = f(\lambda)$$

together with the tip speed ratio

$$\lambda = \frac{2\pi R_{\mathrm{a}} n}{v_1}$$

it is possible, for a fixed blade pitch angle (e.g. at around $\beta = 90$) at different wind speeds v_1 (as a parameter), to derive the power and/or the torque–rotational speed characteristics for wind turbines. These, together with the load torque–rotational speed curve of the generator being driven, determine the running mode and the stability of the entire drive, as shown in Figure 3.19.

If the wind turbine delivers its energy via a synchronous generator to a clock-pulse-generating grid, then at different wind speeds the turbine will be constrained by the generally constant grid frequency to a likewise constant rotation speed of $n/n_1 = 1$.

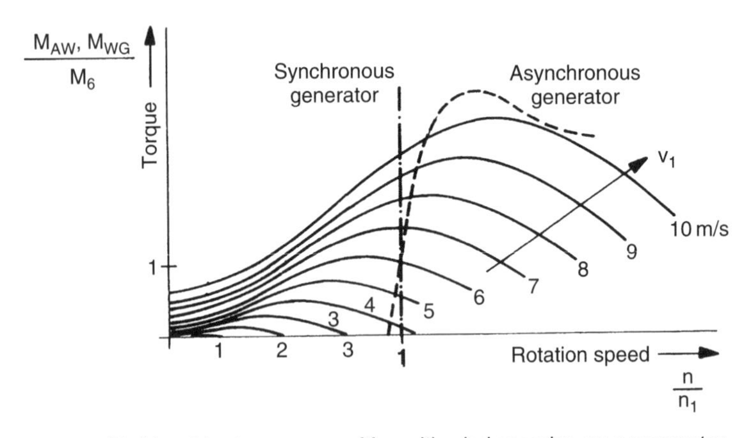

——— Turbine drive torque curve M_{AW} with wind speed v_1 as a parameter

—·—·— Load torque M_{WG} of a synchronous generator

- - - - Load torque M_{WG} of an asynchronous generator

Figure 3.19 Torque–rotational speed relationship of a grid-connected wind turbine (for a turbine drive torque M_6 at a wind speed of 6 m/s)

Similarly, if wind energy is delivered through an asynchronous generator then, when the wind speed increases, the turbine speed will also slowly increase (due to increasing slip). For normally designed machines, however, a change in rpm of only around 1 % (see Figure 3.14) needs to be allowed for.

It should be pointed out here that the stable operating regime of the turbine may lie beyond the breakdown point of an asynchronous generator, should the turbine characteristic fall away more rapidly than that of the generator in question. In consequence, steady-state stability will always be attained when

$$\frac{dM_W}{d\omega} > \frac{dM_A}{d\omega} \tag{3.18}$$

This can be achieved by designing the asynchronous machine accordingly, particularly for the operating range that lies between breakdown and saddle points.

Figure 3.19 shows that sufficient turbine torque to drive the generator is only available at or above wind speeds of about 3.6 m/s for asynchronous machines and 3.8 m/s for synchronous machines. When transmission and generator losses are taken into account, an even higher wind speed must exist before any power at all is delivered to the grid. When the wind speed lies below nominal (minimum or no-load) levels, the machine acts as a motor and drives the turbine.

Under normal operation of wind power plants, continual changes in the wind speed occur. The tip speed ratio changes in consequence, so the coefficients and values for power and turbine torque are not constant either. If the generator is rigidly coupled to the grid, continually changing operating states result.

Under variable-frequency generator operation, the speed of rotation can be freely set within given limits. The turbine's utilization of available wind power may therefore be optimized within the possible range of adjustment. This can be seen from Figure 3.20. By plotting the torque of the turbine against speed of rotation, the power of the turbine can be given for every level of wind speed. The example in Figure 3.20 shows this for $v_1 = 5\,\text{m/s}$. For every wind speed, therefore, an optimal performance may be determined; the corresponding values for the turbine drive torque and speed of rotation are shown in Figure 3.20 as a heavy dashed line. The load torque–rotational speed curve of a self-exciting synchronous generator under variable-frequency operation can be drawn in the turbine characteristic field for various excitation currents I_E. The level of excitation can be used to set the generator load torque according to the turbine drive torque at optimal turbine performance. This will always be attained at the speed of rotation corresponding to the intersection of the curve for the turbine torque at optimal turbine performance with that of the generator load torque for a given excitation current. This can be achieved by using frequency-converter systems with direct current intermediate circuits to decouple generator and grid frequencies.

If the generator frequency can be modified (e.g. by using an intermediate circuit converter or direct converter) then, by choosing the frequency, the speed of rotation may be set and the required torque level obtained. This is plotted in the turbine characteristic field in Figure 3.21 for different operational states of synchronous and asynchronous generators. Coupling this with Figures 3.19 and 3.20 allows a frequency, and thereby a speed of rotation, to be chosen that yields optimal turbine power. In Figure 3.21 this is shown, for example, as the operating state 1. The assumption is that the turbine output power for the chosen frequency is fed to the grid.

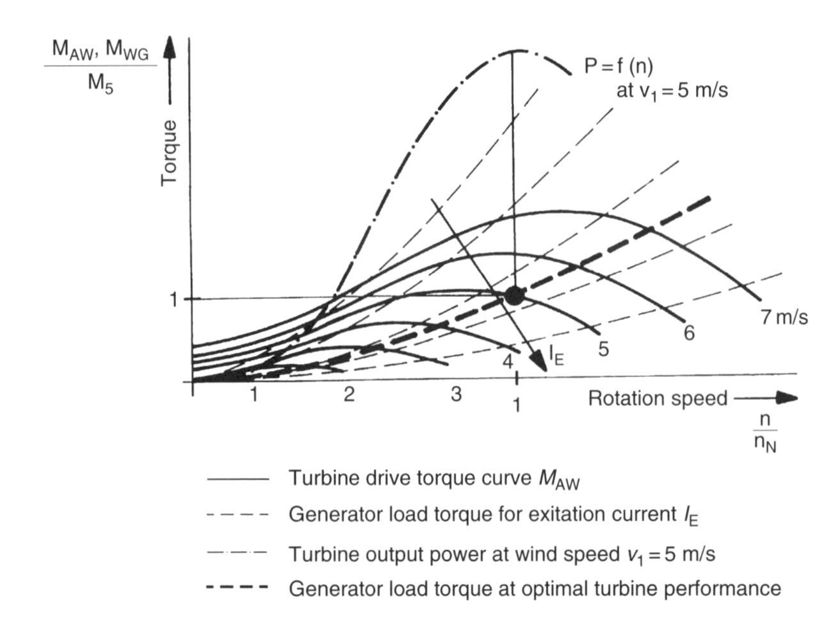

Figure 3.20 Variable-frequency adaptation of a self-exciting synchronous generator to a turbine (for a turbine drive torque at a wind speed of 5 m/s)

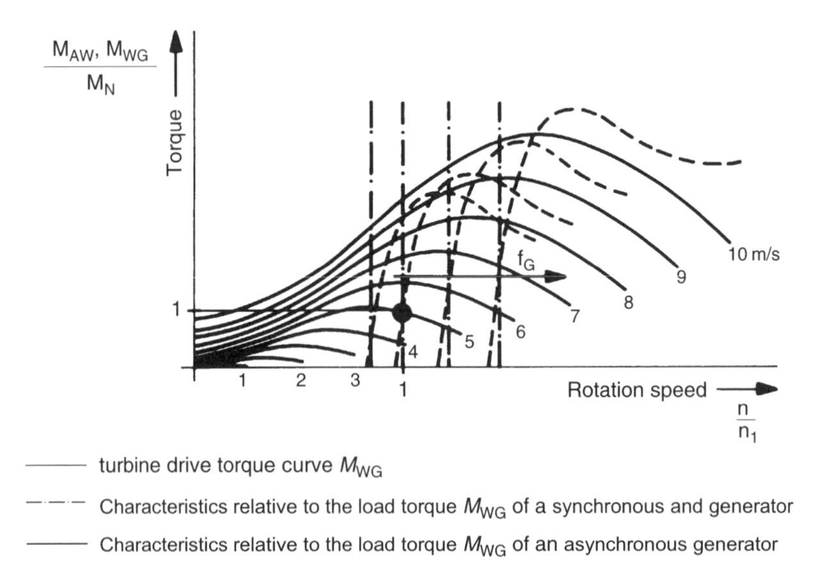

Figure 3.21 Torque–rotational speed relationship and characteristic adaptation by variation of the generator frequency f_G (for a turbine drive torque M_5 at a wind speed of 5 m/s)

To adapt a machine to the optimum wind power, the operating points of the entire system must each intersect the maxima of the turbine performance–rotational speed curves, so that the turbine is running with constant tip speed ratios, as already explained. The speed of rotation of the turbine must therefore adapt to the prevailing wind speed. Controlling a machine for maximal exploitation of wind power therefore implies that, for a turbine of this design, electrical energy might be delivered to the grid, even at very low wind speeds. All the same, transmission and generator losses mean that a certain minimum wind speed is necessary to keep the turbine operating. If the wind speed drops below this so-called start-up speed, the turbine will stop.

In addition to producing the maximum possible performance figures in the range between cut-in and nominal wind speeds, generator frequency adaptation allows deliberate reductions in turbine power beyond this range. In this way,

- overload protection can be provided and
- mechanical effects due to pitch control, for example, anticipated.

This provides the possibility of reducing the mechanical interventions on the turbine system that must be made in response to changing wind speeds, thereby reducing component loading. As mentioned at the beginning of this section, in most wind turbine operating conditions, mainly static torques are determinative. Dynamic torques overlie these in marginal situations. The following considerations show that dynamic torques can far outweigh static torques and in consequence cause effects on the wind turbines that are relevant for design and dimensioning.

3.4.2 Dynamic torque

Immediately after a change in operating conditions due to

- cut-in and cut-out procedures,
- grid disturbances,
- rapid auto-reclosure in the grid or
- deliberate generator short-circuit,

short-time effects in the generator give rise to transient currents and torques. These can be estimated [3.5] and, as far as possible, can be attributed experimental values. Knowledge of such effects is important, especially when dimensioning mechanical (see Section 2.4) and electrical components and determining machine behaviour.

3.4.2.1 Connection procedures

Connection of stator coils to the grid during

- motorized run-up from standstill or
- near synchronous operation

are of particular relevance. From the point of view of the grid, it is necessary to differentiate between the connection of the generator to

- a rigid combined grid and
- a weak isolated grid.

In addition, from the point of view of the generator, differences can be expected between

- unexcited and
- excited

motor state during connection.

Motorized run-up from stationary or from very low rotation speeds is not normally an option for synchronous machines, and is only practised in exceptional cases with asynchronous generators. When calculating the maximum occurring moments in asynchronous machines that are connected to the grid with the rotor stationary, we need to differentiate between torque paths for

- running up and
- locked

machines (Figure 3.22(a) and (b)). The maximum values occur after a half-period. When the machine is locked

$$M_{\text{max}} \approx 2M_{\text{AM}},$$

i.e. the motorized start-up torque is doubled. When the generator is running up

$$M_{\text{max}} \approx 4M_{\text{AM}},$$

i.e. maximum torque is around four times the start-up torque.

At maximum start-up torque [3.7], similar values are obtained (depending upon machine layout) of

$$M_{\text{max}} \approx 2.6M_{\text{K}},$$

i.e. 2.6 times the breakdown torque, with the expected torque peaks increasing with the size of the machine. Somewhat higher values are found using the equation [3.5]

$$M_{\text{max}} = M_{\text{K}} \left(1 + \frac{1}{\cos \varphi_{\text{K}}}\right),$$

where $\cos \varphi_{\text{K}} = 0.3$ to 0.5.

These torque peaks can reach three to four times the breakdown torque. However, maximum values only affect the shaft and drive train very briefly, i.e. for approximately one period with the rotor locked and for approximately two periods with the motor running up.

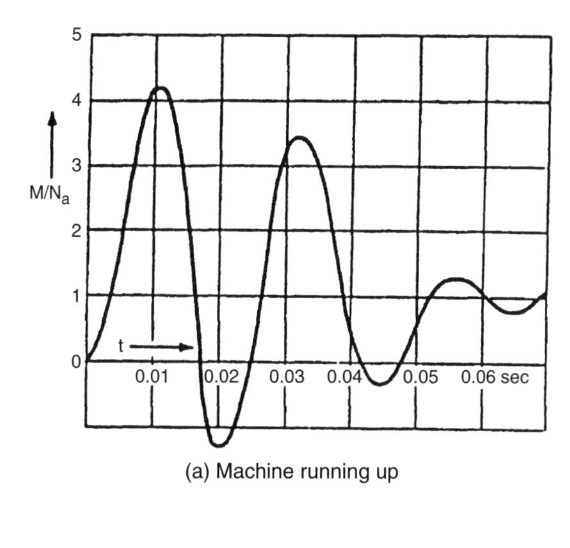

(a) Machine running up

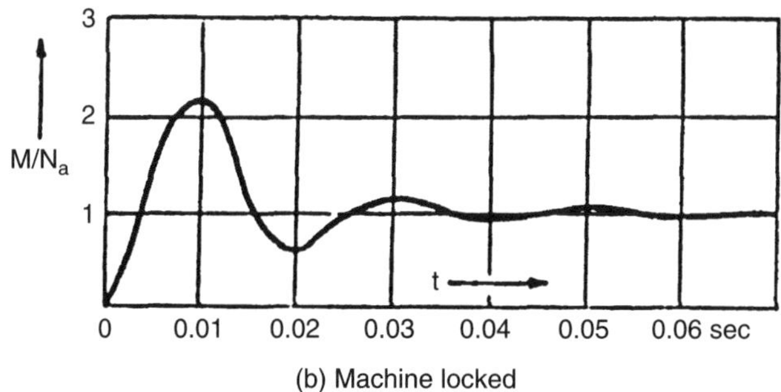

(b) Machine locked

Figure 3.22 Graph showing torque following the connection of a cage motor [3.6]

A reduced component then occurs, which oscillates around the starting torque and dies away after a few further periods.

Synchronous generators coupled directly to the grid are (with a few exceptions) not important for use in wind power plants in practice, and for this reason connection procedures with the machine stationary will not be dealt with in detail here due to their complexity [3.8, 3.9].

If, however, the motorized run-up of synchronous generators needs to be considered, the field winding must be connected via an external resistor with approximately ten times the resistance of the exciter during run-up, as the load voltage would be too high for the field winding insulation with an open-field circuit. Currents that are generated in the starting/damping cage of the synchronous machine give rise to an asynchronous torque. The starting/damping cage, which is usually significantly weaker than in asynchronous machines, therefore develops much lower torques, which have a similar path for nonsalient pole machines as for asynchronous machines.

In salient pole machines the magnetic anisotropy of the rotor caused by the pole vacancy and the incomplete damping cage lead to further deviations. The effects of these are highly dependent upon the design, and cannot therefore be treated as generally valid. Maximum starting torques in relation to a nominal torque [3.8] of approximately

$$\frac{M_{\mathrm{max}}}{M_{\mathrm{N}}} = 1 \text{ to } 3.5$$

can be expected. As already mentioned in Section 3.4.1.2, the use of synchronous generators is basically limited to grid connection via frequency-converter systems. This permits the inrush current and starting torque to be limited to values close to the nominal value with the aid of the power converter, which means that in practice such systems are not usually associated with significantly increased moments in the drive train.

Grid connection at near-synchronous speed is possible in principle for excited or unexcited asynchronous machines, either directly or with the aid of a synchronization device of the type always used in synchronous generators. Low- and medium-output asynchronous generators can be connected to the grid either directly or via power converters, and large machines can be soft-connected using synchronization devices. Increases in the moment of a magnitude relevant to drive train loading can usually be avoided.

The maximum current and torque for direct grid connection near-synchronous speeds are highly dependent upon the degree of saturation of the main and leakage reactance of the asynchronous machine, as well as the timing of the connection (crossover or peak voltage) in all three phases, particularly when the generator is in an unexcited state.

In the case of asynchronous generators for wind turbines, which have lower saturation than motors due to their design, the nominal torque is only exceeded for a short period during connection in the unexcited state. During the first half-period they achieve similar torque peaks to those for stationary connection (see Figure 3.22). The torque drops rapidly, however, to the appropriate steady-state value for the load.

In order to determine reference values, asynchronous generators with outputs of 11 and 15 kW were selected for laboratory investigations. These exhibit behaviour that is characteristic even for larger units. The generators were connected to a rigid public grid and to a weak isolated grid made up of a 28 kVA emergency generating set. The differences between the two configurations could thus be clarified. Due to voltage dips, lower initial current and torque peaks are observed for weak grids than for rigid public grids.

For the unexcited connection of an asynchronous generator at voltage zero, near-identical results are obtained for the cophasal and antiphasal connection states, which are characterized by their normal operation between the grid and machine. For the rigid public grid, the maximum current during the first half-period in the L_1 phase under consideration is as follows:

Cophasal $i_{\mathrm{max}} \approx 8I_{\mathrm{N}}$
Antiphasal $i_{\mathrm{max}} \approx 7.5I_{\mathrm{N}}$

and for the weak grid

Cophasal $i_{\mathrm{max}} \approx 6I_{\mathrm{N}}$
Antiphasal $i_{\mathrm{max}} \approx 5.5I_{\mathrm{N}}$

In the second half-period, however, the opposite behaviour is observed. In the case of cophasal connection, the steady-state current level is approached after the first half-period, whereas in the antiphase connection a value close to that of the first half-period can be determined.

During the first half-period, these phenomena can be attributed to electrical transient processes, particularly those in the highly saturation-dependent magnetizing inductance of asynchronous machines. During the switching process in antiphase, a weakened electrical transient process is observed at the generator coil due to the voltage leap. During the second half-period, however, the mechanical transient process in particular comes into effect. This also terminates in this phase, so for connection in antiphase higher current and torque values should be expected during the second half-period. For the connection of asynchronous machines that have already been excited to no-load voltage, significant differences can be observed depending on the connecting angle between the machine and grid voltage.

In the case of cophasal connection, inrush currents are created that remain within the nominal range and disappear during the first three to four periods. Therefore no significant differences can be determined between rigid and weak grids. By contrast, excited machines that are connected to the rigid grid in antiphase produce approximately double the current peaks of those found in unexcited machines, e.g.

$$i_{max} \approx 15I_N,$$

which decay after three to four periods. The same trend can be observed for a weak grid, but with significantly reduced peaks of

$$i_{max} \approx 11I_N.$$

Figure 3.23 shows the results of investigations on the same asynchronous machines (11 and 15 kW) with the aid of a summary of current and voltage space vector variables plus a calculation of the effective output. The connecting angle between the grid and machine

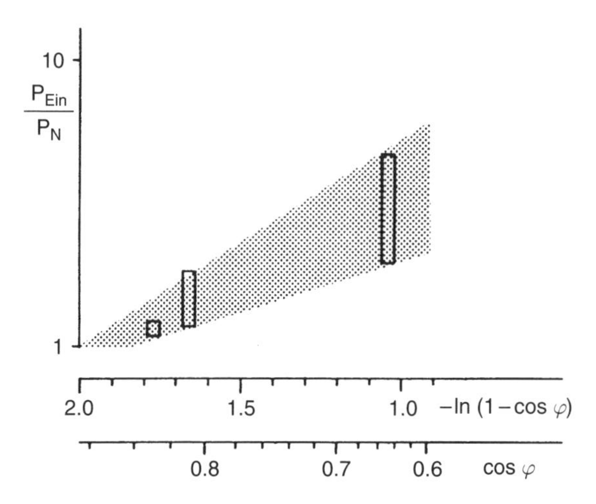

Figure 3.23 Effective inrush power range of asynchronous machines with different saturation levels (connection at synchronous rotational speed)

voltage pointer was varied between 0° and 360°, using approximately 100 intermediate stages. The inrush power exhibits some dependence on the connecting angle; however, from the investigations to date this cannot be unambiguously defined and assigned individual machine parameters.

However, we can indicate the trend for inrush power ranges in relation to power factors during nominal operation on the basis of Figure 3.23. This figure shows that high-power-factor machines have low-making currents with little dispersion due to their low saturation. This is desirable for generator applications because of its beneficial outcome with regard to circuit elements, demands on the winding, foundation loading, etc.

Cophasal connection of excited synchronous machines at near-synchronous rotation speeds gives rise to relationships similar to those described for asynchronous machines. Good synchronization does not cause any significant increase in the moments.

Connection in antiphase in the transient range can be treated as double voltage, as for machine short-circuit, so that the maximum values of the moments described below are doubled. Saturation influences can, however, significantly increase inrush current. This type of loading can be avoided to a large degree by the use of suitable synchronization devices (see Section 3.4.2.4).

3.4.2.2 Generator short-circuit

If an excited generator is suddenly short-circuited, high current and torque peaks occur during the transient process, which die away after a few periods. In synchronous generators with the machine's excitation unchanged in relation to no-load excitation ($U_p = U_1$), the maximum moment of a nonsalient pole machine, as described in Section 3.4.1.2, is found to be

$$M_{\mathrm{K\,max}} = -\frac{3U_1^2}{2\pi n_1 X_\mathrm{d}} \tag{3.19}$$

as a quasi-steady-state variable. For a sudden three-phase short-circuit of the machine the maximum value of the pulsating short-circuit moment, which only lasts for a few periods, is determined by the following equation [3.5]:

$$M'_{\mathrm{K\,max}} = M_{\mathrm{K\,max}} \frac{X_\mathrm{d}}{X'_\mathrm{d}} \tag{3.20}$$

where the transient reactance is characterized by X'_d. Reference values for X_d, X'_d and X''_d are given in Section 3.7.3 for machines on the order of magnitude of 100 kW to a few megawatts.

For nonsalient pole machines it is found that [3.10]

$$\frac{X_\mathrm{d}}{X'_\mathrm{d}} \approx 7$$

and therefore an appropriate multiple of the pulsating maximum value of the quasi-steady-state pull-out torque is found, i.e.

$$M'_{\mathrm{K\,max}} \approx 7 M_{\mathrm{K\,max}}.$$

If a subtransient current occurs due to the damping winding, an even higher maximum moment can be expected during the first period:

$$M''_{\mathrm{K\,max}} = M_{\mathrm{K\,max}} \frac{X_{\mathrm{d}}}{X''_{\mathrm{d}}} \tag{3.21}$$

where X''_{d} represents the subtransient reactance of the synchronous machine. The relationship

$$10 \leq \frac{X_{\mathrm{d}}}{X''_{\mathrm{d}}} \leq 13$$

thus gives a peak value of around 13 times the pull-out torque.

For salient pole machines, transient peak values of an undamped moment can be expected that are approximately 30 % higher than those for symmetrical rotors [3.5], i.e.

$$M'_{\mathrm{K\,max}} \approx 10 M_{\mathrm{K\,max}}.$$

Assumed peak values of

$$M''_{\mathrm{K\,max}} \approx 15 M_{\mathrm{K\,max}},$$

taking damping into account, lie on the side of safety. Due to the design-specific complexities of this process, it is not possible to conclusively derive a simple estimate. A detailed description will therefore not be provided.

No steady-state short-circuit current is created in a three-phase short-circuit with an asynchronous generator. The generator voltage collapses after the initial periods, so no further current flows and no load moment is maintained. During the first half-period, torques of

$$M_{\mathrm{K\,max}} \approx 2 M_{\mathrm{K}} \text{ or } (5\text{–}6) M_{\mathrm{N}}$$

are reached. In two-phase short-circuit, on the other hand, peak values of

$$M_{\mathrm{K\,max}} = (3\text{–}5) M_{\mathrm{K}} \text{ or } (9\text{–}15) M_{\mathrm{N}}$$

must be expected, due to the residual excitation.

Cautious estimates for synchronous generators with direct current intermediate circuits lead us to expect a maximum of four to five times the nominal moment during the first half-period for mains supply in two-phase short-circuit. Thereafter, no more electrical power is supplied to the grid and the torque drops to zero.

3.4.2.3 Grid abnormalities

Grid abnormalities are predominantly caused by grid short-circuits, rapid auto-reclosure, changes in grid voltage and grid frequency fluctuations. We can refer to the descriptions above when considering their consequences.

Single-phase, two-phase and three-phase grid short-circuits in the direct vicinity of the generator are almost identical to generator short-circuits. Grid short-circuits some distance from the generator cause currents to flow across the transformer and conductor impedances that lie between the short-circuit and generator. These currents and the resulting torques are less than the generator short-circuit values, in accordance with the limiting conduction data. The moments given in Section 3.4.2.2 are thus always based on the least favourable assumptions, thus producing safe designs, based on a simulation of extreme situations.

The rapid auto-reclosure familiar to electricity supply companies, in which grid sections become separated from the energy supply for approximately 100 to 500 ms, can, because of the connection between the wind power plant and consumers, lead to a brief, unintentional period in which the entire system operates in isolation. As a state of equilibrium does not usually exist between power supply and consumption, voltage and frequency will drift away from the grid values. In the worst case, the generator may be reconnected to the grid in antiphase after this interruption. Such phenomena are also possible if wind turbines are connected together but do not supply consumers. Because of the connection between direct-coupled asynchronous generators and variable-speed power converter supply systems, it is possible for the asynchronous generators to go over to the consumer or motor mode during the interruption. In this situation, component types and individual plant regulation systems greatly influence the behaviour of turbines and the network as a whole (see Section 4.1.3).

Changes in grid voltage within the normal range do not cause any dominating torque levels. In grids supplied via power converters, rapid grid-frequency fluctuations can lead to significant loading of the drive train and generator of wind turbines. Although values remain well below maximum moments, such frequency variations can have a very negative effect on the lifetime of the drive train, particularly if they occur frequently.

3.4.2.4 Starting and synchronization devices

In order to connect motors and generators in a manner that will protect the drive train and winding, processes that reduce current, and therefore torque, are often used.

The star-delta connection is the simplest and most widespread measure for the reduction of inrush current peaks and the resulting torque shocks and grid reaction for motors in the 10 kW range, which start their run-up at no-load or at low-load moments. This is only possible if the coil is designed for connection voltage in the star-delta mode. Star connection start-up reduces phase voltage and current by a factor of $1/\sqrt{3}$. Electric power and torque drops to around a third of the value for direct connection. Moreover, due to the 42 % reduction in phase voltage, significant saturation effects that would otherwise occur, significantly increasing current, are avoided. This minimizes current and torque peaks.

For the motorized running up of large turbines without blade pitch variation systems, these and similar measures are generally not sufficient. The starting current, which, as shown in Figure 2.55, amounts to around six to eight times the nominal value, will be reduced to around four times this value; however, the running-up time is around three times as long as that for direct connection and varies around values of 10 seconds or more. This means that the grid connection and the switch gear and protection devices must be designed to withstand these relatively short period currents. Therefore, for these procedures, so-called weak wind generators are fitted, either as a separate machine or in the form of a pole-changing coil within the main machine. These are designed for around 20 % of the nominal output of the turbine and run-up at a low rotation speed. This means that the starting current amounts to around the nominal level for the main generator, or exceeds this value only slightly (see Figures 2.55 and 3.33). The available grid capacity can thus by fully utilized, both for running-up and full-load operation.

One further start-up method for wind turbines is the so-called soft start. In this method, gradual voltage and current increases are achieved and running-up currents limited to nominal operating values or similar, by using a thryristor actuator and drive angle alterations. During the starting-up phase, however, increased grid reactions must be expected due to harmonics. As these only occur for short periods and vary greatly due to the drive angle alteration, it is not feasible to compensate for them. Filtering devices are not generally used in this application.

Synchronization devices are common in power station operation for connecting synchronous generators to the grid and were/are used in the few exceptional cases in which wind turbines with synchronous machines are coupled directly to the grid. Synchronization is of much greater significance for wind turbines where large asynchronous generators are connected to the grid.

After running the turbine up and exciting the machine, the generator and grid voltage levels and positions in all three phases are compared with each other, synchronized by small speed and frequency changes and connected when almost synchronized. This keeps transient processes low and to a large extent prevents high drive-train loading and grid reactions.

3.5 Generator Simulation

Referring back to the above descriptions of generator-side steady-state and dynamic torques, simplified simulation models can be developed. These can be drawn upon for in-depth analysis of the interaction between wind turbines and the supply grid, and for the control of these systems. The normal five-coil model for synchronous machines [3.11] is taken into account in this context.

In addition to the previously common simulation procedures based upon PSpice, MATLAB/SIMULINK, etc., new opportunities for the simulation of rotating and static power converter systems will be opened up in the future by PV-Simplorer [3.12]. It will be possible to simulate both conventional (diesel, battery) and renewable (photovoltaic, wind, etc.) power supply systems in graduated dynamic models. The simulation takes into account highly dynamic subtransient processes in the microsecond and millisecond ranges, transient processes in the high millisecond and second ranges and longer-lasting behaviour (including that relating to energy and tariffs) in the ranges of several seconds, minutes and hours. Static and rotating loads are also considered. Furthermore, the use of implicit databases yields significant simplifications compared to previous simulation procedures.

3.5.1 Synchronous machines

Electrical energy supply plants (power stations, diesel generators, emergency generating sets) usually use synchronous generators. If rapid excitation processes are required, these are excited independently via a slip ring or, in the case of brushless machines (which are associated with much greater delays), they are designed to be self-exciting. These systems can be used for both the public grid and isolated operation, as well as standalone supply systems.

Up until now, wind turbines have been fitted with synchronous generators coupled directly to the grid only in exceptional cases. In such cases, measures for the control of mechanical vibrations (e.g. oscillation dampers on the generator foundation) were usually necessary (Section 3.4.1.2). Options for active damping by suitably designed coils are described in reference [3.11].

If short-term transient processes are of interest, the five-coil model [3.12, 3.13] can be used to describe the torque on the generator shaft in relation to the currents and voltages in the stator and rotor for grid operation of synchronous machines. The defined operating ranges can be accurately approximated using a system of differential equations, flux linkages and the mechanical torque–speed relationship or a corresponding logic block diagram. In this context, time constants for the electrical changes for stator and rotor currents lie in the millisecond range, and are thus significantly beneath the relevant times for mechanical processes in the seconds range. Furthermore, the following assumptions are normally made:

- losses are disregarded,
- the saturation of iron is assumed to be linear and
- only the effect of the fundamental wave is taken into account.

Transient, subtransient and saturation-related processes, the consequences of which are described in Section 3.4.2, cannot be adequately taken into account on the basis of these idealized assumptions. Therefore, simplified simulation models have been developed for synchronous machines, which simulate only those operating states and connection processes that are of interest. Thus relevant operating methods are approximated, or at least a trend shown, with justifiable calculation costs.

Results of very simplified simulations of the regulation of effective output for a turbogenerator [3.13] in grid and isolated operation, which only take account of the mechanical inertia of the system, the grid frequency, the turbine speed and the resulting rotor displacement angle ϑ (including power angle) and voltage, can indicate significant deviations from the actual rotor displacement angle. Wind turbines in particular tend to show such deviations due to the changes in excitation that are often necessary because of the varying dynamic loads. Furthermore, interesting special cases cannot be accurately represented.

However, we can use the above simplification to derive structures that allow us to simulate not only steady-state operation but also particularly relevant processes such as connection, grid disturbances and generator short-circuit. Using the structure developed in accordance with Figure 3.24 we can approximate the load moment at the generator M_w (as a negative variable in the power consumption indication system) and the electrical output P_{el} from the difference in speed or frequency between mechanical and electrical rotation arising from the rotor displacement angle ϑ and its sinusoidal quantity ($\sin\vartheta$) in connection with the steady-state pull-out torque M_{KS}.

We can take account of the changes in moment according to the operating or extreme state under investigation by varying the excitation, while taking into account the time delay T_E and maintaining the minimum and maximum steady-state pull-out torques ($M_{KS\,min}$ and $M_{KS\,max}$). Furthermore, dynamic increases in the moment m_{dyn} acting in the short-time range (T_v) can be summarized in relation to the steady-state pull-out torque M_{KS} and damping effects (k_D) in accordance with the machine data in Section 3.4. Electrical and magnetic processes and the consumption by synchronous machines of their own power (e.g. due to excitation) are

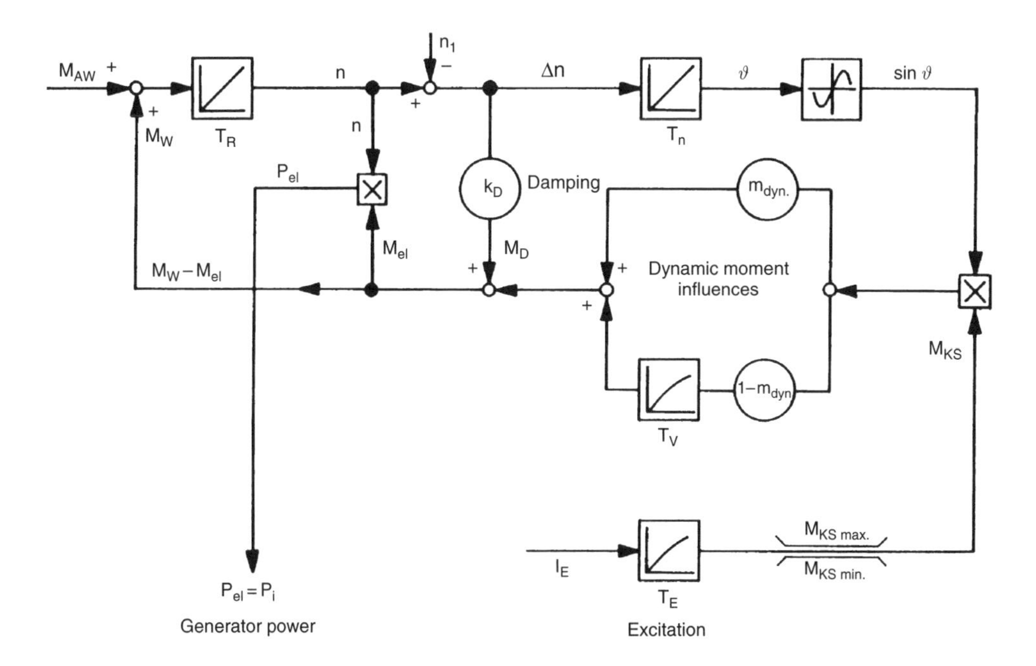

Figure 3.24 Simplified structure for representing the load moment at the synchronous generator

not taken into account here. The structure developed in Figure 3.24 can be used to obtain a good approximation of steady-state and dynamic processes, limited to electrical torque and mechanical effects, with the minimum of calculations.

Figure 3.25 shows a comparison of the simulation results between the five-coil model [3.13, 3.14] and the structure shown in Figure 3.24 for the neutral and overexcited state of the synchronous machine and for moment leaps of half and full nominal value. The reactions of the electrical generator moment and the rotor displacement angle shown display almost identical extreme values and paths for all four load states in both simulations. The structure represented in Figure 3.24 therefore gives the option of reflecting the load behaviour of the synchronous machine to a high level of accuracy, with significantly reduced calculation effort and machine data quantity.

Electrical torque is not determined on the basis of electric currents and magnetic flux: only their effects are – as far as possible – represented analytically or by structural blocks. The model represented here is only valid if the machine does not fall out of step, i.e. as long as the rotor displacement angle does not exceed 90° or the equivalent 1.57 radians. For the simulation of three-phase short-circuit of the synchronous machine, the simplified structure shown in Figure 3.26 can be derived from the results of the five-coil model, where

- the time constant T_D determines the damping of torque oscillations,
- the factor k_U determines the maximum distortion of the generator moment, as well as damping, and
- the factor k_A determines the rate of change of the rotor displacement angle after the generator has fallen out of step.

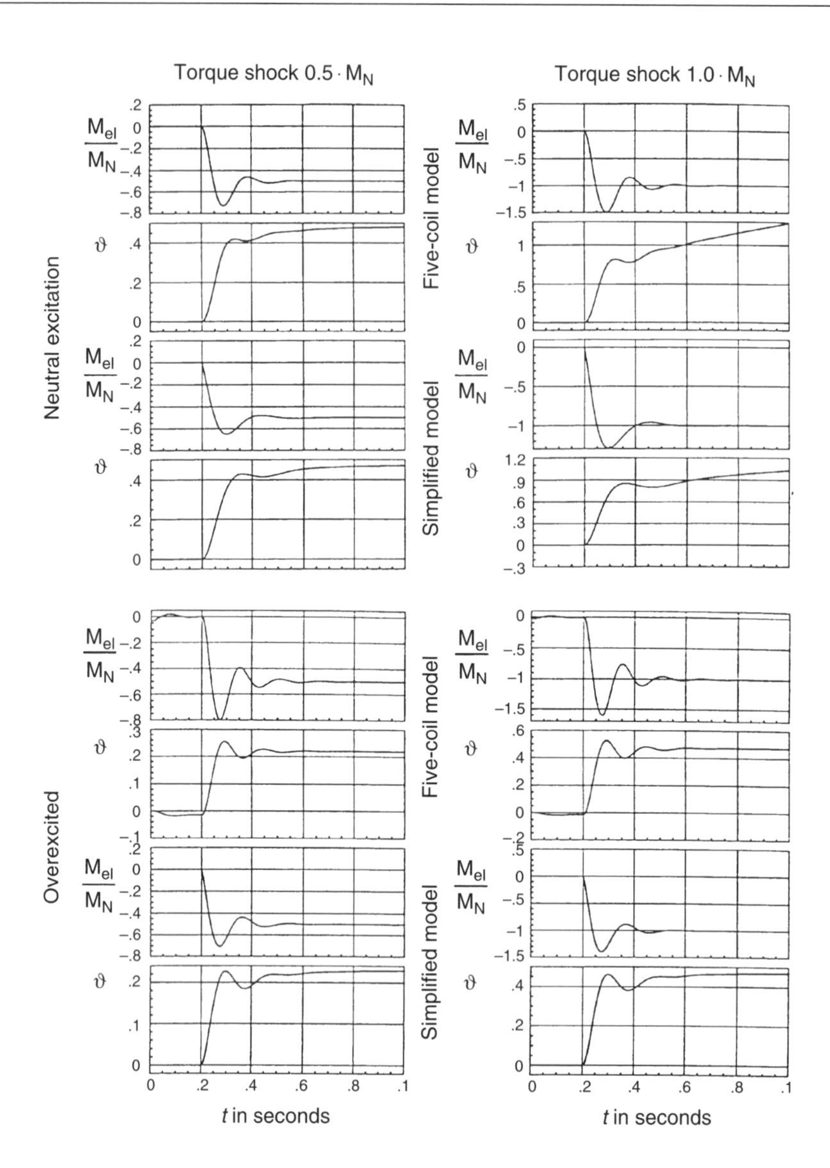

Figure 3.25 Comparison of simulation results (electrical generator moment and rotor displacement angle) between the five-coil model and the simplified model of a synchronous machine as shown in Figure 3.24 (input size of the torque leap from 0 to 0.5 or 1.0 M_N are not represented in this figure)

At the instant of the short circuit, the values k_A and k_U of the dynamic process can be derived from the step-function response. Calculations based on this simplified model show only small differences when compared with the results from the five-coil model as shown in Figure 3.27. This presupposes that only the mechanical component of the system and the generation of electrical moment are considered.

In a similar manner, further representation options are also available for two-phase short-circuit, which will not be discussed further at this point. It is clear that, using this type

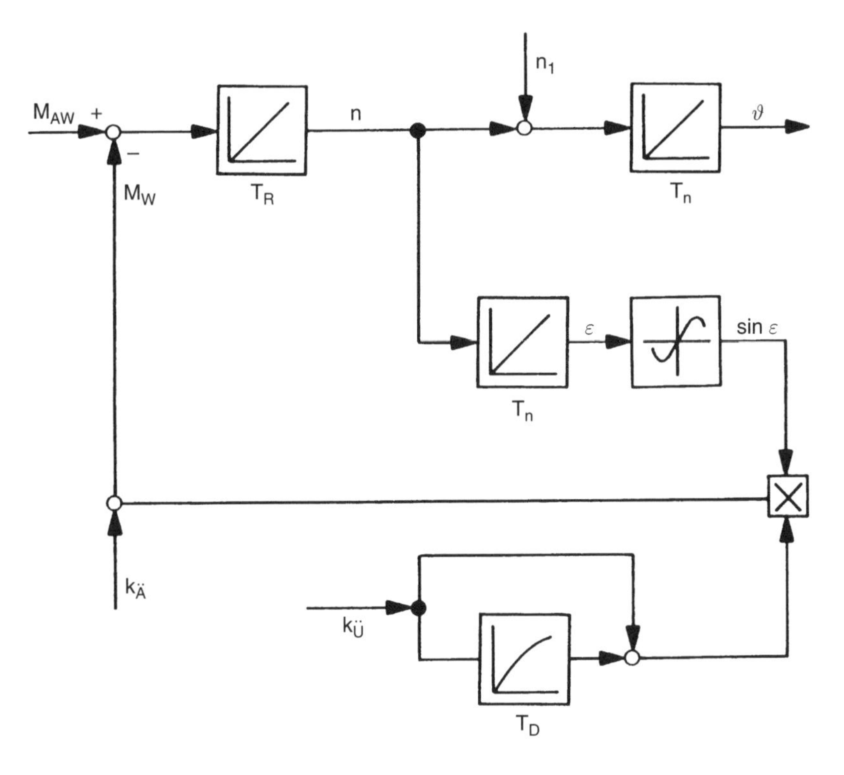

Figure 3.26 Simplified structure for the simulation of generator moment for three-phase connection processes and short-circuits in synchronous generators

of simplified model for specific investigations of torque behaviour, results can be obtained that are a good match for those obtained by the five-coil model, which requires much more calculation effort. Furthermore, by making appropriate modifications to these models, we can also take into account approximations of transient and saturation-dependent processes.

The use of synchronous generators in wind power plants is almost always limited to independent (isolated) operation or corresponding conditions. In grid operation these are achieved by decoupling the generator speed from the rigid grid using power conversion devices (controlled or uncontrolled rectifier, direct current intermediate circuit, previously line-commutated or self-commutated inverter). Such configurations offer a great deal of potential for research and development, e.g. with respect to plant-specific control and operation strategies and with regard to the stability and controllability of synchronous machines when working alongside direct current circuits (see Chapter 4).

3.5.2 Asynchronous machines

The electrical transient processes for asynchronous machines, like those for synchronous machines, can be represented and summarized in a simplified form for the relevant time period that is decisive for mechanical loads. Further aspects specific to the use of asynchronous generators in wind turbines will be dealt with in Section 3.6.

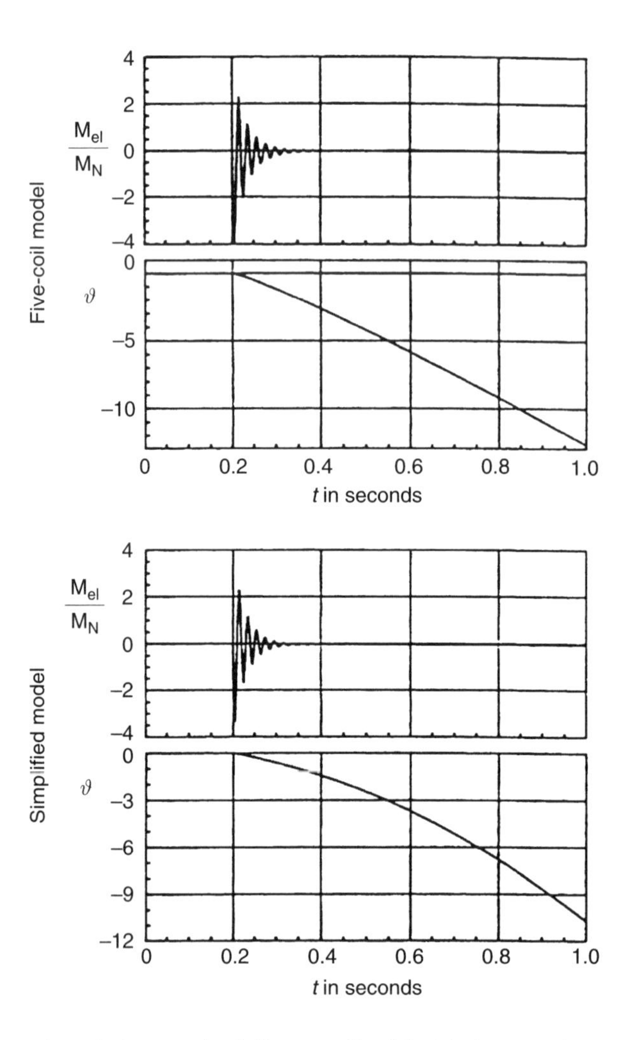

Figure 3.27 Comparison between simulation results (electrical generator moment and rotor displacement angle) between the five-coil model and the simplified model shown in Figure 3.26 for triphase short-circuit of a synchronous machine

Unlike synchronous machines, there is no completely fixed-speed connection to the grid in asynchronous operation. Furthermore, asynchronous generators generally have a damping effect upon exciting oscillations. Because of the variable-slip grid connection of asynchronous generators due to the large inertial mass of the wind wheel, dynamic moments take on a secondary role. Because of this, short-term increases in moment during dynamic processes can be disregarded in the first approximation or, if necessary, as in Figure 3.24, taken into account in the electrical moment M_{el} using the values given in Section 3.4.2.

The asynchronous load moment at the generator shaft can be simulated in a very simplified form using the steady-state torque–speed characteristic of the fundamental wave field. Therefore, in some cases large modifications of the operands (breakdown torque M_k and breakdown slip S_k), must be carried out here to the operands (breakdown torque M_K

and breakdown slip s_K), so that Kloß's equation (3.5) can be used to represent realistically the slip-dependent torque of the most common current-displacement rotors in the form of the steady-state variable

$$M_s = \frac{2M_K}{\frac{s}{s_K} + \frac{s_K}{s}}.$$

Figure 3.28 shows a corresponding structure for the representation of the load moment for the shaft of an asynchronous generator (see Figure 3.13). Influences caused by changes of the grid frequency

$$f_1 = pn_1 \tag{3.22}$$

in connection with the number of pole pairs, p, can be taken into account here by the slip

$$s = \frac{\Delta n}{n_1} = \frac{n_1 - n}{n_1} \tag{3.23}$$

as well as the static moment M_s and the electrical moment M_{el} in the generator output equation. Grid voltage fluctuations are taken into account by the approximate proportionality between the torque and the square of the voltage

$$M_{el} \approx \mathrm{f}\left(U_1^2\right) \tag{3.24}$$

or in connection with the steady-state operand M_s in the creation of electrical moment

$$\frac{M_{el}}{M_N} \approx \left(\frac{U_1}{U_N}\right)^2 \left(\frac{M_s}{M_N}\right). \tag{3.25}$$

Under actual conditions, the additional dynamic moments mentioned above generally give rise to higher mechanical component loading and generator behaviour that is less susceptible to overload. However, shaft loading as described in Section 3.4 is usually increased. The resulting electrical transient processes also have an effect on electrical energy transfer, as discussed below, and on the consumer.

Due to the slip-dependent speed variance of asynchronous machines, dynamic moments are of secondary importance in normal operation. Therefore relevant design options for generators are considered below.

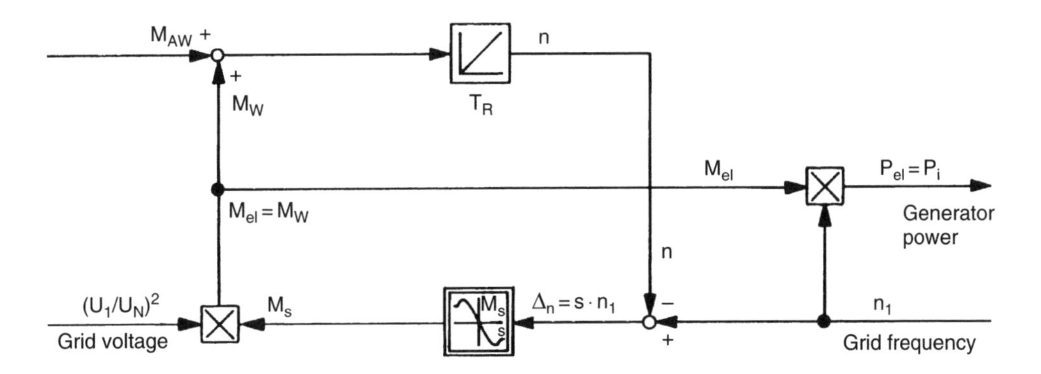

Figure 3.28 Creation of load moment at the asynchronous generator

3.6 Design Aspects

Normal design processes for electrical machines are based primarily on experimental values and the calculations in references [3.14], [3.15] and [3.16]. Currently, these are usually used in a refined form with the aid of computer-aided programmes, and play a key role in the study of electrical machines.

The main dimensions of the machine, i.e. the diameter and length of the stator and rotor laminated core, can be roughly determined based on experimental values for the so-called output coefficient or tangential force figures, taking into account the rotation speed, output, cooling characteristics, slot and pole pitch, which have been predetermined or must be selected. In this context, the machine type, its size or output class, cooling (air, fluid, hydrogen) and the materials used, etc., have a significant influence on the dimensions of the generator. Calculations of the magnetic circuit with flux and field divisions and the determination of individual losses in the electrical, magnetic and mechanical components permit, with the aid of iterative processes, a near-optimal machine design. This process, which has been described in relatively simple terms, turns out in practice to be very costly and calculation-intensive, as the considerations contain a multitude of nonlinear relationships, and a great deal of design experience is required to obtain the desired optimization.

The full calculations upon which the design and construction of electrical machines are based will be omitted at this point, due to the wide-ranging prerequisites for the use of design methods, which lie far outside the scope of this book.

The representation that follows therefore aims to give some pointers and trends for design, based on the discussions in this chapter and on operating experience and simulation calculations for asynchronous and synchronous generators.

3.6.1 Asynchronous generators

Asynchronous generators with cage rotors were by far the most common type of generator for mechanical–electrical energy conversion in the wind turbines manufactured and installed in the past. Generators with slip-ring rotors have, up until now, been the exception. However, developments in power electronics over the last few years have meant that they are taking on increased importance for geared systems. In addition to slip-ring rotor machines with brushes, brushless machines (so-called cascade machines) are also used [3.18, 3.19]. The frequency converter is supplied via an additional auxiliary machine (see Figure 5.27(a)).

Different requirements are imposed on asynchronous generators for use in wind turbines than for those in diesel generators. In diesel units, high no-load losses in the generator can be very desirable from the point of view of operating characteristics. They bring about low motor fouling at only slightly increased fuel consumption. Furthermore, from the point of view of operating behaviour, they bring significant advantages. Torque fluctuations are notably reduced, for example, when loads are decreased, resulting in lower frequency and voltage deviations. In wind turbines, however, no-load losses are generally undesirable, since they make it more difficult to convert low wind speeds into electrical output, and increase the standstill time of the plant.

In a generator based on a good mechanical design the saturation state of the magnetic circuit has a significant influence over no-load losses. High magnetic saturation usually results in high no-load losses. Load variations, however, bring about lower voltage changes with increasing saturation. Furthermore, the permissible current density of the generator (in particular for connection and shock load) is significantly increased due to the high magnetic energy content. The adjustibility and operating behaviour of diesel units can therefore be favourably influenced by high generator saturation. However, operating problems can occur, particularly in low wind-speed ranges, when generators designed in this manner are used in wind turbines.

In wind turbines operated independently, generator excitation normally occurs after turbines have been run up. No-load losses caused in this manner cannot be covered at high saturation in the lower wind-speed range; thus the turbine speed can drop to the point where the generator is de-excited. The reduction of significant losses results in a subsequent increase in turbine acceleration. This pendulum process can repeat itself over and over again in a certain wind-speed range if no action is taken by the plant control system.

When wind turbines are operated as part of a grid, high inrush currents occur, which can trip fuses and significantly load windings (see Section 3.4.2.1), particularly if generator saturation is high. Furthermore, efficiency is sacrificed and generally poor power factors must be accepted. If the standard voltages were changed or adjusted, safety considerations [3.20] could increase in importance in addition to the aspects discussed here.

The behaviour of the induced voltage U_{i} takes on particular importance in a detailed consideration of the processes in asynchronous generators, since it has a significant influence on the degree of saturation of the magnetic circuit. According to Figure 3.7(a), the induced voltage is found to be

$$\underline{U}_{\mathrm{i}} = \underline{U}_1 - (R_1 + \mathrm{j}X_{1\sigma})\,\underline{I}_1 \tag{3.26}$$

Based on the general current circle diagram, normally represented in simplified form as circle I_1, the circle U_{i} shown in Figure 3.29 can be derived for induced voltage. The U_{i} circle intersects the real axis at U_1 near the nominal voltage at no load and at approximately $U_1/2$ when stationary. Both circles are represented in Figure 3.29 for an asynchronous machine designed as a generator with 22 kW nominal output, as an example for larger asynchronous machines operating at a nominal rating for the motor and generator mode. For the sake of completeness, the I_μ circle, which will not be dealt with in detail here, is also drawn. Saturation-dependent influences on the circles, which lead to significant deviations from the circular shape, will not be considered.

The circle diagram (the so-called U_{i} circle) traces the path of the pointer tip for induced voltage. We see a clear increase in U_{i} for a generator's normal load operation when compared with motor operation (compare U_{iNG} with U_{iNM}). Asynchronous machines therefore fall significantly further into the saturation state in generator operation than in motor operation. No-load losses increase and the power factor falls. Both variables therefore take on much poorer values. This is clearly illustrated by the following discussion.

3.6.1.1 Performance characteristics from measurements

In order to demonstrate the operational options and problem formulations, and the energetic effects that are dependent upon machine selection, we will show the results of metrological investigations on five different asynchronous machines of the same size, with

- 11 kW nominal output;
- motor- or generator-type design;

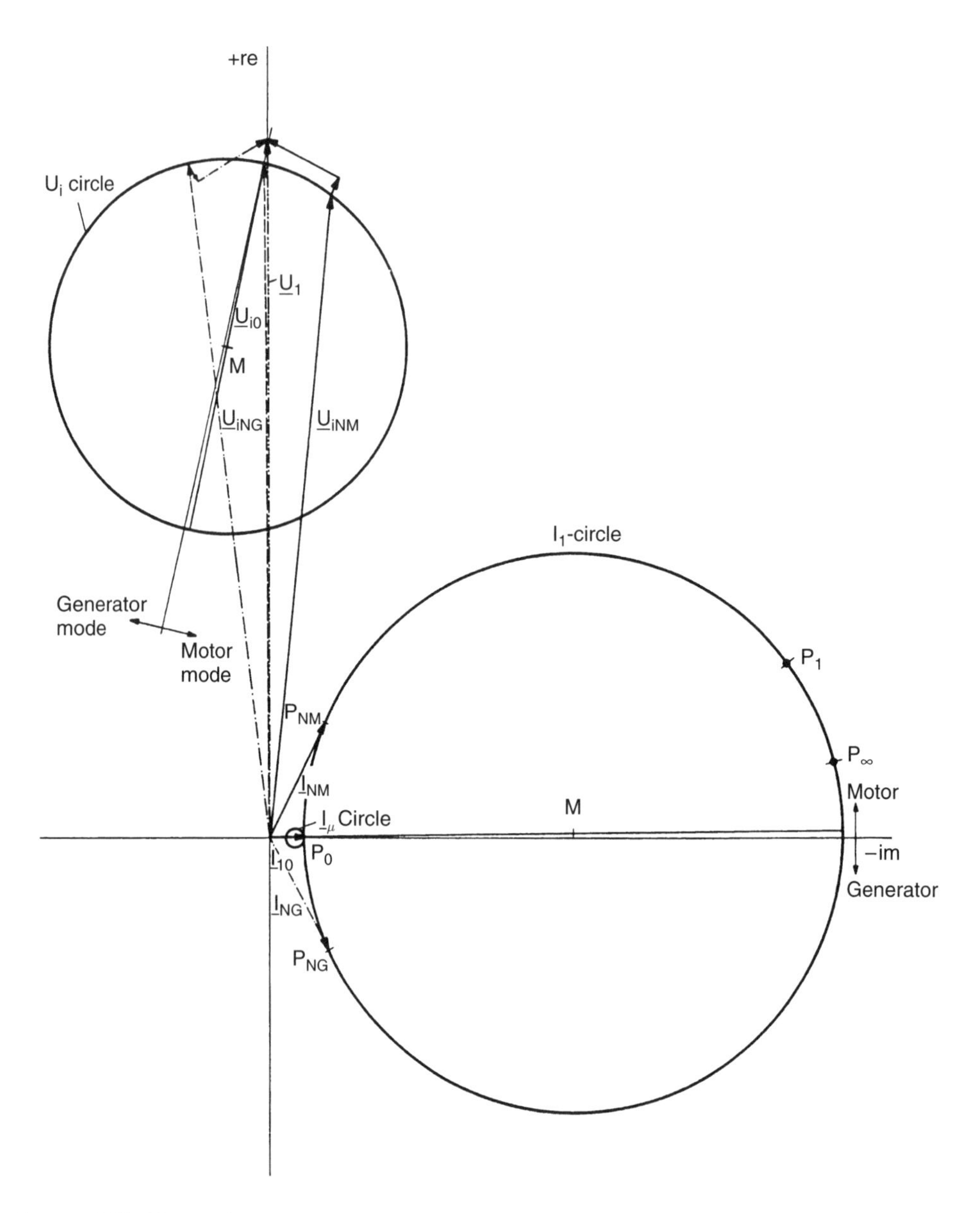

Figure 3.29 Circle diagram of a 22 kW asynchronous machine for stator current (I_1 circle), magnetization current (I_μ circle) and induced voltage (U_i circle)

- output-class-dependent slip value and increased slip values due to
  - — constructional changes in the case of cage rotor machines or
  - — circuitry measures with the aid of additional resistance in the rotor circuit of slip-ring rotor machines;

- for stator voltages between 280 and 500 V or between 320 and 480 V;
- in the electrical output ranges of
  - — generator no-load operation or
  - — motor no-load operation respectively and
  - — up to a maximum of 1.3 times nominal output.

To record the measured values in a stable-temperature, steady-state condition, a warm running phase of around two hours was necessary for each measurement. The rating of the machine was selected such that the characteristic attributes of the large unit class would be accurately represented.

To demonstrate the transient behaviour of individual machines, and in order to specify guidelines for the design of generators specifically for use in wind turbines, there follows a small selection from the comprehensive programme of investigations based on some performance characteristics from a selection of machine types (generator or motor design, slip-ring rotor). The relevant performance characteristics for

- active power consumption,
- apparent power,
- reactive power,
- stator current,
- efficiency,
- power factor and
- generator slip

are represented in the following in relation to

- electrical output power and
- stator voltage.

The grid voltage was adapted to the selected stator voltage with the aid of a regulating transformer. Due to the fact that the power values were measured at various distances, they had to be interpolated for spatial representation in the form of computer graphics before they could be represented in the selected power grid (based upon 1 kW increments).

In Figure 3.30 the measured values are shown for a generator-type asynchronous machine, from the maximum voltage downwards. This particularly illustrates the trends that increase steadily with voltage, which is true for apparent power, reactive power and mechanical power input, and partly so for efficiency. Values that decrease with increasing voltage, such as the power factor and slip, remain difficult to recognize due to being partially obscured. These trends are, however, clearly shown in the subsequent figures, which show voltage increasing.

Figure 3.30(a) and (b) illustrates the increase in apparent and reactive power with increasing stator voltage and output power. The increases with voltage are caused by the effects of saturation in the magnetic circuit and the associated increase in current with rising voltage, particularly at no-load and low outputs. The resulting increases in copper and iron losses in the machine require higher drive input power values, as illustrated by Figure 3.30(c).

As shown in Figure 3.30(d), the minimum current runs from the lowest voltage values (280 V) at no-load to maximum output at approximately 460 V. High electrical output at

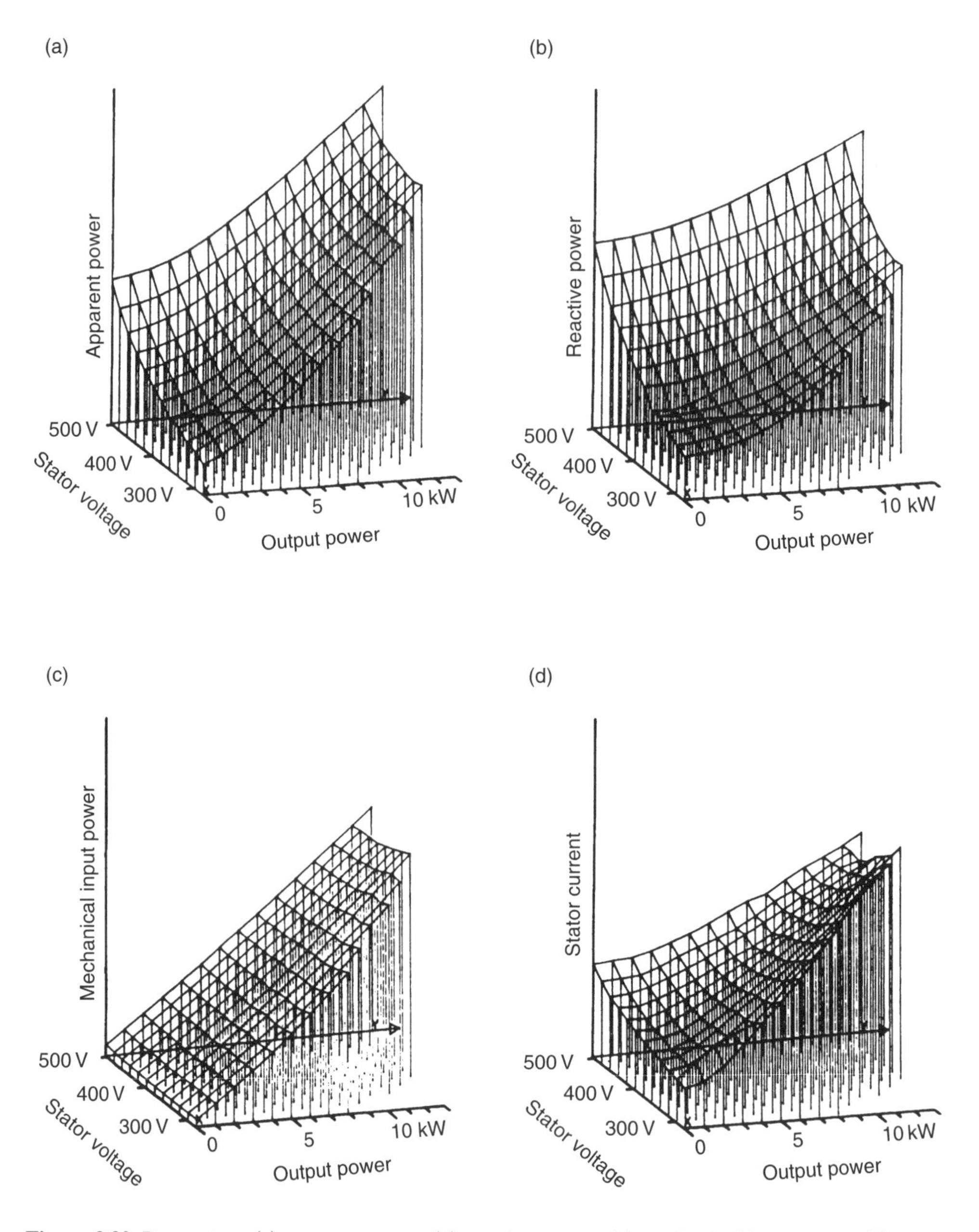

Figure 3.30 Parameters: (a) apparent power, (b) reactive power, (c) mechanical input power, (d) stator current, (e) efficiency, (f) power factor and (g) slip of a generator-type asynchronous machine (11 kW nominal output) as a function of electrical output and stator voltage (results from a series of measurements)

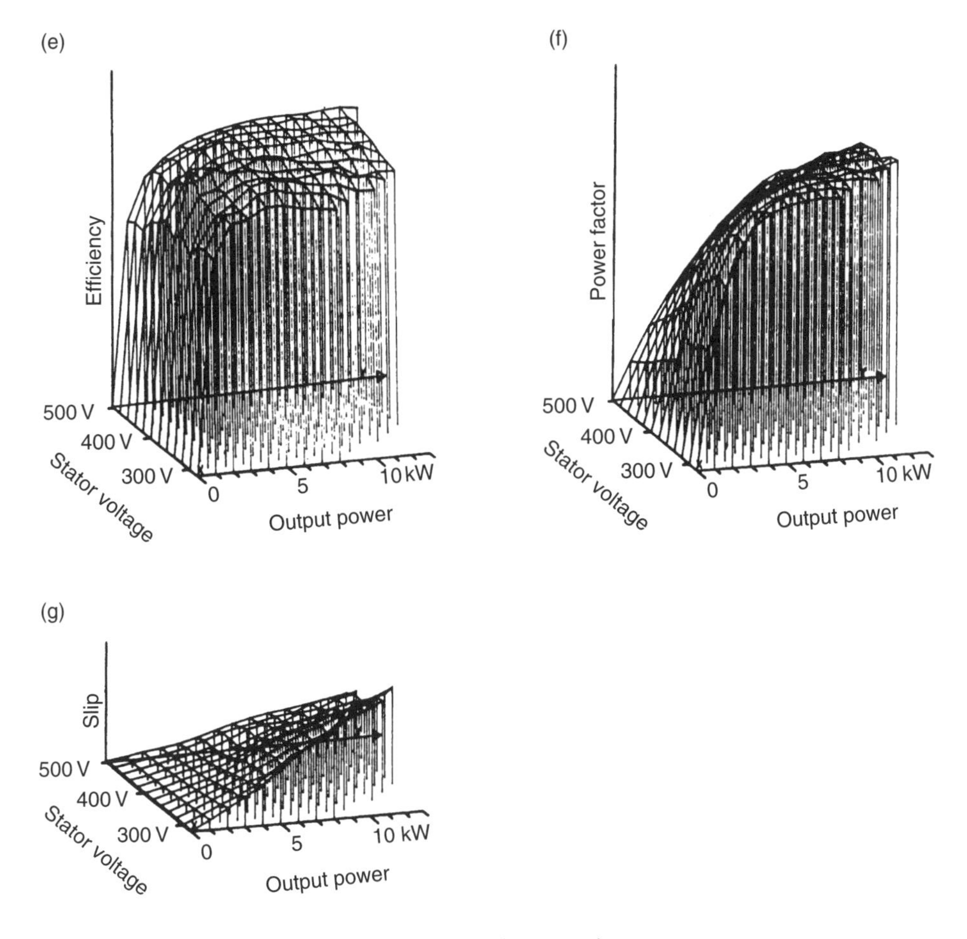

Figure 3.30 (*continued*)

low-voltage values thus causes high stator currents due to output demands. At high voltages, high stator currents are caused by the high level of magnetization due to high saturation.

Figure 3.30(e) shows that efficiency increases progressively at high voltages,i.e. if the machine is designed in the heavily saturated range. Conversely, at low voltages, and thus dimensioning in the unsaturated range, maximum efficiency is achieved at approximately half the nominal output. Higher output values only bring about a small reduction in efficiency in such designs, a characteristic that can be very desirable in wind turbines, which are usually operated at partial load. Furthermore, Figure 3.30(f) shows that for highly saturated machines the desired power factor values typical of generators are only achieved in the overload range. Unsaturated design, on the other hand, gives the maximum power factor at only approximately two-thirds of nominal output. Such machine types also exhibit high slip values, as shown by the slip levels (in the lower-voltage range) in Figure 3.30(g), which, in combination with the torque fluctuations that are always present in the turbine, results in a high degree of elasticity and compliance in the transmission. Output fluctuations are thus reduced and highly fluctuating component loadings in the entire power transmission section are significantly reduced. Measurements on machines of differing designs have shown that, assuming that driving torque fluctuations are similar, generators with slip values four times higher than nominal slip transfer around a quarter of the output fluctuations to the grid compared to a normal design (see Figure 3.15).

Therefore, the generator parameters of a motor-type slip-ring rotor machine will be described below for different slip behaviours, which can be adjusted with the help of additional resistance in the rotor circuit. Figure 3.31 shows a comparison of the relevant parameters for different levels of slip. The machine in question was operated with a short-circuited rotor at approximately 3 % nominal slip or with additional resistors in the rotor circuit, which are selected to achieve 7, 14 or 21 % nominal slip at nominal load and voltage.

Figure 3.31 shows an increase in input power at higher slip values due to increased losses in the rotor circuit, and thus a worsening of the machine's efficiency, particularly in the full-load range. The mechanical input power increases with increasing slip. Stator current and power factor values, on the other hand, show only small differences. They are almost identical for all slip values at corresponding stator voltage and generator load.

Finally, the slip levels illustrate the very strong voltage dependence of these variables and their significant increase as voltage decreases. This trend is highlighted in a very striking manner by the levels at the highest slip values (21 % operating at the nominal rating). It is thereby possible to achieve the desired degree of slip by the selection of

- rotor resistance in connection with
- rotor leakage (see Equations (3.7) and (3.11))

and the specification of

- stator voltage.

As was clearly shown at the beginning of this section with the help of the derived circle diagram for induced voltage, very different behaviours can be expected from the motor-type design compared with generator-type dimensioning. Figure 3.32 shows marked trends and constructive areas of influence, by direct comparison of the relevant parameters.

Mechanical input power and, in particular, stator currents are higher for the motor-type machine than for the generator-type machine across the entire generator operating range. The excessive increase in current at high voltages clarifies the high saturation dependence of the motor. Efficiency and, to an even greater degree, power factor values are much more favourable for machines of generator dimensions, particularly at higher voltages. At low voltages, however, only small differences are apparent. The generator-type machine with 3 % slip compared with 4 % nominal slip of the motor demonstrates significantly more rigid behaviour, due to the more efficient design, resulting in a more rigid grid connection.

In summary, we can state that the asynchronous machines used in wind turbines should be designed to achieve a low degree of saturation. Figure 3.29 illustrates this point, since the generator always achieves higher induced voltage values and thus higher saturation states in normal operation than in motorized operation. This achieves

- a low mechanical input power,
- stator currents as low as possible for the prevailing load state and

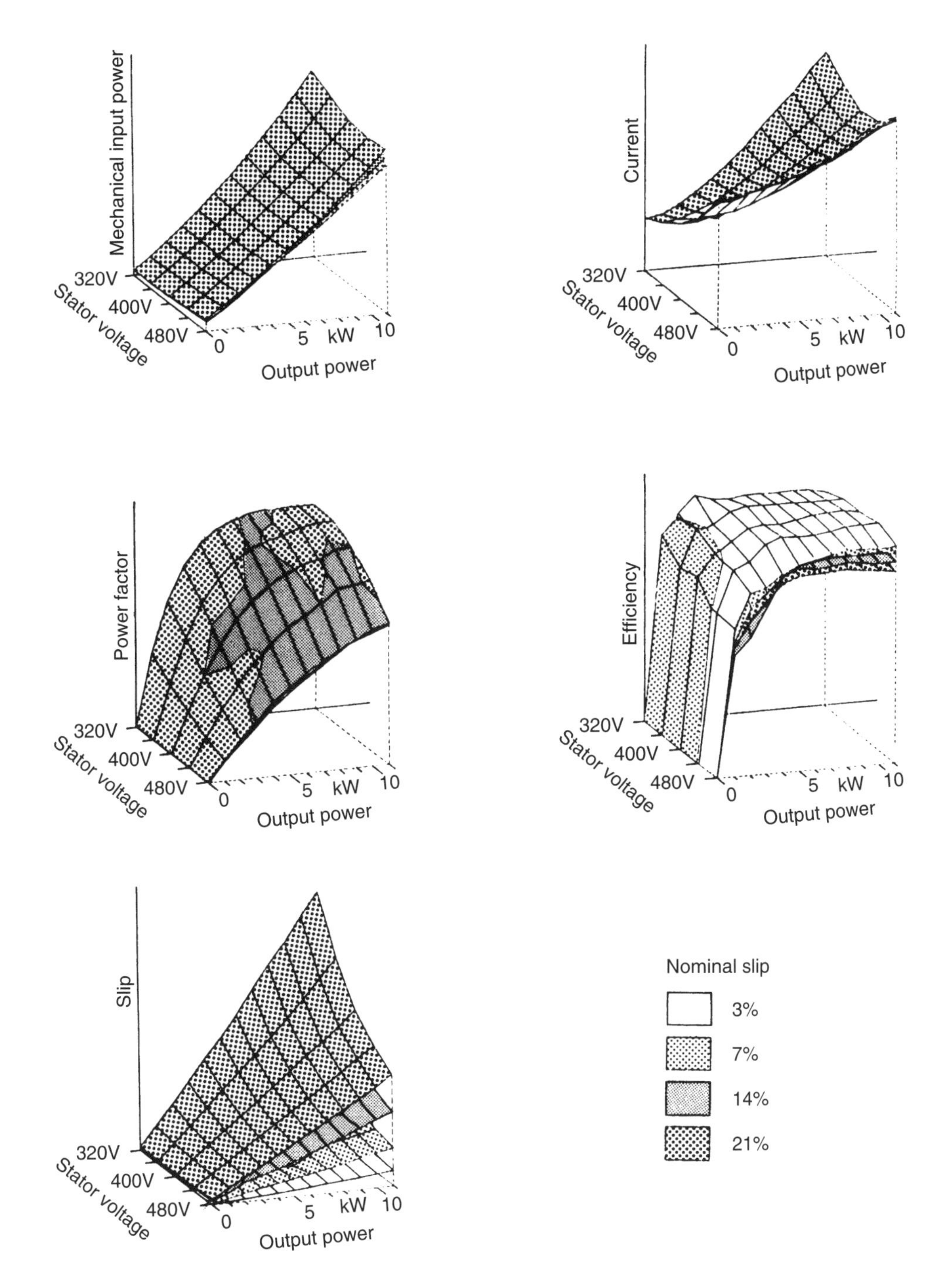

Figure 3.31 Comparison of parameters of a motor-type asynchronous machine with a slip-ring rotor (11 kW nominal output) at different nominal slips in relation to electrical generator output and stator voltage (results from a series of measurements)

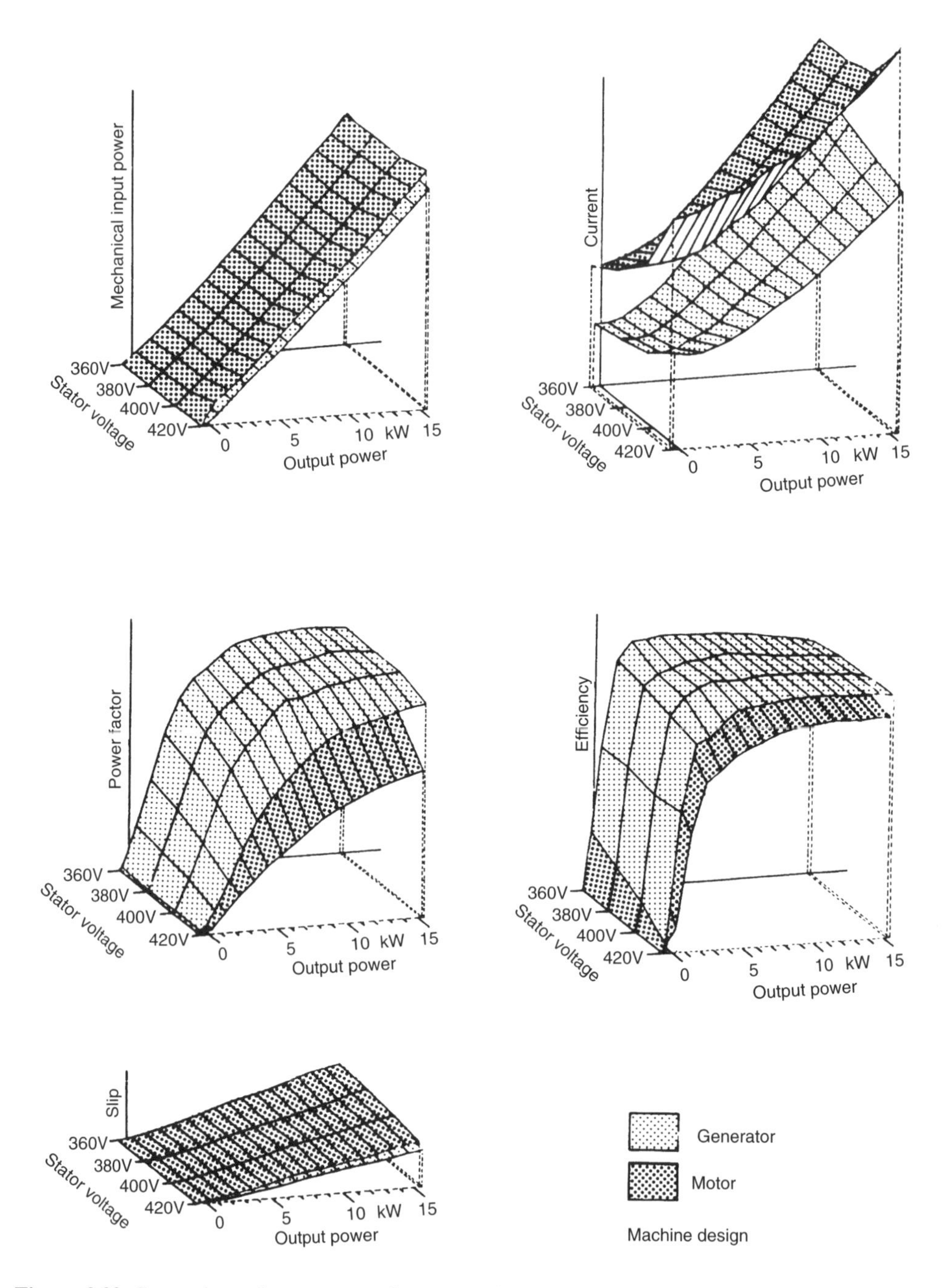

Figure 3.32 Comparison of parameters of motor- and generator-type asynchronous machines (11 kW nominal output) in relation to electrical generator output and stator voltage (results from a series of measurements).

- favourable efficiency and power factor values, both in the partial load range and at high slip.

Altogether, this type of design allows the following to be achieved in asynchronous generators:

- low reduction of efficiency,
- high rotation speed flexibility and
- correspondingly low drive-train loading

due to better flexibility in the turbine rotation system. In the following, we will describe specific designs for operating ranges. The grid connection is described in Chapter 4.

3.6.1.2 Weak wind generators

As mentioned in Section 3.4.2.4, weak wind generators are used for the motorized running up of wind turbines without blade pitch variation. They are designed to achieve 20 % of turbine nominal output at 50 to 75 % nominal speed, which means that the weak wind generators can operate in a favourable output range at low rotation speeds. This permits energy yields in the lower load region to be significantly increased (Figure 3.33). Thus – similarly to the variable-speed operation of plants with frequency converters but with only two fixed speeds – the potential output of the turbine can be more fully exploited. A weak wind generator can either take the form of an additional machine that rotates with the turbine

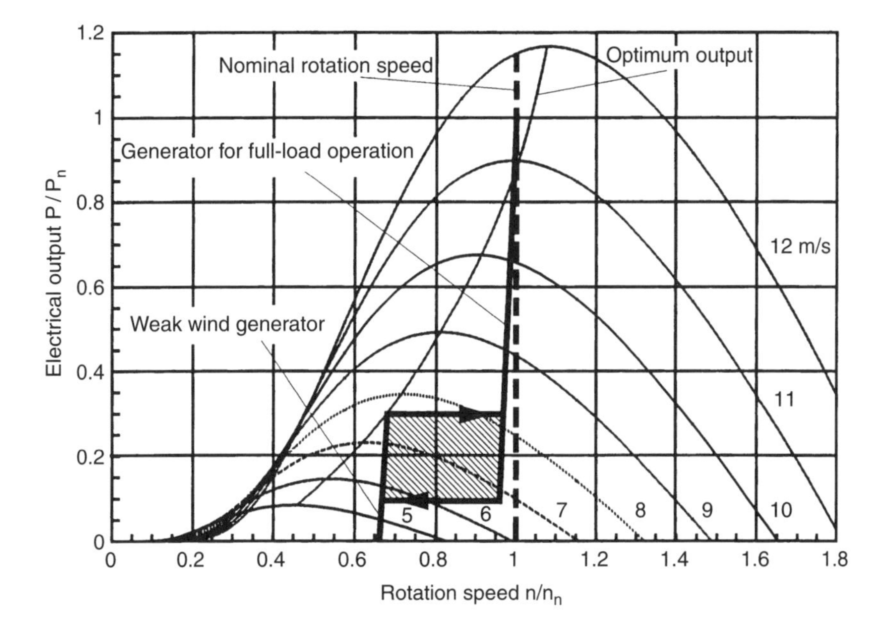

Figure 3.33 Group of turbine characteristics for output as a function of rotation speed, with wind speed being a parameter, and the working range of weak wind and a full-load generator

and has been specially designed for this application or it can be a pole-changing generator designed for weak wind operation at low speeds and for full load at higher speeds.

To avoid a high frequency of changeovers on the borderline between weak wind and full-load speeds, and the associated loads to the drive train and switch gear, an output overlap must be provided between the two operating conditions. The changeover hysteresis shown in Figure 3.33 must be adapted to the plant circumstances and the prevailing local conditions.

3.6.1.3 Drive-train junctions

Drive-train junctions provide further options for the improvement of starting procedures and partial-load operating ranges. In this approach, two to four generators G_1 to G_4 (see Figure 3.34(a) and (b)) are coupled to the gearbox. Transmission loading can thus be reduced and the design of the system significantly improved.

These configurations permit the linking of different features. For example, using a (modular) 250 kW generator, wind turbine outputs of 250 and 500 to 1000 kW can be achieved by fitting one to four machines. This permits significant cost reductions for development, storekeeping, etc. In the four-generator layout, moreover, a specific design is possible for the motorized run-up. Furthermore, by pole changing, the generator in the foreground of Figure 3.34(a), for example, can be operated at different rotation speeds. The weak wind characteristics of the plant can thus be improved. It is also possible to operate all generators in high-efficiency ranges by selecting appropriate individual changeover levels. Depending upon requirements and design goals for the entire system, the option of flexible grid connection can be included, as well as other features that we will not describe in detail here.

3.6.1.4 Variable-slip asynchronous generators

As already described in Section 3.4.1.1, power gradients and grid effects can be reduced at increased slip values by the special design of the machine in accordance with Equations (3.6) to (3.12). As shown in Figure 3.16, significantly better results can be achieved at a more favourable utilization ratio by slip regulation. In this approach, slip values less than 5 to 10 % are generally used, which are controlled by a microcontroller through additional rotor resistors rotating along by using pulse-width-modulated insulated gate bipolar transistor (IGBT) power output elements. With regard to the control dynamics, similar reaction times can be achieved to those in systems with a frequency converter supply, despite the simple layout and low system cost. The operating range of slip-regulated asynchronous generators is not limited to variable-pitch wind turbines. According to reference [3.21], stall-regulated turbines also permit the use of such systems.

The reduction of the transmission and generator load as well as the resulting more favourable flicker values and so-called *K* factors (See Section 4.3) are achieved by design and construction measures. To this end:

- climatic influences due to temperature changes, air density, moisture, etc., must be compensated and
- high intrinsic safety,

(a) Double generator configuration (Adler 25, 165 kW). Reproduced by kind permission of Feus Peter Molley

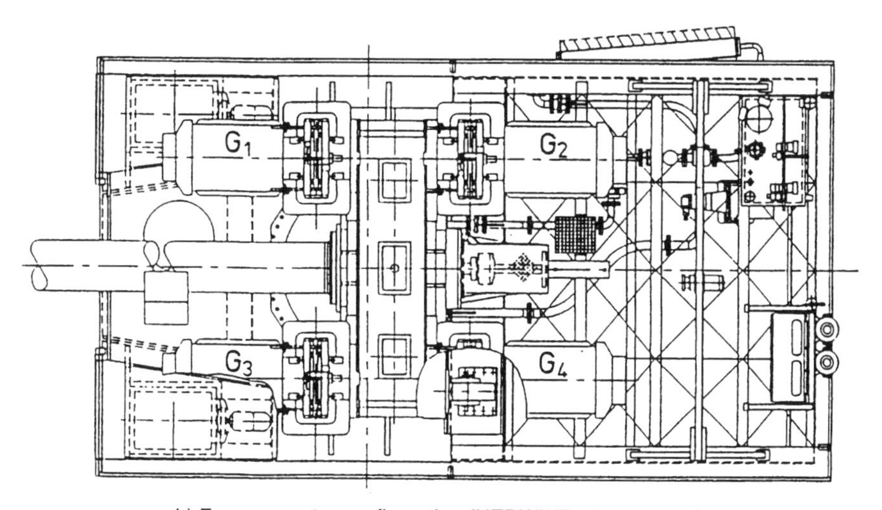

b) Four-generator configuration (NEDWIND 50, 1000 kW)

Figure 3.34 Multiple generator systems

- good control dynamics of the slip-control unit and
- rapid (target and actual value) data transfer guaranteed.

Furthermore, all important operating states of the generator and the power electronics of the higher management of the wind turbine must be made available (by a communication interface).

3.6.1.5 Shaft voltages and bearing currents of three-phase machines in a frequency converter operation

Parasitic effects in machines equipped with frequency converters give rise to currents that propagate through shafts, the running surfaces, rollers and housing of bearings, and which can (depending upon the intensity of the current) cause damage to bearings. Shaft voltages are particularly common in low-voltage, high-output machines, which are used, for example, in the generators of wind turbines [3.22–3.24].

Asymmetries in the magnetic working flux circuit, e.g. caused by ventilation openings, laminated core mountings, magnetic anisotropy of the laminations and asymmetries in the windings, result in a parasitic magnetic flux. This surrounds the rotor, increases with the level of flux per pole and induces voltages in the shaft of 50 and 150 Hz in particular, the so-called classic shaft voltage. Magnetic asymmetries must therefore be avoided in the design and manufacture of the machines. Furthermore, in large generators a bearing can be constructed in insulated form, in order to interrupt circulating currents [3.22]. According to reference [3.24] permissible limit values for shaft voltage in roller bearings lie at 0.5 V.

The use of high-speed IGBT frequency converters further promotes the generation of bearing currents [3.25–3.28]. Pulse-width-modulated a.c. converters are generally designed with a direct current intermediate circuit and generator-side inverter. This connects the rotor phases with positive and negative intermediate circuit potential on an alternating basis. As a result the total voltage at the output of the pulse-width-modulated a.c. converter (in contrast to symmetrical three-phase voltage systems) exhibits a pulse-shaped path that deviates from zero. This so-called common-mode voltage is dependent upon the frequency converter pulse pattern or the rate of rise of the common-mode voltage and can be measured at the neutral point of the generator winding. It is responsible for the bearing currents caused by the pulse-controlled a.c. converter.

Furthermore, in electrical machines capacitive couplings exist between the windings, the stator housing and in particular the rotor, which give rise to potential differences between the rotor and stator [3.29]. These arise in the same direction at both bearings, break down the common-mode voltage and are generally dependent upon the machine geometry. Since the capacitances are very small and lie in the nF range, these potential differences are limited by discharge routes or they lead to discharges that can break down the bearing lubrication film. This results in microscopic marks on the running surfaces of the bearing. These effects can be minimized by a common-mode voltage-minimized pulse pattern. In slip-ring rotors, earthing brushes connecting the stator and rotor are used. Furthermore, the stator housing is connected to the frequency converter via the cable shield in order to guarantee this as a good path for the return of common-mode current [3.24]. Since the impedance of these current routes is often too high, stray earth currents sometimes flow through the generator bearing or the bearings of components that are connected to the generator shaft (e.g. gearbox, sensors). These capacitive earth currents, which can reach levels of up to 10 A, put the bearings at risk and have an unfavourable effect upon electromagnetic compatibility (EMC). This situation can be remedied by the use of voltage-rise filters, known as dv/dt filters, which limit the gradient of voltage rises; these filters ensure the best possible return line for the high-frequency earth currents by providing a two-sided earth shield. Bearing damage to components connected to the generator shaft are avoided by the provision of an insulated coupling or the use of lower-inductance earthing than in the generator. Additional protection is provided by the earthing of the shaft at the non-isolated bearing.

Similarly to the classical shaft voltage mentioned at the start of this section, frequency-converter-dependent circulating currents arise as a result of common-mode current flow in the motor winding due to the occurrence of capacitance between the windings and the laminated core. Asymmetries in current distribution across the machine give rise to a flow that induces pulse-shaped voltages along the generator shaft. These are overlaid upon the classic shaft voltage and can give rise to low-frequency currents. For this phenomenon, too, the steepness of the common-mode voltage can be reduced by the use of a voltage-rise filter, and circulation currents can be weakened or prevented by good insulation of a bearing.

Shaft voltages and bearing currents thus give rise to various damaging effects, which are also overlaid upon one another. To prevent bearing damage, several measures are therefore necessary. To insulate generator bearings, isolated bearings can be used or, if standard bearings are used, an isolating layer can be inserted between the bearing and shaft. The bridging of the non-isolated generator bearings represents an additional preventative measure. Connections that are as short as possible and exhibit low impedance, even at high frequencies, are necessary in conductors, cable shields and earth connections. Furthermore, the use of voltage-rise filters and the design of common-mode voltage-minimized frequency converters are of great importance for avoiding bearing damage.

The following considers synchronous machines, which in combination with frequency converter systems permit variable-speed turbine operation.

3.6.2 Synchronous generators for gearless plants

In addition to the simply constructed and very robust asynchronous machine with a cage motor and the increasingly common double-fed asynchronous generator, the conversion system with a frequency-converter-coupled synchronous machine has also achieved a large market share. For the conditioning of electrical energy and grid connection, load-side rectifiers for intermediate circuits (until the start of the 1990s with line-commutated inverters based upon thyristors and thereafter with line-side self-commutated converters based upon IGBT technology) have gained the greatest importance. Gearless plants dominate.

However, in synchronous generators the entire machine output must be fed through the frequency converter. In double-fed asynchronous generators, on the other hand, the frequency converter output is oriented towards the selected speed variation range or the slip power to be transmitted (e.g. 30 to 40 % of the nominal value).

This type of generator and frequency converter system offers the advantage of decoupling rotor speed from grid frequency over a wide variation range (e.g. (0.5–1.1) n_N). Thus the potential for intermediate storage of kinetic energy in the rotating mass is utilized over a wide range for output smoothing and, in contrast to fixed-speed converter systems, by selecting optimal output operating points for better efficiency (Figure 3.35).

The normal type of synchronous generators for use in wind turbines are – unless more stringent requirements are imposed upon the control speed of the exciter – of a brushless design with an integral exciter. The auxiliary generator thus supplies the field winding of the main generator via a rotating rectifier (see Figure 3.6). In order to increase the subtransient reactance of the machine and to limit the intermediate circuit current in the case of a short-circuit, the generator rotors can be designed without damper windings (compare the equivalent circuit diagram in Figure 3.36(a) and (b)). This can reduce the maximum value of stator current by around 30 % (see Figure 3.37).

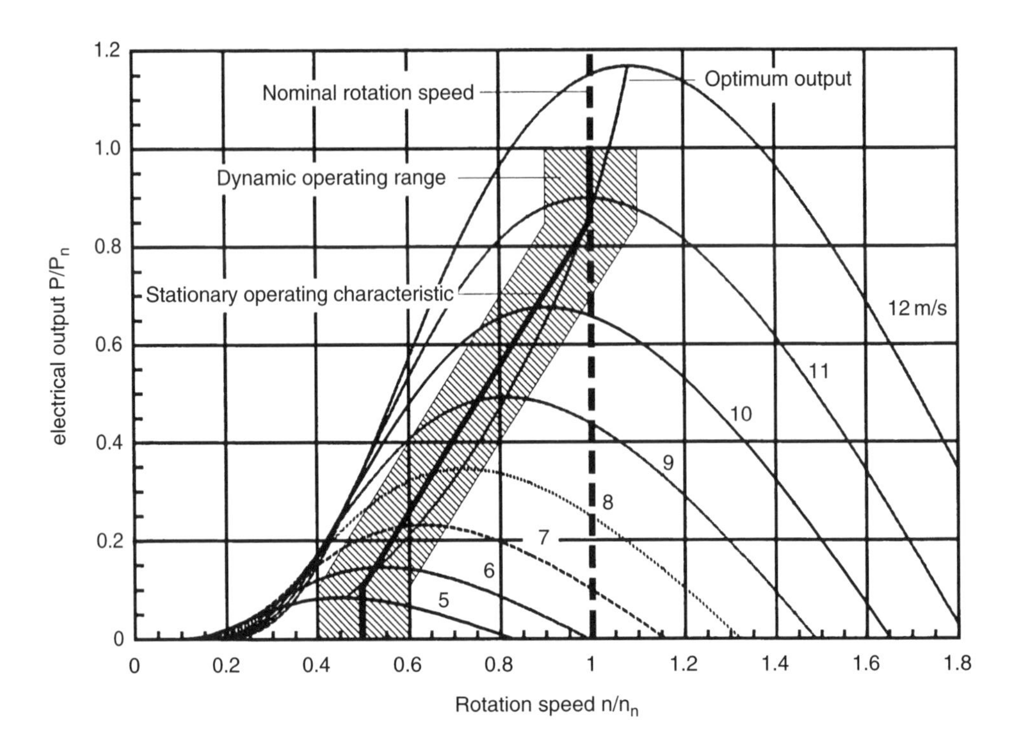

Figure 3.35 Power–speed diagram of a variable-speed wind turbine with steady-state and dynamic working ranges near to optimal output

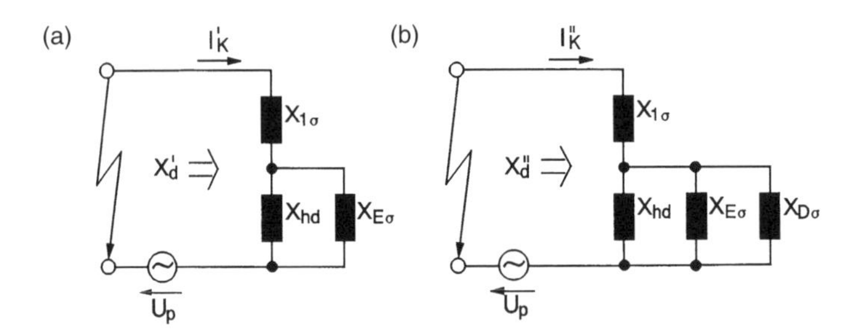

Figure 3.36 Equivalent circuit diagram for a synchronous machine in short-circuit (a) without damper winding and (b) with damper winding

Gearing represents a considerable cost factor in the drive train of a conventional wind turbine. It is also subject to dynamic load shocks, e.g. due to gusts. In addition, gears bring about losses, and require constant monitoring and maintenance during operation. Therefore, since the early days of wind energy technology [1.3], the option of installing plants with directly driven generators with no gears has been considered. Such configurations are in a period of rapid market growth and consolidation. More and more manufacturers are currently

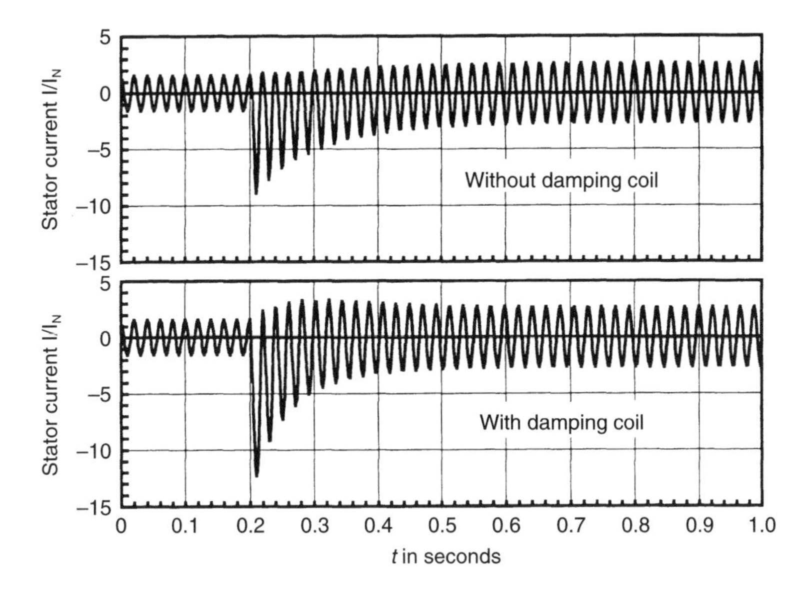

Figure 3.37 Stator current during the short-circuit of a synchronous machine with and without damper winding (simulation results)

going over to developing gearless conversion systems. The electrically excited generators used in these systems have, up until now, not been fitted with an auxiliary generator. The field winding is fed via slip rings.

The new type of configuration must incorporate

- modular power components, in particular
  - a generator and
  - power converter;
- stepped output, with
  - 10 to 100 kW for hybrid plants and
  - 0.5 to 5 MW class for large plants;
- good transportability and
- simple installation.

Furthermore, the entire system must facilitate both

- excellent compatibility for the grid supply and
- the option of connection in weak grids and hybrid plants.

These requirements can usually only be achieved by conversion systems, which

- can be connected to form a grid,
- have voltage and/or frequency controlling characteristics, as well as
- energy optimization and supply-compatible tuning for the control and operation of the plant.

Different system configurations are available that fulfil these requirements. Conventional generator types, which require geared transmissions, will not be considered further in the following discussion, which will concentrate on the new type of configuration.

Gearless synchronous generators can basically be divided into the following lines of development. In particular,

- separately-excited salient pole machines or
- permanent-magnet designs, with
- axial,
- radial or
- transverse air gap

come into consideration. In addition,

- non-salient pole,
- unipolar and
- claw pole designs

are possible. Some of these systems are described in further detail below.

Firstly, it should be mentioned at this point that for wind turbines in the 100 kW to MW class at a normal speed range between 60 and 15 rpm, 100 to 400 poles would be necessary to achieve a generator frequency of 50 Hz. To incorporate this large number of poles and the associated stator windings in the generator around the rotor and stator would require a relatively large machine diameter in the conventional layout, resulting in a correspondingly high component mass for construction elements.

By decoupling the generator rotation speed and grid frequency, lower frequencies can be selected with the aid of frequency converters, thus reducing the required number of poles. Alternatively, a higher number of poles can be achieved using new layout types, e.g. in permanent-magnet machines. Lower generator frequencies, however, generally increase the requirement for smoothing in the direct current intermediate circuit. This can be significantly reduced by a trapezoidal generator voltage pattern (compare Figure 3.38(a) and (b)), which is achieved by the design and operation of the machine at high saturation. This, however, results in higher losses.

Stringent demands are imposed on wind power plants and their components with regard to reliability and service life. Plant availability above 98 % is currently the norm. However, these values can only be achieved with repeatable manufacturing and installation procedures and precise execution. Components and subsystems are therefore manufactured and fitted, as far as possible, at the factory. After testing, the units are transported to the site and installed. Then, the entire system is commissioned. Thus, in the best case, the generator, as

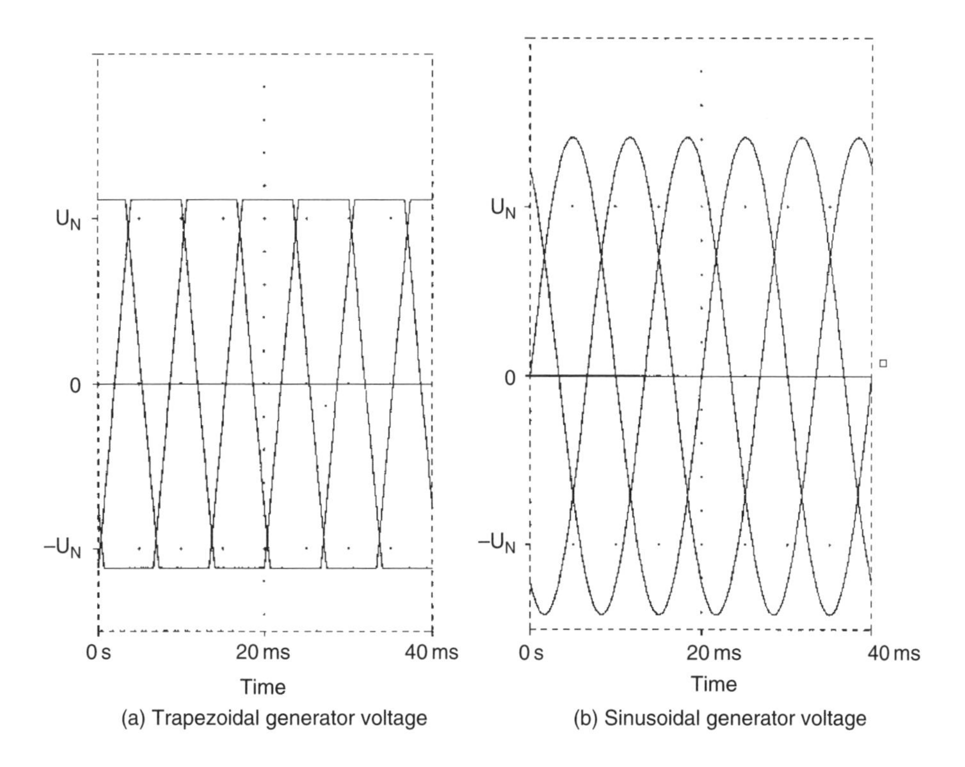

Figure 3.38 Voltage pattern for a six-pulse bridge connection

the main component of a gearless wind turbine, is fully assembled and tested at the place of manufacture and, where possible, transported to the site as a single unit incorporating the nacelle and hub and integrated into the tower head. The maximum dimensions for the transport route must be adhered to.

To facilitate road transport, generators with a housing diameter of around 4 m are desirable. If a good route with suitably dimensioned bridges, underpasses, tunnels, etc., is available significantly larger units can also be transported by road and rail. A generator/nacelle diameter of 5 to 6 metres should, however, not be exceeded.

Based on this preliminary discussion, the numerous design options and limiting factors become clear. The following description is limited to four different systems.

3.6.2.1 Separately excited salient pole machine with radial air gap field

The key components of the Enercon E30, E40, E66 and E112 gearless wind power plants are the synchronous generators driven direct from the turbine. These are separately excited salient pole machines with a radial air gap field, of conventional design and using established materials technology. Two rotors with excitation windings on T-shaped pole cores are shown at the centre of Figure 3.39. Almost 80 poles are fastened on to a yoke ring. The stators, as shown in Figure 3.39, partly wound on the left and almost fully wound on the right, are made up of identical armature segments, which are joined together, overlapping in the

Figure 3.39 Electrically excited synchronous generators for Enercon E40 gearless wind turbines; stator (standing) and salient pole rotor (lying)

direction of travel. The stator windings are located in the slots of the stator laminated core, as shown in Figure 3.39 in the right-hand foreground.

3.6.2.2 Permanent-magnet synchronous machine with radial air gap field

In the alternative form of vertical axis turbines, the so-called H-Darrieus system, gearless systems were initially introduced in the 30 and 300 kW classes. The ring generators conceived for these plants are designed with a radial air gap field and permanent-magnet excitation. They are connected to the grid via a frequency converter, thus permitting variable-speed operation. The plant was originally designed for the supply of electrical power for arctic regions, and is thus suitable for operation at very low temperatures.

The 300 kW prototype shown in Figure 1.31(a) is fitted with a ring generator, which was installed at ground level and had a diameter of 12 m (Figure 3.40(a)). Subsequent plant designs (Figure 1.31(b)), which were manufactured in a short production run, had the generator at the top of a three-legged mast (Figure 3.40(b)). As already mentioned in Section 1.2, however, this type of system has not yet managed to break through into the wind generator market.

Radial air gap generators with permanent-magnet excitation are particularly common in low-power horizontal-axis plants. With this design, it is possible, as in the layout shown in Figure 3.40, to fit the rotor with excitation poles of alternate polarity and the stator with salient poles. This makes it simple to dismantle the stator, e.g. for transport. Furthermore, the grain-orientated structures in the electrical sheet steel can make good use of the flow of magnetic flux to reduce magnetic losses. The insulation of differing phases can also be

(a)

(b)

Figure 3.40 H-Darrieus HM 300 plant with the generator at (a) the mast foot and (b) the mast head

significantly reduced, since no intersections protrude from the end windings. This results in a large generator diameter and high material cost. Compact plant designs are thus not possible.

Smaller pole divisions and diameter, and the associated higher generator voltage frequencies and better utilization of the active materials, can be achieved using split windings in the slots. However, this necessitates more insulation in the stator. A division of the stator can only be achieved by making sacrifices in utilization and increased use of coil segments. For this generator design, the stator diameter should be as small as possible. For machines in the megawatt class, external diameters between 4 and 6 m can be achieved. A range of

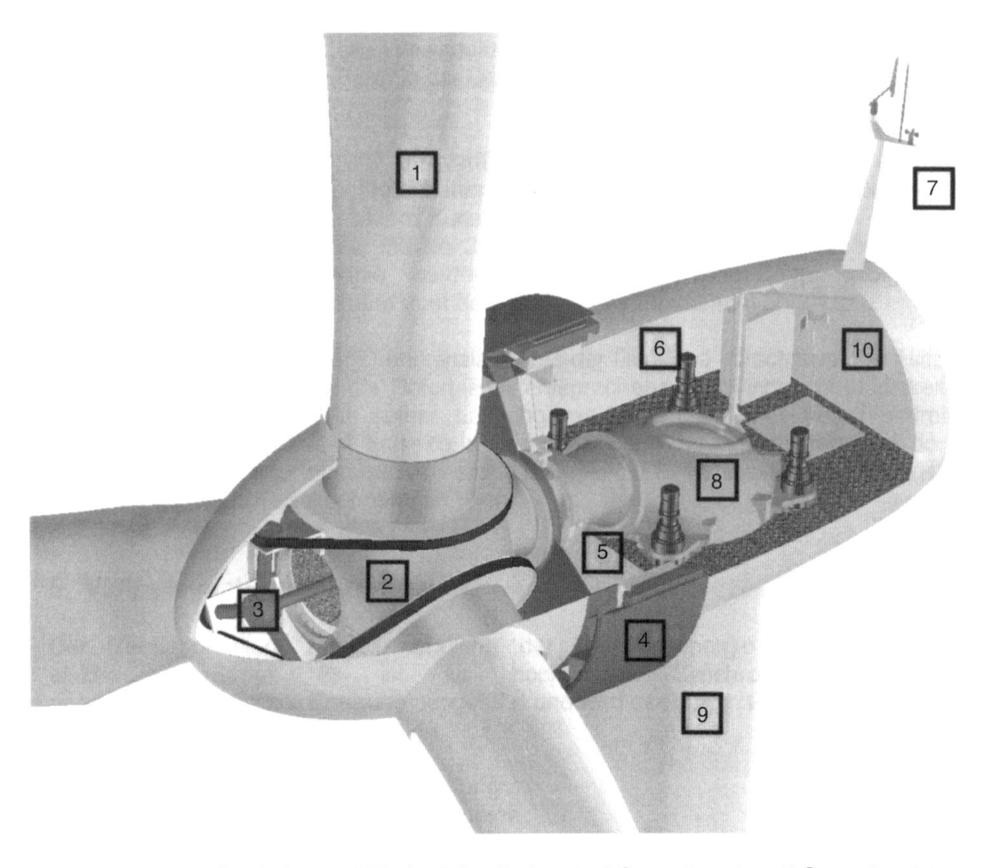

1 Rotor blade 2 Cast hub 3 Blade pitch adjustment 4 Generator rotor 5 Generator stator
6 Azimuth drive 7 Anemometer 8 Machine mounting 9 Tower 10 Auxiliary crane

Figure 3.41 Tower head of the VENSYS 62 gearless turbine (1200 kW) with a permanent-magnet external rotor generator [3.30]. Reproduced by permission of Vensys

single-ring generator modules, e.g. the 500 kW size, could also facilitate favourable layouts here, thus keeping the costs for development, manufacture, storage and spare parts service low.

A further possibility for reducing the generator diameter in comparison with standard machines was used by Prof. Dr F. Klinger (HTW Saarbrücken) in the development of the permanent-magnet generators for the GENESYS 600 and VENSYS 62 wind turbines (1200 kW) (Figure 3.41). In addition to the axial iron or laminated core length, the decisive quantity for the dimensioning of the mechanical–electrical energy converter is the air gap diameter of the generator, which when increased leads to a quadratic increase in power.

Normally, synchronous machines are designed with the rotor running inside a stator. The three-phase windings are, almost without exception, located in the slots of the stator laminated core. Thus, in addition to the air gap diameter that dominates in multipole machines, the slot height, the magnetic return (the so-called armature return) and the mechanical support structure that encompasses all electromagnetically active components determine the external diameter of the generator.

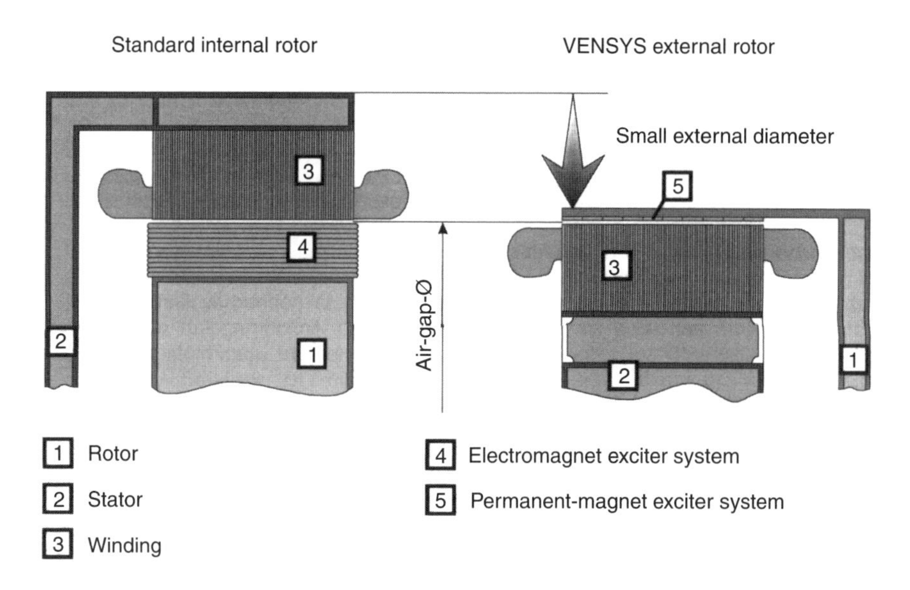

Figure 3.42 Generator diameter comparison between external and internal rotor designs (page 181)

Given the same air gap diameter or the same power, significantly smaller generator diameters can be designed by the use of permanent magnets that are just a few millimetres thick, using the external rotor variants as shown in Figure 3.42. Lower machine masses can thus be achieved, along with the associated advantages during manufacture, transport and installation.

3.6.2.3 Transverse flux machines

Motor and generator designs with magnetic circuits laid out across the direction of motion can mainly be attributed to the work of Prof. Dr Ing. Weh at the Institute for Electrical Machines, Drives and Rails at the Technical University of Braunschweig [3.31]. Figure 3.43 illustrates the difference in flux direction in a laminated core and current flow in one phase of a stator between conventional longitudinal-flux (a) and transverse-flux machines (b).

The principle layout of a two-sided transverse flux machine with permanent-magnet excitation as shown in Figure 3.44 illustrates the significant manufacturing cost of magnetic rotors and stator parts. The cross-section shown in Figure 3.45 shows the complete layout of the machine. Designs based on the reluctance principle, in which very cheap magnetically soft materials are used instead of expensive permanent magnets, are also possible. The advantage of this cheaper variant is traded off against the disadvantage of poorer utilization.

3.6.2.4 Permanent-magnet synchronous machine with axial air gap field

A further option is to design the generator as a permanent-magnet, multiple-pole synchronous machine with an axial air gap field [3.33, 3.34]. Permanent magnets of alternating polarity are fastened to the rotor disc to create an excitation field, which can, as shown in Figure 3.46,

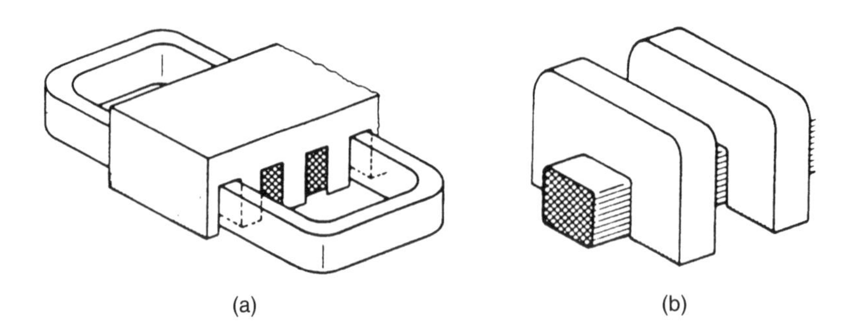

Figure 3.43 Stator laminated core and winding arrangement of (a) a longitudinal-flux machine and (b) a transverse-flux machine [3.32]

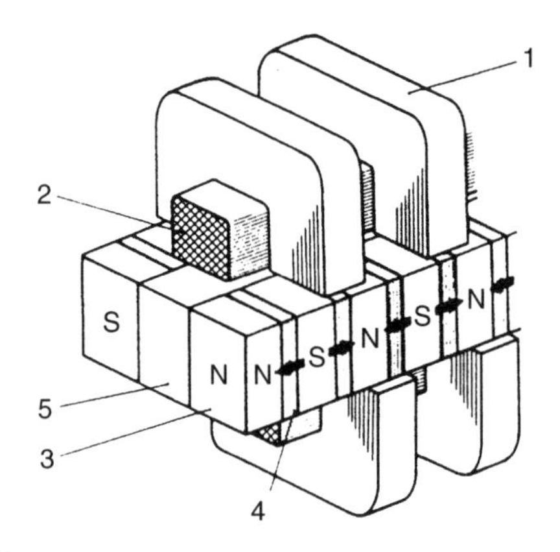

Figure 3.44 Principle layout of a double-sided transverse-flux machine [3.32]

comprise many single magnets or, for large quantities, one or a few individual parts. A large number of poles can be attained using this layout. This means that this type of machine can be designed for very low speeds (e.g. in the 100 kW range at 30 rpm with 100 pole pairs and a grid frequency of 50 Hz).

The ring-shaped stator (Figure 3.47) carries the armature coils. It can be made up of normal alternating current windings with projecting end windings that cross one another or, as shown in Figure 3.48, of several identical single-phase sectors. The winding can be constructed of parallel conduction layers, which may be made of copper or aluminium plates, or suitable preformed coils.

Axial-flux machines have a constant air gap. The electrically and magnetically active component of this type of generator can therefore be made of a few different components,

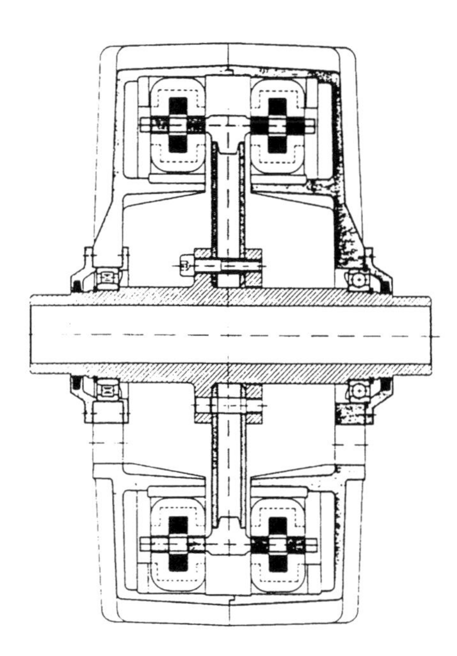

Figure 3.45 Cross-section through the stator and rotor of a transverse-flux machine [3.32]

which can be manufactured and assembled in large quantities. The layered winding, as well as having a high slot space factor, gives compact end windings (Figure 3.47), since only small or zero bending radii have to be observed during manufacture, thus making short connecting routes between the grooves possible, despite the high conductor cross-section. Since, in principle, it is only the section of the conductor within the magnetic field that actively contributes to energy conversion, small end windings give high power density.

Due to its construction in sectors and the use of stacked windings, the cost for the insulation of adjoining conduction layers can be kept low. An insulation layer thickness of a few μm is generally adequate. The slot area can thus be fully exploited for the conduction of electricity. Due to the short end windings, the proportion of the conductor that is inside the magnetic field is very high. These constructional characteristics can lead to a significant reduction in weight compared with other multipole generators. When machine-side pulse-controlled a.c. converters are used, however, the generator phase windings must be protected against partial overvoltages by voltage rise filters.

Due to the space-optimized layout of the windings in multipole generators, higher overall efficiencies can be achieved than is the case for conventional designs. Excitation losses are also eliminated by the use of permanent magnets. The generator therefore has a higher efficiency than electrically excited induction machines, particularly in the partial-load region. Furthermore, the need for maintenance is reduced in this type of generator, since no slip rings and brushes are required to supply the excitation current to the rotor.

The specific labour and material costs for highly permeable permanent magnets (Neodym) lies at around 100 to 150 euros per kilogram, i.e. around ten times higher than aluminium or copper conductors, at 5 to 8 euros per kilogram, or low-loss electric sheet steel and welded constructions, at 5 to 10 euros per kilogram [3.35]. Thus an economical generator design

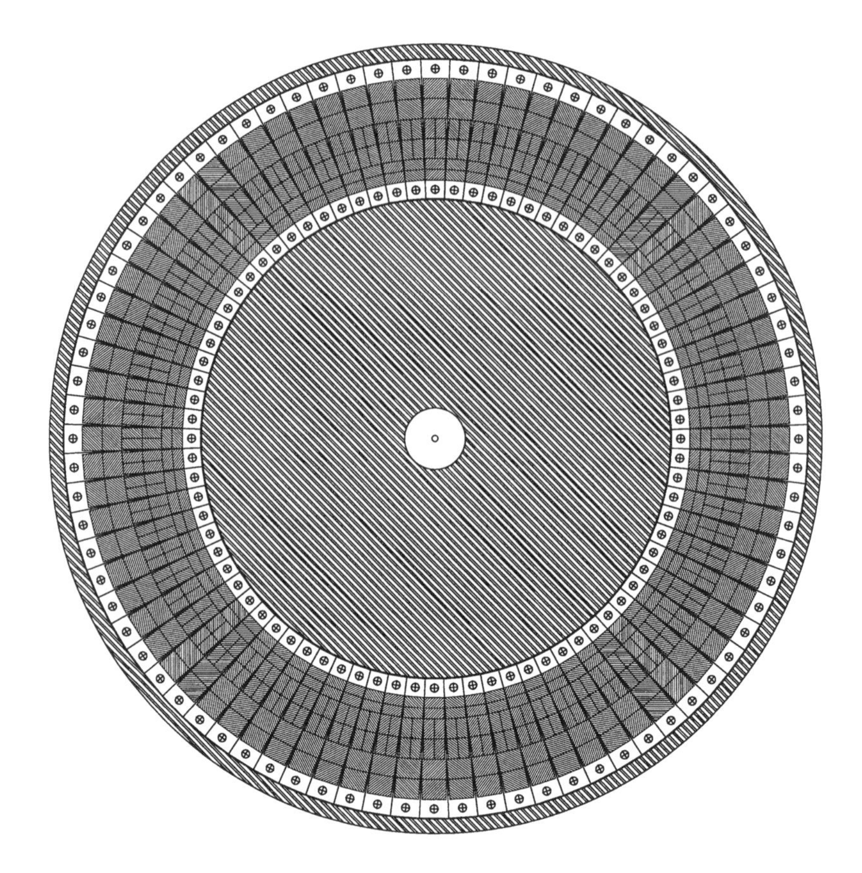

Figure 3.46 Rotor disc of a permanent-magnet multipole machine with an axial air gap field

requires the use of magnetic materials to be minimized. This can be achieved, for example, if the generator is designed to be underexcited. However, step-up converters or the supply of exciter reactive current via the machine-side frequency converter are necessary to maintain the voltage in the intermediate circuit (see Section 4.1).

Permanent magnet generators can only be economically designed with a relatively small air gap due to the high cost of magnetic materials. System configurations with a one-sided air gap thus exhibit enormous axial forces (e.g. approximately 500 kN for a 100 kW generator). These require correspondingly rigid and expensive constructions.

Double air gap arrangements allow the forces to be largely compensated and thereby these disadvantages avoided and the design simplified. This variant (Figure 3.49(a) and (b)) can be designed with both axial and radial air gaps. In the magnetic circuit there are two air gaps connected in series. As a result, the normal components of magnetic force acting upon the rotor are enormously reduced. If approximately 25 % of the normal air gap is maintained on both sides the axial force is reduced to a maximum of around 10 % of the value for a one-sided air gap. Furthermore, due to the modular structure of system groups (Figure 3.49(c)) relatively wide power ranges can be covered by a few construction sizes.

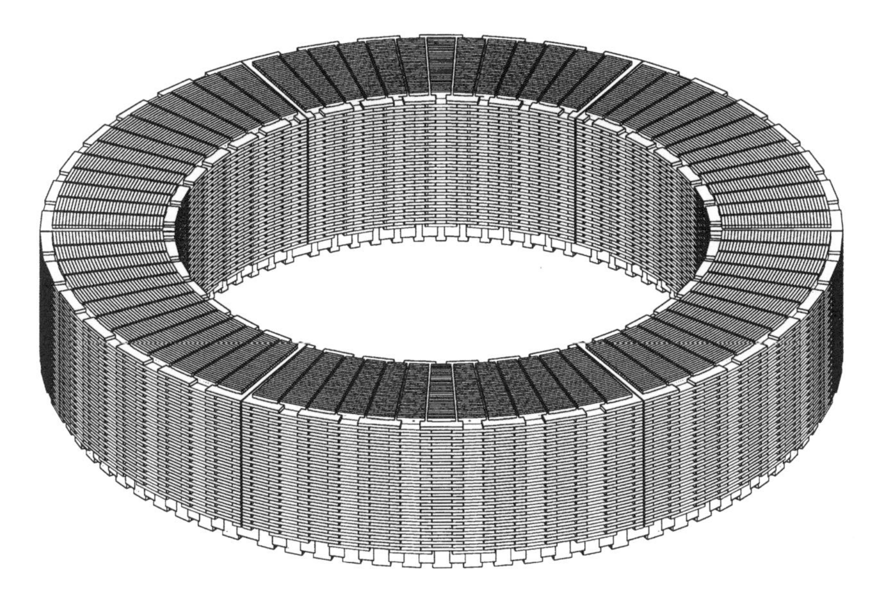

Figure 3.47 Stator with a laminated core and winding from a multipole machine with an axial air gap field

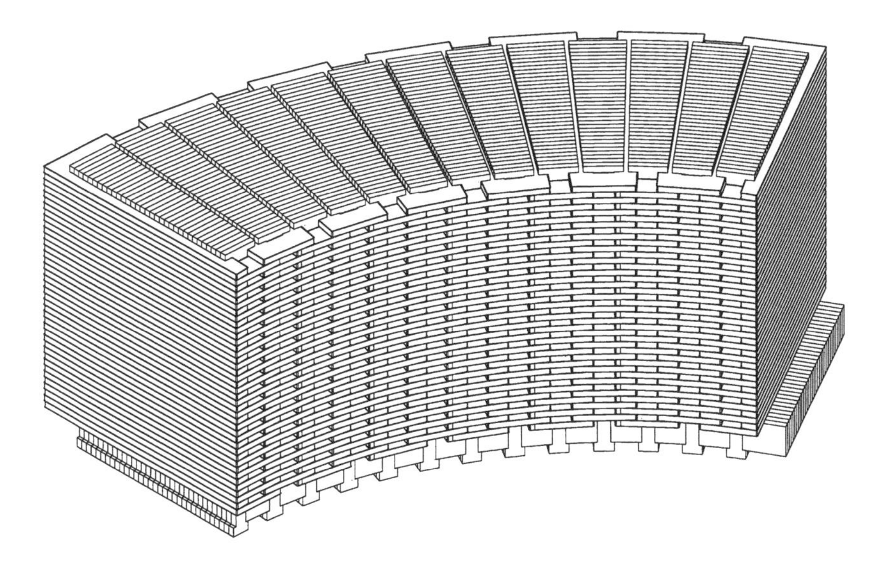

Figure 3.48 Single-phase stator segment (laminated core and stack of coils) of a multipole machine with an axial air gap field

To sum up:

- The 'Enercon concept' with a radial air gap flux and electrical excitation by salient poles has proved itself very well, and currently dominates the market. However, further improvements are still possible using a layout with a high number of poles, etc., which can give a more compact form, for example.

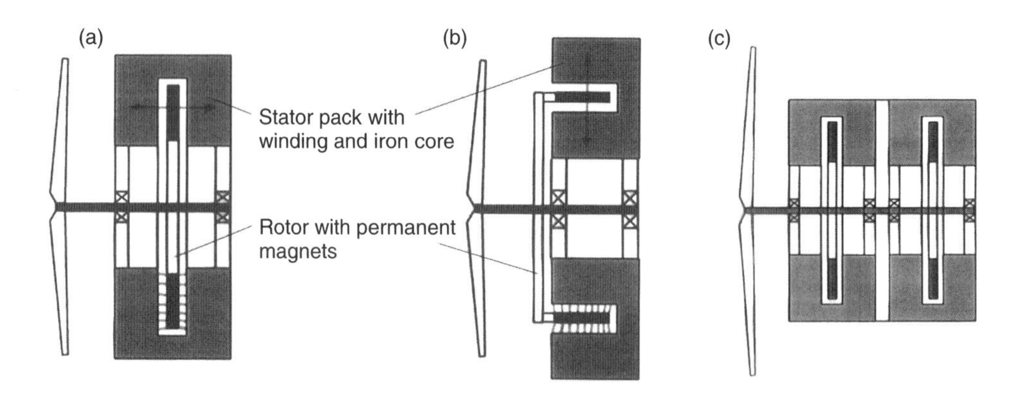

Figure 3.49 Schematic representation of double air gap designs for machines with (a) axial and (b) radial air gaps and (c) modular structure of system groups

- Corresponding designs with permanent-magnet excitation necessitate high expenditure on the frequency converter and protection concept. Double salient pole machines facilitate simple division of the stator into several sectors. This results in larger models.
- Transverse-flux machines are expensive to construct due to their layout. For the permanent-magnet design, the magnet costs are high due to the double air gap. Cheap reluctance machines, however, have a high mass.
- Highly permeable permanent magnets for the excitation of the machines still lead to significantly higher generator costs. At an additional price in relation to electrically excited synchronous machines of 50 euros per kW, for example, and at an exciter power component of 2 %, amortization times are around five to ten years.
- Permanent-magnet multipole generators with the stator divided into sectors, either with an axial or a radial air gap, have some advantages, particularly with regard to machine utilization and expected efficiency. In the case of the axial air gap field, a double stator layout is advantageous, to compensate for axial forces in the rotor. Modular construction lends itself particularly to this type of generator.
- Radial-flux machines with external rotors permit smaller generator external diameters than standard internal rotor variants.
- Encapsulating the electrically active generator components significantly reduces the risks of corrosion and insulation damage. However, correspondingly higher requirements are imposed upon the cooling systems. High-quality vacuum impregnation of the three-phase windings additionally increases operating reliability.

To round off this chapter, the next section briefly describes performance-related data of conventionally designed machines.

3.7 Machine Data

During the planning, design and construction of plants supplying electrical drive and energy, knowledge of experimental values for the electrical machines on the market is of fundamental significance with regard to design, proportioning and economics. For new developments, in particular, manufacturers' and suppliers' ratings are necessary for the design of subsystems. Hence, data overviews are very helpful for the preselection of components and systems, as well as for the creation of preliminary sketches. Important ratings are thus given below as a function of nominal output.

3.7.1 Mass and cost relationships

The mass and cost values in relation to nominal output are shown for asynchronous (Figures 3.50 and 3.51) and synchronous (Figures 3.52 and 3.53) machines for different numbers of poles. The diagrams show that both for synchronous and asynchronous machines the specific mass and cost per kW the machine output falls with increasing size. This trend is explained by the laws of the model. These state that more favourable mass and cost relationships can always be expected for higher-output machines, given the same or similar conditions. If, however, large units give rise to very high costs due to the costs of manufacture, transport, etc., the above relationships could lead to completely different trends. Further differences in mass occur, for example, in asynchronous machines, depending upon whether these are designed with a steel or an aluminium housing.

A comparison of the diagrams in Figures 3.50 and 3.52 shows that synchronous and asynchronous machines exhibit similar mass relationships. On the other hand, the cost

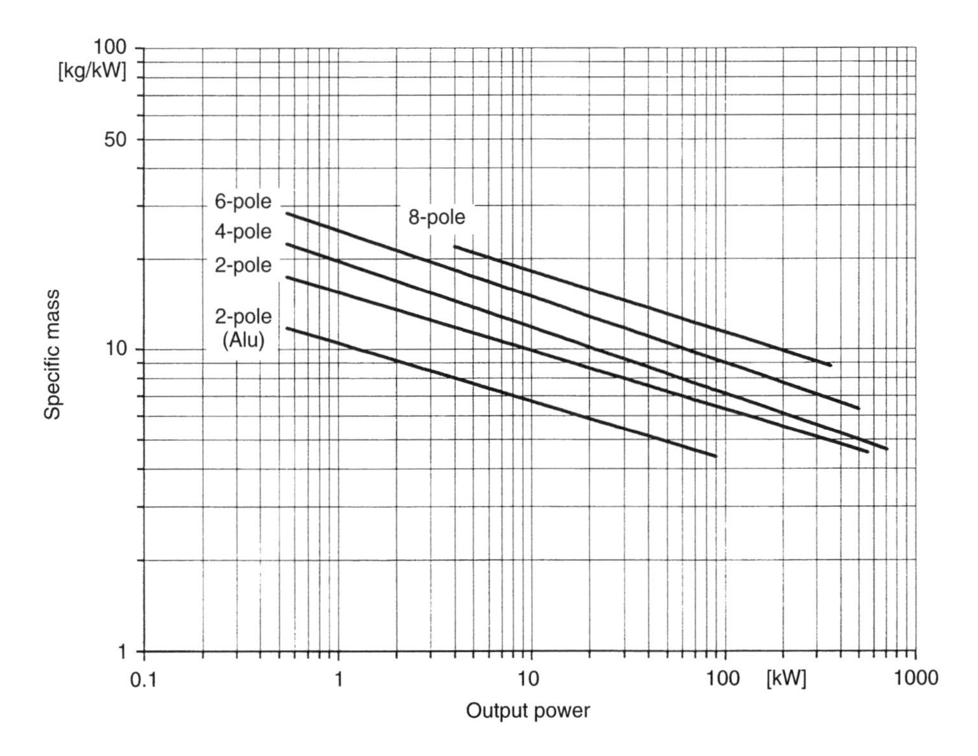

Figure 3.50 Specific mass of asynchronous machines (per kW) as a function of nominal output, with the number of poles as a parameter

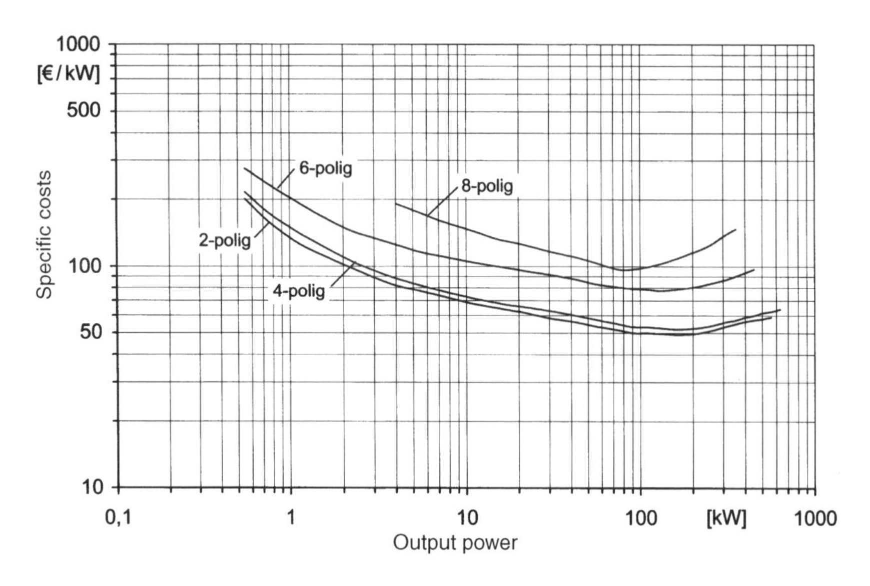

Figure 3.51 Specific costs of asynchronous machines (per kW) as a function of nominal output, with the number of poles as a parameter

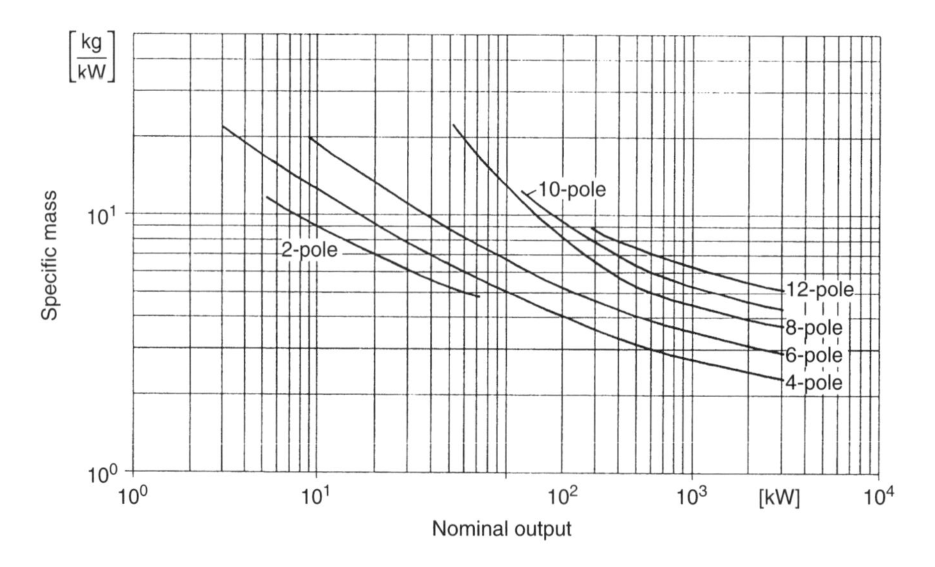

Figure 3.52 Specific mass of synchronous machines (per kW) as a function of nominal output, with the number of poles as a parameter

relationships shown in Figure 3.51 and 3.53 show considerable differences in the lower output range (e.g. 10 kW). The more favourable values for asynchronous machines can mainly be attributed to the simpler design and mass production. In the higher-output range, both designs show roughly the same cost relationships. It is worth noting at this point that

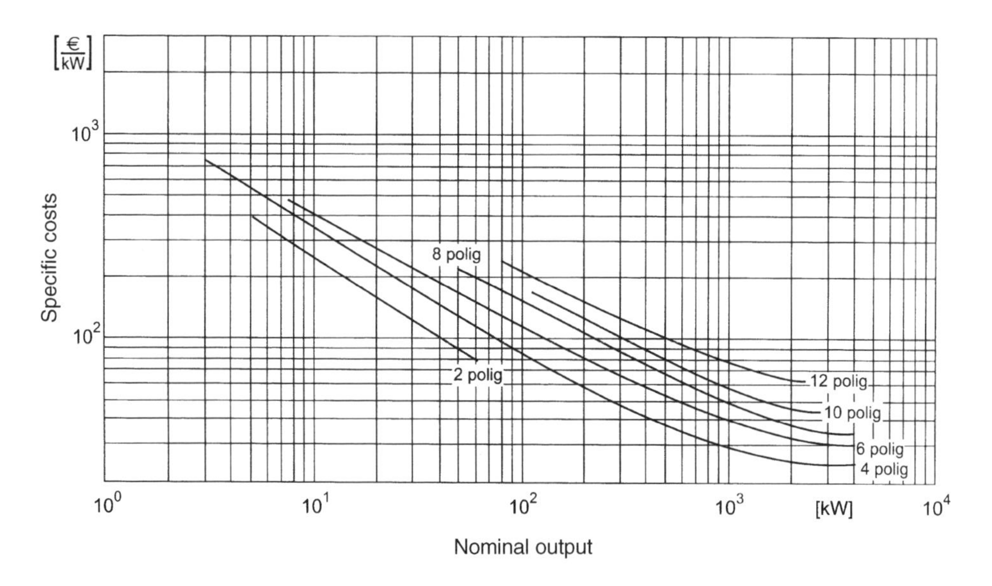

Figure 3.53 Specific costs of synchronous machines (per kW) as a function of nominal output, with the number of poles as a parameter

the actual purchase price can deviate significantly from the values given, depending upon the quantity purchased and discount level. Furthermore, machines with a higher number of poles must obviously have a correspondingly higher mass due to their high torque, which thus leads to a comparable increase in cost.

3.7.2 Characteristic values of asynchronous machines

When asynchronous machines are used, starting currents (Figure 3.54) must be taken into account in the dimensioning of switchgear and protective gear, as well as nominal currents (see Figure 2.55). If no current limiting measures are taken, these must also be taken into consideration in the design of the drive train.

No-load current (see Section 3.6.1) is an important parameter of asynchronous machines. It is determined primarily by the design of the generator, and its uses include the dimensioning of compensation units, if these are designed for no load running of the machine, as shown in Figure 3.12(a). Figure 3.55 gives approximate reference values for no-load current. These can differ significantly from the values shown, depending upon the design and application.

Further reference values are given in Figures 3.56 and 3.57, which show efficiency and the power factor. Both diagrams show that asynchronous machines with a high number of poles generally lie below designs with a low number of poles, both for efficiency and the power factor, because of their larger air gap. Owing to the relatively large end windings of bipolar machines, their efficiency is less favourable than, for example, that of four-pole machines, but generators with a higher number of poles do tend to produce somewhat lower efficiencies.

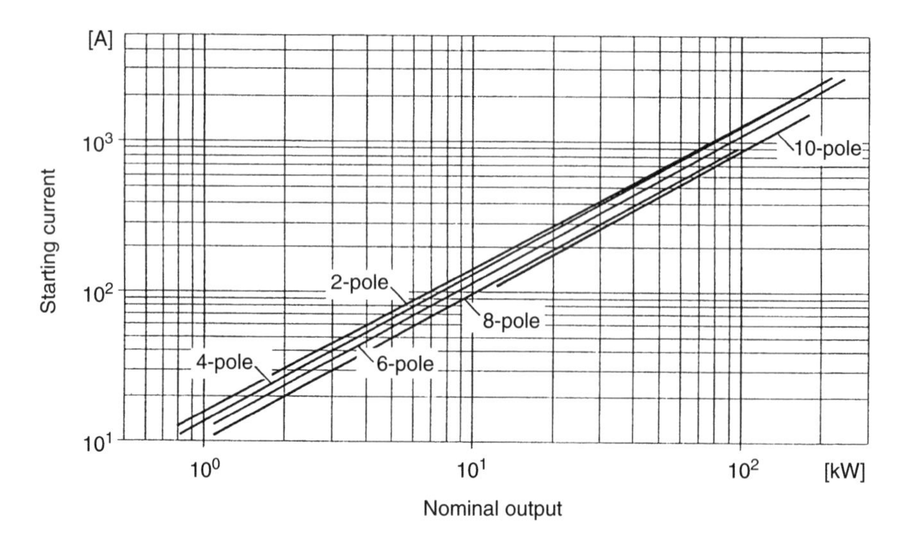

Figure 3.54 Starting current of asynchronous machines (at 400 V) as a function of the power output, with the number of poles as a parameter

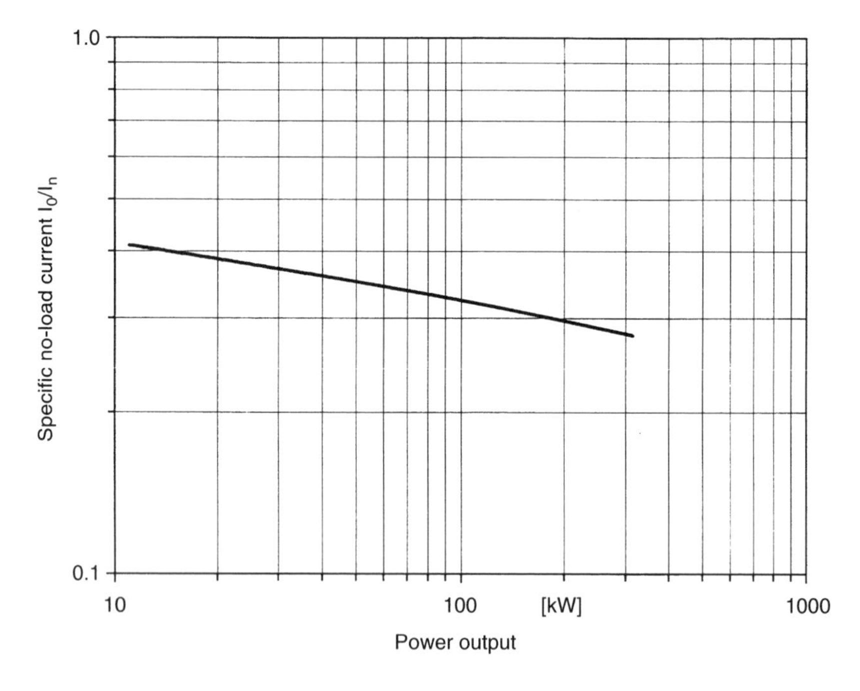

Figure 3.55 Specific no-load current of asynchronous machines as a function of nominal output at 400 V operating voltage

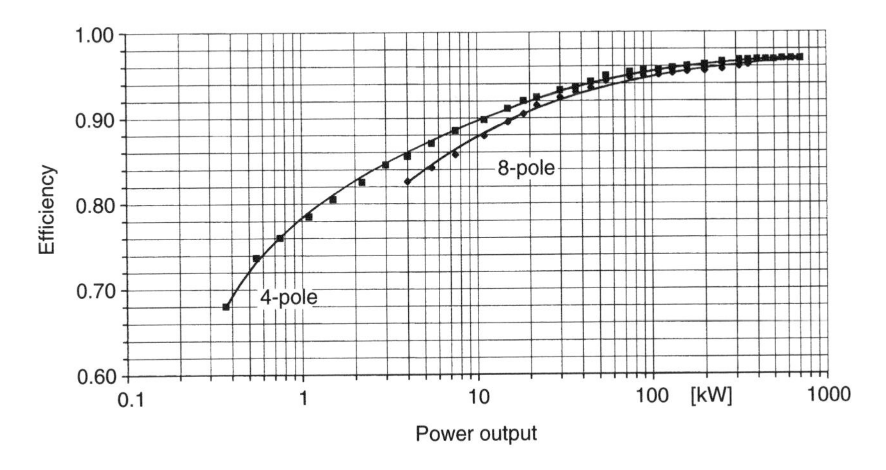

Figure 3.56 Efficiency of asynchronous machines as a function of nominal output, with the number of poles as a parameter

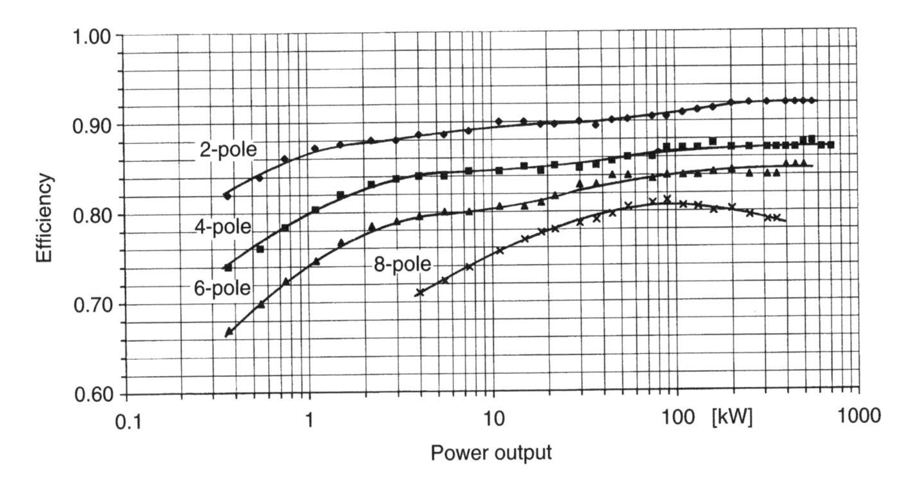

Figure 3.57 Power factor of asynchronous machines as a function of nominal output, with the number of poles as a parameter

As mentioned in Section 3.6, the size of a machine is primarily determined by the inside diameter of the stator. Figure 3.58 gives reference values for the diameter. According to reference [3.36], this is proportional to the fourth power of the machine's output.

3.7.3 Characteristic values of synchronous machines

Reference values for the efficiency of four-pole electrically excited synchronous machines are shown in Figure 3.59. The values are plotted as a function of the utilization P/P_N for nominal outputs between around 100 kW and 3 MW.

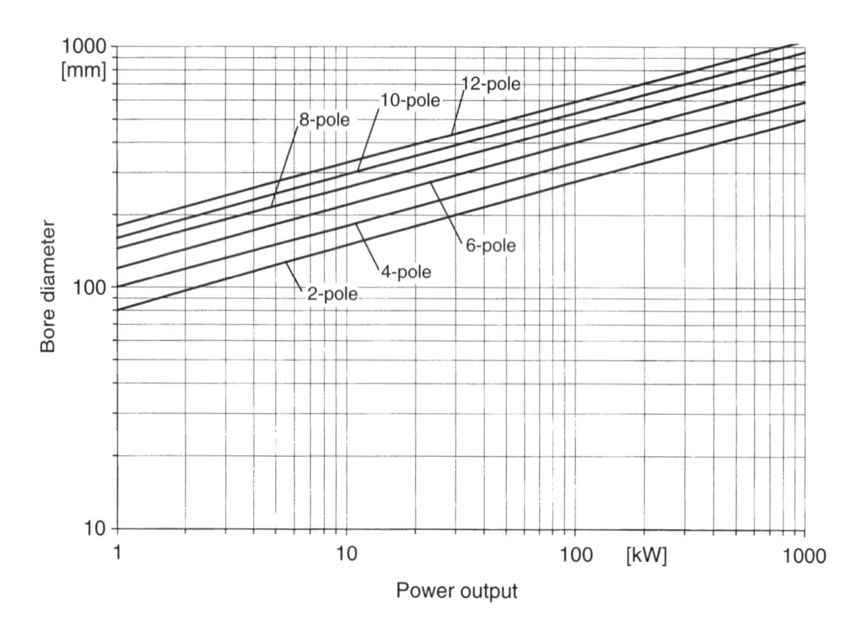

Figure 3.58 Stator bore diameter as a function of machine output, with the number of poles as a parameter

Direct-axis synchronous reactance and its transient and subtransient values are determining variables for the electrical and mechanical components of electrical and mechanical processes in synchronous machines. They are therefore shown in Figures 3.60 to 3.62, based on the nominal impedance $Z_N = U_N I_N$ in relation to nominal output, with the number of poles as a parameter. Reference values can thus be obtained from the diagram and used for initial designs. However, the above-mentioned characteristic values can deviate from the guide values given here, e.g. due to construction size graduations and design variants. The machine values can, however, be obtained from the equations

$$X_d = x_d Z_N, \tag{3.27}$$

$$X'_d = x'_d Z_N \tag{3.28}$$

and

$$X''_d = x''_d Z_N. \tag{3.29}$$

In steady-state operation, the behaviour of currents, voltages and moments of synchronous machines, in particular, is mainly determined by the so-called direct-axis synchronous reactance. As described in Section 3.3, this is made up of the magnetizing reactance in the direction of the rotor field and the leakage reactance of the stator winding. Figure 3.60 shows the increase in specific direct-axis reactance as the nominal output of the machine increases. It also illustrates the fact that generators with a high number of poles show lower direct-axis reactance.

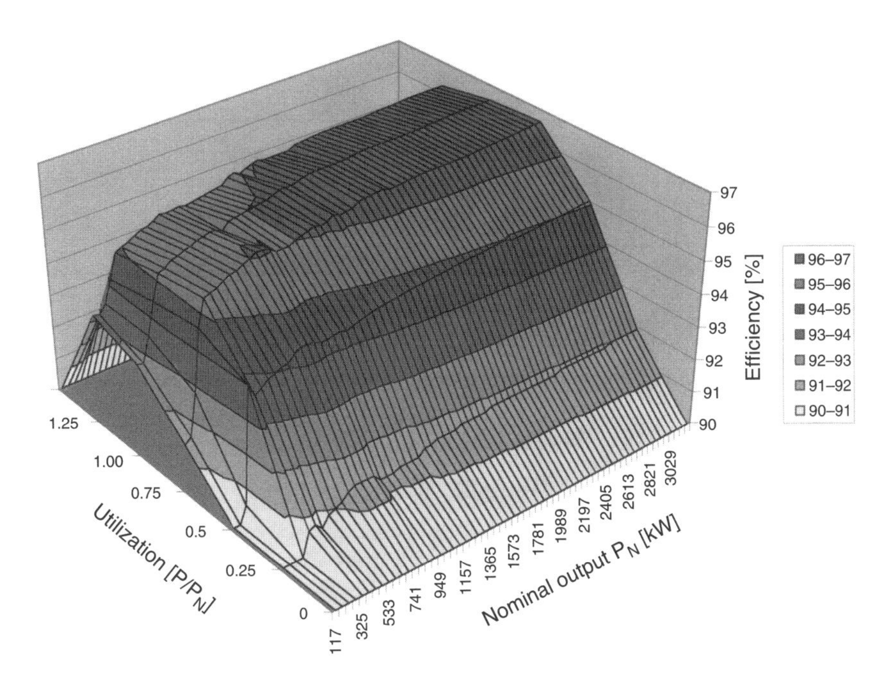

Figure 3.59 Efficiency of electrically excited synchronous machines (four-pole) as a function of output power and utilization (P/P_{N})

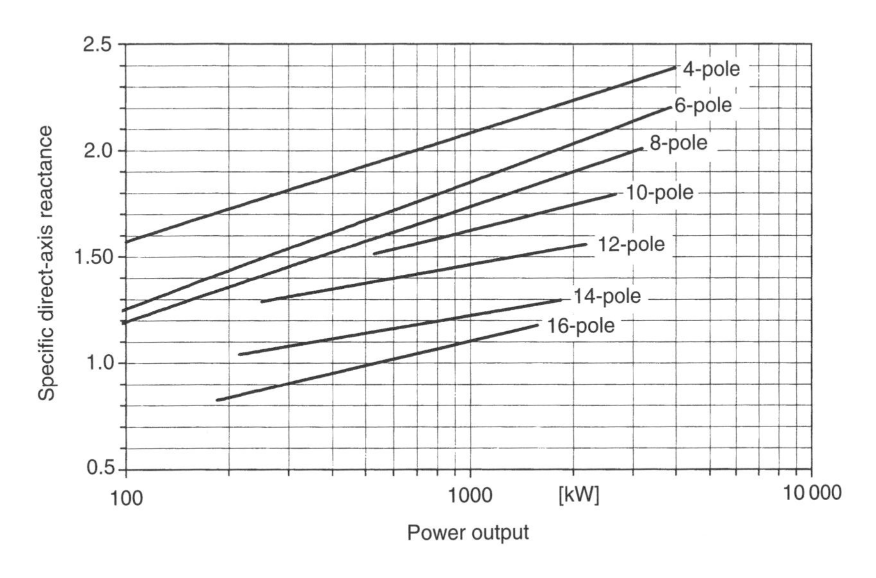

Figure 3.60 Specific direct-axis synchronous reactance of synchronous machines as a function of nominal output, with the number of poles as a parameter

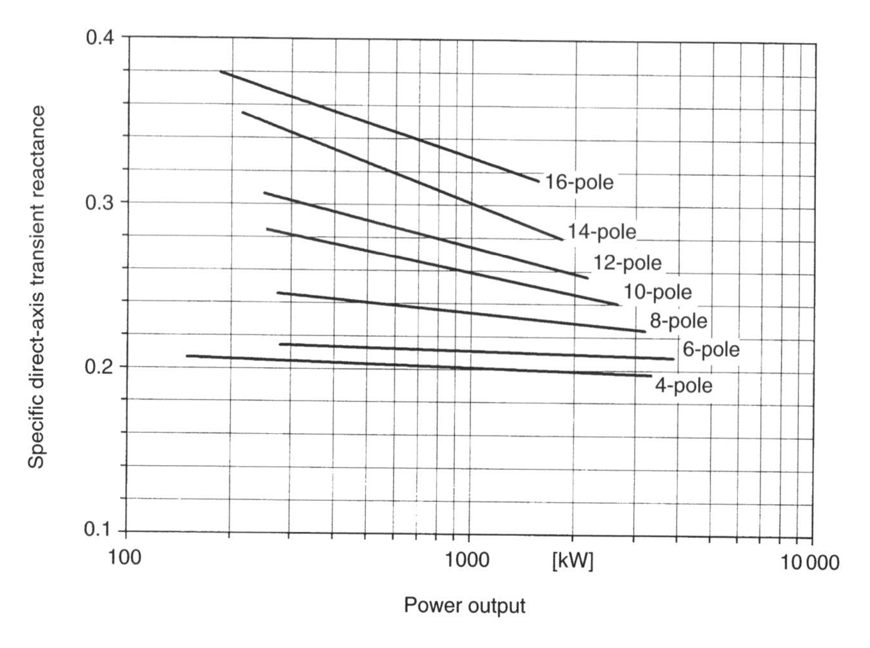

Figure 3.61 Specific direct-axis transient reactance of synchronous machines as a function of nominal output, with the number of poles as a parameter

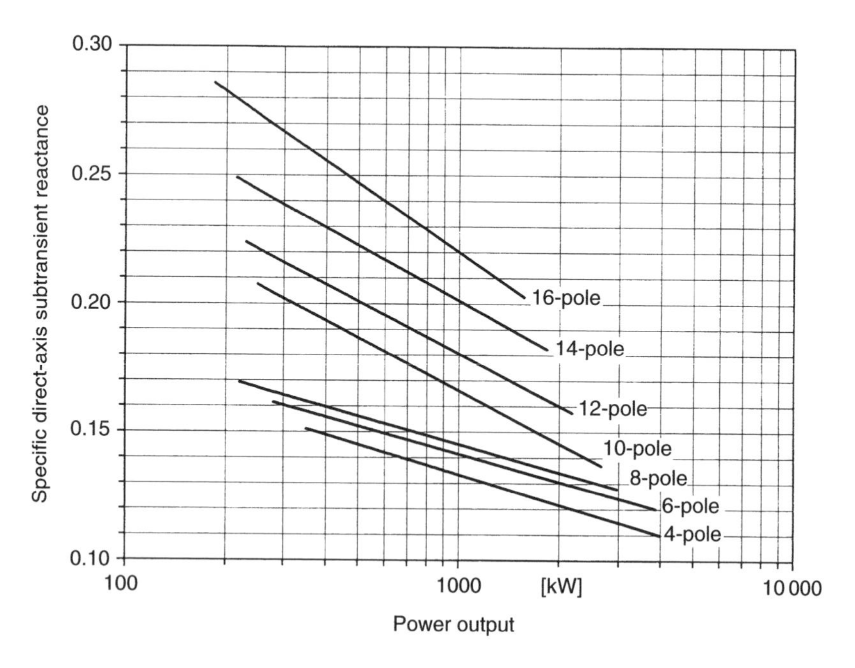

Figure 3.62 Specific direct-axis subtransient reactance of synchronous machines as a function of nominal output, with the number of poles as a parameter

Unlike specific direct-axis reactance, transient and subtransient values decrease in magnitude with increasing nominal output. Higher numbers of poles, on the other hand, give rise to greater values. As shown in the equivalent circuit diagrams in Figures 3.7(b) and 3.36, the synchronous direct-axis reactance is clearly greater than the transient and particularly the subtransient values.

The power conditioning of electrical energy and the connection of generators to the grid will be examined in more detail in the following chapter.

4

The Transfer of Electrical Energy to the Supply Grid

With regard to the transfer of energy to electrical supply installations, we must differentiate between

- systems with limited supply options that either operate in isolation or supply weak grids and
- unlimited capacity connection with the rigid grid.

Wind energy converters should give reliable operation in both areas of application.

Due to its very high capacity (in comparison with the nominal values of the consumers connected to it), the so-called rigid combined grid can be regarded both as an infinitely rich source of active and reactive current and, for the low-level energy supply devices that wind turbines usually represent, as a sink of unlimited capacity with constant voltage and frequency.

Unlike thermal power plants, wind turbines are usually installed at remote sites with limited supply options. Therefore a weak grid connection is often made using dead-end feeders, which are sometimes long. In large wind energy converters and wind farms, supply power can reach the same order of magnitude as grid transfer power, or can even approach its level, which means that mutual influences must be taken into account. Table 4.1 lists, in simplified form, the requirements and equipment needed for the connection of wind energy converters (WECs) to the grid [4.1].

On the one hand the effects of wind energy converters on the grid are determined, on the one hand, by power conditioning and the resulting grid connection. On the other hand, safety

Grid Integration of Wind Energy Conversion Systems, Second Edition S. Heier

Table 4.1 Requirements and equipment for the grid connection of wind turbines

Grid connection	Disconnection point in accordance with DIN/VDE 0105 available to the electricity supply company at all times
Switchgear	Section switch with at least power circuit breaking capacity (in pure parallel operation can be realized by WEC grid protection) Design for maximum short circuit current (WEC, grid) Inverter: connection point on grid side
Protective gear	Synchronous and asynchronous generators: • undervoltage protection, range: $(1.0\text{–}0.7)U_N$ • overvoltage protection, range: $(1.0\text{–}1.15)U_N$ • underfrequency protection, range: 48–50 Hz • overfrequency protection, range: 50–52 Hz Inverter: • voltage protection as for generators • no frequency protection required
Reactive power compensation	Power factor in range 0.9 capacitive to 0.8 inductive plants $\leq 4.6\,\text{kVA}$ per external conductor not required Larger plants: electricity supply company approval necessary Self-commutated inverter: not usually necessary
Connection conditions	Connection only if all external line voltages are present Synchronous generator: • synchronizing device necessary Voltage difference: $\Delta U \pm 10\,\%U_N$ Frequency difference: $\Delta f \pm 0.5\,\text{Hz}$ Phase difference: $\Delta\varphi \pm 10°$ Asynchronous generators • no-voltage connection in range $(0.95\text{–}1.05)n_{syn}$ • for motorized starting: limitation of starting current Inverter: • connection only if the alternating current side is off-load or the conditions for synchronous generators are met
Grid feedback	Adherence to the compatibility indicators for disturbance variables in accordance with DIN VDE 0838/IEC 77A/IEC 61000: • voltage fluctuations and flicker • harmonic currents The operation of ripple control systems must not be impaired
Commissioning	Testing: • disconnection devices • meters • protective gear for one- or three-phase grid failure, frequency deviations, automatic reclosing • fulfilling the connection conditions

aspects and grid protection can be affected by the influence on protective equipment and short-circuit power, and the function of circuitry can be impaired. Furthermore, grid feedback is possible, which can cause changes to harmonics and voltages and can affect grid regulation.

4.1 Power Conditioning and Grid Connection

Power conditioning, as well as energy conversion, represented a decisive milestone in the development of wind energy technology [4.2]. In the 1980s, generator and grid connection designs for wind energy converters were based upon conventional electrical energy supply installations and had rigid grid connection systems. Only high-output pilot plants were constructed and operated as variable-speed units. This configuration did not make a breakthrough until electronic power converters were fitted to the 50 kW class of wind power plants at the end of the 1980s. Development progressed from the cheap six-pulse converters with thyristors through quasi-twelve-pulse circuits to the so-called pulse-controlled converters with semiconductor switches operating in the kilohertz range.

For a few years now there has been a trend away from robust single systems, mainly characterized by stall-controlled turbines with asynchronous generators and direct connection to the grid, towards more expensive units. Synchronous machines, often based on gearless, ring-type designs with controlled or machine-commutated rectifiers, direct current links and self-commutated inverters, are favoured in these machines. Double-fed asynchronous generators, on the other hand, permit similar speed-variation ranges with considerably smaller converter systems in reactive current adjustment ranges equivalent to the converter output. The gears necessary in these machines make it possible for manufacturers of what were conventional wind energy converter systems with fixed-speed, usually stall-regulated turbines and asynchronous generators to make an extremely simple transition to innovative variable-speed mechanical–electrical conversion units.

The decisive advantage of significantly lower converter outputs previously had to be traded off against the considerably greater cost for measurement, calculation and control technology, in particular for field-oriented machine supply. The rapid development and enormous price reductions in these fields are, however, opening up, improving prospects for this technology. The use of such systems, even at a high cost, is justified if, by adjusting the turbine speed to the prevailing wind speed, the compatibility of the plant to the environment and to the grid can be improved, leading to a higher energy output and reduced drive-train loading.

This type of system also requires a converter system that is capable of conditioning the variable-frequency electrical energy from the turbine generator for supply to a grid of (almost) constant frequency and voltage.

4.1.1 Converter systems

Power electronic converters, so-called power converters, are the most common solution for the conversion and control of electrical energy. They are also used to an increasing degree in wind energy converters to adjust the generator frequency and voltage to those of the grid, particularly in variable-speed systems. Wind turbines connected through converters, either fully via the stator or partially via the rotor circuit, are increasingly common, particularly in systems of the megawatt class.

Power converters have significant advantages over the rotating transformers based on groups of mechanical components or mechanical commutators that were common in the past, namely

- low-loss energy conversion,
- rapid operator intervention and high dynamic response,
- wear-free operation,
- low maintenance requirement and
- low volume and weight.

Figure 4.1 shows the options for low-wear energy conversion using power converters. These are defined as follows. Rectifiers convert single-phase or three-phase alternating current into direct current, with the electrical energy flowing from single-phase or three-phase alternating current systems into direct current systems. Inverters convert direct current into single-phase or three-phase alternating current. The energy flows into the alternating-current side. Direct current conversion is the conversion of direct current with a given voltage and polarity for use in a direct current system with a different voltage and possibly reversed polarity. In alternating current conversion, alternating current of a given voltage, frequency and number of phases is converted for use in an alternating current system with a different voltage, frequency and possibly a different number of phases.

The main components of power conversion systems are the power section, with so-called power converter valves, which carries the electrical power, and an electronic signal processing unit, which performs numerous control, protective and regulating tasks. Over the past few decades, great progress has been made in improving the efficiency of both parts of the converter, owing to the rapid development of semiconductor and digital technology. Basic designs are described in more detail in references [4.3], [4.4], [4.5] and [4.6].

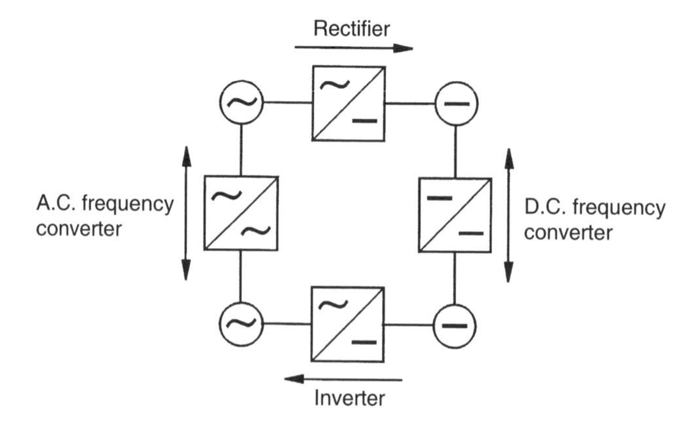

Figure 4.1 Electrical energy conversion by power converters

As wind turbines are almost always fitted with three-phase generators, only three-phase converters are relevant for power conditioning. We shall therefore limit the following discussions to these. We differentiate here between

- direct converters and
- indirect converters.

Direct converters are used particularly for the reduction of frequency. In the case of supply from or to a 50 Hz grid, the operating range of 0 to 25 Hz [4.3, 4.7] is preferred. Developments according to reference [4.8], however, also permit the conversion of frequency of the same order of magnitude. Direct converters require two complete antiparallel power conversion bridges per phase to operate the consumer and supply systems (Figure 4.2). This results in high costs for power gates and control elements.

The conversion of grid frequency f_1 into machine frequency f_2, or vice versa, in a direct converter takes place by the selection of voltage sections from the three phases (Figure 4.3) and by triggering the power converter such that the voltage path after smoothing has the

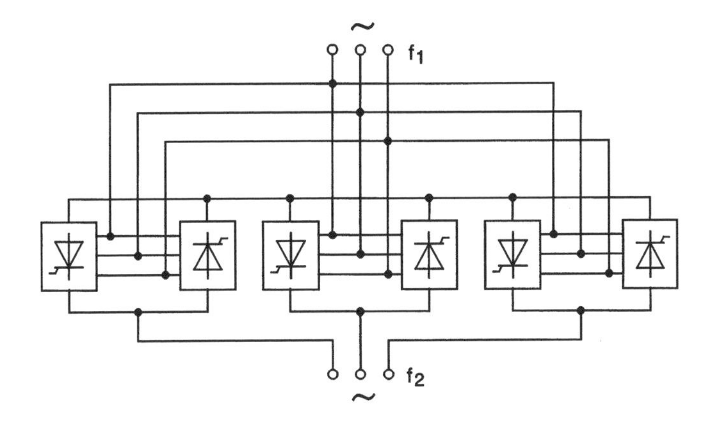

Figure 4.2 Block diagram for direct converters

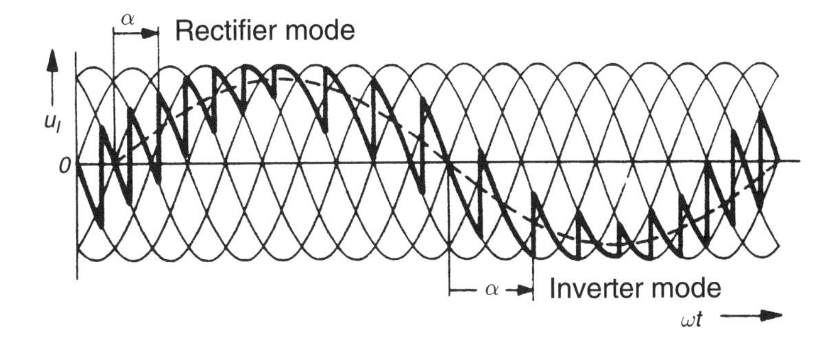

Figure 4.3 Voltage in one phase with a direct converter

amplitude, phase position and frequency required by the machine. Transfer via converter thus allows active and reactive power to be both supplied and drawn. Figure 4.3 illustrates the schematic voltage path and frequency conversion achieved over one phase (U_1). Rectifier and inverter operation can alternate depending upon the load current after a zero crossing of the current with a certain delay for the duration of a half-period.

Indirect converters consist of a rectifier, constant-voltage or constant-current d.c. link and an inverter. A converter with a constant-current d.c. link will be referred to as an *I* converter and one with a constant-voltage d.c. link will be referred to as a *U* converter. Particular characteristics of the link (Figure 4.4) are

- the inductor for current smoothing in the *I* converter and
- the capacitor for voltage smoothing in the *U* converter.

Indirect converters have achieved a clear dominance in energy conversion and the connection of variable-speed wind turbines to the grid. Direct converters have only been used in individual cases to supply the rotor circuit of double-fed asynchronous generators. The following discussions therefore concentrate on the indirect converter. We also briefly describe the fundamental characteristics of power semiconductors and important power converter components.

4.1.2 Power semiconductors for converters

So-called power converter valves are the main components of the power section of converters. They consist of one or more power semiconductors and conduct electrical current in one direction only. These valves generally alternate periodically between the electrically

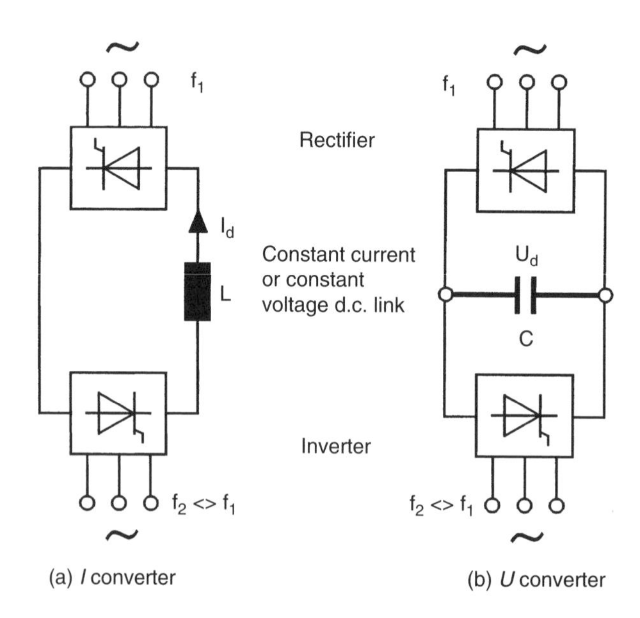

Figure 4.4 Indirect converters

conductive and nonconductive states, and therefore function primarily as switches. As no mechanical contacts have to be operated they can initiate and/or terminate current conduction very rapidly (i.e. in the microsecond range).

Power converter valves can be either controllable or noncontrollable. Noncontrollable valves (diodes for example) conduct in the forward direction and block in the reverse direction. Controllable valves permit the selection of the moment at which conductivity in the forward direction begins. Thyristors can be switched on by their gate and block if the direction of the current is reversed. Switchable thyristors and transistors, on the other hand, can be switched on by one gate electrode and off by a second, or the same, gate.

4.1.2.1 Semiconductor diodes

Diodes consist of positively (p) and negatively (n) doped semiconductor material with a barrier layer between them that ensures current can flow in one direction only. This is possible if the diode voltage is positive. If the current direction and voltage are reversed, the diode becomes nonconducting and blocks the flow of current. Diodes can therefore only be used in uncontrolled rectifiers and for protective and backup functions, e.g. in the form of freewheeling diodes in direct current circuits or similar circuit elements.

In addition to limit values for current and voltage in the forward and reverse directions, and its thermal behaviour, another decisive variable (particularly for protective functions) is conducting-state dynamic behaviour. For the effective protection of semiconductor components, so-called fast-recovery diodes with low stored charges are necessary to protect power converter valves from destruction by overvoltage.

4.1.2.2 Thyristors

Thyristors are semiconductor components with four differently (p and n) doped layers. Conventional thyristors, the gate turn-off (GTO) thyristor, the metal oxide semiconductor-controlled thyristor (MCT) and the Integrated Gate Commutated Transistor (IGCT), are the main types used in converters, and are described briefly in what follows.

Thyristors, unlike diodes, do not automatically go into a conducting state when a positive anode–cathode voltage is applied. The transition from blocking to conducting state is initiated by the supply of a current impulse to the gate, and is known as the ‘firing’ of the thyristor. Once fired, thyristors behave like diodes. They remain in the conducting state as long as a current flows in the positive direction and does not fall below the component’s minimum value, the so-called holding current. If a thyristor is in the off-state, it can be fired by a new current pulse or periodic pulse sequences at the gate.

In conventional thyristors it is not possible to stop the flow of current by intervention at the gate. Switchable thyristors, on the other hand, do permit this. The best-known type is the gate turn-off, or GTO, thyristor. This type of thyristor requires a freewheeling arm for uninterrupted current. Despite the option of switching off and the resulting advantages for control and regulation, GTO thyristors have not yet achieved the market share that was expected for them.

The metal oxide semiconductor-controlled thyristor, abbreviated to MCT, behaves in a similar manner to the GTO thyristor. The MCT can be switched on almost without power by a negative voltage (in relation to the anode) at the gate. A positive gate voltage switches it off, and at the current's zero crossover it automatically switches to block operation.

Integrated Gate Commutated Transistors (IGCTs) represent a further development of the GTO thyristor. They combine the very good conducting-state behaviour of thyristors with the switching capability of bipolar transistors. Their hard gate control means that when they are switched off they go directly from the thyristor mode into the transistor mode, so that no additional protective circuit is necessary. In the reverse conducting, disc-type variant the freewheeling diode that is generally necessary for converter circuits is integrated into the silicon chip. In the asymmetrically blocking variant an external diode is also required. The rate of change of the current is largely determined by the turn-off behaviour of the freewheeling diode and is set by an inductor between the inverter bridge and the d.c. link. Since single switches have no electrical insulation from earth they are mounted on separate coolers. These are electrically isolated. The firing circuits are incorporated into the IGCT. Further tuning is thus unnecessary. The fact that the materials in the discs of the IGCT are not rigidly connected means that they can move slightly in relation to one another. This leads to a high service life with some 100 000 temperature cycles.

The MCT has not yet achieved any significant degree of popularity. However, its further development to higher off-state voltages could lead to it becoming a good alternative to the IGBT [4.9, 4.10] in the middle-output range, which will be briefly described in the next section.

4.1.2.3 Transistors

Transistors are semiconductor components with three differently (p and n) doped layers. Bipolar, metal oxide semiconductor field-effect and integrated gate bipolar transistors are the main types used in converters. As valve components they function exclusively as switches.

Bipolar transistors (BPTs), in their function as power semiconductors, are usually used in the emitter mode. This allows a high level of power amplification to be achieved. Almost like switches, they become conductive when a control current is passed through the base electrode. When switched off, the on-state of the transistor is terminated and the flow of current blocked. In order to achieve a low on-state voltage, and thus low losses, transistors are operated with a relatively high base current. The transistors therefore operate in the so-called saturation range.

Much smaller control currents are needed for Metal Oxide Semiconductor Field-Effect Transistors (MOSFETs) than those for bipolar transistors. MOSFETs can be switched almost without power by voltage control at the gate. This, however, requires that the internal capacitances of the transistor be recharged. Increasing the switching frequency causes more frequent charge reversals and thus higher losses in the driver. MOSFETs are used in the lower-output range at high switching frequencies (see Table 4.2) for switched-mode power supplies and converters, and have advantages over bipolar transistors and IGBTs, particularly at high switching frequencies.

Insulated gate bipolar transistors (IGBTs) combine the advantageous characteristics of MOSFETs and bipolar power transistors. The field-effect transistor at the control input

Table 4.2 Characteristics and maximum ratings of switchable power semiconductors (maximum rating is currently not achieved simultaneously)

	Component					
	BPT	IGBT	MOSFET	MCT	GTO	IGCT
Symbol	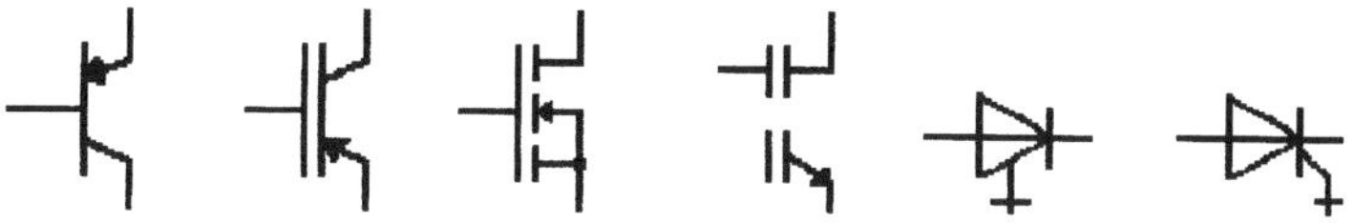					
	Maximum rating					
Voltage (V)	1200	6500	1200	6000	6000	6000
Current (A)	800	3600	700	600	6000	6000
Output (kVA)[a]	480	4000	70	2400	24 000	24 000
Turn-off time (μs)[b]	15–25	1–4	0.3–0.5	5–10	10–25	10–15
Frequency range (kHz)	0.5–5	2–20	5–100	1–3	0.2–1	–2
Drive requirement	Medium	Low	Low	Low	High	High

[a] Ideal switching capacity per switch ($U_{d\ max}\ I_N$).
[b] Including delay times and partial current phase.

facilitates rapid switching at very low driving power. IGBTs automatically limit increases in the current at the output. This results in good overcurrent and short-circuit characteristics. Integrated freewheeling diodes protect the transistor in the off-state mode. Different types of IGBT are used as individual transistors or are connected together in modules of two to six transistors to form bridge connections. Normally, transistors are built into modules with driver switches, protective switches and electrical insulation. IGBTs can be connected in parallel. However, this requires that all transistors exhibit the same thermal behaviour.

The development and availability of new power electronic semiconductor components has given new impetus to converter technology and its application in the field of drive and power engineering. Particularly in the small- and medium-output range, new components (IGBTs) have largely pushed transistors and GTOs out of the market. IGCTs will gain more significance in this output spectrum in the future. Table 4.2 shows the symbols, maximum ratings and characteristics of the switchable power semiconductors described briefly above [4.11] that are currently available on the market.

4.1.3 Functional characteristics of power converters

The main components of power converters are the power converter valves and their electrical connections and trigger equipment. Also necessary are protective elements, energy stores, auxiliary devices and devices for commutation, filtering, cooling and protection, and usually also transformers. It is not possible to describe the many connection options and the complex relationships for the construction and operation of power converters within the framework of the discussion below. This will be limited to the basic three-phase current variants that can be used in wind energy converters. In addition, the most important fundamental circuits will be briefly illustrated, based on the control and timing of converters (Figure 4.5).

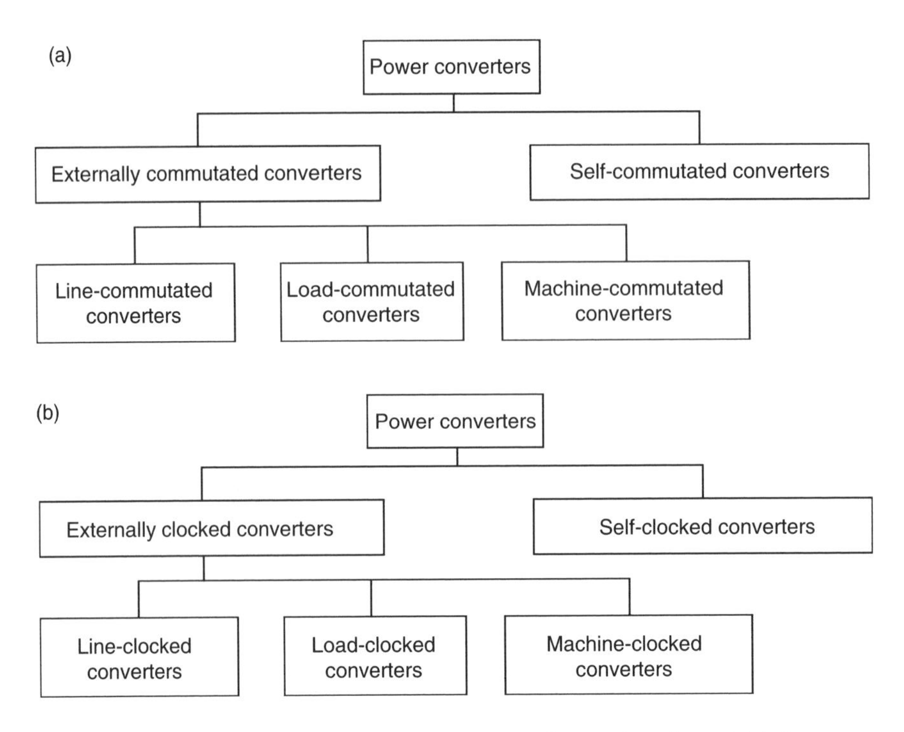

Figure 4.5 Differentiation between power converters according to the origin of (a) the commutation voltage and (b) the elementary frequency

Power converters must be commutated in their voltage and clocked with the corresponding frequency. The origin of the commutation voltage and commutation reactive power at the current transition to another valve is decisive for voltage commutation. Externally commutated power converters operate using natural commutation. They require a grid, load or machine that specifies the voltage and can supply reactive power (Figure 4.5). Self-commutated converters, on the other hand, operate with forced commutation. The required reactive power is provided by capacitors.

The internal function of power converters must also be differentiated with regard to the origin of the clock frequency. Externally clocked power converters take their control pulse from the system with which they work in parallel. Line clocking is the adjustment of the zero-crossings or phase intersections to match those of the grid voltage. Thus the load- or machine-clocked power converter orients itself to the load or machine voltage. Self-clocked power converters have an internal clock generator and are thus not dependent upon external frequency information.

As well as the commutation voltage and clock frequency, the so-called pulse number, the number of nonsimultaneous current transitions (commutations) from one valve to another within one cycle, is an important parameter of power converter circuits. Three and six (Figure 4.6), as well as twelve, pulse connections are normal for three-phase systems. The pulse number is characterized by the number of sine peaks (pulses) of the unsmoothed direct current.

One of the most important and commonest power converter fundamental circuits is the six-pulse three-phase bridge connection. Figure 4.6(b) shows the block diagram of this

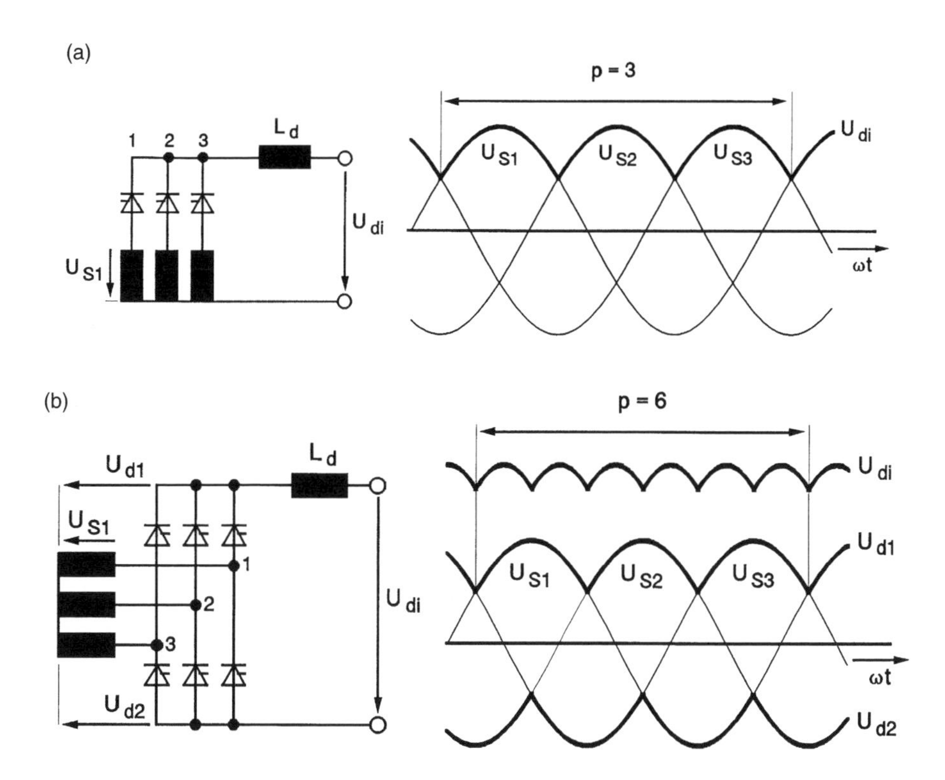

Figure 4.6 Circuit and pulse number of power converter circuits: (a) a three-pulse star connection and (b) a six-pulse bridge connection

circuit with thyristors and the corresponding uncontrolled voltage path. The fully controlled six-pulse three-phase bridge circuit represents the most common variant of line-commutated power converters and of rectifiers for frequency conversion.

Commutation, the transfer of current between the individual valves, can occur in different ways. If the conducting valve is turned off before the next valve is fired then for a brief period no current flows in the connection. As gaps occur in the direct current, this process is known as the intermittent flow. In contrast, it is possible to fire a second valve while the valve to be turned off is still conducting. This creates a temporary short-circuit between two alternating current lines. The current in the valve to be turned off is quickly forced beneath its holding point. This interrupts the short-circuit before the operating current is exceeded. This changeover is known as the commutating operation.

4.1.3.1 Line commutation

Thyristors, unlike diodes, can block positive voltages if they are not fired. By firing the thyristors, it is thus possible to delay the connection or current flow by a specified time $t_v \geq 0$, or by the angle $\alpha = \omega t_v$, compared with the natural point of ignition determined by the grid. Thus three-phase bridge connections can function as rectifiers or inverters. Their operating state is dependent upon the trigger delay angle α. This is the same in all three phases for steady-state operation (Figure 4.7).

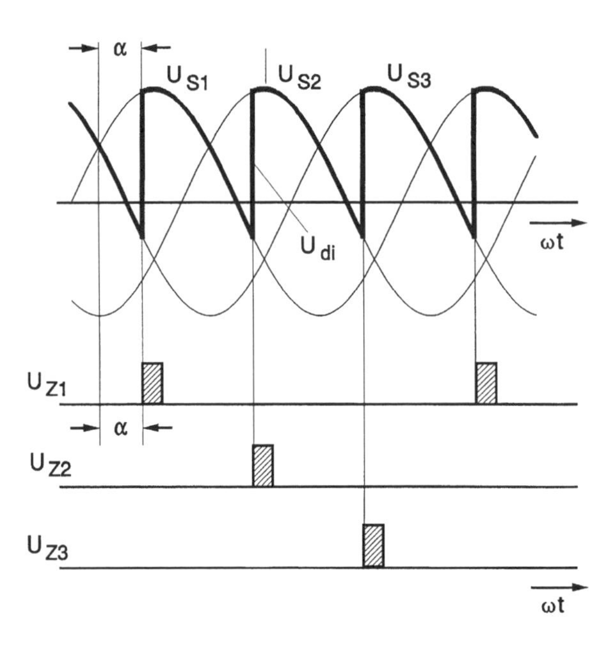

Figure 4.7 Path of unsmoothed direct voltage U_{di} and firing pulses U_{Z1}, U_{Z2}, U_{Z3} and the valve trigger delay angle α

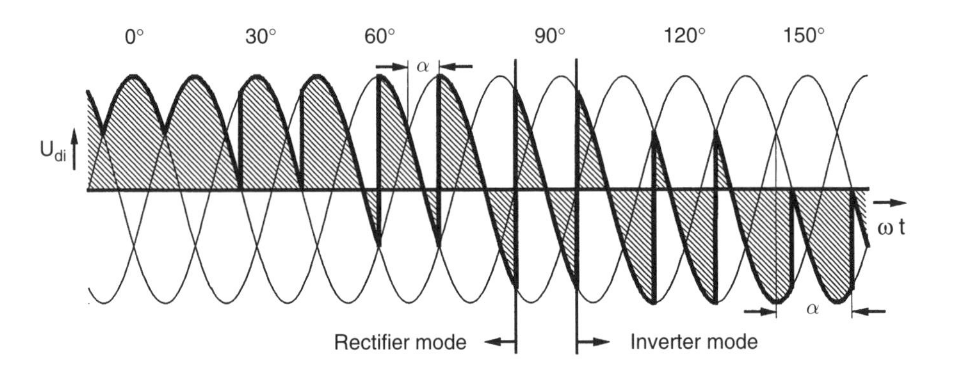

Figure 4.8 Voltages and trigger delay angles of power converters in the rectifer and inverter modes

Thyristors can only allow current to flow in one direction. The voltage direction depends upon the trigger delay angle. In the range $0 \leq \alpha \leq 90°$ the power converter works in the rectifier mode (Figure 4.8). The (arithmetic) mean values of direct voltage and direct current have the same polarity sign. Thus a connected consumer is supplied. For firing angles $\alpha > 90°$, the direct voltage mean value is reversed. The power converter thus operates in the inverter mode.

In principle, the trigger delay angle α can be varied between 0 and 180°. However, before valves can accept a positive blocking voltage, they need a so-called circuit-commutated recovery time. If this is not adhered to, there are still enough charge carriers in the thyristor to reignite a positive voltage immediately after return. The current then commutates back to the valve, which is not completely turned off, and the inverter shoots through. The maximum value of the trigger delay angle is called the inverter stability limit, and in normal operation this lies at around $\alpha_{max} = 150°$. If voltage changes are expected in grids supplied via converters then this value must be reduced accordingly. In the event of 15 % voltage dips, the inverter stability limit lies at approximately $\alpha_{max\,15} = 138°$.

If the converter has to facilitate current as well as voltage reversal, antiparallel valves, as shown in Figure 4.9, are necessary. The black part of the valve is for one current direction and the white part for the other direction. They must also be triggered accordingly.

In the steady-state mode all valves of a bridge conncction operate with the same delay angle, as shown in Figure 4.7. Thus, in the course of one period, each thyristor in the positive and negative bridge half carries the current for one-third of the duration of the period in the triggered state. For the ideal, smoothed direct current (with the aid of inductance L_d as shown in Figure 4.6), the grid current in one phase is found to be rectangular blocks of 120° length, which are positive during the first half-period and, after a 60° interruption, are switched over to the negative side of the alternating current line. This current form deviates significantly from the sinusoidal path. The grid current thus contains strong harmonics. This results in corresponding grid effects (Section 4.3.5).

Line-commutated power converters require commutation reactive power and phase-control reactive power for the control of energy systems. In driving the power converters, the current flow is delayed in relation to the grid voltage by the trigger delay angle α. Due to this phase shift, phase-control reactive power is required, which, if we disregard the commutation reactive power, is roughly proportional to the cosine of the delay angle α. As the phase shift φ of the fundamental frequency current, compared with the adjoining voltage, is approximately equal to the delay angle α, the rectifier is operated close to $\alpha = 0$ and the inverter is operated as close as possible to the inverter stability limit in order to keep the requirement for the fundamental component reactive power low.

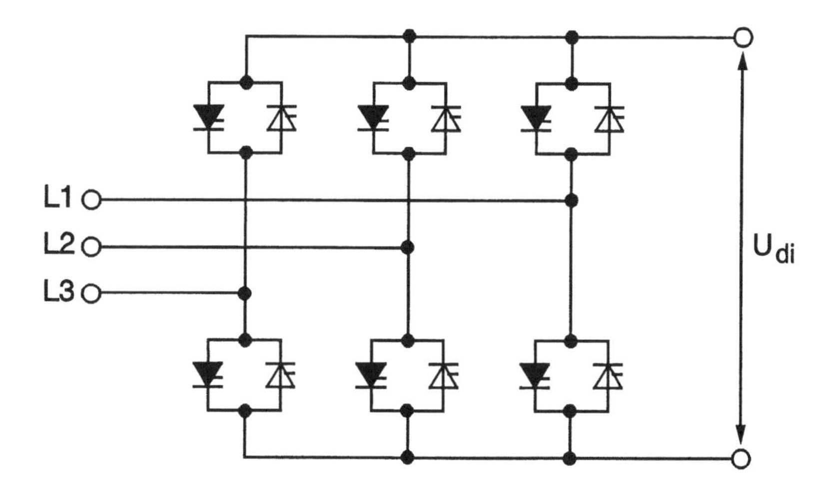

Figure 4.9 Two-way inverter with antiparallel valves

One significant disadvantage of the line-commutated power converter is that intervention for the purpose of regulation is only possible at a few discrete points in time within a cycle. Control commands between these times will only be reacted upon after a delay – a so-called dead time. Their statistical mean is half the period divided by the pulse number. For six-pulse three-phase bridge connections at 50 Hz, the mean dead time is 1.67 ms.

4.1.3.2 Self-commutated systems

In addition to control by the control unit, self-commutated power converters require that the necessary commutations can take place independently of the available back-electromotive force (e.m.f.). The power converter valves must therefore be switchable at any chosen point in time.

When using conventional thyristors, which cannot switch themselves off automatically, a special reset switch is necessary. This disconnects the current-carrying valve at any selected point in time. Reset circuits can have different layouts. They are made up of a commutating capacitor, a GTO thyristor and a commutating inductance, arranged in parallel to the main thyristor.

Self-commutated power converter systems can benefit from the inclusion of transistors. At the appropriate operating frequency, the grid effects and effects on the supplied machines can be made much more favourable. The present market situation, with a clear trend towards transistor converter systems, reflects this. As the self-commutated power converters that come into consideration here are designed with a d.c. link, this layout will be discussed further in the next-but-one section.

4.1.4 Converter designs

The grid connection of wind turbines is basically determined by generator characteristics and permissible grid effects. When using converters, the conversion losses and the extra costs of the technology must be covered by correspondingly higher energy yields. Therefore, low losses and low costs must be the aim for feeding power. Grid characteristics at the point of connection take on a particular significance in this context.

The power supplied to the grid should approximate a sinusoidal path, exhibiting low harmonics. The control and stabilization of the tendency to oscillate, particularly in permanent-magnet synchronous machines, by the appropriate influencing of the generator current is necessary to guarantee reliable plant operation. Moreover, it is desirable to implement grid frequency converters in the form of active reactive power filters (filter consisting of semiconductor devices) or as a unit to support or limit the grid voltage.

Frequency converter systems are used to supply the current generated over a wide range of frequencies by wind generators operated at variable speeds to a constant-frequency grid. Very different demands can be placed on the frequency converter depending upon the generator design. As described in Chapter 3, synchronous machines are mainly used in this application.

Generators with excitation windings allow voltage to be controlled at the terminals or in frequency converter branches by excitation devices. Permanent-magnet machines do not provide this option.

Depending upon the design of the frequency converter, the operation of the generator and the d.c. link can be influenced. This is not immediately possible when uncontrolled

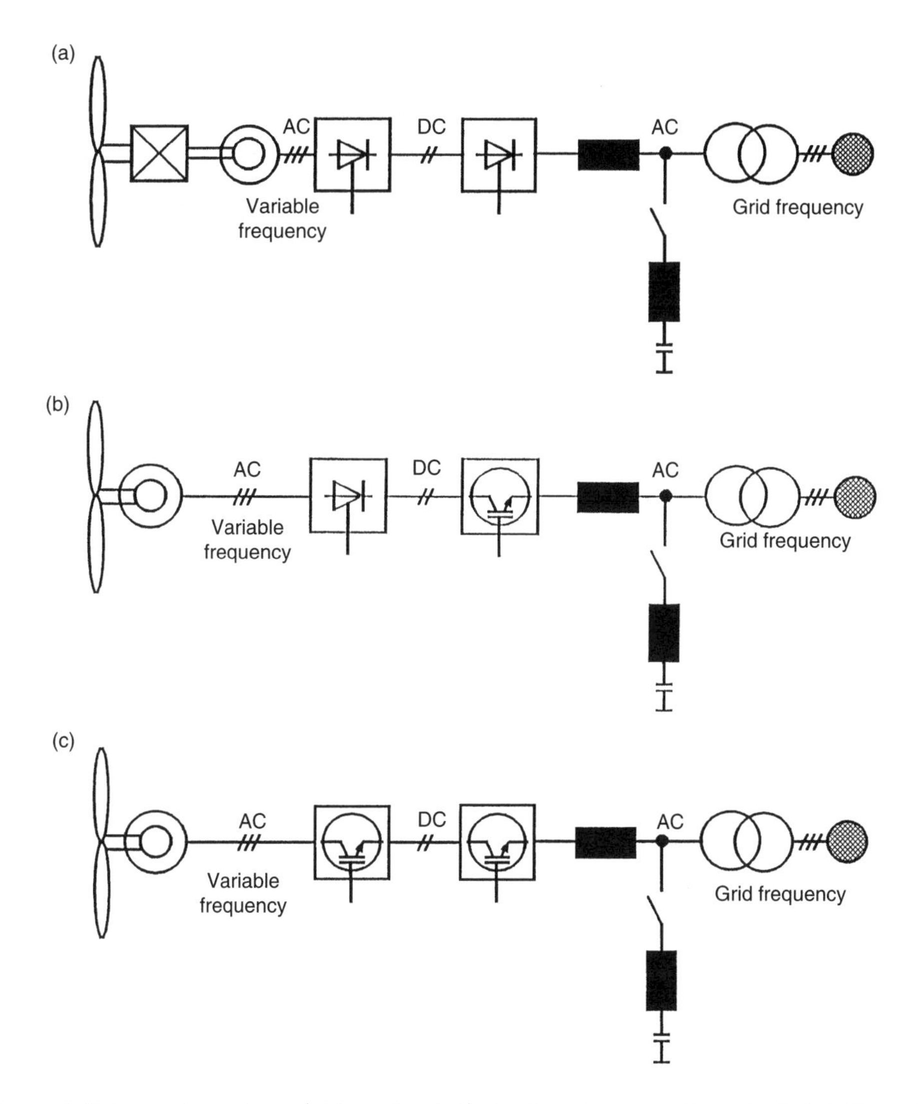

Figure 4.10 Converter systems (with a d.c. link) for the grid connection of wind turbines: (a) an uncontrolled or controlled thyristor rectifier with a line-commutated thyristor inverter; (b) an uncontrolled or controlled thyristor rectifier with a self-commutated switching transistor inverter; (c) a controlled or self-commutated switching transistor rectifier and inverter

rectifier bridges are used (Figure 4.10(a)). Controlled thyristor rectifiers facilitate control interventions, which are able to adapt a power output and energy situation in the d.c. link within a certain range (Figure 4.10(a) and (b)). Using pulse frequency converters, on the other hand, the generator states can be used on the basis of the magnitude and phase of the generator current, to change and regulate the d.c. link voltage (Figure 4.10(c)).

For cost reasons, generators in the 100 kW and megawatt range are designed for output voltage in the low-voltage range (below 1000 V). Dimensioning for between 500 and 800 V is common. Generator designs in the medium-voltage range for offshore use [4.12] facilitate new, economical grid supply concepts. This is only possible if semiconductor components with the required current and voltage data for the selected frequency converter design are available. If this is not the case, several valves or power converter branches, for example, must be connected in parallel (see Figure 4.16). The parallel connection of identical frequency converter units can achieve further operating advantages, as described in Section 4.1.5.5, and grid effects can be minimized owing to the phase differences between the individual power converters.

Further system criteria, determined by the requirements of the generator and grid, must be observed in the design of the frequency converter. Electromagnetic compatibility, system losses, protective measures in normal operation and in the case of failure, and the control of energy flow into the grid are of particular importance here. The concepts and designs that follow relate to the most commonly used type – the indirect converter.

4.1.5 Indirect converter

As shown in Figure 4.11, the power branch of systems for the grid connection of wind energy converters that operate using a constant-current or constant-voltage d.c. link can be subdivided into three subsystems. These will be described briefly in what follows. A good converter design is also based upon the recording of state variables, as well as turbine management and monitoring.

A generator-side rectifier (RE) converts the generator voltage or current to a d.c. link value and controls the generator operation and thus also the wind turbine. Decisive variables here are the turbine and generator driving torque, rotation speed, rotation angle, and generator voltage and current. The d.c. link (DCL) with energy storage enables the grid frequency to be disconnected from the generator frequency. The performance of the d.c. link is influenced by the voltage and current levels within it. The grid is supplied via the inverter. Its operating state is characterized by the voltage, current and frequency, and the resulting active and reactive power values.

Power converters can therefore, depending upon design, be made up of different subsystems. Table 4.3 shows machine-side and grid-side power converters for three-phase

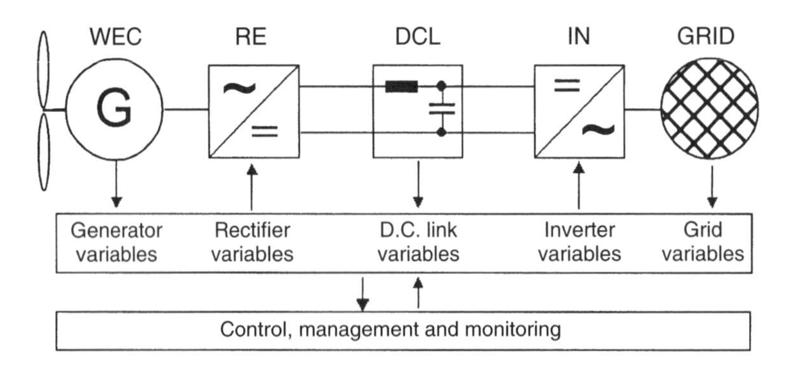

Figure 4.11 Grid supply of the wind energy converter (WEC: turbine, generator) via an indirect frequency conversion system

Table 4.3 Basic circuits of power converters [4.13]

Type	Name	Abbreviation	Energy flow
Rectifier with load-independent direct current	Current-controlled rectifier	CCR	
Rectifier with load-independent direct voltage	Voltage-controlled rectifier	VCR	
Inverter with load-independent direct current	Current-source inverter	CSI	
Inverter with load-independent direct voltage	Voltage-source inverter	VSI	

systems with constant-voltage and constant-current d.c. links, and gives the load-independent values for each. The normal abbreviations are also given. Rectifier and inverter subsystems can be constructed using thyristor or transistor components, and the construction is, in principle, identical for both types. The drive and the pulse pattern [4.12] of the frequency converter are decisive for their function.

A particular characteristic of the frequency converter with a constant-current d.c. link is the high series inductance in the d.c. link, as shown in Figure 4.4(a). As a result, the d.c. link current can be considered to be constant and load-independent during the short commutation period. In contrast to this, the frequency converter based on a constant-voltage d.c. link (Figure 4.4(b)) has a high case capacitance in the d.c. link. This means that the d.c. link voltage can vary only relatively slowly.

The most important rectifier systems that are relevant to the conversion of wind energy will be briefly described below. The supply line between the generator and rectifier is, however, of particular importance here.

4.1.5.1 Supply line between the generator and rectifier

As well as ohmic losses, wiring also has inductive and capacitive reactance components. These affect the transmission of three-phase current. The resistance per unit length R', the inductance per unit length L' in the longitudinal direction, the conductance per unit length G' and the capacitance per unit length C' in the shunt arm in relation to the line length can be represented for a line element, as in Figure 4.12. A long line can be viewed as a corresponding number of line elements connected in series.

In wind energy converters the generator is normally at the top of the tower, but the grid connection and the management, control and frequency conversion systems are located at or in the foot of the tower. Therefore the supply line in plants, which currently have a tower height of up to 120 m, has an influence on the frequency converter.

For the low-loss connection of the generator current to a controlled rectifier with an operating frequency of up to 20 kHz, very steep connection gradients are necessary. This causes high-frequency interference signals in the supply line, which give rise to transient overvoltages and electromagnetic emissions. The sudden increase in voltage brings about polarization effects in the insulation of the winding, which give rise to a capacitive

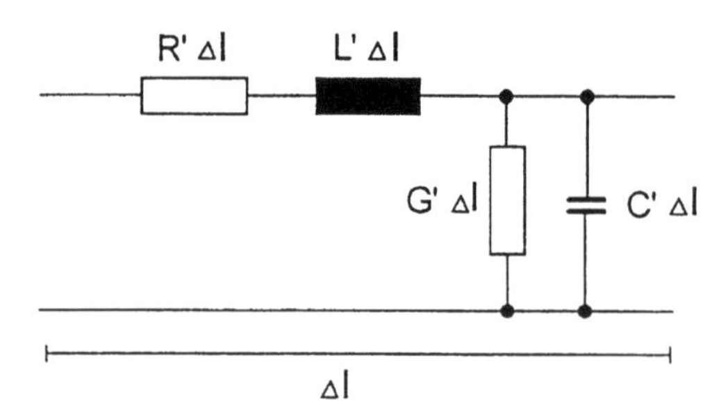

Figure 4.12 Equivalent circuit of a line element

displacement current in the insulation. If no measures are taken to limit this current such as a voltage rise filter, also known as a $\mathrm{d}U/\mathrm{d}t$ filter, it will damage the insulation over the long term.

Overvoltages caused by switching operations can be reduced by using long build-up times. However, this increases switching losses and reduces the clock frequencies that can be achieved. Similar effects can be achieved by reducing the line resonance to low frequencies using inductors at the rectifier input, thus limiting the rate of increase of the current.

The use of *LC* low-pass filters, which allow the maximum generator frequencies to pass undamped and block the rectifier pulse frequency, facilitates better overvoltage limitation. However, this does influence the shape of the generator current.

Moreover, the rectifier pulse pattern can be used to limit the overvoltage caused by long generator supply lines. Predetermined prepulses and afterpulses work against the overshoot in the line, thus creating defined voltage states ([4.14], [4.15]). The additional impulses, however, give rise to increased switching losses.

4.1.5.2 Rectifiers

Electrically excited and permanent-magnet synchronous generators in variable-speed operation supply different frequencies from those of the grid. The rectification of the generator output voltage or its current, and the filtering effect of storage elements in the d.c. link, have the effect of largely disconnecting the generator frequency from the grid frequency. The quality of the d.c. link is basically dependent upon the rectifier type and storage elements used.

The main functions of the rectifier in a frequency converter system, apart from the conversion of variable-frequency electrical energy, is to exert influence over and control the generator and turbine train by controlling the output and the protection and electrical disconnection of the generator in the case of a fault.

Rectifiers can be basically divided into

- uncontrolled diode rectifier bridges,
- diode bridges with series-connected direct current (d.c./d.c.) regulators and
- controlled rectifiers.

Uncontrolled diode rectifier bridges or continuously triggered thyristor rectifier bridges (see Figure 4.6), which exhibit the same normal operating behaviour, can be used if the generators permit voltage regulation by the excitation unit. Thyristors can protect the frequency converter rapidly and effectively from short-circuits in the event of faults by blocking the firing pulses.

When using permanent-magnet generators, there is no excitation effect on uncontrolled rectifiers. Operating behaviour can only be influenced via the power output of the grid-side frequency converter and by mechanically controlling the output or speed of the turbine. Very rapid control interventions are not possible due to the time constants of the control system.

We must also take into account the fact that the combination of synchronous generators with rectifiers under dynamic load can activate unstable behaviour [4.16]. This is caused by the steep current–voltage curve of the d.c. link. This can make the generator tend to vibrate at frequencies below 10 Hz. Stable operation can be achieved using constructional measures on the machine to damp the vibrations or by using controlled rectifiers [4.17].

The combination of an uncontrolled rectifier with a d.c./d.c. converter (Figure 4.13) represents an alternative to the controlled rectifier. The d.c. chopper controller determines the value of the d.c. link voltage by altering the timing (clock signal) relationships. It must be capable of holding the speed-dependent generator voltage or the rectified value almost constant. Using a so-called step-up converter, a d.c. link voltage that is too low can be increased to the level required for the grid-side frequency converter without reducing the utilization of the generator. The disadvantage is that the phase position between the generator current and voltage cannot be influenced. For synchronous generators with excitation windings, this process represents a low-loss option for controlling the d.c. link voltage.

A controlled rectifier with non-switchable power converter valves (see Figure 4.6) can be operated as a system with a load-independent current. The lower dead time of the frequency converter, which is in the millisecond range, compared with the mechanical time constants of generators means that parasitic oscillations and oscillating torques can be prevented by intervention. To maintain a control range in both directions, the firing angle must have an initial value of, for example, $\alpha_0 = 15°$, so that only a correspondingly lower d.c. link voltage can be expected. Thus, both a positive and negative control range is possible. This, however, reduces machine utilization. This type of rectifier can be used for electrically excited and permanent-magnet generators.

Controlled rectifiers with switchable valves can fulfil further demands and desires regarding wind energy converter control and operation. Pulse-width-modulated control of the constant-current or constant-voltage D.C. link facilitates near-optimal generator control by creating a pulse pattern. Along with rapid influence of the current path, harmonics in the generator current can be largely suppressed. Moreover, because of the elementary frequency

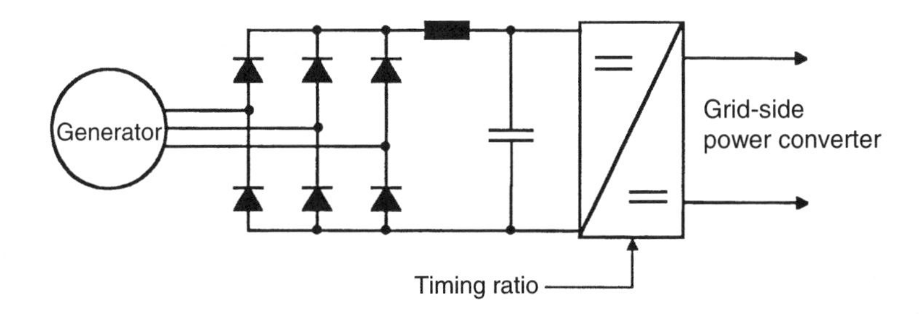

Figure 4.13 Uncontrolled rectifier with a d.c./d.c. regulator

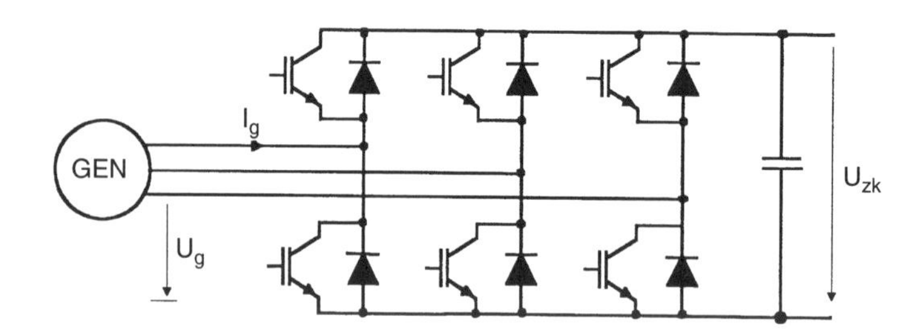

Figure 4.14 Pulse-controlled a.c. converter in the form of a step-up converter

of the rectifier in the kHz range, ripples in the d.c. link are of a relatively high frequency (in comparison with diode or thyristor bridges). The need for storage elements in the d.c. link is reduced accordingly.

In conjunction with the generator inductive reactances, which serve as short-term energy stores, it is also possible to operate the d.c. link as a three-phase step-up converter (Figure 4.14). For this, the d.c. link voltage U_{zk} must always lie above the generator peak value to prevent uncontrolled rectifier operation through the freewheeling diodes. At low speeds, with correspondingly low generator voltages, the step-up converter increases the voltage, thus ensuring the flow of current in the d.c. link.

The shape of the control pulse pattern must be adapted to the generator design. Generators with trapezoidal-induced voltage thus require square-wave currents. For variants with sinusoidal voltage, on the other hand, the pulse pattern is based upon a sinusoidal current reference.

A comparison of losses shows that diode bridges or continuously triggered thyristors exhibit on-state losses only. Both on-state and switching losses occur in the current path of a step-up converter, because of the pulsing at a semiconductor switch, e.g. the IGBT. In addition, the current always flows through several semiconductors at the same time. The switching losses are dependent upon the type of operation and the pulse frequency.

To sum up, we find that controlled rectifiers with switchable semiconductor components have a higher technical cost. However, by using them significant advantages can be achieved, particularly when operating with permanent-magnet generators. They facilitate the adjustment and control of the amplitude and phase positions of the generator current, as well as the resulting torque. Better regulation and protective functions (e.g. in the drive train) can be achieved than is the case with conventional systems, which means that the use of controlled rectifiers in d.c. link converters is very advantageous, despite the higher losses.

4.1.5.3 Thyristor inverters

With the development and introduction of thyristors, the power converter made its breakthrough in the form of controlled rectifiers and line-commutated inverters, which will be briefly considered here. They have found a broad field of application in drive and energy technology, and for a long time dominated the market for d.c. link converters. In six-pulse inverter bridges (see Figure 4.15(a) and (b)), however, the high harmonics that typically occur are very disruptive to the grid owing to the significant deviation of line current blocks from the sinusoidal form. Costly filter devices, mainly for the very strong fifth and sixth

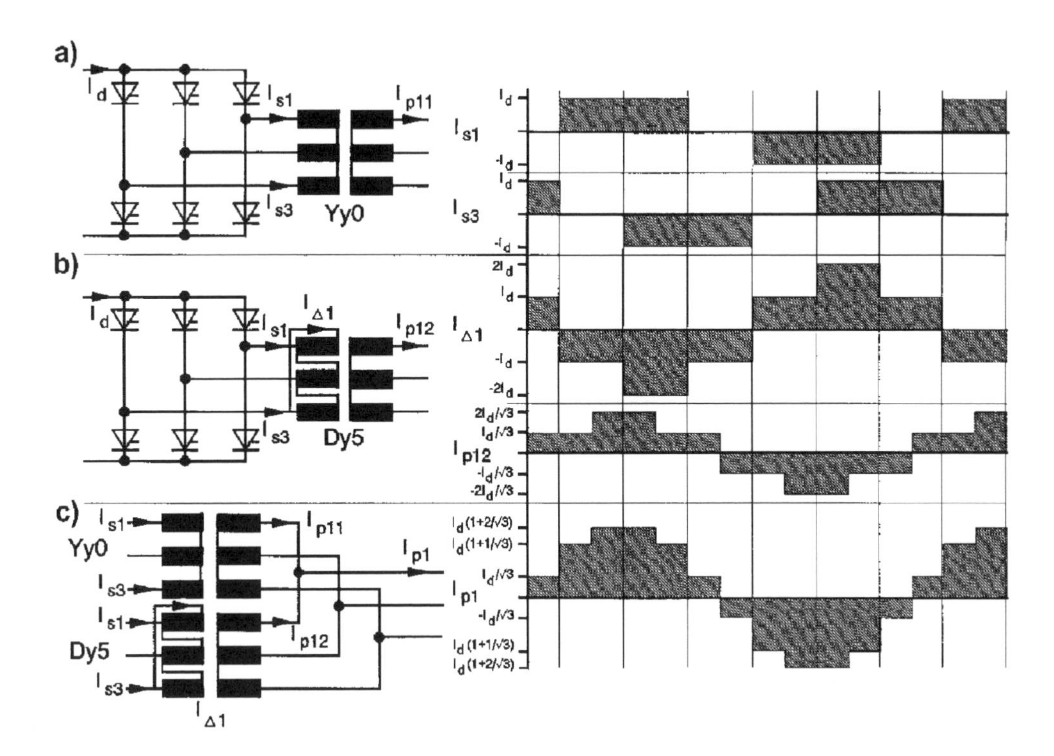

Figure 4.15 Connection and current path of inverters: a six-pulse inverter with (a) star and (b) delta transformer coupling and (c) a twelve-pulse design with the transformer circuit as in (a) and (b)

harmonics, must also be installed. It is also impossible to adjust the frequency and phase position of the line current. These are predetermined by the grid and supply system, and are dependent upon the firing angle.

Significantly better grid compatibility can be achieved with the twelve-pulse design. This is normally achieved by connecting two six-pulse systems in parallel, which are then brought together, via a magnetic circuit, by one common transformer with two secondary windings, phase-shifted by 30° (Figure 4.15(c)). This brings the line current much closer to the sinusoidal form. The fifth and seventh harmonics almost disappear compared with the six-pulse design. Only the significantly reduced amplitudes of the eleventh and thirteenth harmonics and the subsequent weakened components must be taken into account.

These quasi-twelve-pulse frequency converters, however, generally necessitate specially designed generators and grid transformers (for connection at the medium–high voltage levels) with two separate windings in the three-phase system (Figure 4.16).

4.1.5.4 Pulse-controlled inverters

Favourable conditions, in particular with regard to grid effects, are created by the use of self-commutated frequency converters, particularly in designs that operate with a pulse modulation process: so-called pulse-controlled inverters. The cheap availability of efficient power converter relay valves and the progress in signal processing meant that, just a decade after a short

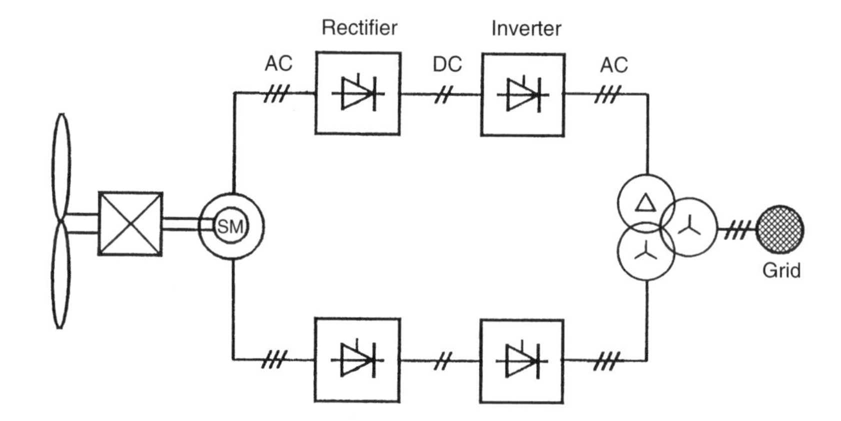

Figure 4.16 Twelve-pulse grid supply via two six-pulse frequency converters connected in parallel

introductory period, self-commutated frequency converters based upon IGBT technology represented the state of the art due to the impetus provided by wind power technology. Higher outputs can also be achieved by parallel connections, as shown in Figures 4.16 and 4.21. At the moment, 5 MVA frequency converter systems are already being planned and constructed for wind energy converters. Conventional and GTO thyristor inverters currently still dominate the market for high outputs in the high megawatt range. New market opportunities are opening up for IGCT semiconductor switches. They offer advantages in terms of switching and service life. It is anticipated that IGBT modules offering a further increase in frequency converter output will soon be available. The discussion below is limited to these.

As a pulse-width-modulated (PWM) inverter with a load-independent voltage, the circuit shown in Figure 4.17 achieves rapid regulation due to its high pulse frequency. Six electronic power circuit elements (IGBT) with an integrated freewheeling diode can be used for this. This circuit permits highly dynamic regulation and the supply of near-sinusoidal currents.

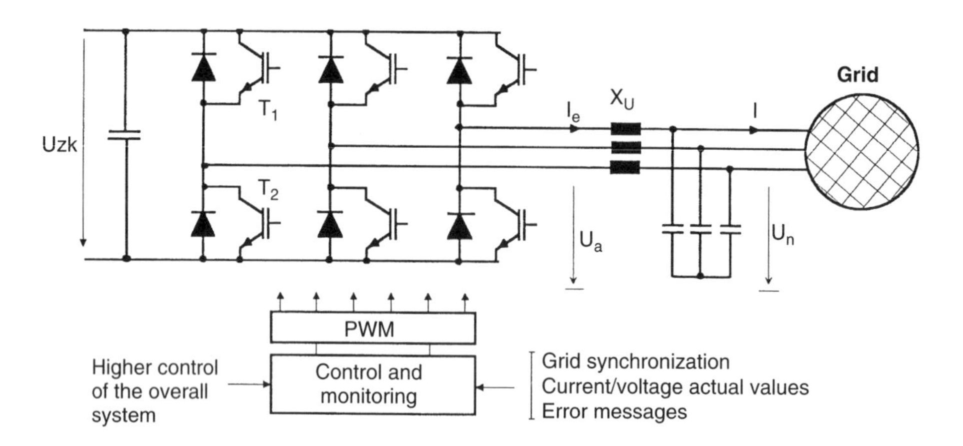

Figure 4.17 Block diagram of a pulse-controlled a.c. converter connected to the grid via a constant-voltage d.c. link

Moreover, both the supply of pure active power and the compensation of reactive power from consumers and other generators in the grid or in grid branches can be selected.

The transistors are switched by the control device, which is synchronized with grid voltage U_n, such that the grid current I or I_e has a near-sinusoidal path. The inverter inductors, with a series reactance X_u, limit the rate of change of the current. The difference between the momentary voltage values at the power converter output and the grid determines the flow of current. The combination of inductors with series-connected capacitors represents a low-pass filter, which serves to reduce the harmonics of the current.

The characteristics of the pulse-controlled a.c. converter are, as mentioned above, principally determined by the controlling pulse pattern. Figure 4.18 shows the path of line current I, which is regulated within the tolerance band ΔI, and illustrates the influence of the switching cycles of the transistors T_1 and T_2 of a bridge branch (Figure 4.17). The desired value can be preset as a sinusoidal variable in the middle of the tolerance band, synchronized with the grid voltage.

The transistor T_1 is switched on during a positive increase in current. If the upper limit of the tolerance band is reached, T_1 switches off and T_2 switches on. The current drops to the lower limit, which initiates another changeover.

Current continuously oscillates between the maximum and minimum value in relation to the desired value. The tolerance band therefore determines the required timing. As shown in Figure 4.18, the switching rate can be freely adjusted as a function of current and the selected tolerance band. In other procedures the switching rate can also be preset. The desired values, the trigger procedure and the resulting switching rate, however, influence the total harmonic distortion (Figure 4.19). High-quality requirements with regard to output current thus result in higher losses in the frequency converter.

The value and phase angle of the output current of a self-commutated pulse-controlled inverter can be freely selected by regulation within certain design-dependent limits. Its use in wind energy converters thus permits a contribution to grid support in grid operation.

Along with the pure active power operation, the current can also be adjusted to lead and lag in relation to voltage, as shown in Figure 4.20. The supply of capacitive and inductive reactive power is thus possible. Due to its short access time, this type of plant can, at high operating frequencies, be used for dynamic reactive power compensation, as well as energy supply. If necessary it can take on an active filtering role.

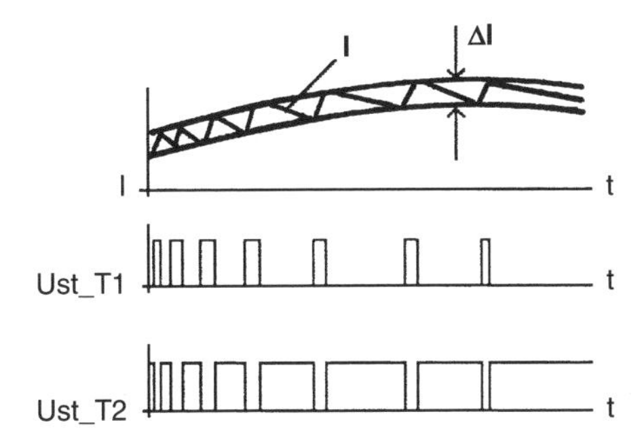

Figure 4.18 Tolerance band regulation of current and control signals for a bridge branch of a pulse-controlled inverter

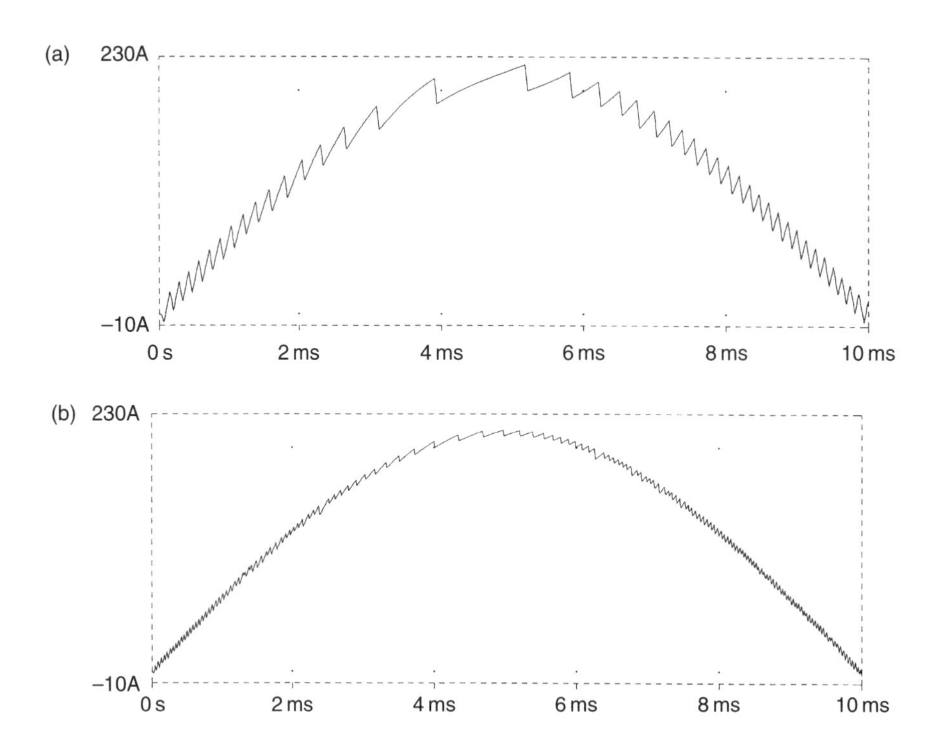

Figure 4.19 Graph of output current of a pulse-controlled a.c. converter: (a) with a large tolerance band $\Delta i/I_{\mathrm{eff}} = 15\,\%$; (b) with a small tolerance band $\Delta i/I_{\mathrm{eff}} = 5\,\%$

If the clock frequency is drawn from an internal clock generator, systems with pulse-controlled a.c. converters can also help to stabilize the grid frequency, particularly in weak isolated grids. The connection of wind turbines via pulse-controlled a.c. converters thus has clear advantages with regard to high grid utilization.

4.1.5.5 Parallel operation of frequency converters

The maximum output of conversion and transmission systems can be limited by the available components. Thus, for example, when they were introduced the use of IGBT pulse-controlled inverters with individual components was limited to systems up to around 300 kW nominal output. For wind energy converters in the 500 kW class, output therefore had to be split. However, further advantages can be achieved by such junctions (branching points) of the energy circuit. For example, one frequency converter junction can be disconnected in the partial-load range. On the other hand, if one part of the frequency converter group fails, the wind energy converter can continue to operate at around half the nominal output. The superposition of frequency converter harmonics operating in parallel reduces the grid influence by partial or extensive cancelling out. The use of identical modules for several output classes makes development, manufacture and storage, as well as service, much easier and reduces costs.

There are various options for the parallel operation of frequency converters. Figure 4.16 shows the normal connection in conventional thyristor frequency converter technology, which is also used for switching transistor systems. This requires a six-phase design of the

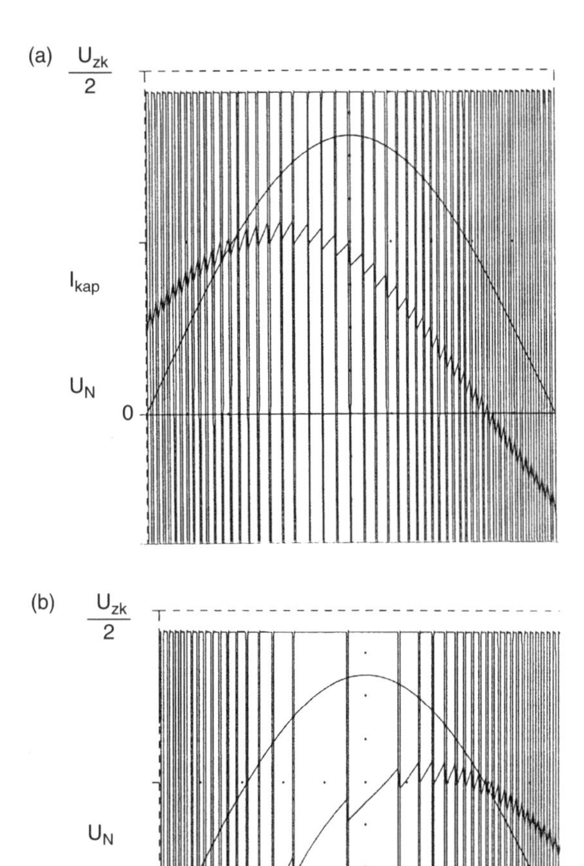

Figure 4.20 Current and voltage path of a pulse-controlled inverter for (a) capacitive and (b) inductive supply

generator winding. The parallel frequency converter units are thus electrically separated, giving favourable conditions for high reliability of operation and low grid influence. Further variants are possible with three-phase generators, with the energy flow divided on the three-phase side or in the constant-current d.c. link. In addition to magnetic coupling by means of partial transformers on common iron cores, the electrical energy can also be brought together on the three-phase side by bus section reactors in all six phases or connection to the grid in pairs via centre tapping of three reactors. In the case of electrically coupled units, increased cost is incurred in the splitting of the current flow in order to prevent the overloading of one frequency converter branch.

4.1.5.6 Frequency converter concepts

Frequency converters have the task of conditioning the electrical energy from wind generators and supplying it to the grid. The influence on the grid should be kept as low as possible. The characteristics of the supplying generator and the grid have a decisive influence on the way in which frequency converters are constructed. Advantageous functional characteristics and favourable types of behaviour in components and subsystems can be taken into account in the design of the frequency converter.

Active and reactive power control with rapid intervention, low grid influence, etc., can be achieved using pulse-frequency converters in the grid with IGBT valves and a constant-voltage d.c. link. High operating frequency, a low triggering requirement and the capacity to disconnect overvoltages are decisive advantages of IGBT frequency converters. By contrast, the low pulse frequencies of GTO thyristors, which are also available in significantly higher power ranges, bring about a poorer-quality feed current.

Figure 4.21 shows the frequency converter design for wind energy converters that have generators with sinusoidal synchronous internal voltage. The constant voltage d.c. link is supplied by the generator via a controlled rectifier. This is current-controlled, so the value and phase of the generator current is determined by the triggering of the rectifier. By phase-shifting the generator current $\underline{I}_{\text{Generator}}$ or $\underline{I}_1$ in over- and underexcited operation, as shown in Figure 3.10, the voltage level at the generator terminals can be altered and matched to the d.c. link. The rectifiers and inverters are of largely similar construction.

For generators with trapezoidal-induced voltage, the rectifier must be designed differently, because the generator only carries current in two winding phases at the same time. Moreover, the output voltage must be stepped up by a step-up converter for current to flow in the d.c. link. The six-pulse bridge shown in Figure 4.22 functions using the generator winding inductances as storage reactors in step-up operation. In this system the rectifier must be placed right next to the generator, since high-frequency oscillations down the long line through the tower could cause disruptive overvoltages.

Figure 4.23 shows the layout of a frequency converter with an uncontrolled rectifier and series-connected step-up converter in the d.c. link. This type of system was used in gearless wind energy converters with generators that require electrical excitation and exhibit a trapezoidal voltage path. The two variants described previously are, however, preferred in the new systems.

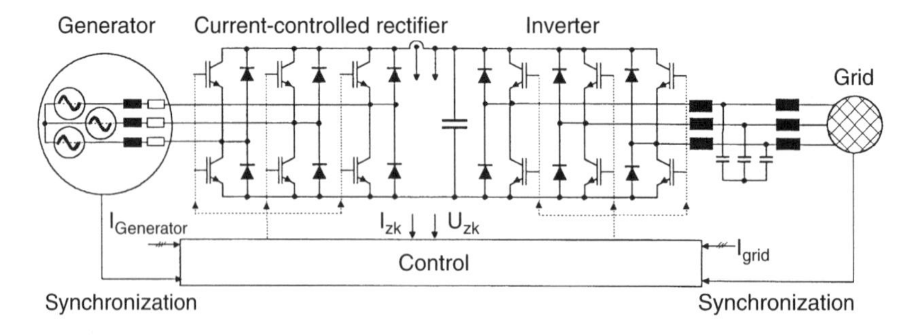

Figure 4.21 Frequency converter concept for an electrically excited or permanent-magnet generator with sinusoidal voltage

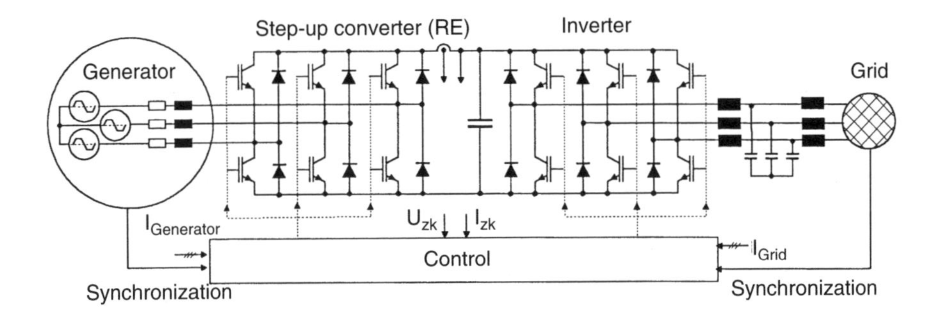

Figure 4.22 Frequency converter design for generators with trapezoidal voltage and a step-up converter in the rectifier circuit

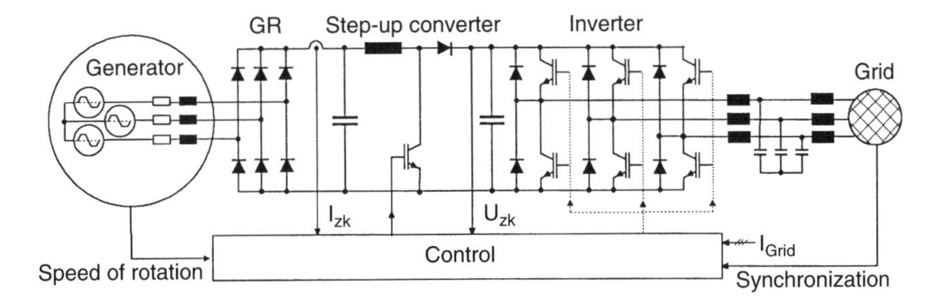

Figure 4.23 Frequency converter design for generators with trapezoidal voltage, a square-wave current path and a step-up converter in the d.c. link

4.1.6 Electromagnetic compatibility (EMC)

Semiconductor valve switching on and off processes in frequency converter systems cause interference emissions over a wide signal spectrum. These must be limited to such a level that other electrical devices can be operated in the vicinity without problems [4.18]. Furthermore, the function of wind power plant frequency converters must not be disrupted by low levels of external interference that could give rise to dangerous operating states.

Interference signals are transmitted by different routes. This transmission can occur by an electrical connection via the impedances between the (emitting) interference source and the (receiving) potentially susceptible equipment. Moreover, inductive and capacitive connections via magnetic and electrical fields are also possible. In addition, electromagnetic waves can be emitted and received by radiation.

The ‘law of electromagnetic compatibility (EMC) of devices’ has been in force (implemented) in Germany since November 1992. Since January 1996, the emission of interference and immunity to interference have been regulated by law. The term ‘device’ includes all electrical equipment, systems and components, and therefore includes wind turbines and their frequency converters. Different EMC limits apply, depending upon the power range and application.

Frequency converter units for wind power plants and their regulation and control components are electromagnetic sources. Their connection and supply lines carry current pulses of high clock frequency. These pass out of the range of the protected junction boxes, etc., via generator and grid lines. The frequency converter supplies the grid with current afflicted by harmonics. This is distributed over all voltage levels and causes a voltage drop at grid impedances. The shape of the grid voltage is distorted by the influence of harmonics. The EMC standards DIN VDE 0838 for general application, IEC 61000-3-6 or VDE 0838-6 for medium-voltage and high-voltage grids and IEC 77A/169/CDV or VDE 0838-4 for low-voltage grids relate to harmonics (VDE 0838-7 and IEC 77A/136/CDV or VDE 0838-5 relate to voltage fluctuations and flicker in the medium-voltage and high-voltage grids or low-voltage grids) and set down the permissible limit values. These can be adhered to by the creation of a pulse pattern giving as near to sinusoidal current as possible. The frequency converter timing gives rise to high-frequency current components. Compact filters in the grid line can reduce these to permissible values.

Rectifiers generate current pulses in the supply line on the generator side of the frequency conversion system. In uncontrolled operation, only current peaks with a relatively low pulse frequency occur during the charging of the smoothing capacitors. Controlled rectifiers create interference signals due to the pulsing of current into the generator supply line. These can be significantly reduced by a spatial arrangement in which the rectifier is located directly adjacent to the generator. This allows the electrical energy to be transmitted to the grid-side power converter via the direct-voltage level. Filters at the rectifier input also facilitate a reduction in interference signals. Depending upon the ratio of interference frequency to generator frequency, however, filters lead to reduced rates of current rise.

Broadband emissions of interference signals, caused by computing and control units, can be limited by fitting these devices in switchgear cabinets. Mutual interference of these components must also be avoided to ensure the function of the system as a whole.

High-frequency radiation can induce disruptive voltages in lines that are not screened (e.g. control lines). This can be compensated by input filters and blocking capacitors in the circuits, as can interference originating from grid supply lines.

Direct or indirect lightning strikes are potentially very hazardous for components, and can occur in the wind turbine, the wires or the surrounding area. The incorporation of lightening arrestors and a risk-minimizing wiring layout, e.g. not parallel to the lightning conductor and earth wiring, very effectively limit the damage that occurs.

The wide range of electronic devices in industry, business and households necessitates that all devices and equipment keep within predetermined interference limit values. To avoid complicated and expensive measures for the limitation of interference, possible effects during operation must be taken into account during the system development phase. The supply, transmission and use of electrical energy can thus be achieved safely.

4.1.7 Protective measures during power conditioning

Faults in the grid and generator or the failure of components may give rise to voltages or currents in the frequency converters of the wind energy converter that are so high that functional groups can be destroyed. Suitable measures are therefore necessary to protect the turbine and grid from damage. This requires the constant monitoring of all relevant variables. Moreover, control devices such as rectifiers and inverters must recognize deviating plant conditions and initiate suitable protective measures.

4.1.7.1 Generator side

A short-circuit between the input terminals of the rectifier creates an overcurrent in the generator. The sharp increase in current or the collapse of voltage must be recognized as a fault by the frequency converter controller. The tripping of fuses or safety switches terminates the generator-side short-circuit. The control system must turn down the grid-side power converter and the plant management must then brake the plant mechanically. Measures for the prevention of bearing and shaft currents have already been described in Section 3.6.1.5.

4.1.7.2 Frequency converter

Rapid switching processes in the power converter give rise to significant overvoltages in inductive devices. These must be limited to protect electronic components. Protective circuitry [4.19] made up of resistors, capacitors and diodes acts against harmful voltages and converts the energy stored in inductive reactances into heat.

4.1.7.3 D.C. link

In the event of a short-circuit in the d.c. link the voltage in the d.c. link collapses. The rectifiers and grid-side frequency converters supply the short-circuit via the freewheeling diode, which means that intervention is no longer possible. Therefore the grid and generator must be disconnected from the frequency converter by the system management.

4.1.7.4 Grid side

The regulations and guidelines of the power supply companies and manufacturers' associations must be adhered to for the supply of wind energy to the public grid [4.20–4.22]. According to these regulations, frequency converters on the grid side must recognize voltage or frequency changes to prevent unintentional isolated operation. In the case of overvoltage and undervoltage outside the stipulated limits or rapid auto-reclosing in the grid, wind turbines and their frequency converters must be disconnected from the grid (within 100 ms in the event of overvoltage and in 3 to 5 s in the event of undervoltage). Systems for the recognition of these faults can be integrated into the plant control system or designed as external units.

Grid-side shorts to earth and short-circuits lead firstly to an increase in current and secondly to a grid voltage dip in one or more phases. This gives a clear indication of a fault. Immediate control interventions, or an immediate blocking of the power converter, cut off the supply to the short-circuit. Disconnectable switch components have significant advantages for protective functions, due to their short intervention time. In particular, when using IGBT frequency converters, the advantages already mentioned come to bear due to their limiting characteristics. As well as being influenced by control and power conversion devices, the grid connection must also be protected by fuses.

In permanent-magnet generators a grid failure leads to an increase in generator and d.c. link voltage. This must be limited by intervention at the rectifier or in the d.c. link until mechanical countermeasures can be taken and the system brought into a safe operating condition. The inertia of the rotating masses counteracts a rapid increase in rotation speed, but also hinders a rapid shut-down. Overvoltages caused by the effects of lightning on the generator supply lines or the grid connection can be effectively limited by lightning arrestors on the rectifier input and inverter output.

Up until now we have largely based our discussions concerning the electrical energy supply, the planning of energy distribution and the layout of grid protection devices on the assumption of central electricity suppliers and the distribution of energy to decentralized end users. However, if decentralized wind turbines are integrated into the supply structure possible changes must be taken into account, particularly with regard to protective devices.

4.2 Grid Protection

Grid protection encompasses measures to protect against excessively high currents and voltages in the individual supply levels that could damage components and devices. Measures to protect against overvoltage are familiar and therefore will not be discussed further.

Wind turbines supplying the distribution system can impair the function of (overvoltage) fuses, hamper the coordination of reconnection devices, cause undesired isolated operation in the case of auto-reclosing or grid failure and thus put previously operational safety devices out of action.

4.2.1 Fuses and grid disconnection

If grid faults occur between a grid fuse and the associated reconnection device, e.g. due to a short-circuit, the generator can continue to supply current to the load side. Under realistic conditions, the currents generated by synchronous and asynchronous generators are not normally capable of tripping the existing fuses during the first two cycles [4.23].

Asynchronous generators only supply a small continuous short-circuit current ($I_k \approx 0$) in a three-phase short-circuit, since the field excitation is weakened. The worst case is a two-phase short-circuit [4.24–4.26].

Due to their high run-up time constants (see Figure 2.70), wind turbines with asynchronous generators that are run up by the grid in motor mode require electrical connections to be designed in accordance with the starting currents that will occur. Fuses and wiring must be dimensioned for five to eight times the nominal current. In small (by today's standards) wind energy converters the power and speed of the generating system is often graduated in that the running up of the wind turbine is governed by a small generator, designed for weak winds at low speeds (see Section 3.6.1.2), which reaches the nominal current of the generator dimensioned for full-load operation during the running-up period. This arrangement offers very favourable grid connection possibilities. The demands placed upon the drive train are also much lower with this running-up procedure.

The use of synchronous generators usually necessitates modifications to the reconnection device [4.24–4.26]. For the medium-voltage level, isolating switches should be used that can disconnect the decentralized generators within two cycles, i.e. significantly quicker than isolating switches on the transformer stations (>10 periods) [4.27].

The additional energy supplied as a result of the connection of wind turbines increases the short-circuiting power in the supplied grid branches, which will be described in more detail below. This can lead to the breaking power of connected devices (fuses, isolating switches, etc.) being exceeded, which may cause significant damage and reduce the safety of a system. However, it also increases the capacity and productivity of the grid.

4.2.2 Short-circuiting power

The short-circuiting power of a machine, a grid branch or a grid can be defined as the apparent power that will be converted at the machine or grid impedance in the case of a short-circuit.

When determining and assessing grid feedback, the special grid short-circuiting power is largely the defining factor for obtaining permissible or limiting values. S_k can be obtained for a node from the simple relationship

$$S_k = \frac{U_N^2}{Z_k}, \tag{4.1}$$

where U_N is the nominal voltage and Z_k the short-circuit impedance between the source and the grid node under consideration, which can be defined using phase values or characteristic three-phase values. Using this simplified assumption, the specified limit values can be safely adhered to in estimates, even in unfavourable cases.

Unlike the assessment of grid feedback, the initial symmetrical short-circuit power of rotating machines (subtransient value)

$$S_k'' = \frac{cU_N^2}{Z_k} \tag{4.2}$$

forms the basis for the design and rating of protective devices. The supplying nonmeasurable initial voltage during the first period must be taken into account here. For this purpose, an appropriate magnification factor c is introduced (DIN VDE 0102). We assume that

- $c = 1.1$ for high-voltage and medium-voltage grids and
- $c = 1.05$ for low-voltage grids.

Using this magnification factor and by multiplication by the so-called withstand ratio $k = 0.1$ to 2.0, the greatest possible short-circuit current is taken into account in the design of protective devices. This is based upon the fact that, for the generator supply, the short-circuit current is not usually driven by the terminal voltage at the level of the nominal value, but the somewhat higher induced voltage.

Figure 4.24 shows the grid connection of a currently standard wind turbine in the 0.5 to 3 MW range. Its connection point (CP) is generally on the low-voltage side (400 to 960 V). The supply connection to the point of common coupling (PCC) – to which, in addition to the grid, consumers are also connected or could be connected – is via a transformer (0.63 to 4 MVA) and a medium-voltage cable primarily in the 20 kV level, which has ohmic, inductive and capacitive components [4.22].

The capacity and productivity of the grid is characterized by the short-circuiting power at the connection point S''_{kCP}. This depends upon the resistance and reactance components of the connecting transformer and connecting cables (R_{L+T} and X_{L+T}) and the short-circuiting power at the point of common coupling S''_{kPCC} of the higher-level grid. The equivalent circuit in Figure 4.25 illustrates these relationships and forms the basis for the following calculations and representations.

For further discussion we have selected a transformer at the connection point (S_N = 2.5 MVA, $u_k = 6.8\,\%$, $U_N = 20/0.4\,\text{kV}$, $P_k = 20\,\text{kW}$) and overhead power transmission lines

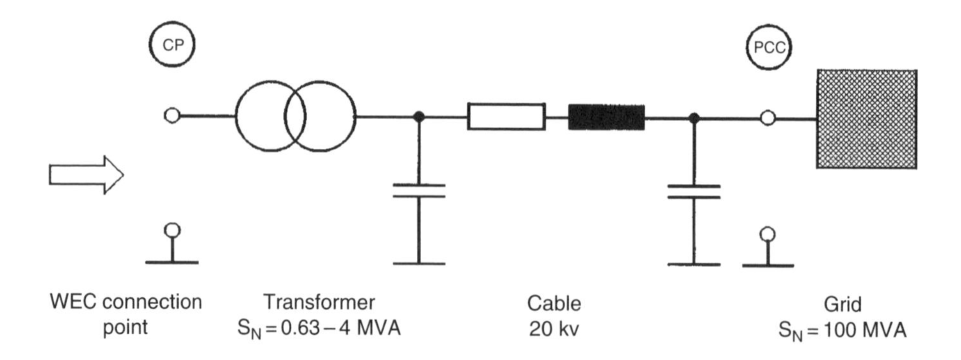

Figure 4.24 Low-voltage-side grid connection of a wind turbine

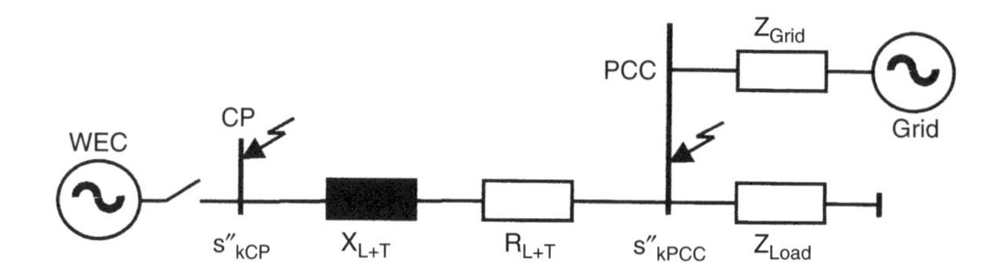

Figure 4.25 Equivalent circuit to determine the short-circuit power S''_{kCP} at the connection point

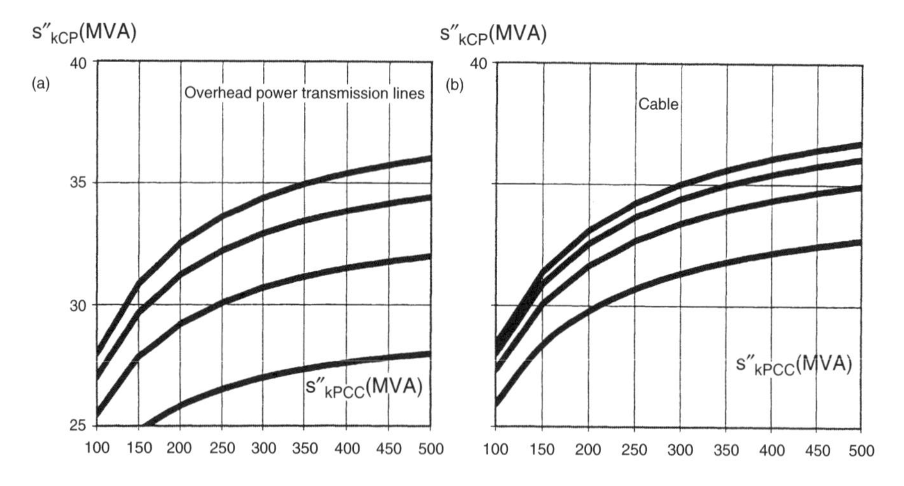

Figure 4.26 Short-circuit power at the connection point in relation to the short-circuit power at the point of common coupling for a 2.5 MVA transformer over 1 to 10 km connections using (a) an overhead power transmission line and (b) a cable

(Figure 4.26(a)) or cables (Figure 4.26(b)), each with a cross-section of 120 mm^2. The calculation results for the short-circuit power at the connection point are thus shown in relation to the short-circuit power at the point of common coupling. Furthermore, we have assumed distances of 1, 2.5, 5 and 10 km between the connection point and the point of common coupling, these being connection distances that are relevant to practical situations. We thus find resistance to reactance ratios of $R/X = 0.14$ (1 km) to 0.25 (10 km) for the overhead power transmission lines and $R/X = 0.14$ (1 km) to 0.33 (10 km) in the case of cable connection. Furthermore, it should be noted that the impedance ratios at the connection point, and thus the value of its short-circuit power, is dominated by the characteristic values of the grid transformer.

The diagrams in Figure 4.26 show the clear influence that the short-circuit power at the point of common coupling and the length and type (overhead power transmission line, cable) of connection to the connection point have on its short-circuit power. An economical increase in the short-circuit power at the point of common coupling from, for example, 100 to 200 MVA is associated with a significant increase at the connection point for all cable and overhead power transmission line connections. Doubling this value from 250 to 500 MVA, on the other hand, brings about significantly lower changes, particularly if long lines are used. This comparison illustrates, among other things, the possibilities and limitations of grid amplification measures at the point of common coupling for increasing the short-circuit power at the connection point.

Furthermore, the influence upon the relationships at the connection point by the selection of supply lines in the medium-voltage range must also be taken into account. A comparison of Figure 4.26(a) and (b) shows that for short connections the type of line and – as also shown by further calculations – its cross-section (within permissible limits) have only a slight influence on the short-circuit power S''_{kCP}. However, at greater distances the short-circuit power falls most when overhead lines are used. Different conduction materials (aluminium, copper) and the cross-section of lines and cables have, as shown by further calculations, less effect. The short-circuit power in the grid falls from the point of common coupling with increasing line length. However, if energy suppliers are connected, it increases again in the direction of the grid connection point.

The impedance of generators and transformers is usually very high compared with the impedance of the line. The distribution of generators thus has no dominant influence on short-circuit power along the line [4.23]. To simplify the determination of short-circuit power, as an approximation the generators can be combined as one characteristic point that is selected to assist the calculation as much as possible. In a short-circuit of rotating machines, significantly altered impedances – compared with normal operation – are, in effect, what influence the subsequent short-circuit power.

4.2.3 Increase of short-circuit power

Like motors, synchronous and asynchronous generators in operation increase the short-circuit power of the grid [4.28]. The calculations that were normally used for short circuit power up until ten years ago disregarded machines that were not constantly connected to the grid. This could lead to false estimates with regard to the dimensioning of disconnection devices and to the grid effects that could be expected.

When calculating the initial impedance Z_{k} that is effective in a short circuit, the fact that the rotating electrical machines will exhibit different impedances in their subtransient time response compared with normal operation must be taken into account. The simplified short-circuit equivalent circuit diagram in Figure 4.27(a) without and Figure 4.27(b) with a connected wind energy converter can serve as a generally valid starting point. The influences

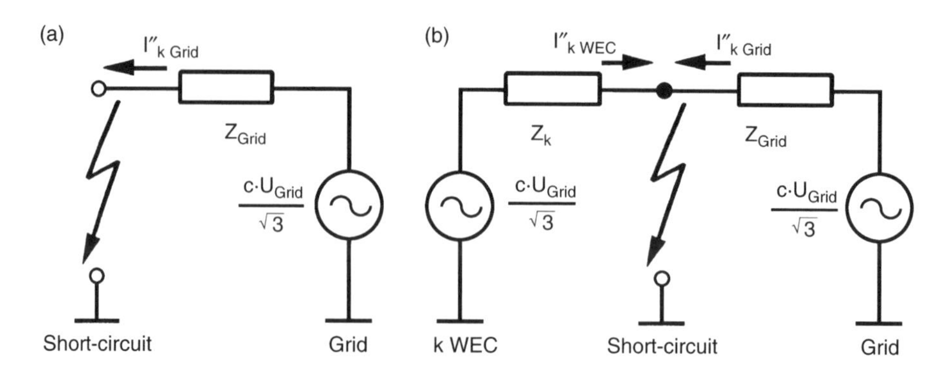

Figure 4.27 Equivalent circuit diagram for a generator in (three-phase) short-circuit (a) without a wind energy converter and (b) with a wind energy converter

due to winding and earth capacitances in the case of a short-circuit, which are very low, are disregarded. The short-circuit power contribution to the grid by rotating machines is independent of whether the machine is working in the motor or generator mode at the moment of the short-circuit.

In synchronous machines the subtransient direct-axis reactance can form the basis for the determination of the short-circuit reactance X_k in the equivalent circuit diagram:

$$X_k = X''_d = x''_d \frac{U_N^2}{S''_N}. \tag{4.3}$$

In this equation x''_d represents the ratio X''_d/Z_N (see Section 3.7.3). To estimate the maximum values of short-circuit currents, however, the lower saturated value of subtransient direct-axis reactance $X''_{d\ total}$ must be taken into account. Depending upon the machine size, the ohmic component R_k can be assumed to lie between 5 % of short-circuit reactance for high-output, high-voltage machines and around 15 % for low-voltage machines.

In asynchronous machines the effective subtransient reactance in the subtransient time period is determined by the subtransient impedance

$$Z_{an} = \frac{U_N}{I_{an}\sqrt{3}} = \frac{I_N}{I_{an}} \frac{U_N^2}{S_N} = \frac{I_N}{I_{an}} Z_N. \tag{4.4}$$

This also characterizes the short-circuit of the machine, for which the leakage reactance of the stator and rotor winding $X_{1\sigma}$ and $X'_{2\sigma}$ respectively and the corresponding winding resistances R_1 and R'_2 are decisive machine parameters (see Section 3.3), where

$$X_k = X_{1\sigma} + X'_{2\sigma} \tag{4.5}$$

and

$$R_k = R_1 + R'_2. \tag{4.6}$$

The ratio of starting current to nominal current lies, in accordance with reference [4.26] and Figures 2.55 and 3.54, approximately in the range of

$$5 \leq \frac{I_{an}}{I_N} \leq 7,$$

where the active component makes up around 10 to 20 % of the short-circuit reactance, depending upon the machine variant.

These discussions show that rotating electrical machines can contribute a multiple of their nominal output to the increase in the short-circuit power of the grid owing to their significantly reduced impedance in a short-circuit. Disregarding the relatively low active component, we can roughly estimate the magnification factor and the increase in short-circuit power for short-circuits near the terminals in synchronous machines from

$$u_{\text{k SM}} = \frac{1}{x''_{\text{d}}} \qquad \text{and} \qquad S_{\text{k SM}} = u_{\text{k SM}} S_{\text{N SM}} \tag{4.7}$$

and for asynchronous machines from

$$u_{\text{k ASM}} = \frac{I_{\text{an}}}{I_{\text{N}}} \qquad \text{and} \qquad S_{\text{k ASM}} = u_{\text{k ASM}} S_{\text{N ASM}} \tag{4.8}$$

For a short-circuit at some distance from the generator, the additional line and transformer impedances between the generator terminals and the short-circuit point must be taken into account. Depending upon the distance and impedances, correspondingly lower magnifications will result.

In accordance with reference [4.22], if the short-circuit currents of an in-plant generation unit are not known, we can use the following as an estimate for the effective values:

- eight times the sum of generator nominal currents for synchronous machines,
- six times the sum of generator nominal currents for asynchronous machines and
- the sum of generator nominal currents for generators with power converters.

For long-term (three-phase) short-circuits, i.e. those that exist over several cycles, significant differences exist, depending upon whether supply is by a synchronous or an asynchronous generator [4.29]. Synchronous machines can supply a short-circuit continuously if excitation is supplied independently of the grid, but if the generator excitation current is drawn from the grid the short-circuit leads to a collapse of the terminal voltage. Asynchronous generators, on the other hand, cannot supply a short-circuit over a long period and so the sustained short-circuit current in this case is equal to zero. The increased short-circuit power

$$S''_{\text{k inc}} = S''_{\text{KN}} + S''_{\text{KM}} \tag{4.9}$$

is thus found from the sum of contributions from the grid and the connected machines or wind energy converters.

A more precise method for determining short-circuit power, or changes to it, is possible by calculating the short-circuit current according to the superposition principle [4.30]. This procedure is particularly suited to the conditions in the case of branched structures. The calculation for a multiple-supplied grid with a corresponding number of sources takes place with the same number of calculations as for single-feed grids. Thus the usual processes can be used for grid calculation. The partial results for current and voltage can be superposed and summed up at each grid or supply point of interest to determine the state variable.

Line-commutated power converters, unlike rotating electrical machines, do not contribute significantly to the increase in short-circuit power. Designs with nondisconnectable valves cannot be immediately switched off in the event of a short-circuit. They can only be relieved of current after a time delay via intercepting circuits. Thus this type of inverter contributes to the initial short-circuit current of the grid.

In normal operation a line-commutated inverter (in the ideal case) supplies square-wave current to the three-phase grid. These currents have a sinusoidal fundamental component. They are predetermined by the grid, which provides the commutation voltage and commutation reactive power for the power converter valves. Since the grid voltage collapses in the case of a short-circuit, the power converter can no longer commutate. The short-circuit current is therefore no longer periodically separated, but flows on in the phase that led when the short-circuit was initiated. Line-commutated inverters thus cannot contribute to the grid symmetrical short-circuit current. The short-circuit current flowing before regulation only contributes to the sudden short-circuit value. This, however, is disregarded in the calculation of short-circuit power.

Self-commutated pulse-controlled inverters can maintain their voltage even in the case of a short-circuit and continue to provide sinusoidal current to the grid for as long as the feed capacity of the system permits. However, since they work with a modulation frequency in the kHz range, intervention is possible within a few microseconds in the case of a fault. Therefore pulse-controlled inverters can be regulated so quickly upon recognition of a short-circuit that they make no contribution even to the initial symmetrical short-circuit current of a grid.

Self-commutated pulse-controlled inverters are thus an ideal grid link, increasing the output capacity of the grid by the frequency converter feed component in normal operation and protecting the grid from any faults.

4.2.4 Isolated operation and rapid auto-reclosure

If the grid is disconnected, the function of protective devices can be impaired by the continued operation of wind power plants, leading to uncontrolled isolated operation, e.g. in disconnected dead-end feeders. Voltage-regulated synchronous generators and self-commutated pulse-controlled inverters are capable of creating grids in the order of magnitude of their capacity. They can therefore, within certain limits, support the grid in the case of voltage failure or maintain a supply in the case of a grid interruption. Synchronous machines coupled direct to the grid are, however, almost never used in wind power plants. Self-commutated pulse-controlled inverters are currently used in variable-speed wind energy converters both in synchronous and asynchronous generators.

Asynchronous generators, which predominate, and synchronous machines with a grid-commutated frequency converter supply, which were commonly used until just ten years ago, do not have this capability. However, in connection with turbine regulation they can supply the grid with active power and support the grid frequency within the range of their capacity. The combination of this type of wind power plant, particularly with plant-determined and consumer-specific reactive-power compensation equipment, maintains the grid in the case of interruption (similarly to synchronous generator supply), with frequency and voltage drift usually being dependent upon the load state.

As well as intentionally induced auto-reclosing described, lasting for about half a second, isolated operation over a longer period is possible. This is the case if wind turbines are disconnected from the grid but continue to supply their power into the fault-free line section and if the momentary output of the installed wind turbines can come close to covering the grid load on the isolated line. Possible reasons include

- opening the lines for maintenance reasons,
- correcting a fault before reconnection of the grid voltage and
- three-phase cut-off due to a single-phase short-circuit in the grid when wind turbines continue to supply in the fault-free phases.

In wind farm configurations and in the case of separate installations, wind power plants with different equipment can supply limited grid sections in combination. When turbines that have asynchronous generators directly coupled to the grid work together with plants that are connected to the grid via frequency converters, grid interruption can lead to various equilibrium states, depending upon the wind turbine control system. If insufficiently high loads are present in the disconnected part of the grid, continued energy supply by variable-speed turbines connected through frequency converters can lead to motor-mode operation of the asynchronous generators. The wind turbines coupled to them can thus be accelerated up to the aerodynamic braking mode.

Grid interruption in the case of the combined operation of wind turbines that are all fitted with asynchronous generators but have different nominal and operating slip values, which is the case for wind turbines of differing size or generator designs, will cause the turbines with smaller slip values to be driven in the motor mode at the rotary frequency of the generators with high slip values. Reconnection to the grid after isolated operation can, in the worst case, occur in phase opposition, which then leads to a high inrush current in the grid and generator, as well as extreme loading of the drive train. Asynchronous generators that have been disconnected from the grid can usually be reconnected without any danger of damage if the voltage has dropped to below a quarter of the nominal voltage in the grid [4.31].

In three-phase short-circuit, almost the same short-circuit time constants can be expected in the vicinity of the generator terminals as at other points on the line, because the total impedance of the fault and the line is usually small in comparison with the generator impedance. For generators in the 100 kW range, these time constants lie in the 100 ms range.

In the case of dead short-circuits, extra capacitors for reactive power compensation and grid filter reactances only have a small influence on time constants, which means that the voltage in the disconnected subsystem will have decayed significantly after only a short interruption. Therefore, no damage can be expected during the reconnection of asynchronous generators after a short-circuit. Asynchronous generators carrying approximately a half-load can, however, in combination with compensation devices, maintain their voltage at above 25 % of the nominal voltage during a short interruption [4.23], thus causing high currents and torques upon reconnection, which may damage circuitry and the drive train.

Wind turbines with synchronous generators and direct grid connection, which up until now have only occasionally been used, usually have a torsionally elastic shaft. Turbines in the MW range can achieve shaft torsions of a few degrees at nominal output. If the load is released (e.g. due to auto-reclosing) then the shaft relaxes. The generator rotor thus accelerates according to its flywheel effect component (k_T) at the rotor time constant T_R (see

Section 2.4) and the gear transmission (e.g. between 1:50 and 1:100 in plants in the MW range). This leads to torsion at the generator, which can lead to a full rotation of the generator shaft due to mechanical effects, and can even lead to two rotations in four-pole machines, due to electrical effects. An overshoot can lead to almost double these values [4.32]. Owing to the large component of the rotor time constant of $(1 - k_T)T_R$, which depends upon the wind wheel (see Sections 2.4 and 2.5), the turbine speed alters very little. The variation in the generator rotor speed, on the other hand, is large. It leads to oscillations around the synchronous speed.

To avoid unfavourable connection conditions (e.g. in phase opposition) the reconnection of wind turbines that have synchronous generators after load shedding or short-circuit should only take place after successful synchronization.

4.2.5 Overvoltages in the event of grid faults

In three-phase lines with distributed supply generators, resonance effects can appear in single-phase short-circuits. Due to the combined effect of capacitive compensation devices and the inductances of coupled generators, and in connection with user and grid filter reactances, overvoltages can occur in the fault-free phases.

Calculations and measurements [4.24], which we shall not describe in more detail here, for this type of grid fault in the medium-voltage range, in which the capacitance of the zero system is connected in series with the inductances of the positive and negative sequence system, yielded overvoltages that reached around twenty times the nominal voltage at low load. Experiments yielded values approximately ten times the nominal voltage. At no load or low load, the single-phase short-circuit current is thus sufficient to extend the electric arc of a fault and to convert a brief fault into a long-term fault if the generator is not disconnected from the grid before reconnection after auto-reclosing. Surge-Voltage Protectors (SVPs) can limit this type of overvoltage and must be capable of absorbing the resulting energy, which in the case of resonance can lie significantly above the values expected in normal operation.

4.3 Grid Effects

The connection of wind turbines to the electricity supply grid affects the grid [3.1]. These interactions between turbines and grid include

- changes in short-circuit power,
- output variations,
- voltage fluctuations and possible flicker effects resulting from these output variations,
- voltage asymmetry,
- harmonics,
- subharmonics and
- other interference emissions.

The investigations described here are mainly limited to output variations and harmonics. The other subdomains will only be briefly outlined.

4.3.1 General compatibility and interference

Public grids must be protected from the disruptive effects caused by wind turbines. Overvoltage, short-circuit and generator protection serve this purpose. In the case of voltage and frequency deviations from normal operation, rapid disconnection from the grid must be ensured. Motor-mode operation should only be permitted for brief periods. The reactive power requirement of wind turbines must be kept within the limits of the grid-specific power factor limits.

The starting current should be kept as low as possible whether the generator is connected with the turbine stationary or at operating speed (see Sections 3.4.2.1 and 4.2.1), in the former case to protect the components of the wind energy converter (from the electrical circuitry through the mechanical drive train and mountings for the generator and gearbox to the turbine) from high shock loads and in the latter case to prevent emitted interference and grid voltage dips.

4.3.2 Output behaviour of wind power plants

Apart from connection procedures, emergency disconnections and other changes of condition, wind turbines in normal operation at partial or full load normally give rise to electrical output variations in the range of 10 Hz to 1 Hz, depending upon the system design, as well as fluctuations over longer periods. Output fluctuations are determined by their amplitudes and rates of change and are caused mainly by periodic and nonperiodic wind-speed gradients. These are influenced by the distribution of air flows over the entire turbine area and the transmission behaviour of the drive train and generator, which can be varied by changes to rotor angle displacement in synchronous machine systems and by slip variation in asynchronous machine systems.

4.3.2.1 Short-time behaviour of a wind farm

In order to quantify the output behaviour of wind turbines when connected together, investigations were carried out in the short-time range at the Westküste wind farm [4.33] with a measuring duration of 1 minute and a sampling frequency of 5 Hz. This showed that in normal operation, e.g. at medium wind speeds of approximately 12 m/s, variations occurred between approximately 10 and 15 m/s, with the wind direction remaining almost constant (see Figure 4.28(a) and (b)). In the 50 kW range (30 and 55 kW respectively), for largely fixed-speed turbines with asynchronous generators coupled directly to the grid, only slightly higher output fluctuations and output gradients can be expected than for variable-speed units with synchronous machines, rectifiers and inverters (cf. Figure 4.28(c) and (d)). In the case of variable-speed turbines, more favourable values can be achieved by control engineering measures.

However, in larger wind turbines (particularly those in the megawatt range) there are fundamental differences. A smaller slip range in larger asynchronous machines, and thus a more rigid grid connection, can bring about higher output fluctuations. On the other hand, higher rotor time constants, and the resulting improved smoothing effect, in variable-speed systems bring about increased output smoothing and permit constant power output, e.g. at wind conditions below the nominal range.

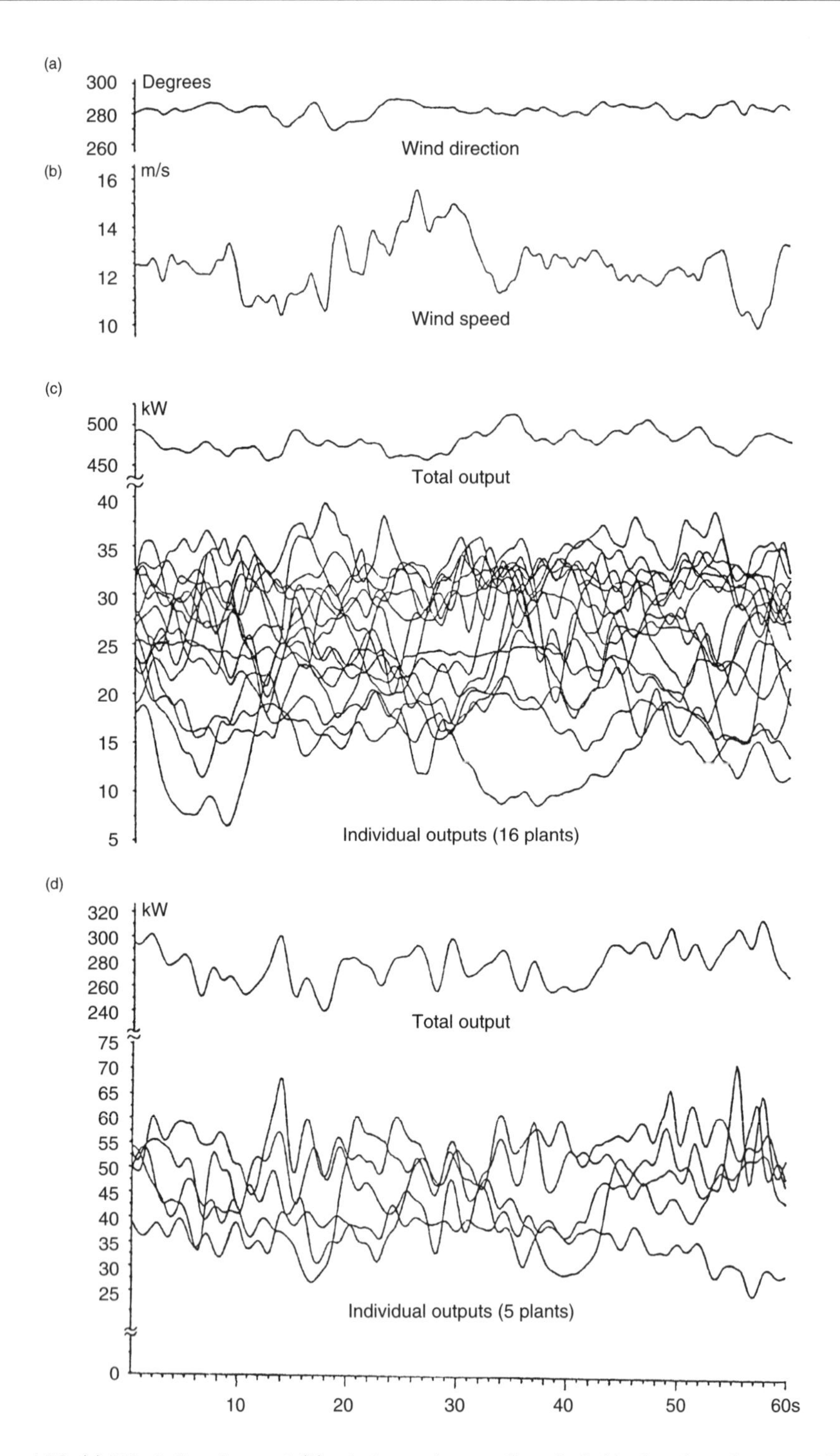

Figure 4.28 (a) Wind direction and (b) wind speed, as well as individual and total output of plants connected at (c) fixed speed and (d) variable speed at the wind farm

The total output of all fixed-speed or all variable-speed turbines as shown in Figure 4.28(c) and (d), measured at the appropriate feed transformers, illustrate the smoothing effect and its dependency upon the number of turbines. Therefore, based on the output behaviour of the individual turbines as shown in Figure 4.29, the different total outputs shown in Figures 4.30 and 4.31 will be considered below, including a consideration of the influence of the turbine configuration. Figure 4.29 clearly shows the very different output behaviour of individual wind turbines in relation to

- wind speed and
- turbine location (for geometry, see Figures 4.30 and 4.31).

The fluctuation range for each turbine over a measuring period of 1 minute is represented in the form of a rectangular plate, with output on the vertical axis and wind speed represented in the different sections. Maximum and minimum values are represented by the top and bottom of the side of the cuboid plates, the standard deviation is characterized by the dotted area and the mean value is represented by a thick line in the inner region. The electrical output values were measured at the same time for each individual turbine; the wind conditions were detected by an anemometer and direction gauge in an undisturbed airflow area at the outer corner of the wind farm close to turbine 7. This wind measurement was used for all turbines, divided into three categories of 'low wind', 'average wind' and 'strong wind', related to a nominal speed of $v_N = 11.9\,m/s$, at which the wind turbines reach their nominal output of 30 kW. The values for individual plants can be determined by projecting the values of individual plates on to the side and lower datum levels. In the 'low-wind' range, for example, an output fluctuation of (0.02 to 0.38) $P_N(P_N = 30\,kW)$ is detected for the first turbine. Standard deviations and mean values can be determined in the same manner.

Consideration of the wind-speed ranges represented shows that output fluctuations are high, particularly at speeds just below the nominal operating speed. Fluctuations are also more pronounced in the back rows (clear from the turbine number). In the lower partial-load range and in the nominal-load range, when operating at above nominal wind speeds, on the other hand, only small output variations occur.

Output smoothing is strongly dependent upon installation geometry and the associated local wind conditions. Wind farm geometry is represented in Figures 4.30 and 4.31 with the prevailing wind direction at the lower edge of the figure. In order to show the differences, the smoothing effect for wind turbines with a fixed-speed grid connection is shown in columns (see Figure 4.30) or rows (see Figure 4.31) for comparison. In addition, the simultaneously measured individual outputs are summed in the step-by-step calculations according to the addition diagram in Figure 4.30. Each calculation is represented by the fluctuation width, standard deviation and mean. For example, in the lower wind-speed range, when only eight plants were operating, based on the 1st plant, the total output is shown for the 1st and 14th plants, the 1st, 14th and 2nd plants, etc., up to the 1st, 14th, 2nd, 9th, 10th, 12th, 7th and 20th plants. Only the plants marked in black were in operation when the measurement was taken. The values in Figure 4.31 were obtained in a similar manner. The dependence of output changes on wind speed or its fluctuation range is particularly clear here.

A comparison of Figures 4.30 and 4.31 shows, owing to the reduced fluctuation range, a clear smoothing of output, particularly in the high wind-speed ranges, i.e. during operation close to nominal output. Moreover, if the output is summed in rows, the row step changes

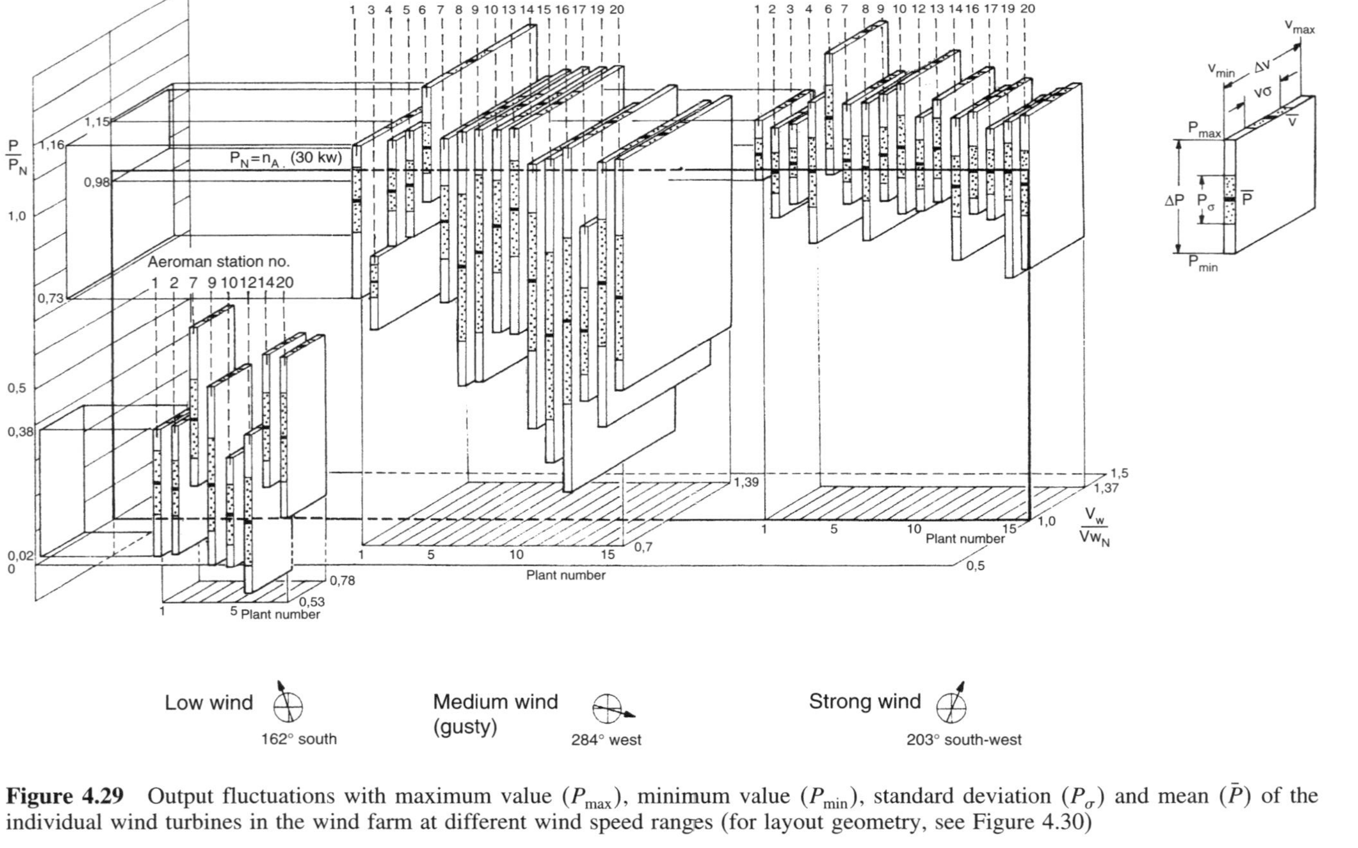

Figure 4.29 Output fluctuations with maximum value (P_{max}), minimum value (P_{min}), standard deviation (P_{σ}) and mean ($\bar{P}$) of the individual wind turbines in the wind farm at different wind speed ranges (for layout geometry, see Figure 4.30)

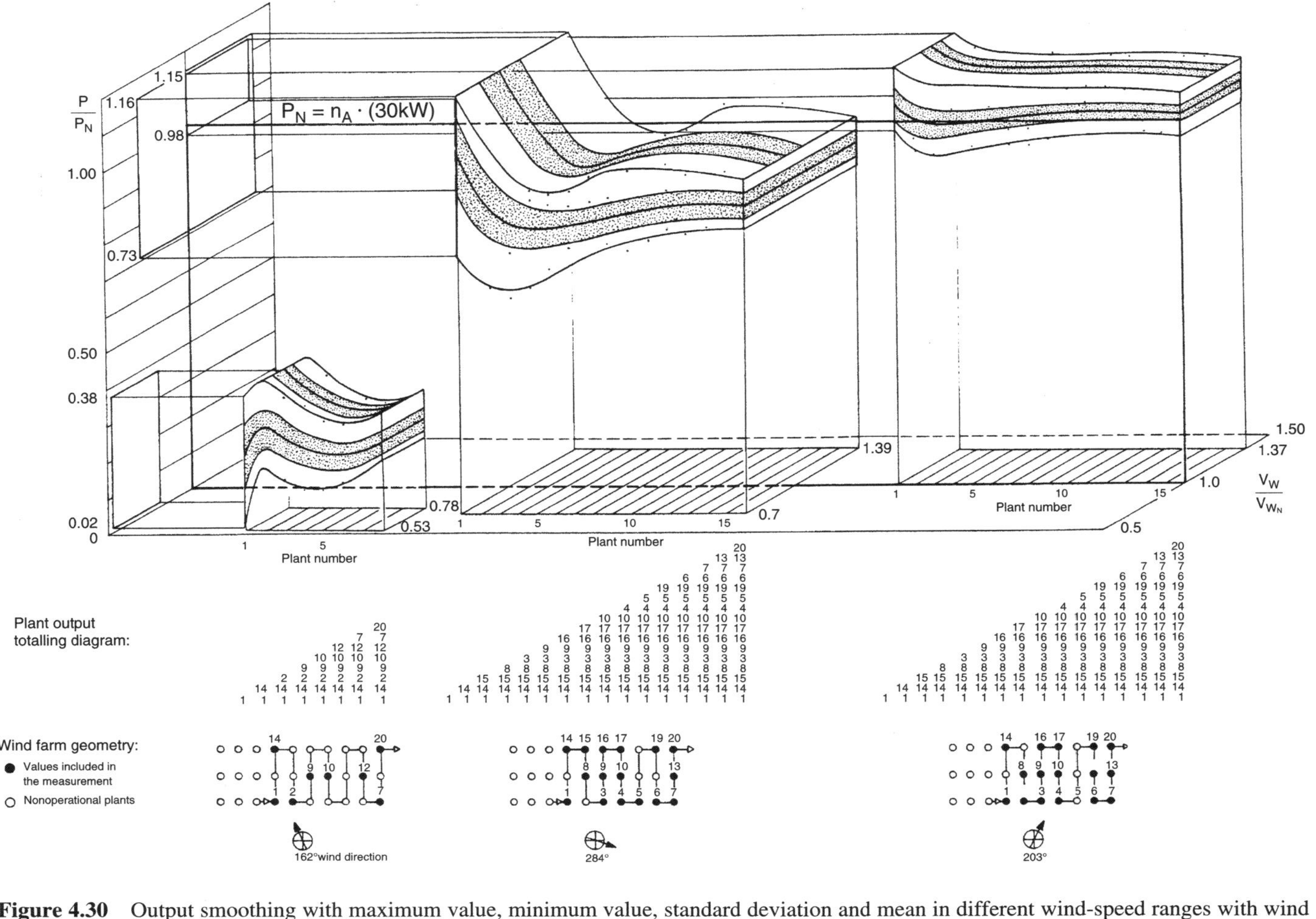

Figure 4.30 Output smoothing with maximum value, minimum value, standard deviation and mean in different wind-speed ranges with wind farm output added in columns

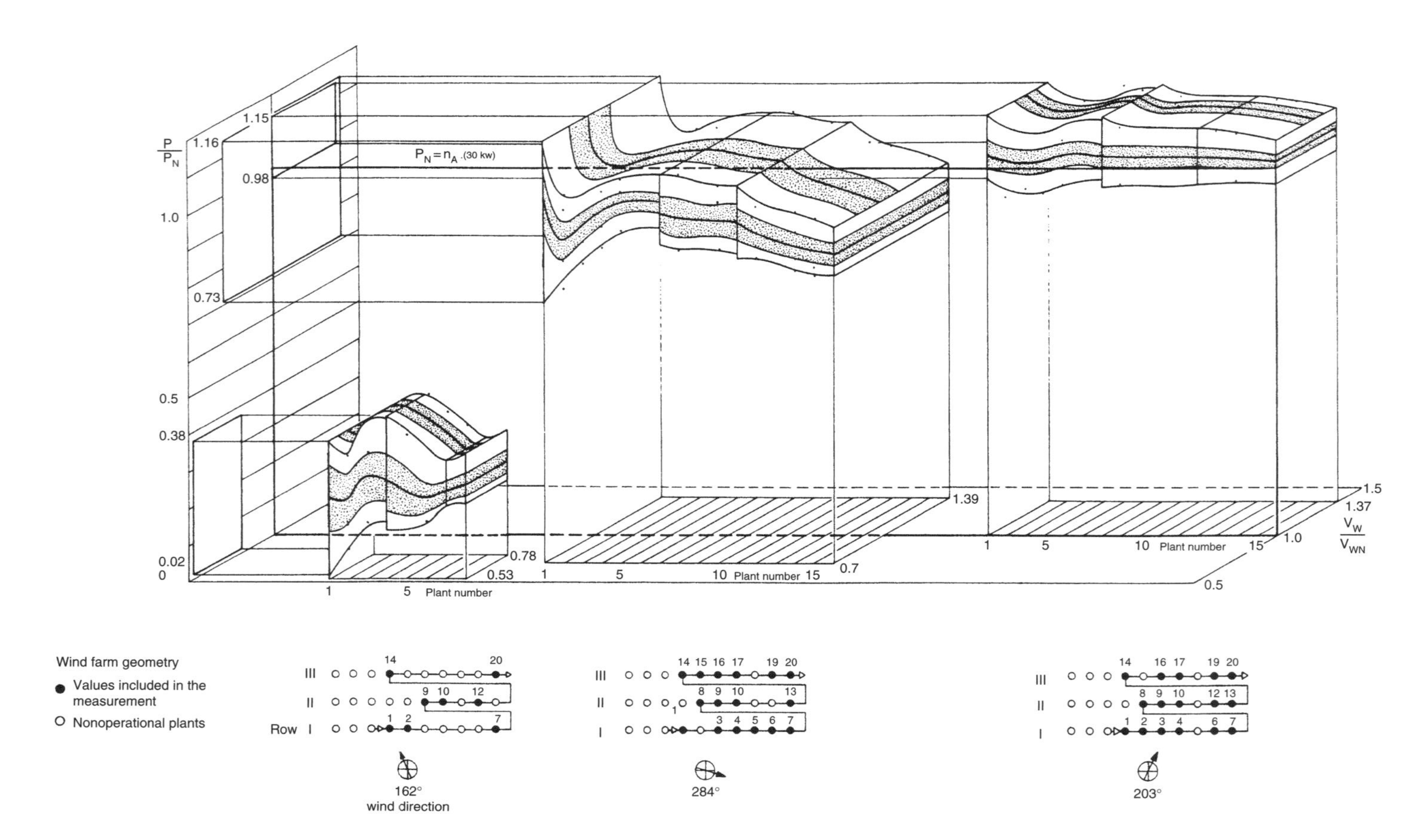

Figure 4.31 Output smoothing with maximum value, minimum value, standard deviation and mean in different wind-speed ranges with wind farm output added in rows

are found again in the fluctuation range and standard deviation. If columns are added, such transitions disappear. As shown in Figure 4.28, the outputs of individual turbines yield a total output supplied into the grid with a much lower fluctuation range. The fluctuation ranges of individual turbines are influenced by the position of the turbine in the wind farm, as shown in Figure 4.29. Both methods of addition demonstrate that by connecting around six plants together, a final value for the fluctuation range of approximately ±6 % of the mean value can be achieved. Due to the spatial offsetting of the individual plants, the airstream meets each turbine at a different time. This yields the largest smoothing effect at the highest individual turbine fluctuations, which occur at low and medium wind speeds.

Due to the wind-speed conditions and wind-speed fluctuations prevailing here, these considerations are valid for short-time investigations. The layout geometry and timing of a wind event must be taken into account here; e.g. in this case a 600 m wind farm depth and 10 m/s wind speed were found in an observation period of 60 seconds. Transferring the results to larger areas and correspondingly longer time periods is only possible to a limited degree. Further measurement results for a layout of plants on a correspondingly larger area with longer measuring periods are discussed below.

4.3.2.2 Long-time behaviour of supply areas

In the VEW Waldeck-Frankenberg supply area in Germany, 33 wind turbines of nominal output between 150 and 500 kW are installed in the Diemelsee area at distances of up to 6 km apart, i.e. around ten times the distance considered above. The total output of the wind farm is 10 MW, of which 70 % of the output is provided by turbines with asynchronous generators with a rigid connection to the grid and 30 % by variable-speed systems with gearless drive train, synchronous generator and pulse-controlled frequency converter. Figure 4.32 below shows plots of the total wind farm energy over time. Figure 4.32(a) shows a plot of the total output of all plants over one hour, Figure 4.32(b) shows this for one day and Figure 4.32(c) for one month in March 1995. Figure 4.32(d) shows the output diagram of the wind farm during the first half of 1995. These figures illustrate the limited output contributions of wind power in the supply area considered here. Furthermore, by breaking the figures down into the time periods that are of interest, relatively precise output gradients can be determined and measures for output smoothing derived. These results can provide characteristic values for the selected time period.

Precise statistical or time-dependent predictions can, however, only be made using the appropriate methods. This requires large amounts of data and long time periods (preferably several years), to obtain reliable results. These can be obtained for the whole of Germany, or for wind-rich regions, within the framework of the 250 MW wind program of scientific measurement and evaluation, which has been set up for a ten-year measurement period. The results of these investigations [4.2, 4.34–4.36] are shown in Figure 4.33 for the North German region. The power duration curves for individual turbines illustrate the duration of the output in question, which is dependent upon local wind conditions; e.g., a turbine at Fehmarn achieves its nominal output for around 1000 hours and operates for approximately 8000 hours in a year (8760 hours). Turbines in the North German area, however, never all reach their nominal output simultaneously. Their energy contributions range over the entire year and 10 % of the nominal output is supplied for more than 6000 hours. In addition to the

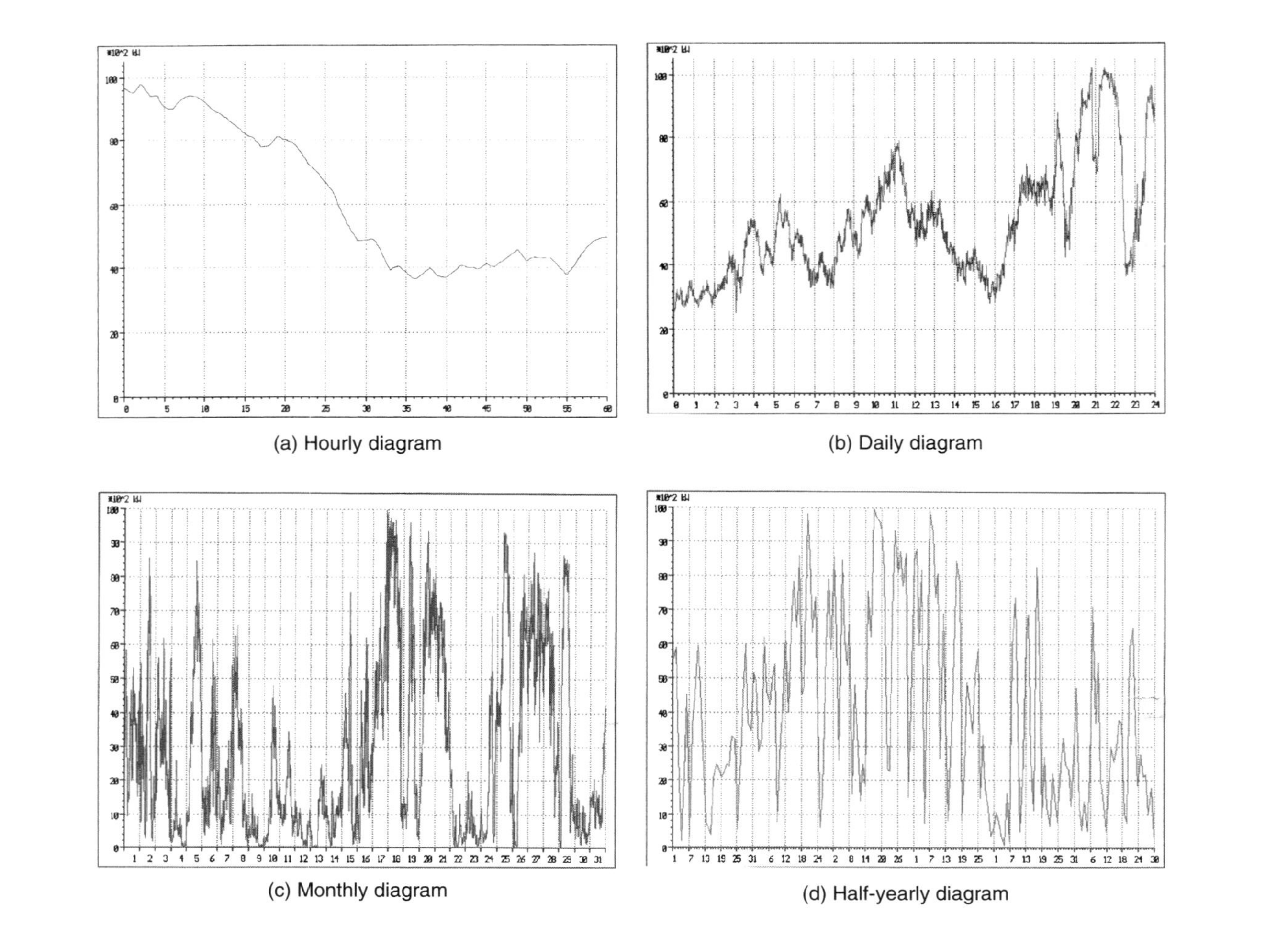

(a) Hourly diagram

(b) Daily diagram

(c) Monthly diagram

(d) Half-yearly diagram

Figure 4.32 Graph of output of a 10 MW wind farm (VEW Waldeck)

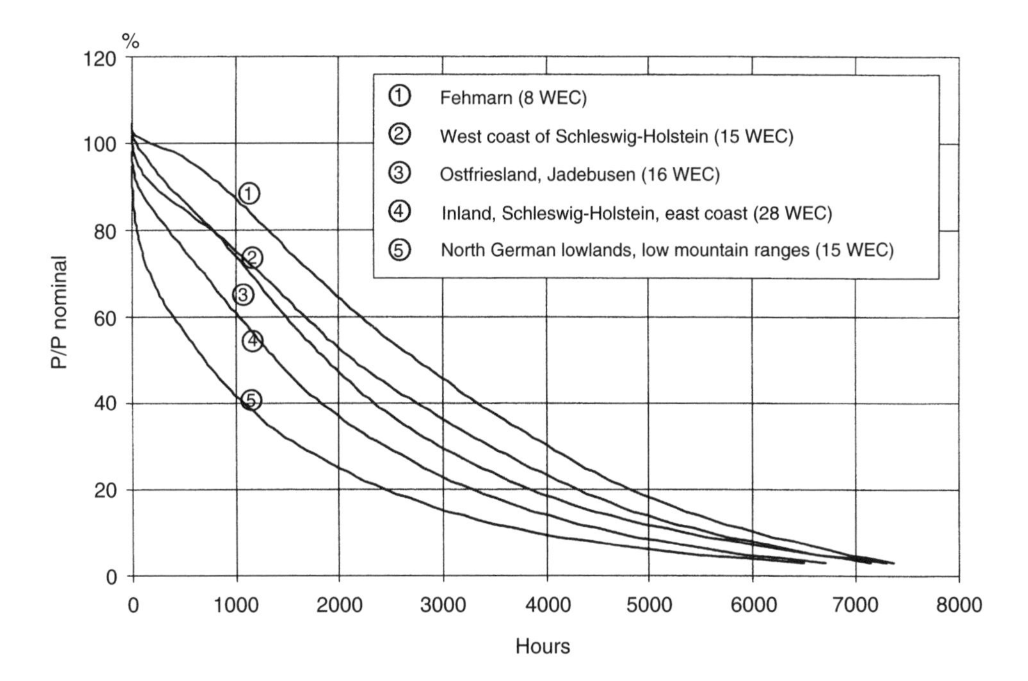

Figure 4.33 Power duration curves of individual wind turbines and a combined group of turbines in the North German area. Reproduced by kind permission of ISET

evaluation of measured data, predications can also be made based upon the above-mentioned program, and these will be briefly described in the following.

4.3.2.3 Wind power predictions

The German interconnected network grid has the greatest installed wind power capacities anywhere in the world. The power contributed by wind generators exceeds the consumed values at off-peak periods in some areas of the grid. Wind power thus plays quite an important role in terms of the operation of grids, load control and generation schedules [4.37–4.41].

In addition to power station failures and stochastic load variations, unpredictable changes in the supply of wind generated power is one of the most common reasons for the use of expensive compensation and control power in power system management. Current consumption over time – the so-called load schedule – can be predicted relatively precisely for the near future by means of the prediction procedures currently in use. Great importance must therefore be attributed to achieving as precise a prediction as possible of wind power, which is subject to weather-dependent stochastic processes and is supplied to the grid in fluctuating form [4.42–4.46]. In particular, precise knowledge of the wind power available on the following day, for example, is of great importance in the planning of generation schedules.

Backed by the German Ministry for Finance and Technology, the Institut für Solare Energieversorgungstechnik (ISET), in cooperation with the power supply utilities, weather

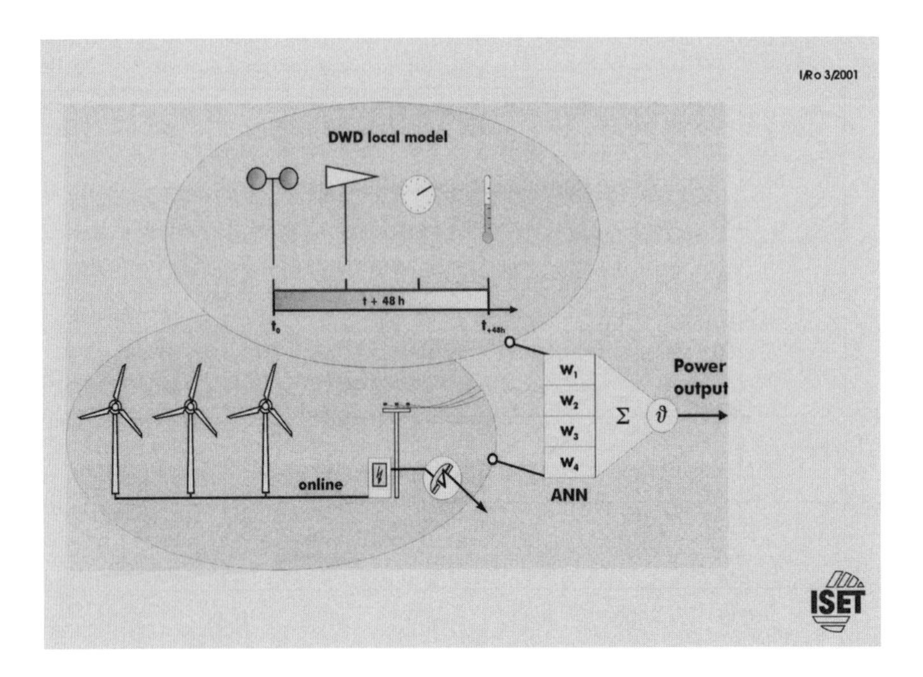

Figure 4.34 Wind power measurement and weather forecasting

service, etc., has developed a numerical model for the prediction of wind power. This model is based upon

- a prediction of wind speed and direction for representative selected locations,
- the transformation of the predicted data to take account of local conditions using a three-dimensional atmospheric model (Figure 4.34),
- determination of the associated wind power with the aid of an 'artificial neural network' and
- extrapolation of the wind power (Figure 4.35) to the total supply in the power supply area in question using an on-line model.

The prediction model supplies the graph of expected wind power over time in the supply area for up to 48 hours in advance. This was achieved by installing measuring equipment at 16 representative wind farms or farm groups. Measured wind and power data obtained in the past was used to improve the predictions produced by neural networks and to help us learn about the relationships between wind speed and wind farm output [4.47]. Figure 4.36 shows the good correlation between measured and forecast wind power by the association of power stations. A further refinement of the procedure will improve the accuracy of the predictions still further. The forecasting model, when implemented at the power distribution board of the power supply grid, will permit enormous progress to be made in the drawing up of a precise load schedule and the cost-oriented planning of power station use, and will lead to overall cost savings in the operation of power stations and of the supply grid.

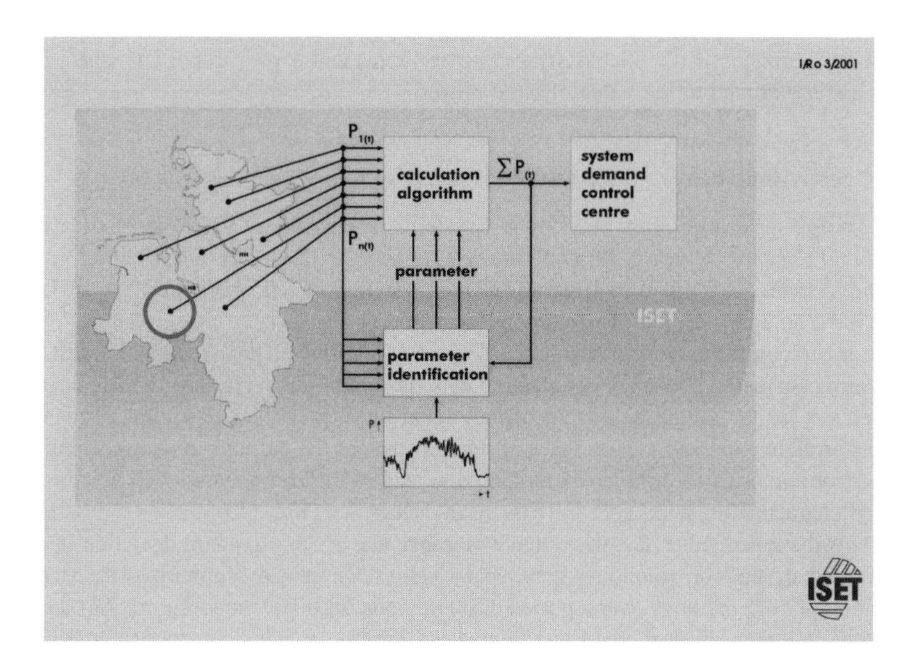

Figure 4.35 Representative wind farm power output measurement and forecasting for supply areas

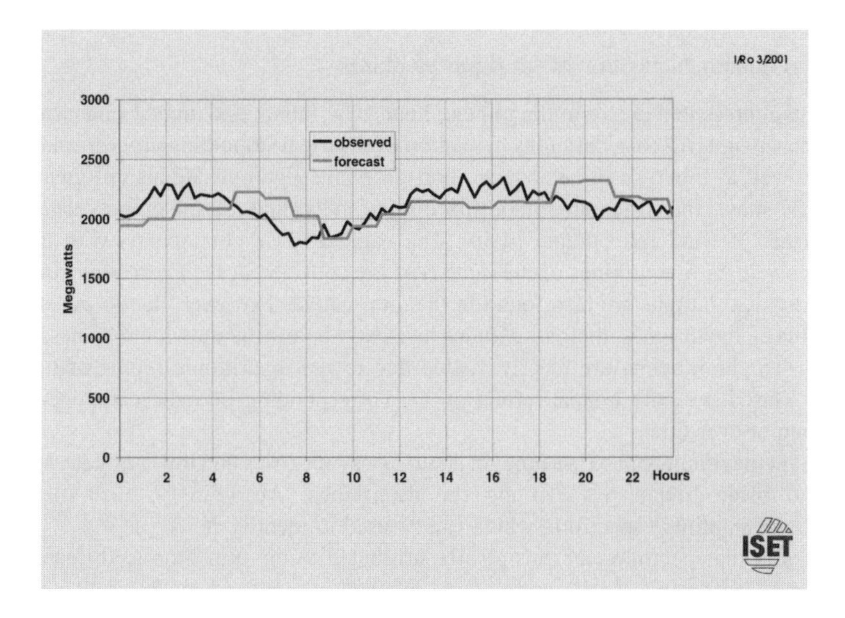

Figure 4.36 Comparison between wind power measurements and 24 hour forecasts over one day

In view of a large-scale use of wind power the expected power compensation possibilities (in particular for fossil fuel electricity generators) represent a very important aspect for the future and are crucial to the value of wind power.

4.3.2.4 Short-time behaviour in a small isolated system

Small energy generation units, e.g. hybrid systems, are normally made up of only two to three wind turbines, with a lead acid battery as a short-term storage and a diesel unit as a back-up system. Such plant configurations are used in isolated operation, e.g. on islands, which have no connection to the interconnected grid [4.48].

Due to its coastal location, the meteorological conditions at the Westküste (west coast) wind farm considered above are comparable with the conditions on islands, which means that the readings for the output behaviour of individual turbines and small groups of plants can be transferred to isolated network systems. In this context, the output behaviour of individual turbines, of two or three turbines in combination, and of different layouts will be considered at three characteristic wind speeds with mean values $\bar{v} = 7.7$, 11.6 and 14.6 m/s. Here $\bar{P}$ represents the mean value, ΔP the fluctuation range and P_σ the standard deviation of output and Δv and v_σ the corresponding wind-speed values. Table 4.4 summarizes the statistical evaluation of 5 Hz readings over a period of 60 seconds for different configurations and flow directions. A comparison of characteristic values illustrates increased output smoothing with larger numbers of turbines and increasing distance between turbines. The greatest smoothing effects are found, as expected, for flow normal to the wind turbine rows. Similar effects were achieved for a triangular shape. In the medium wind-speed range particularly high output fluctuations are found for individual turbines, and good smoothing effects are shown for two or three turbines [4.49].

4.3.2.5 Frequency behaviour of wind power plants

The results represented above in the minute, hour, day, month and annual time scale can be described as long-term considerations in relation to electrical smoothing processes in the hertz range. As well as this type of behaviour, short-time investigations are also of great interest. Figure 4.37 shows the scaled amplitude spectrum of individual turbine outputs together with the spectrum of wind farm output [4.50]. The value at 0 Hz corresponds with the output (active) power. As five turbines of the same type are considered, the alternating components of the individual outputs are also found in the total output. However, due to the smoothing effect between the turbines, their amplitudes lie clearly below the individual values. It should be noted that the maxima are clearly visible due to the logarithmic representation in the diagram. They have only a small effect on the corresponding periodic components in the path of output over time.

Dominant maximum values are brought about by asymmetries and tower effects. Moreover, tower and blade frequencies can also be determined. Accordingly, each plant has a characteristic amplitude spectrum, which can be used to identify it.

Changes in the spectrum can thus also be attributed to the transition to different system behaviours. This permits faults and defects in plant components to be recognized early, which means that this type of frequency spectrum can be used for fault prediction (see Section 5.7.5) [4.51].

Table 4.4 Characteristic values for the output behaviour of wind turbines for different turbine numbers (1, 2, 3), distance between turbines, wind speed and wind direction

	Wind conditions in m/s		
Arrangement	Low	Medium	High
$\bar{v}$	7.7	11.6	14.7
Δv	2.95	8.13	4.35
v_a	0.7142	1.53	0.81
$\bar{P}/P_N$	0.35	0.74	0.97
$\Delta P/P_N$	0.44	0.64	0.44
P_a/P_N	0.118	0.17	0.09
$\bar{P}/P_N$	0.22	0.96	1.02
$\Delta P/P_N$	0.24	0.15	0.26
P_a/P_N	0.068	0.035	0.052
$\bar{P}/P_N$	0.38	0.89	1.04
$\Delta P/P_N$	0.36	0.42	0.216
P_a/P_N	0.098	0.102	0.046
$\bar{P}/P_N$	0.16	0.87	1.03
$\Delta P/P_N$	0.205	0.458	0.148
P_a/P_N	0.046	0.109	0.029
$\bar{P}/P_N$	0.29	0.767	1.031
$\Delta P/P_N$	0.196	0.45	0.175
P_a/P_N	0.06	0.132	0.040
$\bar{P}/P_N$	0.16	0.95	1.022
$\Delta P/P_N$	0.2	0.15	0.23
P_a/P_N	0.046	0.031	0.05
$\bar{P}/P_N$	0.28	0.99	1.025
$\Delta P/P_N$	0.22	0.24	0.184
P_a/P_N	0.05	0.052	0.042
$\bar{P}/P_N$	0.22	0.91	1.037
$\Delta P/P_N$	0.28	0.25	0.22
P_a/P_N	0.075	0.05	0.043

Long-term and short-term changes in wind turbine output can affect the voltage of the grid that they feed, as considered below.

4.3.3 Voltage response in grid supply

The effects of wind turbines on the grid and the power that can be drawn or supplied are determined primarily by the type of grid connection and voltage levels. Changes and periodic fluctuations in voltage and flicker effects must be taken into account, as well as any asymmetries.

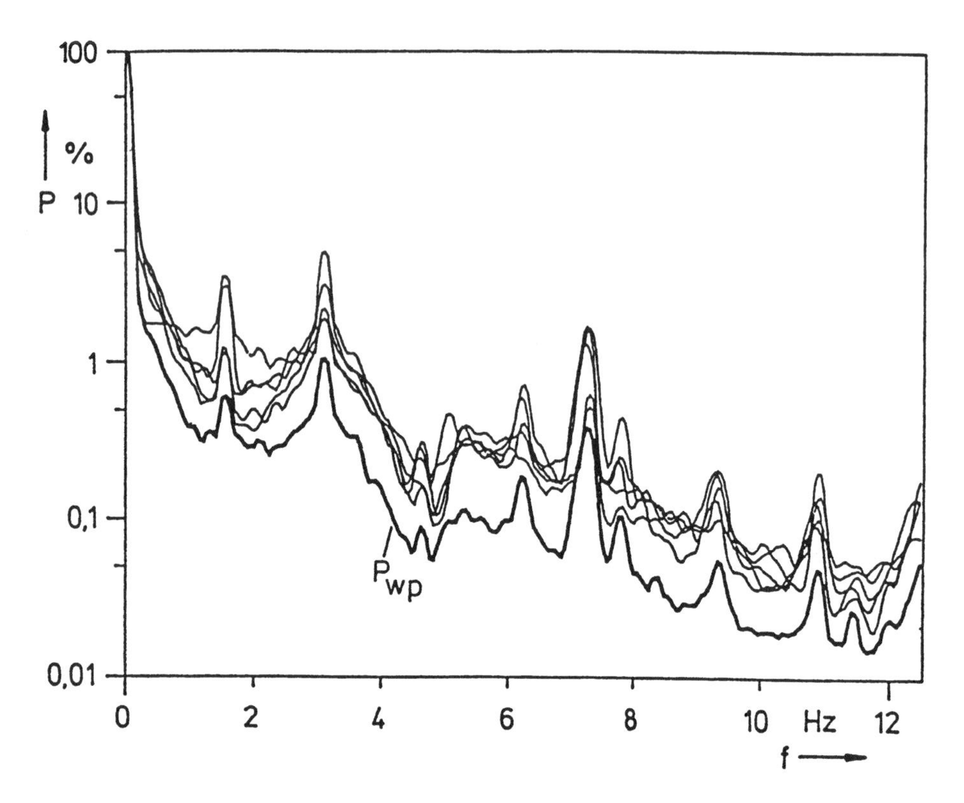

Figure 4.37 Scaled amplitude spectra of the output of five individual wind turbines and wind farm output P_{WF} [4.50]

4.3.3.1 Voltage levels

The electrical components of wind turbines are designed for low voltages up to 1000 V for cost reasons. System configurations in the medium-voltage range represent the exception as yet. As turbine sizes increase, however, these could become more important. For turbines, connection to the grid takes place at the low-voltage level, represented in Figure 4.38 as the lowest block. Systems of the 100 kW to the MW class, on the other hand, supply to the medium-voltage range via a transformer, as shown in Figure 4.24. For large wind farms (e.g. 50 MW) a high-voltage grid connection (middle block in Figure 4.38) can be created in an economically viable manner, and is thus a much cheaper grid connection option. The 220 and 380 kV levels are not currently an option for the grid supply of wind turbines. These will, however, become increasingly important for the connection of large-scale offshore wind farms, which in the future will match the size of conventional power stations at several hundreds of megawatts.

4.3.3.2 Voltage asymmetries

Only three-phase generators with symmetrical windings are used in grid-operated wind turbines. In machines in full working order, asymmetric effects on the grid during supply can thus be disregarded in normal operation.

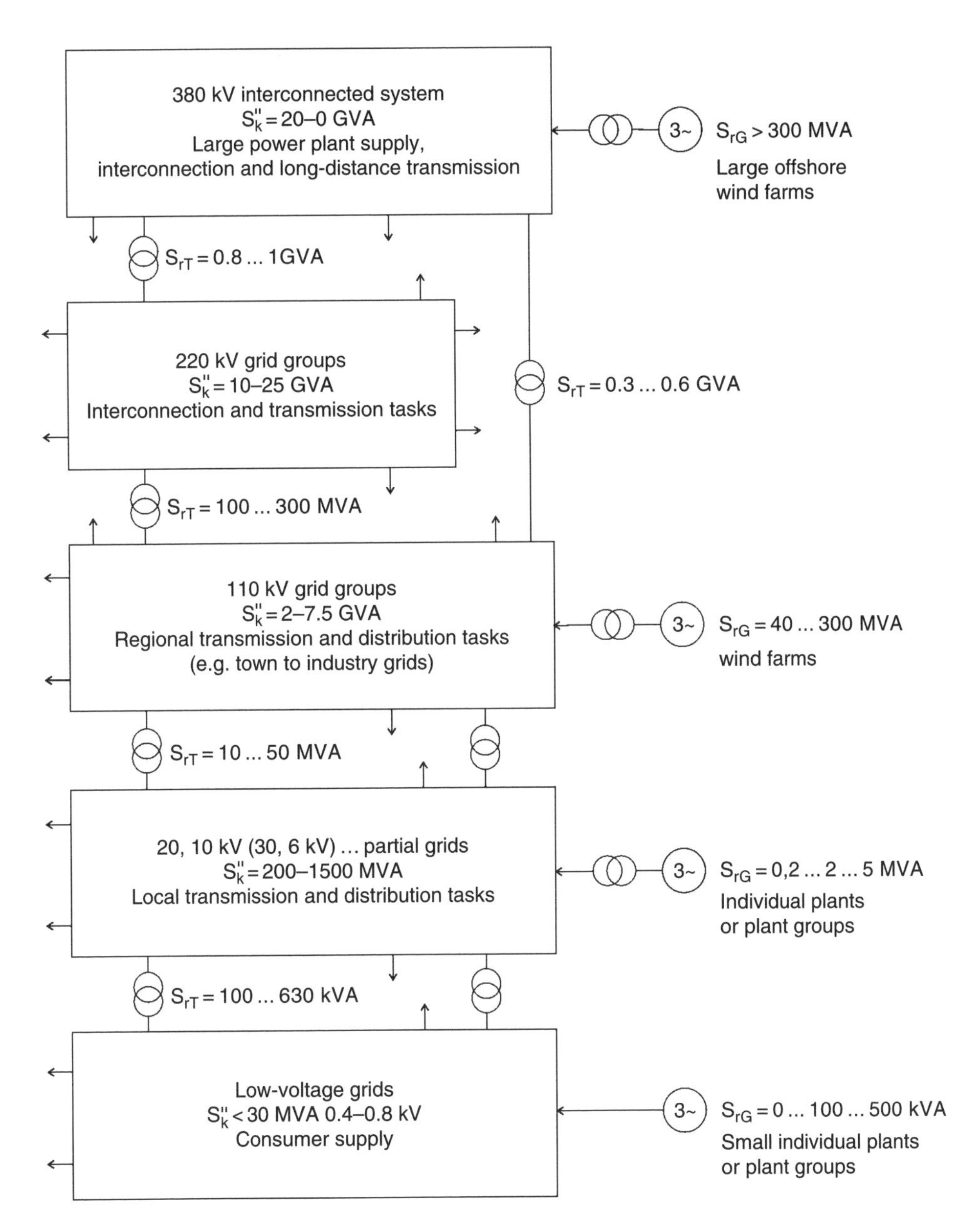

Figure 4.38 Voltage levels and outputs of energy supply grids

Asymmetric loads in the grid can, however, impair the function of lines and transformers as well as that of synchronous and asynchronous generators. Asymmetries can, for example, be caused by the failure of individual thyristors in bridges or asymmetric frequency converter triggering. Asymmetries cause currents to flow in the countersystem. These give rise to a field rotating against the excitation field at twice its speed. This causes additional losses, particularly in the rotor, which reduce the machine's load limit. Local asymmetries can be taken into account, if they are known, in the dimensioning of the generator.

4.3.3.3 Voltage changes, voltage fluctuations and flicker

If a large proportion of the grid load is supplied by wind turbines, output variations due to wind-speed changes can cause voltage fluctuations and flicker effects in normal operation, as described in the preceding section. These are mainly determined by the path of the apparent power of the supply over time, its relation to grid short-circuit power and the corresponding phase angle. Harmonic power can be disregarded here. Voltage changes can occur in specific situations, e.g. as a result of load changes, load or generator connection or release, switching between generator levels, wind-speed fluctuations, tower effects, etc. [4.52].

In the case of severe output fluctuations, the standalone operation of small wind turbines in low-output isolated grids and the connection of large turbines to low-output points in the combined grid can lead to voltage changes. These can be expected in particular in the case of generators connected to the grid at a fixed speed. Large asynchronous generators, in particular, pass output changes caused by wind-speed fluctuations directly on to the grid because of their low slip values [4.53].

In a flexible-speed connection between the wind wheel and grid, on the other hand, a significant smoothing of output can be achieved by the short-time intermediate storage of energy in the rotating masses of the drive train (see Section 2.4). As described in Section 2.5, wind turbines exhibit higher rotor run-up time constants with increasing size and nominal output (Figure 2.70). Larger turbines can thus achieve significantly better output smoothing using variable-speed operation, particularly in the short-time range (Hz range), than smaller units. The speed regulation range is also a contributory factor to the degree of smoothing, with a large speed variation range being more capable of suppressing output fluctuations. Start-up currents do not play a role for turbines coupled via frequency converters, as they contribute less than 1 % of nominal current and can be run up. Short-time voltage changes and flicker are therefore of secondary importance in this type of turbine [4.54].

Small- and medium-sized wind turbines can normally only significantly influence the grid when connected in large groups. The connection of many units, which can exhibit a high level of output fluctuations individually, nevertheless brings about better smoothing of the total output as the number of turbines increases (see Section 4.3.2). Therefore, in normal operation, no harmful voltage fluctuations can be expected [4.55].

The standalone operation of large wind turbines with high output fluctuations can lead to voltage changes in weak grids. The output fluctuations of large units should therefore be as low as possible. This can be achieved by variable-speed turbine and generator systems owing to the large rotor run-up time constant associated with the high centrifugal mass.

Similar grid effects can also be expected for small wind turbines that supply very low-output isolated grids. Due to the low rotor time constant, smoothing effects comparable to those of large turbines can be achieved for small units by increasing the speed adjustment range [4.56].

In contrast to the guidelines aimed mainly at consumers [3.1], in determining the voltage changes caused by the connection of wind turbines the specific requirements of generating systems with regard to the direction of so-called 'voltage drop' in transformers and lines must be taken into account (see Figure 4.39) [4.57, 4.58].

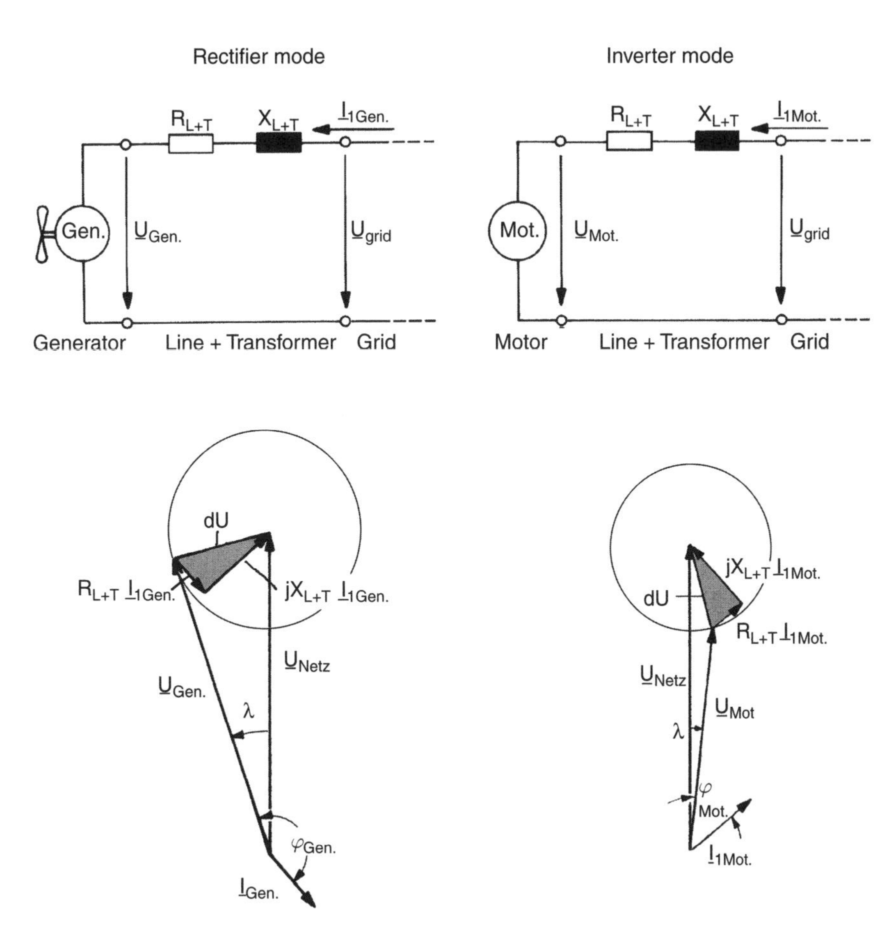

Figure 4.39 Equivalent circuit diagram and vector diagram for generators/motors at the grid

This representation shows that in the motor mode, motor voltage $\underline{U}_{\text{Mot}}$ can generally be determined with adequate precision as the difference between the contribution of grid voltage $\underline{U}_{\text{Grid}}$ and that of voltage drop $\text{d}U = (R_{\text{L+T}} + jX_{\text{L+T}})\underline{I}_{\text{1Mot}}$. The vector diagram for generator operation, on the other hand, shows that grid voltage $\underline{U}_{\text{Gen}}$ and generator voltage $\underline{U}_{\text{Grid}}$ have roughly the same values. However, the voltage drop here causes a clear phase difference between the two voltages, the so-called line angle λ. It is thus clear that the generator voltage can exceed or fall below the grid voltage, depending upon the position of the generator current I_{gen} and the ratio of ohmic to inductive components $R_{\text{L+T}}/X_{\text{L+T}}$ in the lines (overhead power transmission lines, cables) and transformers.

Voltage change, voltage rise

Normally, when power is supplied load components are low in relation to the supply power. Figure 4.40 illustrates in simplified form the grid configuration and associated voltage relationships for the following calculations. Based upon the grid voltage U_1 or the nominal value U_{N} at the higher Grid Point (GP), the corresponding value U_{PCC} is found at the point

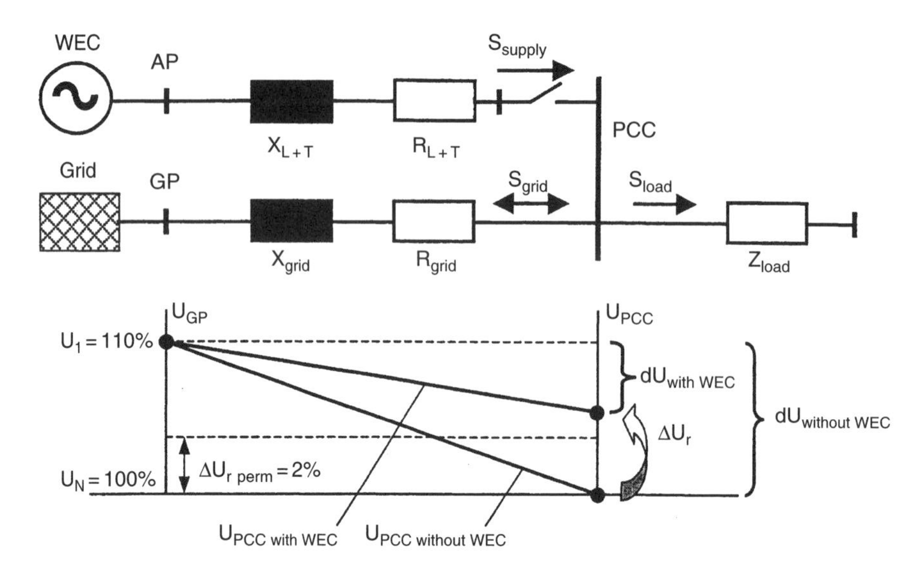

Figure 4.40 Equivalent circuit diagram and voltage change at the point of common coupling of a grid with and without wind energy converters

of common coupling (PCC) given the voltage drop dU with or without supply by a wind turbine at the Connection Point (CP). Here ΔU_{r} characterizes the voltage rise and $\Delta U_{\mathrm{r\ perm}}$ characterizes its permissible values (e.g. 2 % in the medium-voltage network).

Figures 4.41 and 4.42 show the relative change in voltage at the point of supply to the grid in relation to the power factor (cos φ capacitive or cos φ inductive) [4.59, 4.60] if the supply power $S_{\mathrm{supply}} = 1, 2, 4$ and 10 % of the grid short-circuit power S_{k} or, in accordance with Equation (4.13), the reciprocal $K = k_{\mathrm{Kl}} = S_{\mathrm{k}}/S_{\mathrm{supply}} = 100, 50, 25$ or 10. This illustrates the influence of supply ratios on voltage changes particularly well. This calculation is based upon a ratio of active to reactive components for transformers and lines of $R_{\mathrm{L+T}}/X_{\mathrm{L+T}} = R/X = 1(\mathrm{a}), 0.5(\mathrm{b}), 0.25(\mathrm{c})$ and $0.1(\mathrm{d})$. A value of R/X of 1(a) characterizes a weak grid connection and a value of 0.1(d) characterizes a strong grid connection with correspondingly large (a) or small (d) voltage changes at different short-circuit power ratios ($K = S_{\mathrm{K}}/S_{\mathrm{supply}}$).

Figure 4.42 shows, firstly, specifically the ratios according to Figure 4.41 for a pure effective power supply at cos $\varphi = 1$ and, secondly, the ratios for a voltage-neutral supply, at d$U = 0$ for ratios of grid short-circuit to supply power $K = S_{\mathrm{k}} = S_{\mathrm{supply}} = 10, 25, 50$ and 100 at ratios of effective to reactive components $R_{\mathrm{L+T}}/X_{\mathrm{L+T}} = R/X = 1(\mathrm{a}), 0.5(\mathrm{b}), 0.25(\mathrm{c})$ and $0.1(\mathrm{d})$. The range of voltage change is limited here to the permissible value of 2 %. It is clear from this that a near-voltage-neutral effective power supply is only possible in transmission links with low ohmic components. Voltage changes can thus be reduced or avoided by a supply that is matched to system and grid data, even in the event of power fluctuations.

In the integration of supply systems that permit the phase angle φ_{Gen} to be freely adjusted (e.g. synchronous generators, pulse-controlled inverters) it is thus possible to influence the voltage of grid points. Inductive supply leads, according to the above figures, to a reduction in voltage. Capacitive supply, on the other hand, leads to a voltage rise.

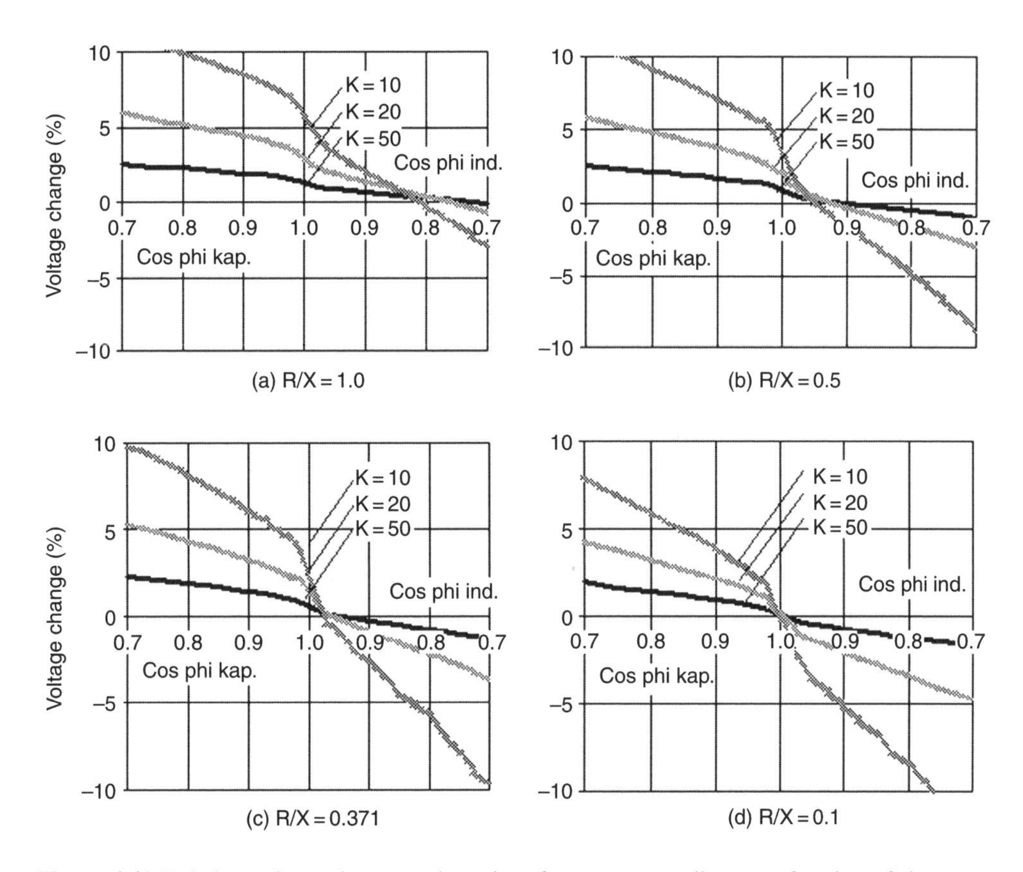

Figure 4.41 Relative voltage change at the point of common coupling as a function of the power factor at different supply ratios $K = S_K/S_{supply}$

Furthermore, the ratios between supply power and load power at the supply point or point of common coupling should also be taken into consideration. According to Figure 4.40 we find

$$U_{PCC} = U_1 - \frac{\left(P_{load} - P_{supply}\right) R_{grid} + \left(Q_{load} - Q_{supply}\right) X_{grid}}{U_1}, \tag{4.10}$$

where

$$S_{grid} = S_{load} - S_{supply} = \left(P_{load} - P_{supply}\right) + j\left(Q_{load} - Q_{supply}\right). \tag{4.11}$$

In the same way, the voltage for the point of common coupling U_{PCC} can also be found in accordance with Figure 4.40, taking into consideration the resistance and reactance components of the line (Z_L). Figure 4.43 illustrates the supply-dependent and load-dependent voltage changes at the point of common coupling for the (in practice often striven for) case of supply at $Q_{supply} = 0$ or $\cos\varphi = 1$. The representations relate (a) to strong grids ($R_{grid}/X_{grid} = 0.1$) with low voltage changes at large power differences and (b) to weak grids ($R_{grid}/X_{grid} = 1.0$) with significantly greater effects. Accordingly, the right-hand side of the

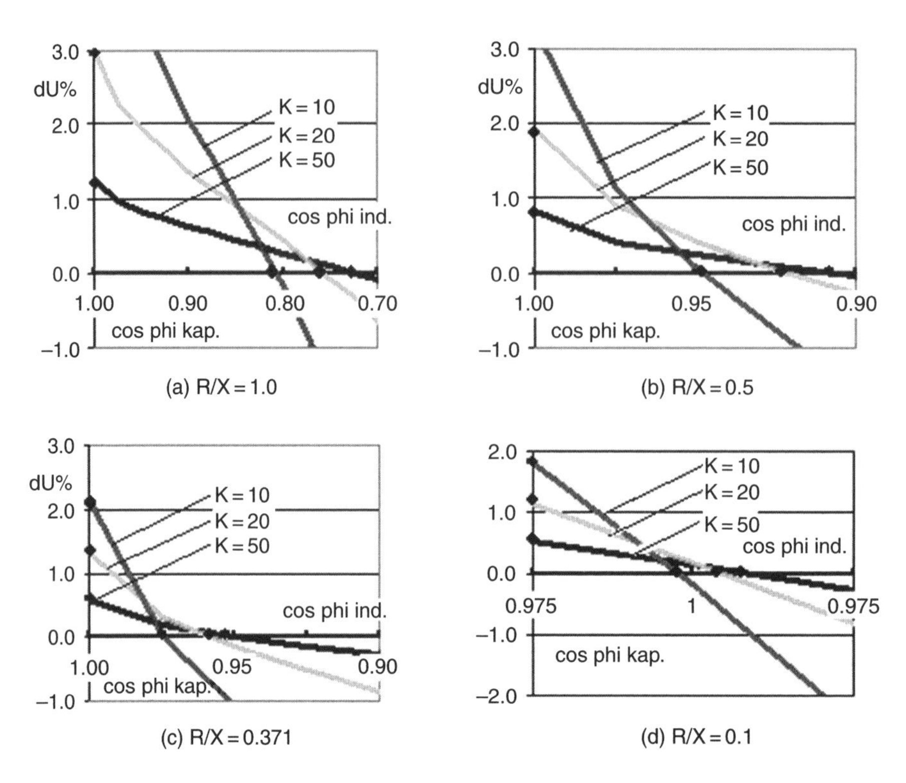

Figure 4.42 Determination of the relative voltage change dU with an effective power supply ($\cos\varphi = 1$) and of the power factor $\cos\varphi$ at a voltage-neutral supply (d$U = 0$)

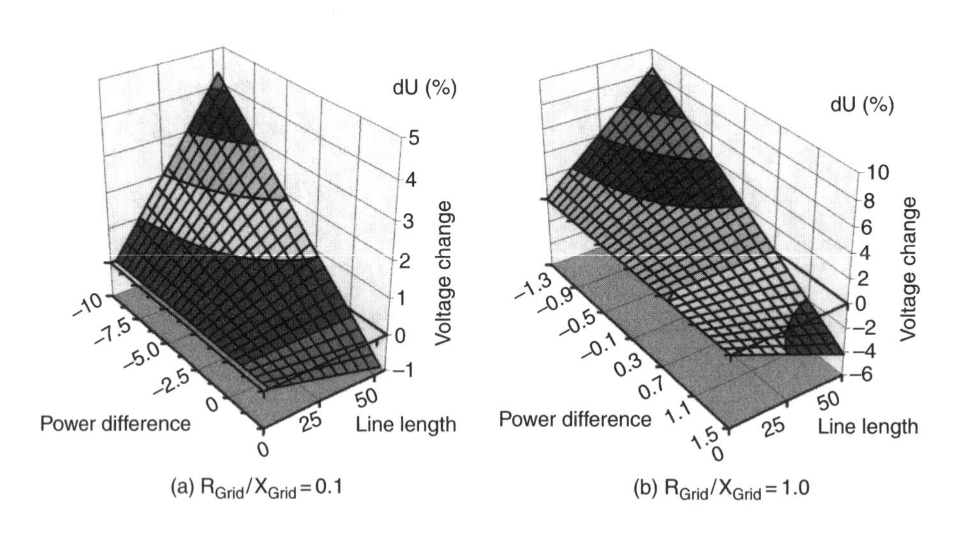

Figure 4.43 Relative voltage change at the point of common coupling as a function of the power difference ($P_{\text{load}} - P_{\text{supply}}$ in MW, supply and reference pure active power) and the line length (in km) at different line ratios $R_{\text{grid}}/X_{\text{grid}}$

voltage area is changed according to reactive power. This property of the grid at the point of common coupling thus permits voltage changes to be kept within the permissible range ($\Delta U_{PCC} \leq 2\,\%$), for example, by a suitable choice of the ratio of effective to reactive power supply.

The voltage change can be tuned at the supply point of the wind energy converter (connection point), the grid connection to the consumer (point of common coupling) or at a higher grid connection point by voltage regulation in the grid (see Section 4.6 on 'grid control'). The last-mentioned requires the monitoring of voltage readings and their transmission from the reference point to the supply point. In this manner, voltage fluctuations at the higher-voltage point, which can occur particularly in weakly loaded grids due to a reversal of the energy flow direction caused by the supply of wind power, can be minimized.

Figure 4.44 shows the voltage deviation in relation to the supply current angle for the motor and generator mode of a machine in a weak grid with a short-circuit power of 8 MVA and, for comparison, a stronger grid (32 MVA). This clearly shows that a generator in the grid configuration selected here can supply inductively at around 18° and a motor can be operated capacitively at 23°, in order to maintain grid and machine voltage at the same level (e.g. $dU = 0$) [4.29]. Away from these supply or reference angles, large changes in voltage are found at low short-circuit power and small effects are found for strong grids. This neutral-voltage angle and the corresponding power factor are dependent upon the ratio of active to reactive components of the transformer and line for the transmission links.

Evaluation criteria and limit values for voltage changes at the point of common coupling are cited in reference [3.1] with 3 % recommended for low-voltage ranges and 2 % for medium-voltage ranges. This includes the provisions according to EN 50 160 as well as the

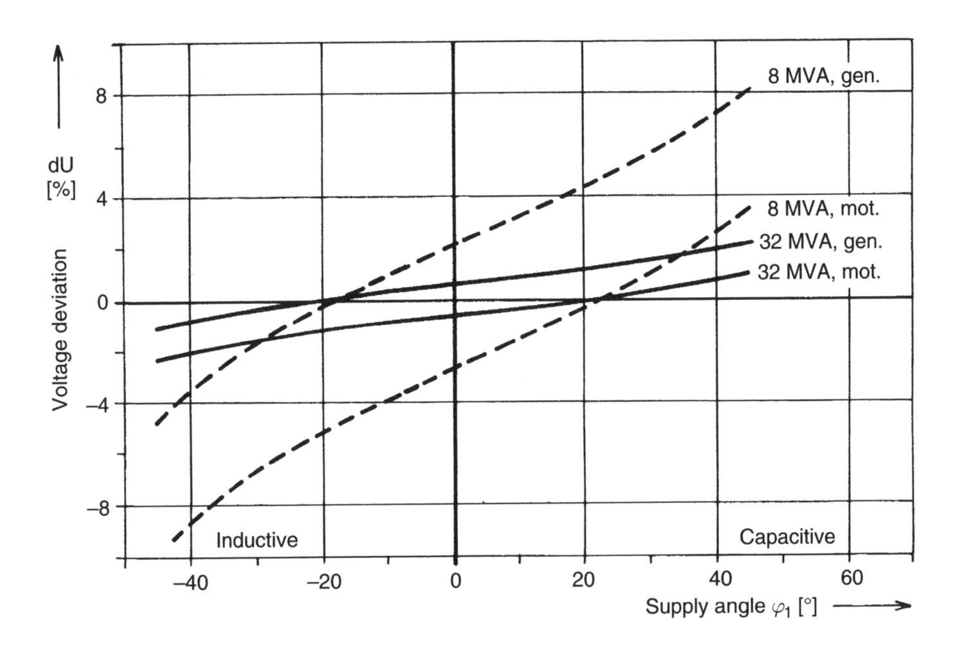

Figure 4.44 Voltage deviation as a function of the supply angle, with grid short-circuit power as a parameter

technical connection regulations of the power supply company. An assessment must take place in cooperation with the power supply company.

According to reference [4.22] the voltage rise at the point of common coupling can be approximated as a function of the maximum apparent power of the turbine $S_{A\,max}$, the grid impedances R_{kV} and X_{kV} at the point of common coupling and the phase angle φ of the in-plant generator according to

$$\Delta u_a = \frac{S_{A\,max}\,(R_{kV}\cos\varphi - X_{kV}\sin\varphi)}{U^2}, \tag{4.12}$$

where U is the nominal voltage of the medium-voltage grid.

The assessment of voltage changes is usually based on the ratio of grid short-circuit power to maximum turbine output, the so-called short-circuit power ratio

$$k_{kl} = \frac{S_{kV}}{\sum S_{A\,max}} \geq 50, \tag{4.13}$$

where S_{kV} represents the short-circuit power at the point of common coupling. This evaluation only applies for radial networks and leads to an absolute consideration which supplies a good approximation for motors and other consumers. In the generator mode, however, there can be significant deviations from the results according to Equation (4.12), as shown in Figure 4.36.

According to reference [3.1], the magnitude and direction of the active and reactive components in accordance with the vector diagram (Figure 4.36) can be included in the evaluation of the voltage changes. This generally yields lower grid effects.

If the grid impedance angle ψ_{kV} and the grid short-circuit power S_{kV} at the point of common coupling are known, the voltage rise is simple to calculate, taking into account the supply angle φ, by

$$\Delta u_a = \frac{S_{A\,max\,1\,min}\cos(\psi_{kV} + \varphi)}{S_{kV}}. \tag{4.14}$$

Here

$$S_{A\,max\,1\,min} = \frac{P_{NG}\,p_{1\,min}}{\lambda} \tag{4.15}$$

represents the maximum turbine mean value of apparent power for the duration of one minute, P_{NG} represents the nominal output of the generator, λ represents the power factor, which generally corresponds to the fundamental value ($\lambda = \cos\varphi$), and $p_{1\,min}$ represents the maximum relative active power over one minute (e.g. $p_{1\,min} = 1.3$). We will therefore not go into more detail on this subject at this point.

In addition to voltage changes related to turbine and primary available output, switching processes, tower effects, etc., lead to aperiodic or periodic voltage changes and flicker, particularly in fixed-speed wind turbines.

Switching-dependent voltage changes

Voltage changes due to the connection and disconnection of individual systems and generator systems may not generally exceed

$$\Delta u_{max} = k_{i\,max}\frac{S_N}{S_{kV}} \leq 2\,\% \tag{4.16}$$

if switching processes take place no more frequently than once every 1.5 minutes. At very low switching frequencies the power supply companies may permit greater values. In Equation (4.16), S_N represents the nominal apparent power of a single unit and $k_{i\,max}$ represents the ratio of start-up to nominal current of the generator. If no figures are available, a maximum value of $k_{i\,max} = 8$ can be assumed for asynchronous generators, for example. If they are connected at close to their synchronous speed $(0.95n_1 \geq n \geq 1.05n_1)$ a value of $k_{i\,max} = 4$ can be assumed. In the case of ideally synchronized synchronous generators and inverters, by contrast, calculations can be based upon $k_{i\,max} = 1$.

The 'grid-dependent switched current factor' $k_{i\psi}$ should be provided and proven by turbine manufacturers or included in testing reports supplied with the wind turbine. This figure evaluates the level of current and its path over time during the switching process and is quoted as a function of the grid impedance angle ψ in the testing report. This allows the fictitious 'equivalent voltage change'

$$\Delta u_{eq} = k_{i\psi}\frac{S_N}{S_{kV}} = \frac{k_{i\psi}}{\lambda} \times \frac{P_N}{S_{kV}} \quad \text{where} \lambda = \cos\varphi \tag{4.17}$$

to be determined, which again may not exceed a limit value of 2 %. The switching processes of several generators should be staggered over time. In the event of maximum voltage changes the interval must be at least 1.5 minutes and at half the permissible value it may be a maximum of 12 seconds.

4.3.3.4 Flicker

Periodic and aperiodic fluctuations in grid voltage cause corresponding changes in brightness in filament bulbs, which causes a 'flickering' of the light. Due to the high sensitivity of the human eye to changes in brightness, deviations of the voltage supplied to consumers from the steady-state value (usually nominal voltage) must be limited. An assessment plan, evaluation criteria and permissible limit values, which characterize the subjective impression of brightness changes, are given in detail in reference [3.1].

In lighting engineering, brightness variations are called 'flicker'. Thus flicker must in the future be related to the voltage deviations that cause it. When discussing flicker disturbance factors, we must differentiate between short-term mean values

$$A_{st} = P_{st}^3 \tag{4.18}$$

in a 10 minute interval (with the short-term flicker level P_{st}, which will not be considered further here) and the 2 hour mean value A_{lt} that prevails in the long term,

$$A_{lt} = P_{lt}^3 = \frac{1}{12}\sum_{i=1}^{12} A_{st\,i} \leq 0.1, \tag{4.19}$$

where the long-term flicker value is

$$P_{lt} = \sqrt[3]{\frac{1}{12}\sum_{i=1}^{12} P_{st\,i}^3} \leq 0.46. \tag{4.20}$$

EN 50 160 allows $P_{st} = 1.0$ (Europe wide). The voltage deviations due to flicker are thus made up of 12 sequential short-term values P_{st} over a period of 120 minutes. Furthermore, P_{st} is proportional to the value of the voltage change. A_{st} is inversely proportional to the repeating rate of the voltage.

For wind turbines the turbine flicker coefficient c and the flicker phase angle $\varphi_f > 0$ are quoted in test reports or the manufacturer's data. These turbine-specific values, along with the nominal power of the individual turbine S_N and the short-circuit power at the point of common coupling S_{kV} according to

$$P_{lt} = c \frac{S_N}{S_{kV}} |\cos(\psi_{kV} + \varphi_f)| \tag{4.21}$$

determine the flicker generated by the plant at the point of common coupling.

When several individual turbines and wind farms are connected, the long-term flicker level $P_{lt\,i}$ should be determined separately for each turbine and the resulting value for the flicker disturbance factor calculated using

$$P_{lt\ res} = \sqrt{\sum_i P^2_{lt\,i}}. \tag{4.22}$$

If we are considering n identical plants, the resulting flicker disturbance factor can be calculated as follows:

$$P_{lt\ res\ estimated} = \sqrt{n}\, P_{lt} = \sqrt{n}\, c \frac{S_N}{S_{kV}}. \tag{4.23}$$

If the limit value is exceeded, the flicker-related phase angle φ_f according to

$$P_{lt\ res} = P_{lt\ res\ estimated} |\cos(\psi_{kV} + \varphi_f)| \tag{4.24}$$

can be taken into account.

The above discussions show that disruptive voltage fluctuations can be minimized, or even avoided altogether, in all configurations by suitable turbine design and by control engineering intervention [4.58]. If necessary, measures for voltage support can be taken (see Section 4.6), either by rotating or power electronic phase shifters or directly by turbine frequency converters.

The desired speed variations in wind turbines can be achieved in the case of fixed-frequency coupling by changing the gear transmission or in fixed transmission by adjusting the generator frequency with the aid of electronic power frequency conversion systems. However, the grid compatibility of these components with regard to harmonics and subharmonics, which will be described briefly below, is of particular importance in this context.

4.3.4 Harmonics and subharmonics

Wind energy can make a useful contribution to the supply of electrical power if high levels of power can be achieved over the usable land area. Existing or new grid connections must be used as cost-effectively as possible. This can be achieved by connecting several or a great many wind turbines together to form wind farms. With the power concentrations found here, however, the compatibility of the supplying turbines with the grid must be ensured, since the grid's capacity is limited. Clear differences are evident here, depending upon the design of the system and the construction of components. Therefore, tendential relationships and problems that can be expected in the connection of different configurations will be examined below.

The combined effect of the units can give rise to

- smoothing or weakening effects or
- amplifying influences

on the harmonics and subharmonics in the grid, and these effects must be assigned particular importance. A basic discussion of circuitry and the design of power converters, generators, etc., including likely grid effects, has already been given in the preceding sections. Therefore, we will limit the following discussion to some comments on fundamental oscillations and harmonics.

Harmonics are sinusoidal oscillations of currents and voltages, the frequencies f_ν of which are integer multiples of the fundamental frequency of the grid. These occur if currents or voltages have a nonsinusoidal form. They can be caused in electricity supply grids by

- consumers with nonlinear impedance characteristics,
- saturation influences in electrical machines and
- power electronic equipment.

The following effects in particular should be highlighted:

- functional disturbances in power electronic equipment that is not resistant to jamming,
- premature ageing of dielectrics and insulation materials, which may for example, lead to the destruction of capacitors,
- disturbance of audio-frequency ripple control systems and
- possible faults in protective devices.

Periodic nonsinusoidal paths of deformed currents and voltages over time (see Figure 4.16) can be considered to be the superposition of the fundamental with different harmonics, and may be represented, for example, for voltage in the form of an equation

$$u(t) = U_0 + \sum_{\nu=1}^{\infty} U_\nu \sin(\omega_\nu t + \varphi_\nu). \tag{4.25}$$

Similar equations can be derived for current. Here U_0 represents the direct-voltage component, U_ν the amplitude, ω_ν the angular frequency and φ_ν the phase displacement angle of the νth harmonic, where

$$\omega_\nu = 2\pi f_\nu = 2\pi \nu f_1 \tag{4.26}$$

characterizes the multiple of fundamental frequency f_1 and the corresponding ordinal number

$$\nu = \frac{f_\nu}{f_1} \tag{4.27}$$

Table 4.5 Compatibility levels $U_{\nu VT}$ in the medium-voltage grid for harmonics up to the ordinal number $\nu = 25$ (EN 6100-2-2 or VDE 0839 Part 88, IEC 77A)

Odd values of ν that are not divisible by 3									
Ordinal number ν	5	7	11	13	17	19	23	25	>25
$U_{\nu VT}(\%)$	6.0	5.0	3.5	3.0	2.0	1.5	1.5	1.5	$0.2+1.3\times 25/\nu$
Odd values of ν that are divisible by 3									
Ordinal number ν	3	9	15	>15					
$U_{\nu VT}(\%)$	5.0	1.5	0.3	0.2					
Even values of ν									
Ordinal number ν	2	4	6	8	10	>10			
$U_{\nu VT}(\%)$	2.0	1.0	0.5	0.5	0.5	0.2			

characterizes their integer multiples. These parameters can be found mathematically by a Fourier analysis and using measuring techniques with the aid of frequency-selective evaluation.

Sinusoidal oscillations with frequencies that do not correspond to integer multiples of the basic frequency f_1 are called subharmonics. This type of oscillation is mainly caused by frequency converters, but can also be caused by rotating electrical machines. For differentiation from the integer ordinal number ν of harmonics, the noninteger factors

$$\mu = \frac{f_\mu}{f_1} \tag{4.28}$$

are used to characterize the subharmonics. The subharmonic frequencies mainly occur as upper and lower sidebands of harmonic frequencies and can be lower than the fundamental frequency (i.e. $\mu < 1$).

Ensuring the quality of supply in the public grid necessitates the restriction of harmonics and subharmonics to a grid-compatible level. EN 61000-2-2 stipulates the so-called compatibility level $u_{\nu VT}$ as limit values for individual harmonic frequency voltages, which may not be exceeded, according to the equation

$$\frac{U_\nu}{U_1} = u_\nu \leq u_{\nu VT} \tag{4.29}$$

for the harmonic levels U_ν in relation to the fundamental U_1. U_ν and U_1 represent root-mean-square (r.m.s.) values. For the sake of simplicity, U_1 can be replaced by the nominal value of the grid voltage for the electrical supply grid. The compatibility levels for harmonics are listed in Table 4.5; for subharmonics, a limit value of

$$u_{\mu VT} = 0.2\,\%$$

is generally specified.

The compatibility levels are valid in the low-voltage and medium-voltage grid, and must be maintained at the so-called point of common coupling and the point of supply. They facilitate the evaluation of individual harmonic levels.

Table 4.6 Total harmonic currents permissible at a medium-voltage grid, related to short-circuit power, that are generated by directly connected turbines [4.22]

Ordinal number ν, μ	Permissible specific harmonic current $i_{\nu,\mu\ \mathrm{perm.}}$ (A/MVA)	
	10 kV grid	20 kV grid
5	0.115	0.058
7	0.082	0.041
11	0.052	0.026
13	0.038	0.019
17	0.022	0.011
19	0.018	0.009
23	0.012	0.006
25	0.010	0.005
>25 or even number	$0.06/\nu$	$0.03/\nu$
$\mu < 40$	$0.06/\nu$	$0.03/\nu$
$\mu > 40$[a]	$0.18/\nu$	$0.09/\nu$

[a] Integer and noninteger within a bandwidth of 200 Hz.

In wind turbines with frequency converters, the harmonic currents should be documented by the manufacturer (e.g. during type testing). According to reference [4.22], with only one point of common coupling in the medium-voltage grid the permissible harmonic currents found are from the specific values $i_{\nu,\mu,\ \mathrm{perm.}}$ according to Table 4.6 multiplied by the short-circuit power at the point of common coupling, i.e.

$$I_{\nu,\mu\ \mathrm{perm.}} = i_{\nu,\mu\ \mathrm{perm.}} S_{\mathrm{kV}}. \tag{4.30}$$

If several turbines are connected at the point of common coupling, the permissible harmonic currents for a turbine are found by multiplying by the ratio of turbine apparent power S_{A} to the power that can be drawn or the planned supply power S_{AV} at the point of common coupling, according to

$$I_{\nu,\mu,\mathrm{A\ perm}} = I_{\nu,\mu\ \mathrm{perm.}} \frac{S_{\mathrm{A}}}{S_{\mathrm{AV}}} = I_{\nu,\mu\ \mathrm{perm.}} S_{\mathrm{kV}} \frac{S_{\mathrm{A}}}{S_{\mathrm{AV}}}. \tag{4.31}$$

When several wind turbines of the same type are connected together the turbine output is replaced by the sum of the individual turbine outputs

$$S_{\mathrm{A}} = \sum S_{\mathrm{nE}},$$

i.e.

$$I_{\nu,\mu,\mathrm{A\ perm}} = I_{\nu,\mu\ \mathrm{perm.}} \frac{S_{\mathrm{nE}}}{S_{\mathrm{AV}}} = I_{\nu,\mu\ \mathrm{perm.}} S_{\mathrm{kV}} \frac{S_{\mathrm{nE}}}{S_{\mathrm{AV}}}. \tag{4.32}$$

In turbines that are not of the same type this value represents only an upper estimate according to reference [4.22].

Medium-voltage to low-voltage voltage transformers that do not transmit a zero phase-sequence system are normally used in public supply grids. In this type of grid connection, the values given in Tables 4.5 and 4.6 for the nearest order can be taken as the basis for harmonics with ordinal numbers not divisible by three.

One measure of the distortion of grid voltage is represented by the harmonic distortion factor. In this context we must differentiate between the harmonic distortion factor k_u, as defined in the past by DIN EN 61000-2-2, and the total harmonic distortion k_{THD}. The harmonic distortion factor gives the ratio of the actual value of the oscillation without the fundamental to the actual value with the fundamental for voltage, according to

$$k_u = \sqrt{\frac{\sum_{\nu=2}^{50} U_\nu^2}{\sum_{\nu=1}^{50} U_\nu^1}} = \sqrt{\frac{U_2^2 + U_3^2 + U_4^2 + \cdots}{U_1^2 + U_2^2 + U_3^2 + U_4^2 + \cdots}}. \tag{4.33}$$

Total harmonic distortion, on the other hand, relates the actual value of the oscillation without the fundamental to the fundamental of voltage, according to the equation

$$k_{THD} = \sqrt{\frac{\sum_{\nu=2}^{50} U_\nu^2}{U_1^2}} = \sqrt{\frac{U_2^2 + U_3^2 + U_4^2 + \cdots}{U_1^2}}. \tag{4.34}$$

The two factors are the same at low values of up to around 20 %. From 30 %, small differences are found, e.g. where $k_{THD} = 0.5$, $k_u = 0.45$, or where $k_{THD} = 1$ the corresponding value of k_u is 0.7 [4.61].

In practice, owing to the limited bandwidth of the grid transmission elements, the spectrum of harmonics is considered up to an ordinal number $\nu = 40$ or up to a frequency $f_\nu = 2\,\text{kHz}$.

Of all the different methods of connecting wind turbines to the grid, line-commutated frequency converters in particular create harmonics. These can be attributed to the approximately square-wave grid currents and the commutation grid voltage dips. However, such systems were only used in wind turbines up until the end of the 1990s. Today IGBT frequency converters are used, which due to their high switching frequencies can supply near-sinusoidal currents.

An inverter of pulse number p generates – if we disregard commutation effects and assume an ideally smoothed direct current in the d.c. link – a harmonic spectrum with the ordinal numbers

$$\nu = kp \pm 1, k = 1, 2, 3, \ldots. \tag{4.35}$$

The amplitudes or r.m.s. values of the current

$$I_\nu = \frac{I_1}{\nu} \tag{4.36}$$

decrease according to the ordinal number ν. For a six-pulse inverter, the 5th and 7th harmonics are thus the strongest. These do not occur, however, for a twelve-pulse inverter, which means that the ordinal numbers 11 and 13 are the first harmonics. All the subsequent harmonics

(17, 19, 23, 25, 29, 31, etc.) exhibit correspondingly lower amplitudes. Taking commutation effects into account, harmonics of other ordinal numbers also occur in practically measurable spectra. In addition, some levels deviate from the values given in Equation (4.20). The emphasis is placed upon the realization of practical findings in the descriptions below.

For wind turbines with asynchronous generators and direct connection to the grid, harmonics have been measured in the normal form up to over 2000 Hz and their amplitude spectra are represented logarithmically up to the 40th harmonic. This showed that a greater number of turbines caused lower amplitudes of harmonics and subharmonics, in particular those of a low order (cf. Figure 4.45).

This phenomenon can be attributed to an increase in the capacity of the grid due to the increase in

- short-circuit power and
- the filtering effect of the wind turbine with the capacitive compensation system,

which increase with the increasing number of turbines. Increases of individual harmonics (here the 11th), which sometimes occur, can be reduced, if this is necessary, by the design of the magnetic circuit and the coil to achieve the desired values.

In contrast to this, turbines coupled to the grid via six-pulse thyristor power converters usually exhibit different behaviour with regard to harmonic effects. For a larger number of converters, and thus higher power components in the grid supply, the amplitudes of harmonics are higher over the entire frequency spectrum, as shown in Figure 4.46.

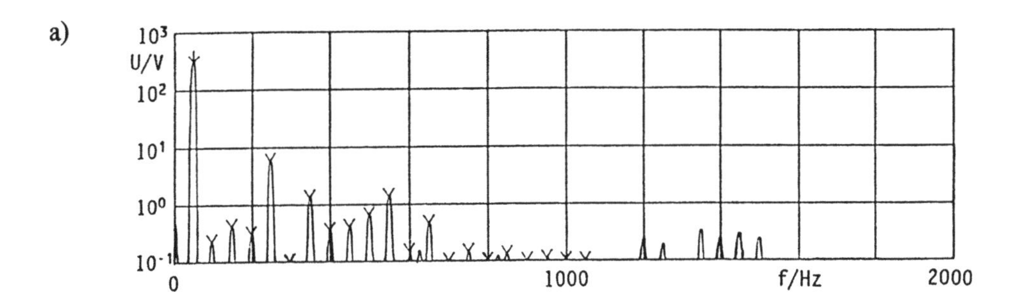

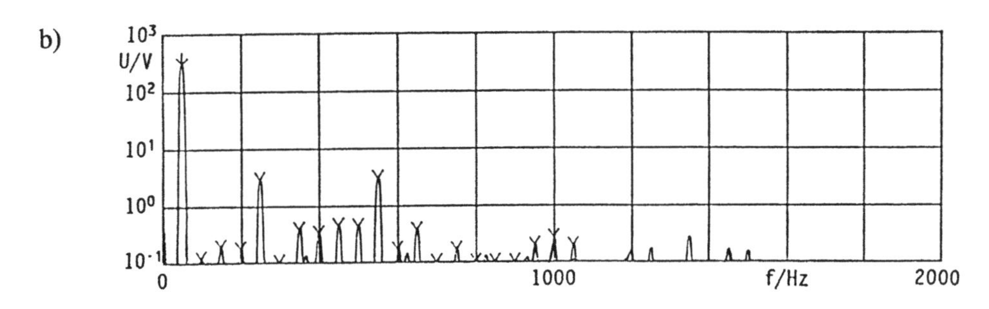

Figure 4.45 Voltage spectra of wind turbines with asynchronous generators and direct connection to the grid: (a) in solo operation (one turbine) and (b) in combined operation (18 turbines)

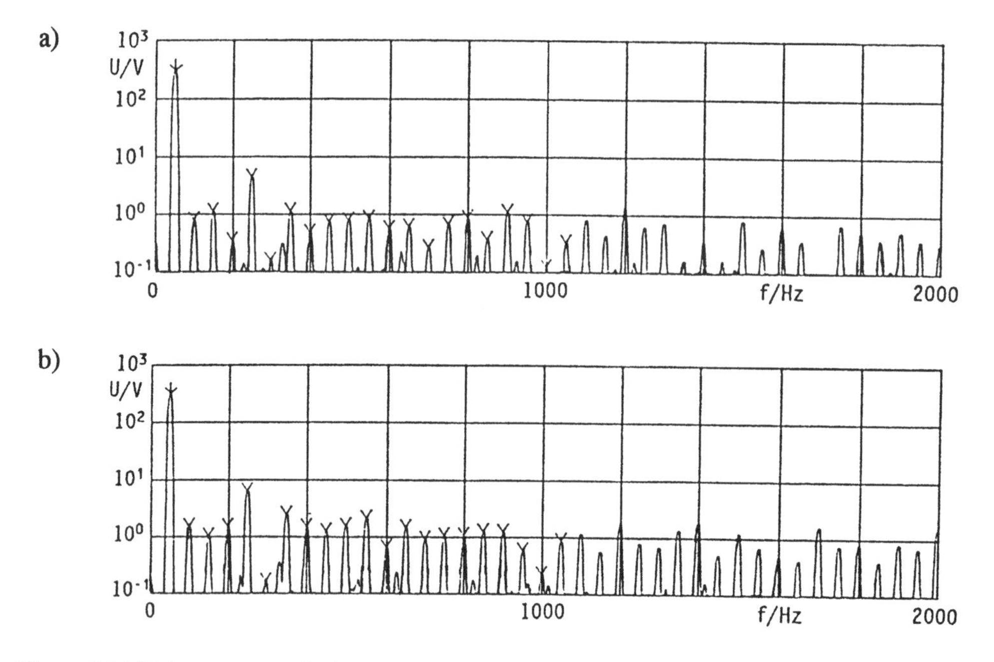

Figure 4.46 Voltage spectra of wind turbines with variable-speed generators and grid connections via line-commutated power converters (a) in solo operation (one turbine) and (b) in combined operation (five turbines)

We therefore find that for wind turbines that supply their energy directly to the grid from asynchronous generators the harmonic component is reduced as the number of turbines, and thus the output, increases. On the other hand, for turbines fitted with variable-speed generators and line-commutated power converters, a more marked influence on the grid by harmonics can be observed as the number of turbines, and thus the output, increases.

The contrasting behaviour of different types of coupling can, as shown in the following discussion, be used to reduce the grid effects caused by harmonics, when different systems are connected together.

The measurements in Figure 4.47 show that by the combination of wind turbine systems that have variable-speed synchronous generators and are connected to the grid via a power converter with those that have fixed-speed asynchronous generators and a direct connection to the grid, a clear reduction in harmonic and subharmonic amplitudes can be achieved, thus improving voltage characteristics (Figure 4.48). Upon closer inspection, lower commutation effects can be seen in Figure 4.48(b). This type of configuration therefore allows grid effects to be significantly reduced.

Data collated using random samples on states at different wind speeds and fluctuations at corresponding times and grid conditions show the following trends, as illustrated in Figure 4.49, for the harmonic distortion factor, this being the characteristic variable for harmonics. In general, the path of the harmonic distortion factor k_u and the total harmonic distortion k_{THD} for low turbine numbers up to around $n_A = 20$ can be approximated in the form

$$k_{THD} = k_{THD0} + k_{THD1} \ln(1 + k_{THD2} n_A), \qquad (4.37)$$

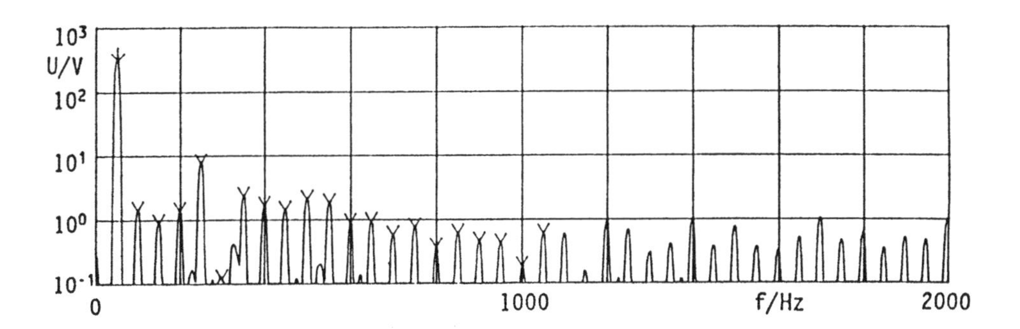

Figure 4.47 Voltage spectra for the combined operation of wind turbines with different grid connections (nine synchronous generators connected via power converters and two asynchronous generators coupled directly to the grid)

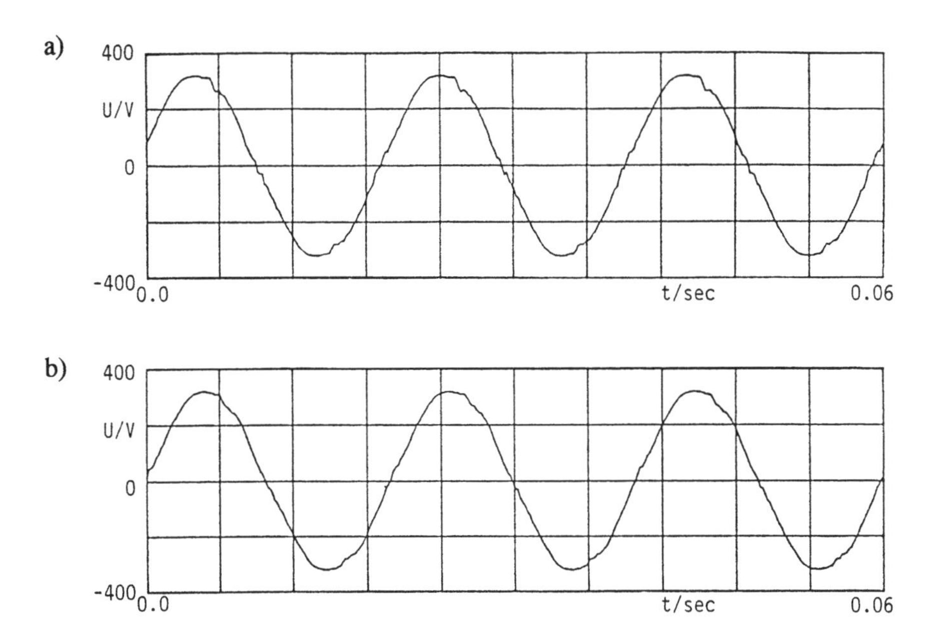

Figure 4.48 Graph of the grid voltage in phase L_1 for supply by (a) five power-converter-coupled turbines (275 kW) and (b) nine power-converter-coupled turbines (320 kW) in combination with two turbines with asynchronous generators (330 kW) coupled directly to the grid of around the same total output

where k_{KHD0} represents the initial value as the axis intercept, which is highly dependent upon the grid short-circuit power and the grid state. Typical values are

$$k_{\mathrm{THD0}} = 1.5 \text{ to } 3\,\%.$$

The gradient is represented by k_{TDH1}. In wind turbines supplied via power converters, this usually takes on positive values, e.g.

$$k_{\mathrm{THD1}} = 0.85$$

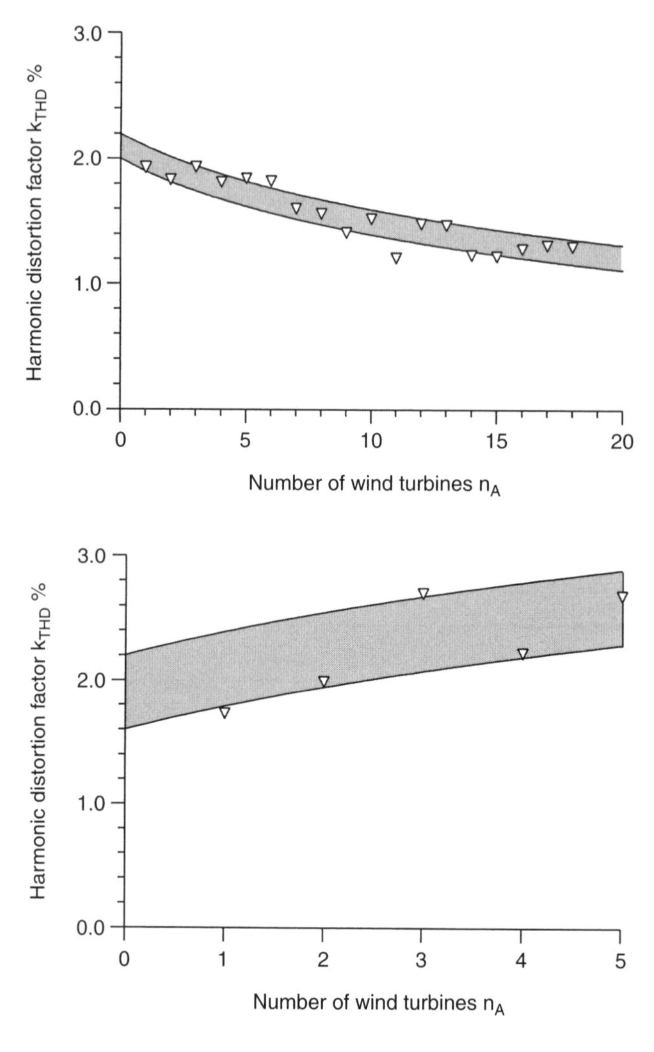

Figure 4.49 Harmonic distortion factor as a factor of the number of wind turbines for (a) asynchronous generators coupled directly to the grid and (b) synchronous generators operating through rectifiers, constant current d.c. links and inverters

and in turbines with generators coupled directly to the grid it usually takes on negative values, e.g.

$$k_{\mathrm{THD1}} = -0.55.$$

The factor k_{THD2} influences the elongation of the approximation function (Equation (4.37)) for values in the range

$$k_{\mathrm{THD2}} = 0.2 \text{ to } 0.4.$$

Pulse-controlled inverters are used for variable-speed systems in the current generation of wind turbines. Due to the relative high-frequency switching processes in the kHz range in power technology, currents and voltages can approximate the sinusoidal form very well by the selection of pulse pattern or so-called bandwidth control. Harmonics with low ordinal numbers are thus largely avoided. In the evaluation of compatibility levels the fixed or variable pulse frequencies of such systems generally lie at the upper end of the range of ordinal numbers and thus usually exhibit low amplitudes. If necessary, these can be reduced by the use of compact filters.

4.4 Resonance Effects in the Grid During Normal Operation

If the grid is considered at the point of connection of one or several wind turbines, the total impedance of the configuration is decisive for behaviour during supply, and this can be determined based on the prevailing combination of ohmic resistances, inductances and capacitances in the supplier, distributor and consumer systems. Depending upon the quantity, type of connection and impedance level of these individual elements, self-resonant frequencies can be determined for the part of the grid under consideration, which can cause rises in the current

$$I_\nu = \frac{U_\nu}{Z_\nu} \quad \text{where } Z_\nu \to 0 \tag{4.38}$$

or the voltage

$$U_\nu = I_\nu Z_\nu \quad \text{where } Z_\nu \to \infty \tag{4.39}$$

in connected components when excited. Such excitation can be brought about in the electrical supply grid by nonlinear suppliers and consumers, e.g. due to saturation effects in transformers, reactors and rotating electrical machines and in particular by power converter units, because in addition to the fundamental frequency these systems also give rise to higher-frequency components due to nonsinusoidal current and voltage paths.

The harmonic content in the grid can be kept low by the appropriate design of electric machines and can even be reduced if machines are selected specifically to achieve this (see Section 4.3.4). Therefore, in addition to power electronic consumer systems, resonance excitation in the grid can be particularly attributed to supply via power converters, which was common in wind turbines and controllable drives with line-commutated frequency converters until approximately ten years ago.

Taking the example of the Westküste wind farm at the beginning of the 1990s, the equivalent circuit diagram shown in Figure 4.50 [4.62] is derived from the block diagram in Figure 4.51. Using a program developed for this purpose [4.63], the impedance path can be graphically represented up to the 40th harmonic, as shown in Figure 4.52, for the point of connection to the harmonic source, shown here as a wind turbine with frequency converter supply.

Figure 4.52 illustrates noncritical operation in all frequency ranges. Relevant resonance points with local maximum impedance lie above the 13th harmonic (650 Hz), which means

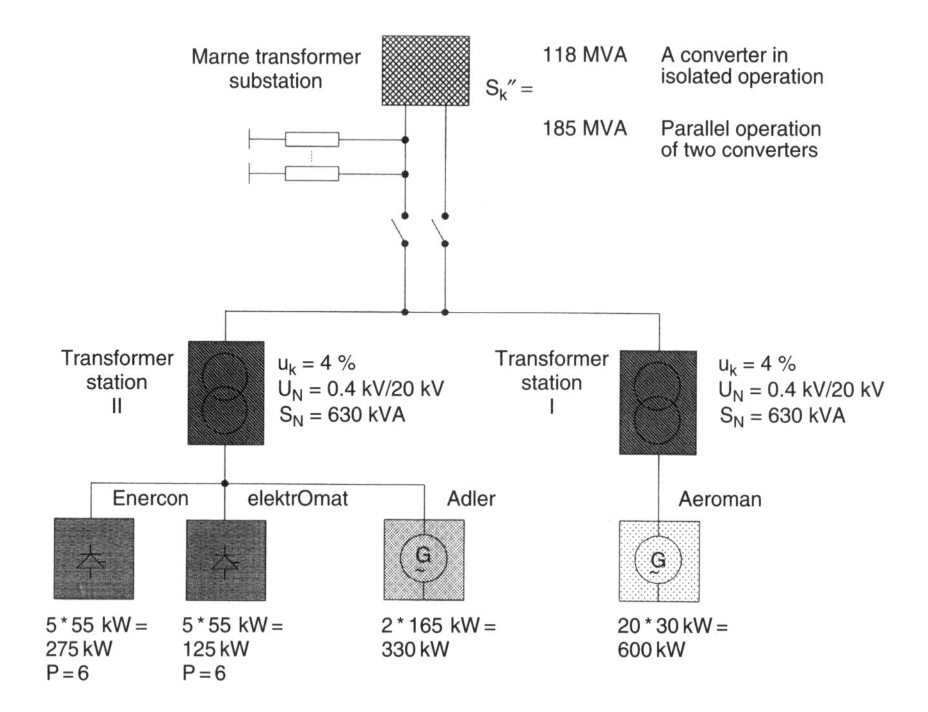

Figure 4.50 Equivalent circuit diagram of the Westküste wind farm [4.62], where 1 to 20 indicate electrical sections of the wind turbine with a compensation device (1990)

that at most one-thirteenth of the maximum value of the fundamental component of the current would be effective. Thus voltages capable of causing damage to components do not occur at any frequency.

The impedance path calculated in Figure 4.52 is shown for a short-circuit power of the Marne grid of $S_K'' = 185\,\text{MVA}$. If the Westküste wind farm were to supply a grid with around a quarter of the capacity, the impedance path shown in Figure 4.53 would prevail. This clearly illustrates that at lower short-circuit power the resonance point is moved to lower frequencies (above the 11th harmonic) and leads to higher impedances and thus a greater voltage drop at the point under consideration. With decreasing short-circuit power, therefore, operation becomes increasingly critical.

If the grid is connected via overhead power transmission lines, the stronger inductive behaviour of the overhead power transmission line slightly increases the resonant frequency compared with cables at the same maximum impedance. If individual wind turbines with asynchronous generators and compensation units are disconnected from the grid, small changes occur in the impedance path, as shown in Figure 4.54. Only in the transition between the connected groups can clear differences be recognized due to different line lengths.

Thus, for example, if the first turbine is connected with the corresponding grid section, the high maximum resonance at 1741 Hz ($\mu = 34.82$) is considerably reduced. Similarly, this trend is evident for the connection of the fifth turbine with its grid branch at the mean ordinal number ($\mu = 23.20$) and for the connection of the first turbine of the last group of

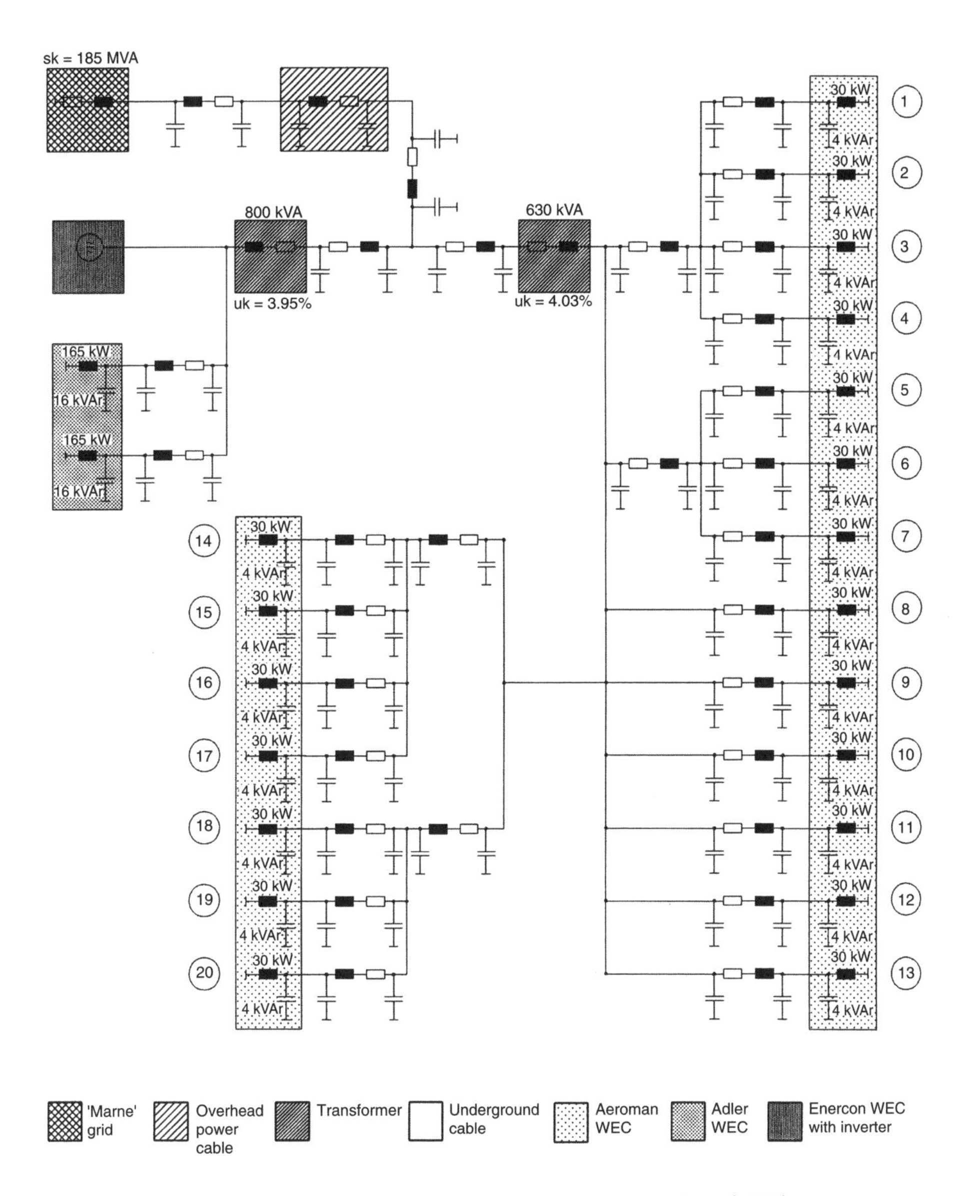

Figure 4.51 Block diagram of the Westküste wind farm (1990)

three (18th turbine) at the lower resonance point ($\mu = 14.04$). In contrast to this, if the stub cable is connected with the 14th turbine, a further resonance point ($\mu = 17.20$) is created.

In a summarized comparison between 20 turbines connected to the grid and their complete disconnection, the relevant impedance values are found to be only slightly changed. The slightly marked resonance point of 20 turbines at 675 Hz ($\mu = 13.50$) completely disappears when all turbines are disconnected. In contrast to this, the slightly higher value at 843 Hz ($\mu = 16.86$) is moved insignificantly to 830 Hz ($\mu = 16.60$). The next highest resonance

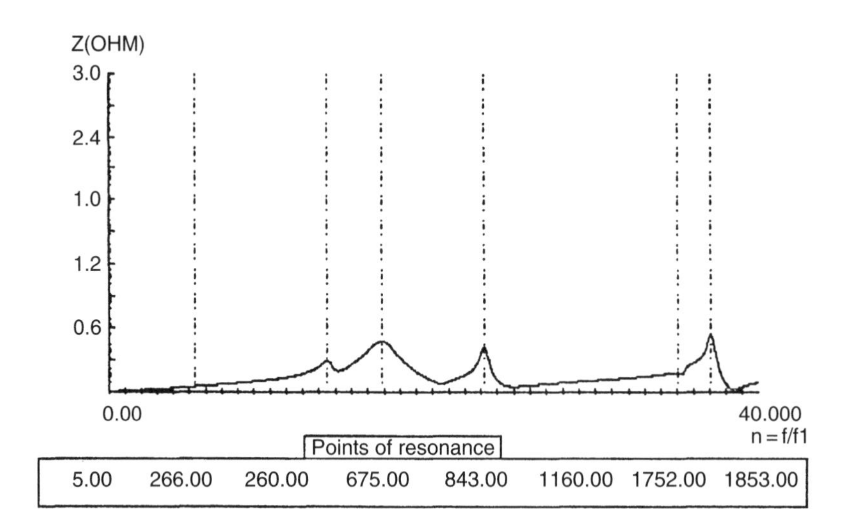

Figure 4.52 Impedance path of the network at the Westküste wind farm with a short-circuit power $S''_{K} = 185\,\mathrm{MVA}$

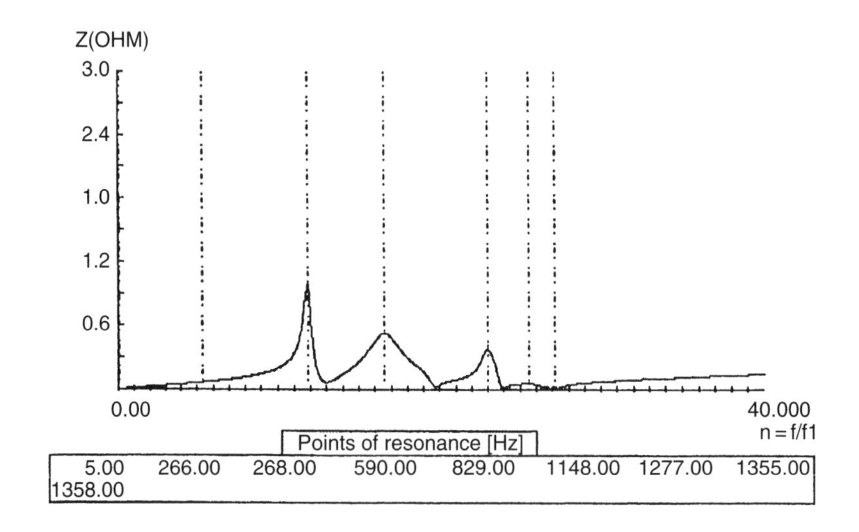

Figure 4.53 Impedance path in a weak grid with a short-circuit power $S''_{K} = 50\,\mathrm{MVA}$

point at 1160 Hz ($\mu = 23.20$) is retained at 1158 Hz ($\mu = 23.16$). The highest value, which occurs in the range up to the 40th harmonic, has not, however, moved significantly from 1853 Hz ($\mu = 37.06$), being 1741 Hz ($\mu = 34.82$).

If approximately the same cable impedances are summarized in the equivalent circuit diagram shown in Figure 4.50, the simplified structure in Figure 4.55 is the result. The resulting impedance path as shown in Figure 4.56 corresponds very well to Figure 4.52. Only at the maximum value, in the region of the 37th harmonic, which is of little relevance

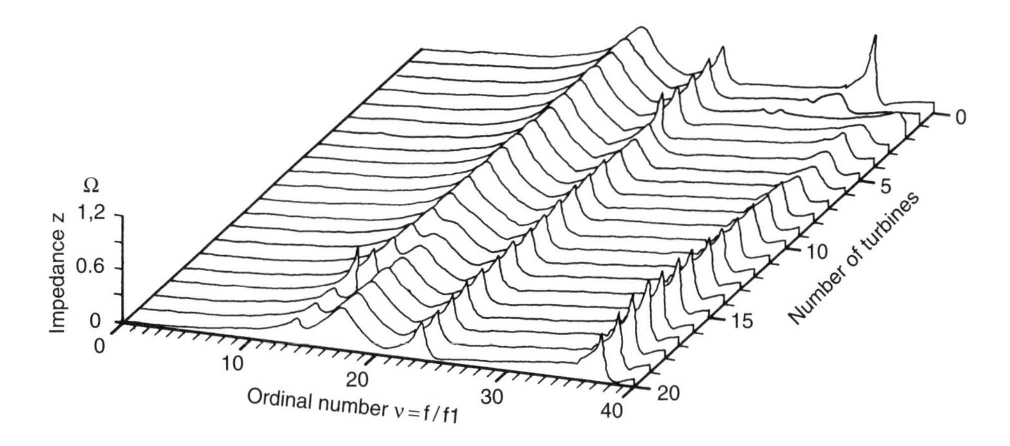

Figure 4.54 Impedance path of the grid at the Westküste wind farm ($S''_K = 185\,\text{MVA}$) as a function of the ordinal number and number of plants (0 to 20 plants of 30 kW each)

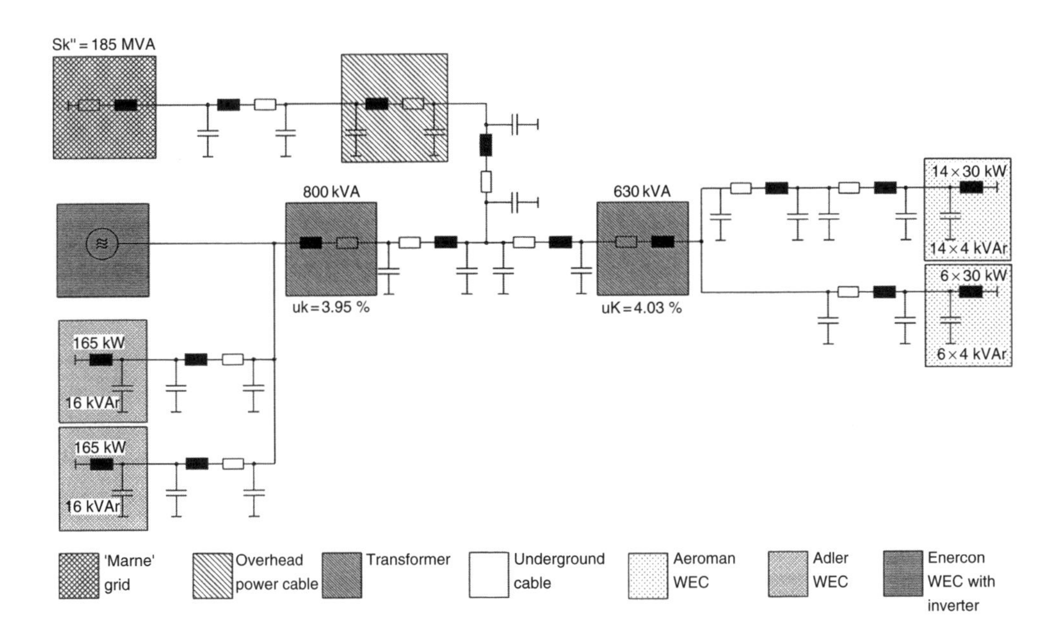

Figure 4.55 Simplified equivalent circuit diagram for the Westküste wind farm

due to the high ordinal number and thus low amplitude, the deviation can be recognized. This type of approximation is largely adequate for determining any resonance points that may occur and could take on critical values.

Further discussions below will cover attempts at remedial measures against resonance effects so that, taking into account the grid control, noncritical operation of wind turbines connected to the grid can always be achieved.

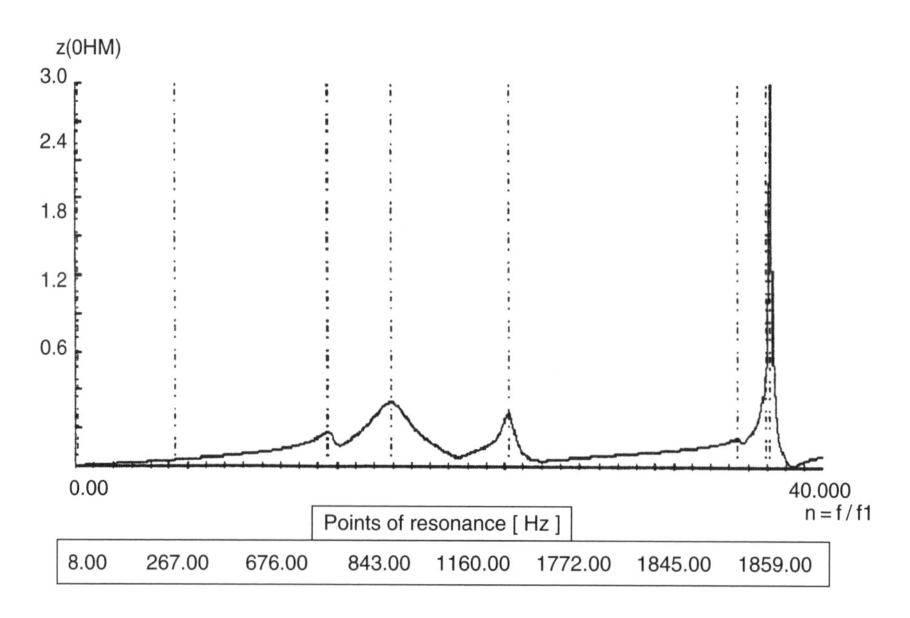

Figure 4.56 Impedance path of the simplified network at the Westküste wind farm

4.5 Remedial Measures against Grid Effects and Grid Resonances

The use of electronic power conversion systems, as well as bringing technical and economic advantages such as good controlability and high efficiency, is also associated with disadvantages, in particular due to grid voltage distortion. The effects are dependent upon the type of power semiconductor, the connection type, the operating state of the power converter and the prevailing grid parameters at the point of connection.

The operation of powerful frequency converters in connection with weak grids leads to an increase in harmonic voltages in the grid, which could have negative effects on the connected consumers. To prevent operating faults and system failures, the harmonic currents supplied by inverters must be kept low, preferably at the point of origin. Systems specifically designed to create low grid effects can be used for this purpose, or alternatively the harmonics (and voltage changes) that exist can be reduced by passive circuits and active intervention. Such devices are known as filters and grid support devices. The discussion that follows will be mainly limited to filters. These are connected in parallel at the grid supply point. Higher-frequency currents flow away via the filter unit and can thus be kept from the grid.

The design of the harmonic filter, as well as requiring precise knowledge of the grid structure and all its parameters, also calls for the specification of the permissible harmonic current at the point of connection. This so-called compatibility level is predetermined by the responsible power supply company.

4.5.1 Filters

Filters are generally made up of capacitors and inductors. Ohmic resistors have the task of reducing high impedances where filters have parallel resonances, or extending the frequency

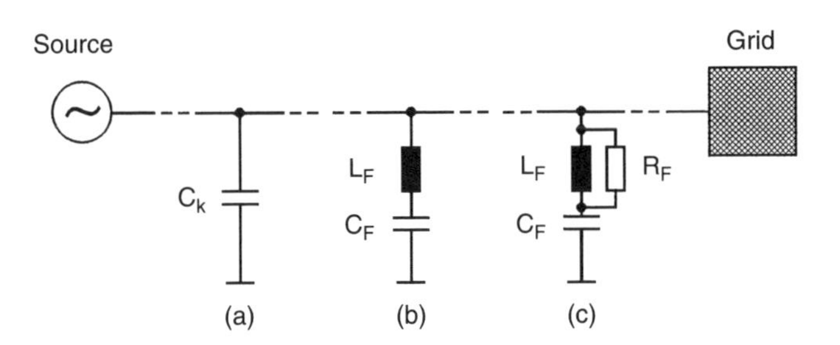

Figure 4.57 Equivalent circuit diagram of compensation and filter devices: (a) power factor correction capacitor, (b) harmonic absorber filter and (c) wide-bandpass filter

range in which filters exhibit low impedances. Only the main features will be considered in the following (Figure 4.57). Complex systems, e.g. double high-pass systems, will not be covered.

Even if capacitors are not used as harmonic filters, their behaviour must still be assigned great importance. Capacitor banks consist of one or more capacitors connected in parallel. They have the task of compensating for the inductive reactive power of motors and generators (e.g. by the capacitive reactive power Q_c) and improving the power factor (Figure 4.57(a)). Due to their impedance behaviour according to the simple equation

$$Z_c(f) = \frac{U_c^2}{Q_c} = \frac{1}{2\pi f C_k}, \tag{4.40}$$

increasing frequency and rising capacitance C_k bring about lower impedances Z_c. This has the result that harmonic currents are led away but fundamental components are almost blocked. Thus power factor correction capacitors also have a filtering effect.

The variables mentioned in the intermediate equation characterize the reference conductor voltage U_c and the compensating reactive power Q_c. As networks usually display inductive behaviour at high frequencies, the grid impedance increases with increasing frequency. Another point to be considered is that current harmonics provide an additional load on the capacitor banks. This results in warming and a reduction in insulating capacity if the capacitors are dimensioned for the fundamental power. This effect can be countered by designing the capacitors for around double the initial voltage.

Six-pulse inverters supply relatively high currents into the grid at 250 and 350 Hz. Due to their high impedance at low frequencies, a capacitor bank can only filter part of these harmonics. An increase in capacitor output would increase the filtering effect. However, the generator or inverter would be significantly overcompensated at its fundamental frequency, thus leading to a highly capacitive supply.

Better filtering can be achieved by the series connection of a capacitor and an inductor – a so-called harmonic absorber. These have infinitesimally small impedances close to their resonant frequency

$$f_{res} = \frac{1}{2\pi}\sqrt{\frac{1}{L_F C_F}}. \tag{4.41}$$

By the selection of filter inductance L_F and filter capacitance C_F as well as compensating output, the harmonic absorber can be dimensioned such that the currents near to the resonant frequency f_{res} (e.g. 250 Hz) can be absorbed. Below the resonant frequency, the filter device has capacitive behaviour and, above it, inductive. Thus small compensating outputs lead to a steep resonance curve; high compensating outputs, on the other hand, bring about a shallow path.

Harmonic absorbers are best used where currents are to be filtered within a narrow frequency band. The fundamental component compensation output and the capacitance of the filter circuit can thus be kept low.

If, on the other hand, currents in a wide frequency band are to be removed from a grid, wide-bandpass filters are required. They consist of a harmonic absorber in which an ohmic resistor is connected in parallel to the inductor (Figure 4.57(c)). This parallel connection has the result that at high frequencies the inductance of the resistor is bypassed and the filter impedance is kept low over a broad frequency band. The resonant frequency of the broadband filter (i.e. the frequency at which the filter exhibits the lowest impedance) is determined as for the harmonic absorber according to Equation (4.41). The behaviour of the broadband filter is characterized by the capacitor at low frequencies and by the ohmic resistor at high frequencies.

4.5.2 Filter design

Taking the higher-voltage grid structure and the permissible compatibility level into account, a total impedance of the grid and filter unit must be achieved by filter design and the selection of components at which harmonics are kept within the range of the permissible compatibility level. We must take into account here the fact that critical operating ranges can occur due to parallel resonances between the filter unit and the grid and among the filters themselves. Moreover, filters must be robust and insensitive to parameter fluctuations such as changes to grid short-circuit power and load, and must always fulfil the necessary connection conditions.

Capacitors for reactive power compensation without an inductive unit can cause resonances and the associated voltage overshoot close to the harmonic source. In the grid areas that contain power converters, inductive-capacitors must therefore be used, whereby the resonant frequency must be selected below the first harmonic frequency.

Filter units are usually laid out in several stages. This makes it possible to tune individual filter levels to the same and different resonant frequencies. In principle, a filter circuit is laid out for the lowest harmonic frequency. Further filters for higher harmonics are dimensioned in order of increasing ordinal numbers. The resonant frequency is selected to be somewhat lower than the frequency of the harmonics to be removed, in order to avoid overvoltage and excessive capacitor currents. By using appropriate levels, the capacitive reactive power in the range of given power factors can be adapted to different load situations.

If filters with different resonant frequencies are used, a parallel resonance is created by their connection in parallel. This occurs at a frequency at which the inductive reactance component of the one filter and the capacitive reactance of the other are equal. Maximum impedance occurs at this point. Harmonic currents that occur in the region of an impedance peak can be reduced only slightly or not at all. If two filter systems (harmonic absorber, wide-bandpass filter, etc.) of different frequency designs are used, the parallel resonance must be positioned either between two harmonics or in a frequency range in which the supplied harmonic currents lie significantly below the permissible values.

It should also be noted that filter levels not only produce parallel resonances between themselves but also produce one or more parallel resonances with the higher-voltage grid, possibly resulting in excessive increases in impedance. This is the case if filter and grid impedance have the same magnitude but opposite signs, e.g. if the filter impedance is capacitive and the grid impedance has the same value but is inductive, or vice versa.

A precise analysis of current and voltage distribution at all occurring frequencies allows the actual loading of individual components in all switching states of a filter to be determined and the components to be designed accordingly. It should be noted here that the total r.m.s. values of voltage

$$U_{rms} = \sqrt{U_{1rms}^2 + U_{2rms}^2 + U_{3rms}^2 + \cdots} \tag{4.42}$$

and current

$$I_{rms} = \sqrt{I_{1rms}^2 + I_{2rms}^2 + I_{3rms}^2 + \cdots} \tag{4.43}$$

are made up of the sum of all the respective individual values. Thus relatively expensive simulation programmes are necessary for exact dimensioning. The voltage endurance of capacitors and the total r.m.s. current are of central importance for inductance. However, harmonic current loads, e.g. the fifth and seventh harmonics of capacitors, and the harmonic voltages of inductors must also be taken into account.

4.5.3 Function of harmonic absorber filters and compensation units

Figure 4.58 illustrates the function of the filter and compensation unit of a relatively small village supply system with a maximum output of approximately 70 kW. The frequency spectrum and voltage path (in phase L_1) are shown at around half the nominal load without a filter in Figure 4.58(a) and with a filter unit connected in Figure 4.58(b) [4.49, 4.56, 4.64]. It is not of primary importance here whether the harmonics are caused by the consumers, the photovoltaic system, the battery or the wind turbine.

All the uneven components in the spectral function shown in Figure 4.58(a) are reduced in Figure 4.58(b) to a few (six-pulse) harmonic components typical of a power converter, leaving a near-sinusoidal voltage path. This clearly illustrates the function of the filter unit.

Different conversion systems exhibit large differences with regard to harmonics, depending upon grid connection. In the case of asynchronous generators coupled directly to the grid, grid effects do not generally increase with the number of generators. On the contrary, harmonics and subharmonics that are already present in the grid are usually weakened, as shown by the measurements in Figure 4.43(a).

In the case of turbines with a frequency converter supply, by contrast, a higher harmonic component is supplied to the grid as the number of turbines increases or output rises (see Figure 4.43(b)). Six-pulse line-commutated inverters give rise to significantly greater grid effects, particularly at the fifth and seventh harmonics, than twelve-pulse inverters.

Similar effects are achieved by combining wind turbines that supply the grid via frequency converters with turbines fitted with asynchronous generators coupled directly to the grid. This is due to the increased short-circuit power and the filter function of generator and compensation reactance, which result in significantly lower grid effects (compare Figure 4.59(a) and (b)). The associated voltage paths are represented in Figure 4.48.

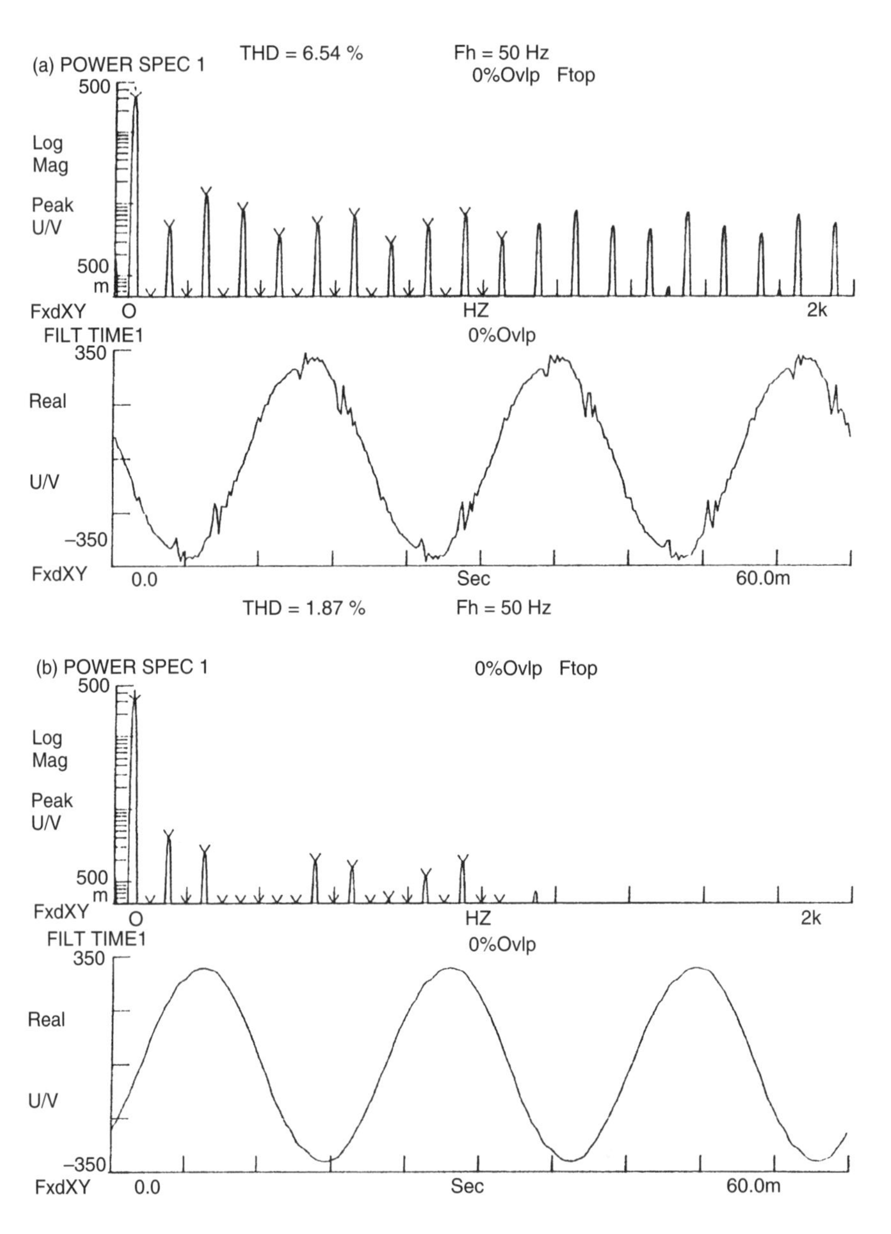

Figure 4.58 Frequency spectrum and voltage path of a village power supply system (70 kW): (a) without and (b) with a filter and compensation unit

4.5.4 Grid-specific filter layout

When large wind turbines supply weak grid areas, e.g. dead-end feeders in wind-rich coastal areas, powerful grid effects must be expected. This is particularly the case if a high-output wind turbine supplies its energy into the grid via a frequency converter with controlled

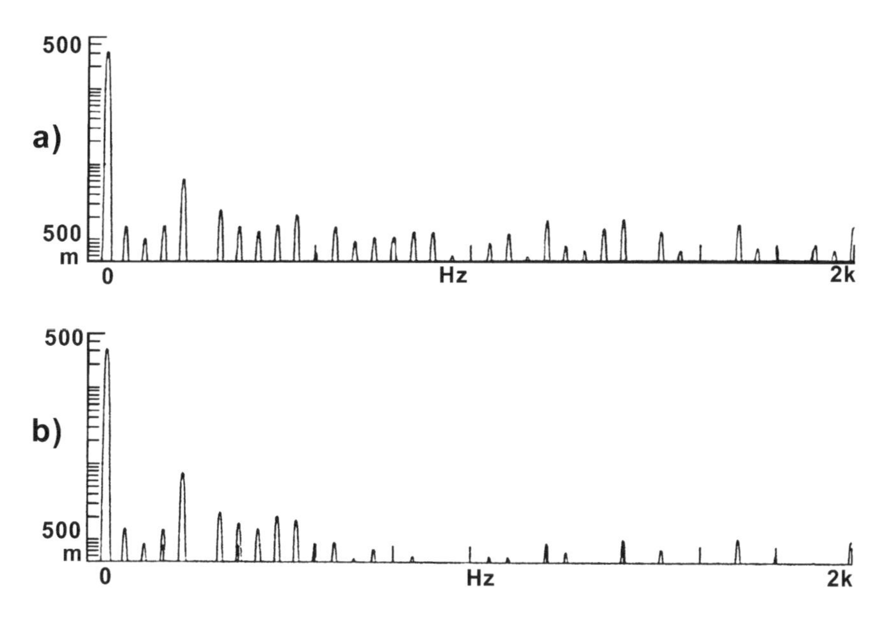

Figure 4.59 Frequency spectrum for (a) 275 kW input power via a six-pulse inverter and (b) 320 kW input power via a six-pulse inverter and 330 kW input power via an asynchronous generator coupled directly to the grid

rectifier, constant current d.c. link and a line-commutated six-pulse inverter in a bridge connection.

When a 1.2 MW turbine was erected, the current grid situation at a coastal location gave rise to a new dimension of grid effects, which could be largely attributed to grid resonances (see Section 4.4) [4.65]. Test runs showed a high level of harmonics in the vicinity of the grid resonance point at the 25th ordinal number, despite the grid filter installed, which was designed according to conventional methods as a harmonic absorber for the fifth and seventh harmonics. This phenomenon cannot be attributed primarily to the effects of the wind turbine, but is basically caused by the grid configuration. The turbine could have been operated without any problems if it had been connected at a different grid connection point with greater grid short-circuit power. Such effects of grid resonance have been observed in other dead-end feeders recently, and will increase in importance with increasing grid loading. The descriptions thus clearly show the limits to the effectiveness of filters.

In order to meet the compatibility level stipulated by the electricity supply company at the grid power interchange, it is necessary to design filters specifically adapted for the entire electrical system. These must be adapted to take into account the characteristics of the power converter, the dead-end feeder and the higher-voltage grid structure. This requires precise knowledge of the following grid and turbine data:

- magnitude and angle of the network's impedance at all frequencies occurring at the point under consideration, which is generally the point of supply to the public grid;
- magnitude and angle of impedance at all frequencies of the network occurring at the infeed point of the harmonic source;
- level of supplied harmonic currents;

- level of permissible harmonic voltage at the point under consideration or at the grid point of supply and the permissible power factor range;
- impedances of other electrical equipment between the infeed point and the point of supply.

In filter design, the frequency path of filter impedance should be such that the permissible harmonic levels are not exceeded at the grid point of supply. These levels, in combination with grid impedance, determine the permissible currents. If a source of harmonics supplies large currents, the filter unit must accept the residual current. Thus, for every frequency that occurs a corresponding filter impedance can be calculated.

Figure 4.60 shows a sample excerpt from the calculated current–voltage distribution in the wind farm, obtained while all generators on the 165 kW turbines were in operation. The frequency-dependent current and voltage values at all nodes, components or connection points (e.g. marked by the arrow for node 54 at 850 Hz) can be determined from this for all operating states with and without the filtering unit. The levels can thus be evaluated and corrected by filter adjustment.

Based upon the above example, it was possible to demonstrate the functionality of the procedure using this type of filter layout. Figure 4.61 shows the harmonic voltage readings after the installation of the filter unit, compared with the levels calculated in the simulation and the permissible values. It is evident from this that the measured and calculated indicators match well for all ordinal numbers that are not divisible by three. Significant deviations can be observed at multiples of the third harmonic. These are due to the assumption of

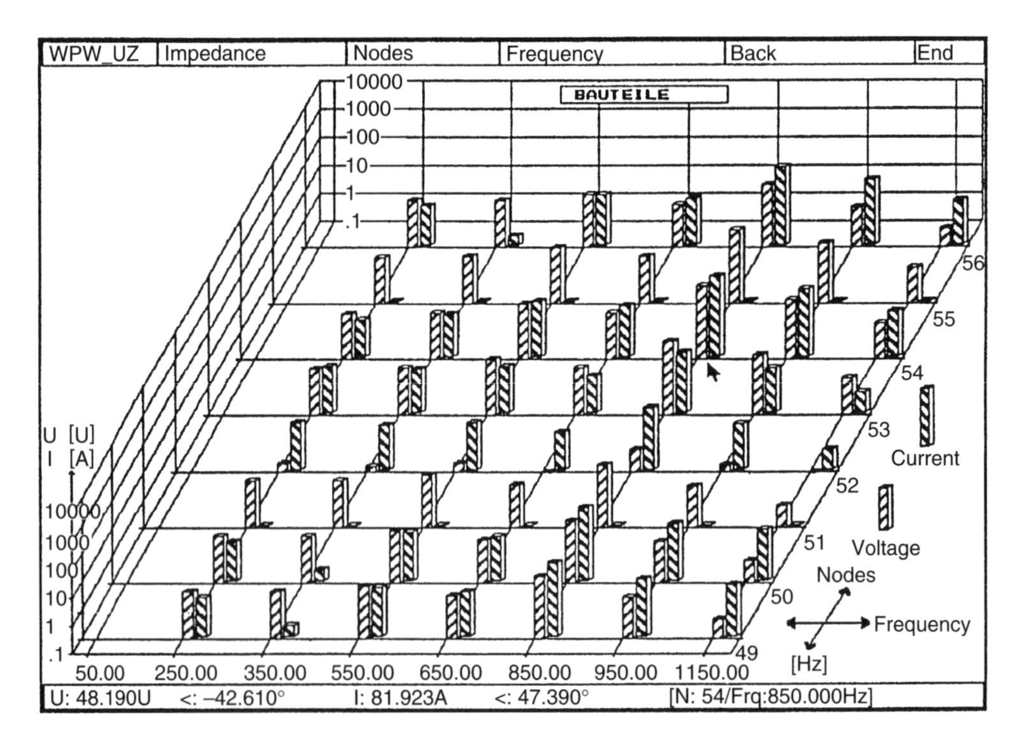

Figure 4.60 Current–voltage distribution in the wind farm with all turbines operating

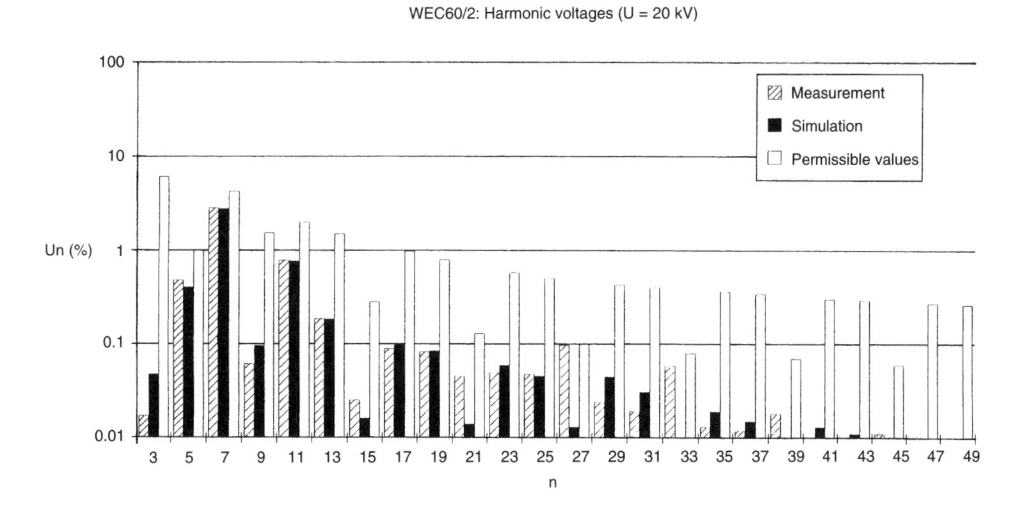

Figure 4.61 Measured, calculated and permissible relative harmonic voltages for a 1.2 MW inverter output with a grid-specific filter (20 kV levels)

mean grid harmonics, which clearly has only limited validity in this grid branch. If precise knowledge exists about these grid distortions, or they have been previously determined, all the inconsistencies shown here can be prevented and values exceeding the limit values avoided.

For grid connections that are susceptible to grid resonances, filter systems can be designed using computerized simulations that allow the local grid connection conditions to be fulfilled. Thus low filter unit costs can be achieved and high grid loads attained.

4.5.5 Utilizing compensating effects

The discussions in Sections 4.3.3, 4.3.4 and 4.5.3 show that low grid effects can be achieved by a favourable combination of differently designed wind energy converters [4.59]. Firstly, asynchronous generators with inverter systems, which produce a filtering effect and increase the short-circuit power, can be used to reduce harmonics. Secondly, the grid-supporting characteristics of units with frequency converter inputs can help reduce the voltage changes or other grid effects caused by turbines connected rigidly to the grid.

A further method of reducing grid effects is possible in the case of the installation of several wind turbines with a frequency converter supply by adjusting their inverters so that they function with different trigger angles in relation to the grid. By an appropriate selection of d.c. link voltage or transformer voltage for two turbines, the trigger angle of the two inverters (Figure 4.62(a)) can be adjusted so that the critical harmonic (here, for example, the 29th) of both systems occurs half a cycle apart, and is thus largely cancelled out (see Figure 4.62(b)) [4.64].

For inverter 1 the inverter stability limit is assumed to lie at the point where the d.c. link voltage remains the same during a voltage dip of 15%, which corresponds to a nominal trigger delay angle of $\alpha_{1N} = 138°$. According to Figure 4.63, a nominal trigger delay angle

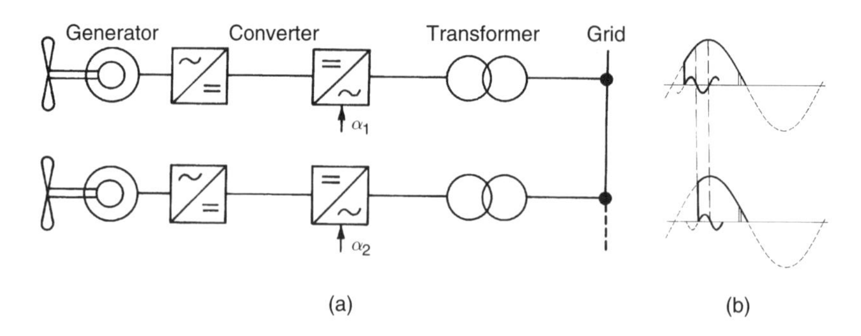

Figure 4.62 (a) Block diagram and (b) harmonic cancellation of two wind turbines with different inverter trigger angles

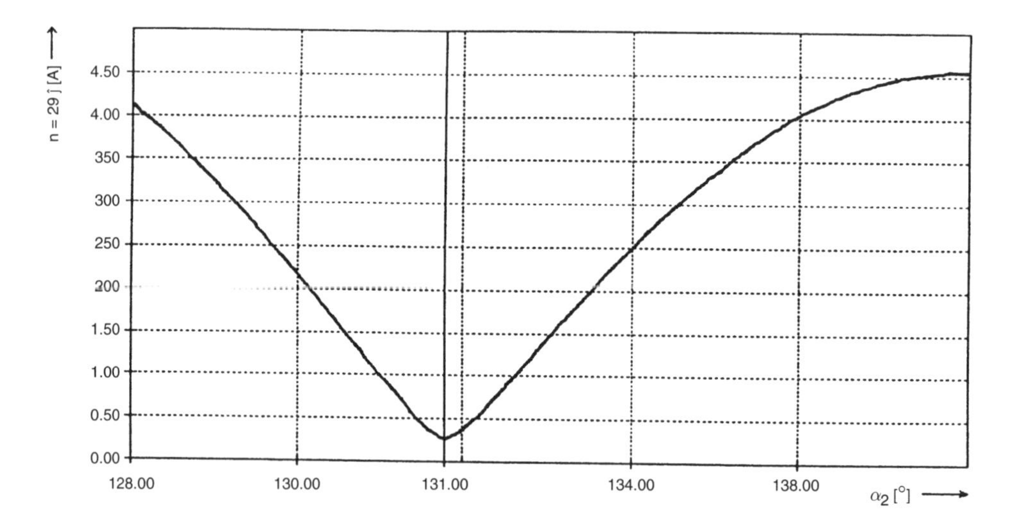

Figure 4.63 Current amplitude of the 29th harmonic as a function of the trigger delay angle α_2 (simulation result)

of $\alpha_{2N} = 131.79°$ is found for inverter 2 for the minimum amplitude of the 29th harmonic. The amplitude spectrum of the superposed grid current for a 1.2 MW plant (value calculated according to Figure 4.64) illustrates the significant reduction in all harmonics between the 11th and 49th ordinal number. Therefore, by using an appropriate system layout, grid effects can be significantly reduced using this process, both for operation at nominal ratings and at partial load, without incurring additional construction costs. Laboratory investigations with line-commutated inverters confirm the effectiveness of this process (Figure 4.65).

The available grid capacity can be utilized well by the formation of a turbine cluster if the turbines are designed with a suitable generator, grid connection and power regulation system, and different turbines are combined in a manner appropriate to the location.

In low-power grids with a proportion of the power supplied by wind power, particularly large system perturbation can be expected. Practical investigations have shown, however,

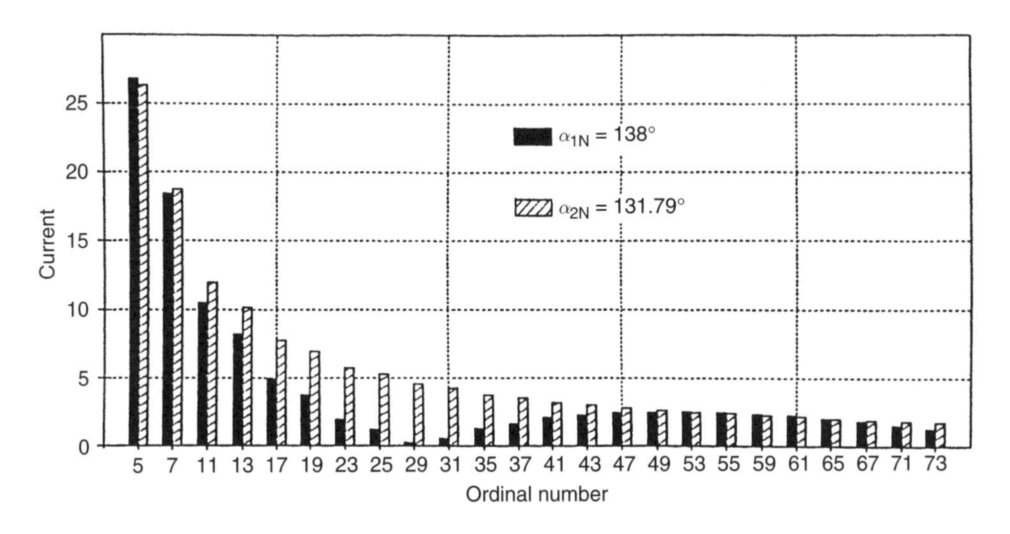

Figure 4.64 Amplitude spectrum of the grid current at the same (striped) and different (black) power converter trigger delay angles (simulation results)

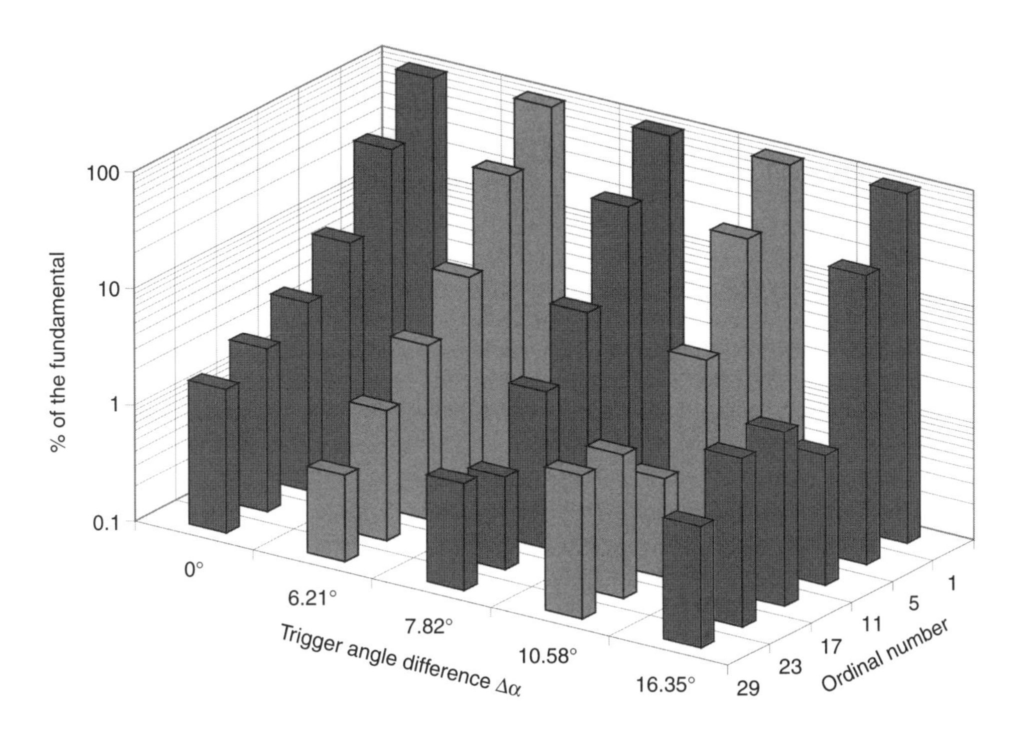

Figure 4.65 Amplitude spectrum of the grid current (measurements) as a percentage of the fundamental for cophasal supply, $\Delta\alpha = 0°$, and trigger difference, $\Delta\alpha$, for the reduction of the 29th harmonic, $\Delta\alpha = 6.21°$, the 23rd harmonic, $\Delta\alpha = 7.82°$, the 17th harmonic, $\Delta\alpha = 10.58°$, and the 11th harmonic, $\Delta\alpha = 16.35°$

that by using an appropriate layout of the grid and its components, wind power can contribute up to 100 % of the supply. Any system disturbance can be avoided by the use of devices to influence the grid characteristic in connection with static or rotating phase shifters, batteries with reversible converters, grid filters, compensation units and grid regulators. In this manner, any existing grid effects plus those that can be expected as a result of the installation of additional turbines can be reduced. Expensive measures to reinforce or extend the grid can therefore be largely avoided.

4.6 Grid Control and Protection

Power station turbines and generators are generally fitted with proportional speed regulators. As shown in Figure 4.66, the speed–power curves of individual generators can be used to derive a steady-state frequency–power characteristic $\omega = f(P_E)$ for the total generator active power P_E of a grid, and this characteristic has a much flatter path than the characteristic of the largest individual machine. For the load P_L, including the power station's own consumption and the loss in the grid, an approximately linear steady-state frequency–output characteristic $\omega = f(P_L)$ is found, the gradient of which is determined by the total characteristic of the consumers. The steady-state operating point in the grid is found at the intersection of the two characteristics at frequency ω_0, where the supplier output just covers the consumer component and the grid losses, i.e. $P_E(\omega_0) - P_L(\omega_0) = 0$. Due to the flat characteristics, each individual generator works with a virtually fixed-frequency grid.

If a load-specific displacement of grid frequency occurs, the producer with a frequency-dependent power characteristic is called upon to maintain the frequency according to the drooping characteristic by output changes. As well as this so-called primary control, frequency and coupling output are monitored and adjusted by so-called secondary control and load management, which will not be considered further here.

4.6.1 Supply by wind turbines

In contrast to conventional power stations, wind turbines, like all renewable power generation systems, usually supply the entire energy offered by the wind into the grid, regardless of the grid state. Moreover, starting and stopping procedures are not usually centrally coordinated.

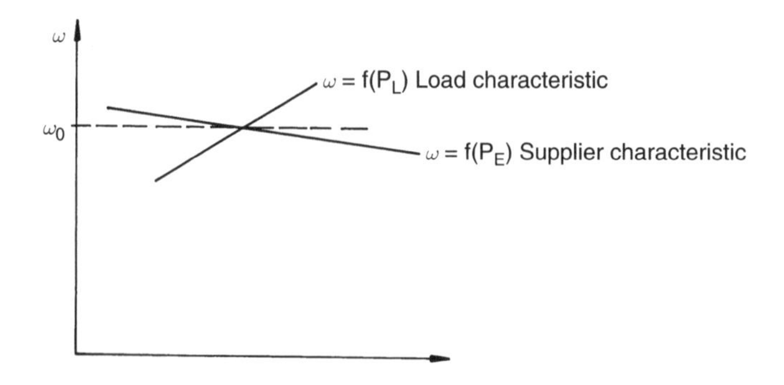

Figure 4.66 Supplier and consumer characteristics of a grid [3.12]

In low-power sections under 10 % this does not cause any serious effects. In Schleswig-Holstein roughly 20 % of the electricity supply is already covered by wind power.

In the near future, however, it is expected that energy contributions from wind power will reach, or even exceed, power consumption values at low-load times in the North German coastal and inland region and low mountain ranges, particularly in windy conditions. Grid effects and measures necessary for grid regulation can take on considerable importance here in an uncontrolled wind energy supply.

Up until now, grid management has taken place based upon the assumption that small power plants, and in particular wind turbines, contribute nothing to the maintenance of grid parameters. They are largely viewed as being only negative consumers, who can be connected as desired. To prevent the formation of asynchronous islands in the grid that are difficult to control, decentralized installations are separated from the grid early, when large suppliers (conventional power stations) fail. Therefore even higher proportions are lost, which could support the grid if there were enough wind.

Initial investigations of output fluctuations [4.66] and interchange power [4.67] in wind farms and grids [4.68] were based upon computerized simulations using the least favourable assumptions. The results of measurements on an existing wind farm with a spatially offset arrangement of individual wind turbines, as discussed in Section 4.3.2, lead us to expect only small output gradients for total output owing to the time sequence of wind and output events. Changes to the grid load can be divided into variations due to the time of day, which usually occur slowly and are for the most part predictable, and changes that occur by chance.

There are fears that the extreme fluctuations in the supplied wind power will lead to problems in grid control or necessitate expensive compensatory measures. Research results from the scientific measurement and evaluation programme, however, refute this. According to reference [4.69] the greatest occurring power fluctuation from one hour to the next in the supply areas of E.ON and RWE amounted to around 20 % of the total installed nominal power, with a probability of occurrence of 0.01 %, i.e. less than once a year.

Even with today's level of meteorology, local wind speeds and directions and their gradients can be reliably predicted to a large degree. This forecasting could be significantly refined and improved in the future to meet the needs of the electricity generation sector. Therefore, with a more precise knowledge of wind turbines and their behaviour and location, more reliable predictions could be made about output over time (see Section 4.3.2.3). With such predictions, methods could be developed in the future that permit long-term grid supporting measures to be taken, replacement turbines to be put in operation in anticipation or consumers that are not at the time absolutely necessary to be switched off.

4.6.2 Grid support and grid control with wind turbines and other renewable systems

The value of wind power can be decisively increased if it is capable of contributing to grid support. Thus wind power could be transformed from a negative consumer into a grid-supporting quantity. The previously somewhat negative image of this environmentally friendly power source will be significantly improved and its long-term prospects guaranteed.

In wind turbines and other small power stations, synchronous or double-fed asynchronous generators with a grid connection via self-commutated pulse-controlled converters are being increasingly used as the mechanical–electrical conversion system. Such electronic power

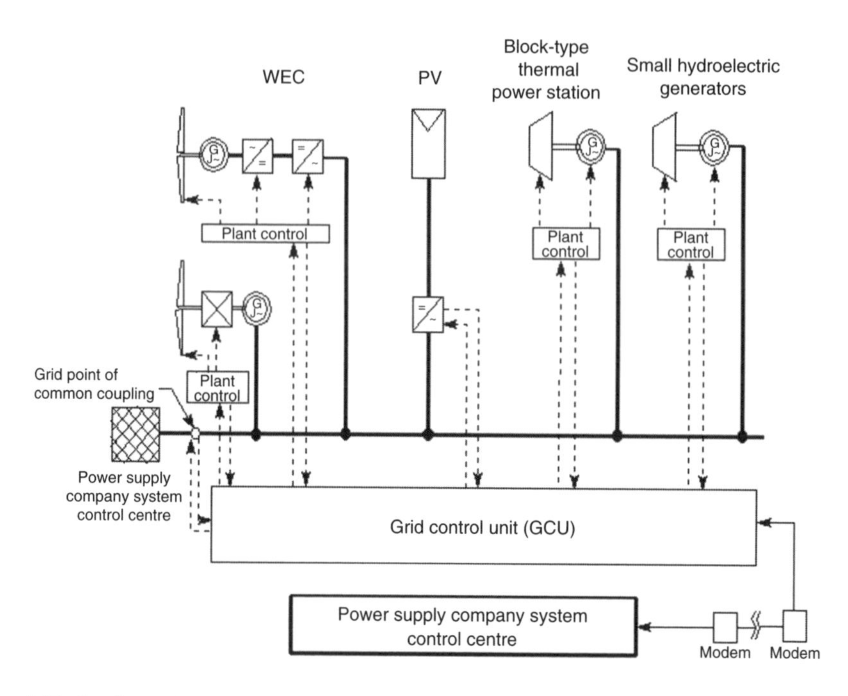

Figure 4.67 Configurations and components for grid control with (spatially separated) regenerative power supply systems

conditioning systems, which have also become standard in photovoltaic systems, offer diverse conventional power stations similar intervention options with regard to the transmission of power to the grid, such as the adjustability of voltage and reactive power and the control of the supply of active power. These technical possibilities, however, remain largely unused to date.

Within the framework of a new research and development project, a concept for grid regulation has been developed and realized in hardware and software. The practical investigations were carried out at various locations in Germany and on the Canary Islands. Thus different grid configurations in combined and separate grids and different meteorological and geographic conditions could be taken into account.

It is the task of a grid regulation unit to match the operating behaviour of renewable, decentralized generation plants (wind power and photovoltaic generators, etc.) with that of conventional power stations and to ensure that they participate actively in grid support. Figure 4.67 shows the configuration and the components for grid regulation. Before practical realization, comprehensive preliminary studies were carried out. To this end, the structure of the subnetworks relevant at the time were recorded and comprehensive simulation calculations (load flow, voltage changes, etc.) were performed, in order to be able to estimate possibilities for control interventions and their effects upon the grid.

The effects of five supply systems of the 500 kW class upon an approximately 23 km long 20 kV grid, connected to the 110 kV level via a 40 MVA transformer with a short-circuit power of around 320 MVA, will be illustrated by the following descriptions, which first of all consider the situation without control interventions. Figure 4.68 shows the differences of

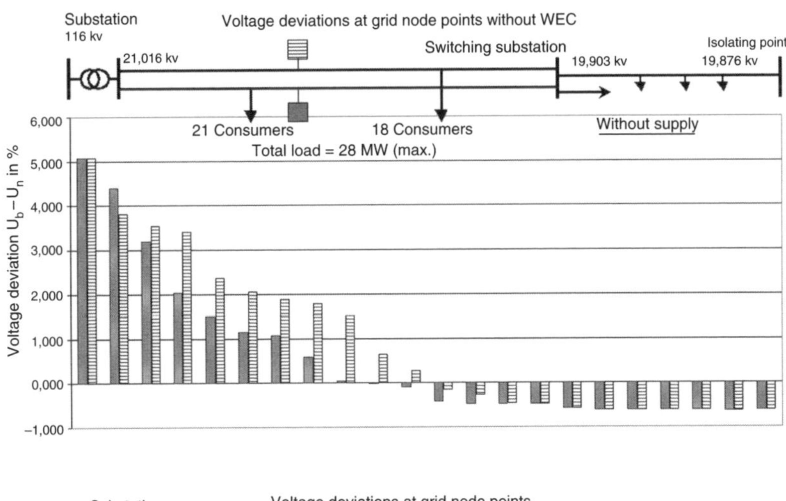

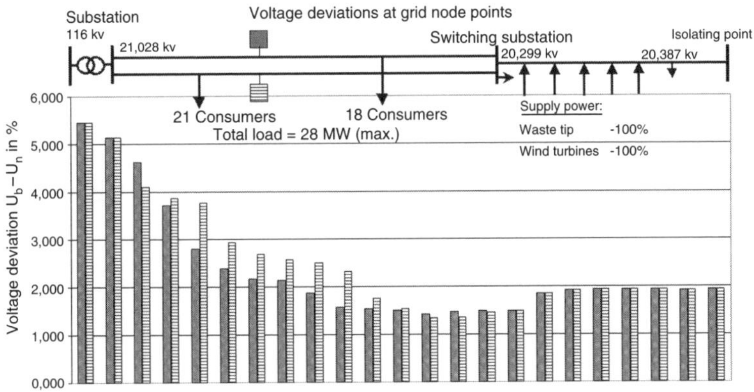

Figure 4.68 Voltage graphs in a 20 kV grid branch of the combined grid (a) without and (b) with maximum supply by renewable systems in the dead-end feeders

voltage paths along the line without (a) and with (b) maximum supply. This shows that in all operating ranges the voltage deviations at the point of common coupling remain below 2 %. However, the voltage change in the grid branch without supply amounts to almost 6 % and at maximum supply only around 3 %. Furthermore, Figure 4.69 also shows the differences in the flow of active and reactive power and in the utilization of the transmission system. Whereas in the grid range under consideration the active and reactive power always flows from the transformer substation to the grid disconnection point when there is no supply (a), the power flows are sometimes reversed if there is supply (b) depending upon the load case, with active and reactive power sometimes flowing in opposite directions to each other. By control interventions, however, it is possible to achieve the desired power flows and favourable grid loads. Furthermore, it is clear that even where supply is 10 % of the consumed power, load is reduced by a good 10 %, particularly in those grid areas with high utilization. As a result the transmission losses (which will not be considered further here) can be correspondingly reduced.

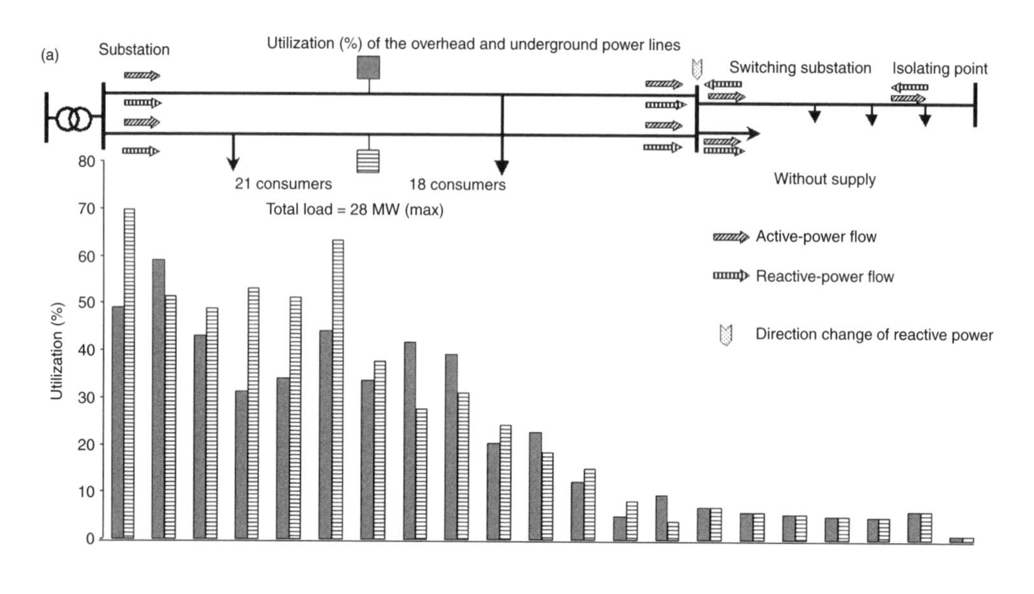

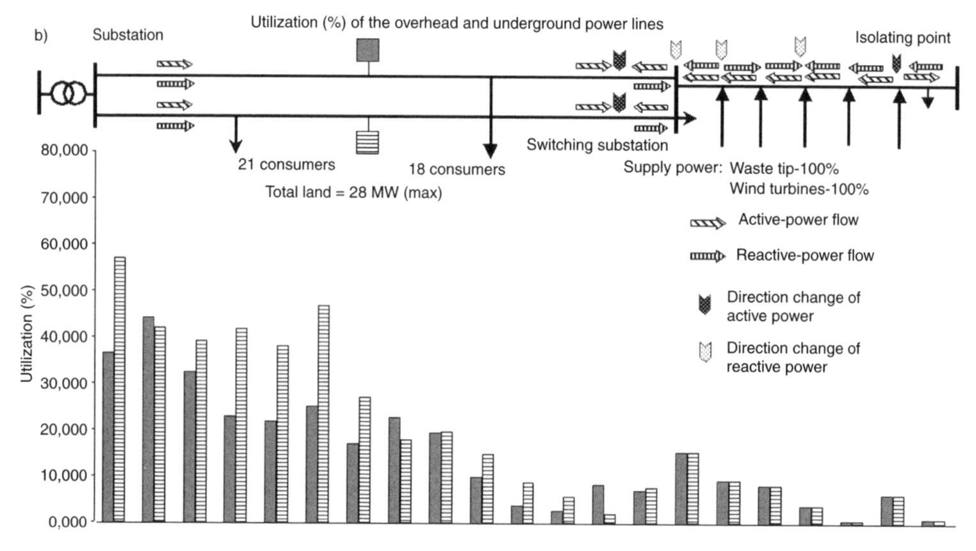

Figure 4.69 Capacity utilization of the transmission systems with active and reactive power flow (a) with no supply and (b) with maximum supply by renewable generation systems

Suitable measurement, conversion and control systems are necessary for control interventions for the support of grid voltage or the adjustment of reactive power. We will not go into further details at this point regarding the various design options that are usually selected specifically to suit the power supply company and component and turbine manufacturer. Detailed descriptions on this subject based upon the above-mentioned EU project are given as examples in references [4.70] to [4.74].

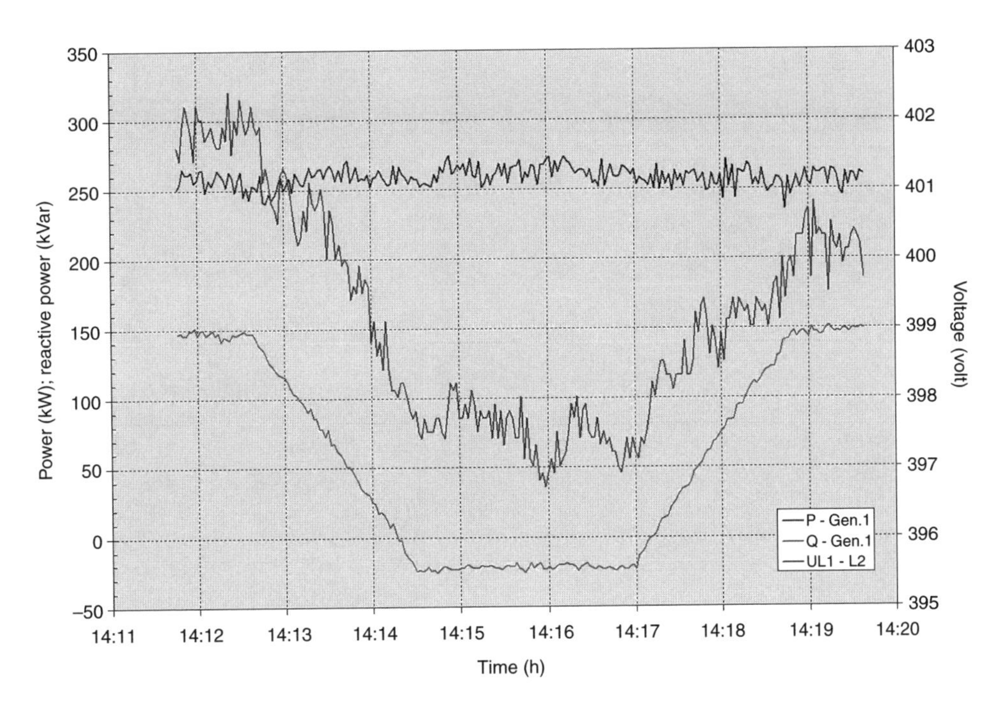

Figure 4.70 Variation in reactive power of a 0.5 MVA power supply system and effects upon the voltage of a weak grid branch

The results of measurements on the grid branch sketched out at the start will be briefly characterized. Figure 4.70 shows, at an approximately constant supply of active feed of 250 kW, the voltage change (with approximately 1 V superimposed voltage fluctuation) from around 402 to 398 V, i.e. a reduction of approximately 1 % (in relation to the nominal value), caused by ramp-shaped reactive power variation of 150 kvar (overexcited, inductive supply) to −25 kvar (underexcited, capacitive supply). Even when the reactive power increases, the voltage rises back to almost its original value. As the measurements show, the voltage changes at the point of common coupling in the grid are approximately half the level of those at the point of supply.

At roughly the same reactive power variations, significantly greater effects were found in separate networks. For example, a voltage change of around 3 % was measured in a separate network with around 500 MW installed power, and a voltage change of around 6 % was measured in a 100 MW grid [4.65]. These results illustrate the particular effectiveness of the procedure in weak grids. However, it is also clear that individual supply systems of the 0.5 MW class are not capable of holding branches in the combined grid at a constant voltage.

The grid supporting function of the grid control unit (GCU) shown in Figure 4.71 is illustrated on the basis of Figure 4.72. In a selected voltage range, e.g. between 20.3 and 20.7 kV, the control unit does not intervene (0900 to 1200 hours). If the voltage falls the reactive power is increased and if the voltage rises the reactive power falls, or is run in the capacitive feed range.

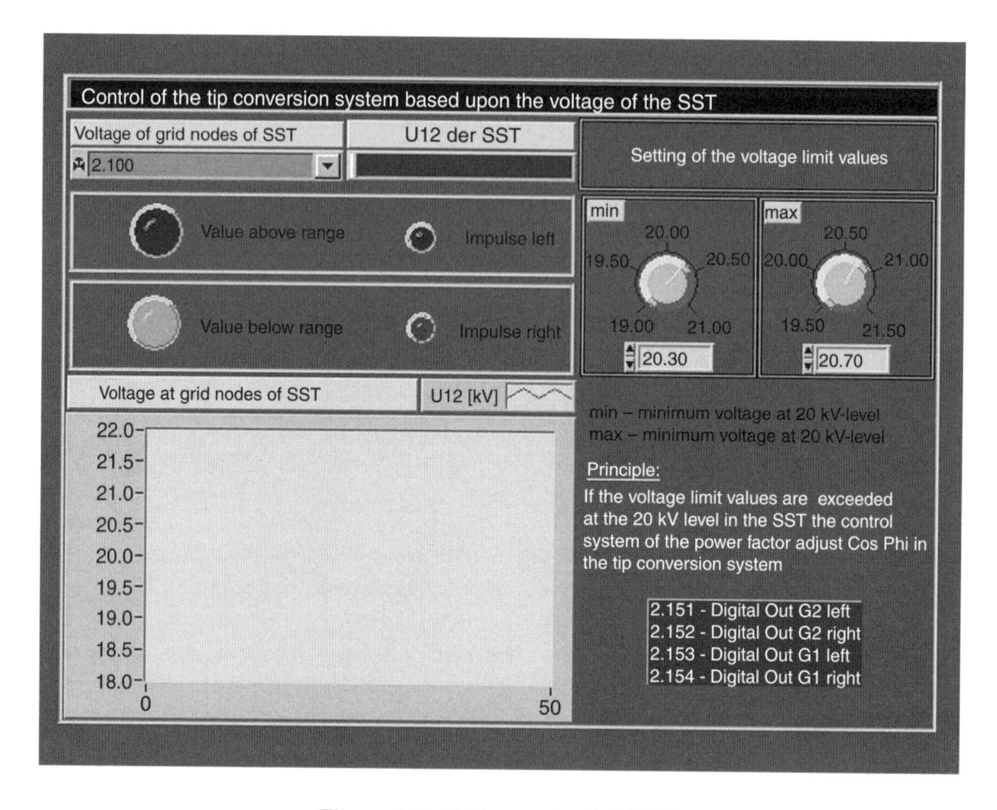

Figure 4.71 Grid control unit (GCU)

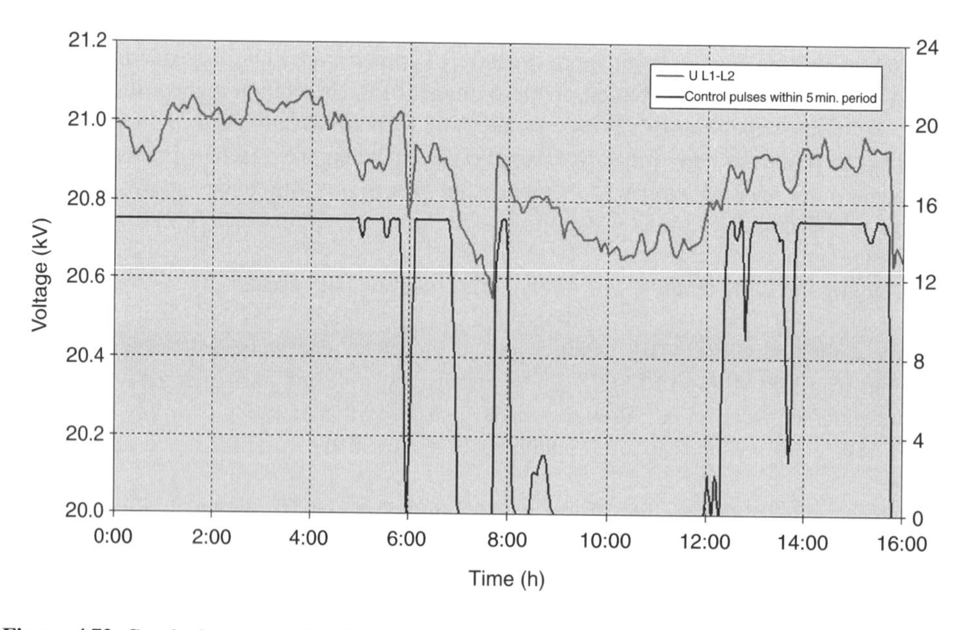

Figure 4.72 Graph showing grid voltage and reactive power control interventions by the grid control unit for voltage stabilization

4.7 Grid Connection Rules

When operating wind turbines in parallel with the public supply grid – as mentioned in the introduction to Chapter 4 (Table 4.1) – the valid international and national standards and directives apply. The grid quality characteristics and limit values in the low-voltage and medium-voltage grid according to Table 4.7 from reference [4.7] that customers and suppliers on the low- and medium-voltage transfer point of the grid may expect can be assumed here. The quality characteristics of the grid are usually evaluated statically and specified for the frequency values of 100 and 95 %. A differentiation can be made here between relatively slow frequency and voltage changes, flicker, voltage asymmetries and harmonics.

Furthermore, voltage-related events should be taken into account. These are determined on the basis of the number of times limit values are exceeded during the monitoring period. As shown in Table 4.7, these cover rapid switching-related voltage changes, voltage dips, short-time interruptions, long-time interruptions and grid-frequency overvoltages.

Electromagnetic compatibility is dealt with in IEC 61 000. A differentiation is made here between conducted disturbances (which occur up to several 10 kHz) and nonconducted disturbances that dominate at higher frequencies (DIN VDE 0 838 to 0 847), which we will not deal with further. Table 4.8 gives a brief overview of the international (IEC) and national classifications (DIN, VDE) for conducted disturbances. Wind turbines are dealt with internationally in IEC 61 400 and nationally classified (in Germany according to VDE 0 127). These standards and directives include the classification according to Table 4.9. Since 1992, the Fördergesellschaft Windenergie (FGW) has issued directives specifying, for example, clear measuring procedures with which comparable data can be determined for different wind turbines. As shown in Table 4.10, they cover the evaluation of the energy yield calculations, grid integration and noise emissions.

According to the Energie-Wirtschafts-Gesetz (Energy-Economy-Laws) (EnWG) of 1998 the operator of the transmission grid is responsible for the organization and operation of the interconnected system. The operator of the transmission grid was therefore made responsible for the safety and reliability of the transmission system. Furthermore, they have to guarantee the quality of the power supply. Moreover, all grid users and participants in the deregulated electricity market must be allowed nondiscriminatory access to the transmission grids and their use. To this end, all grid users must adhere to minimum technical requirements described in the *GridCode 2000* directive. The rules compiled by the Verband der Netzbetrieber (Association of Grid Operators) (VDN) are oriented towards the requirements for uninterrupted interconnected operation in the national framework and take into account the international specifications for grid operation of the Union für die Koordinierung des Transportes elektrischer Energie (UCTE).

In the framework of the GridCode [4.76]:

- the *MeteringCode 2004* VDN gives directives and change procedures,
- the *TransmissionCode 2003* gives grid and system rules of the German transmission grid operators and associated appendices and sample data,
- the *DistributionCode 2003* gives rules for access to distribution grids and
- the *GridCode* gives cooperation rules for the German transmission grid operator.

New rules have been in force since 1 August 2003 regarding the connection to the E.ON electrical transmission grid [4.77], which deviate for the first time from directives for low- and medium-voltage ranges and provide for grid support. These rules set down minimum

Table 4.7 Grid quality characteristics and limit values in accordance with DIN EN 50 160 and the VDEW directives for in-plant generation. From reference [4.75]

Grid quality characteristic	Averaging time	Reference value or limit value			
		EN 51060 LV grid	EN 50160 MV grid	VDEW: EEA at the LV grid	VDEW: EEA at the MV grid
Grid frequency (at synchronous connection to interconnected grid)	10 s	50 Hz + 4–6 % (100 % value) 50 Hz ± 1 % (99.5 % value)	50 Hz + 4–6 % (100 % value) 50 Hz ± 1 % (99.5 % value)	— —	— —
Grid frequency (without synchronous connection to interconnected grid)	10 s	50 Hz ± 15 % (100 % value) 50 Hz ± 2 % (95 % value)	50 Hz ± 15 % (100 % value) 50 Hz ± 2 % (95 % value)	— —	— —
Slow, quasi-stationary voltage changes	10 min	U_n ± 10–15 % (100 % value) U_n ± 10 % (95 % value)	— U_c ± 10 % (95 % value)	$\Delta u_a = 2\,\%$ —	$\Delta u_a = 2\,\%$ —
Long-time flicker intensity	120 min	$P_{lt} = 1$ (95 % value)	$P_{lt} = 1$ (95 % value)	$P_{lt} = 0.46$	$P_{lt} = 0.46$
Voltage asymmetries	10 min	$U_g/U_m < 2\,\%(3\,\%)$ (95 % value)	$U_g/U_m < 2\,\%(3\,\%)$ (95 % value)	—	—
Total harmonic distortion ($v = 1$–40)	10 min	THD = 8.0 %	THD = 8.0 %	—	—
Harmonics	10 min	Harmonic voltage (95 % value)		Permissible harmonic current related to S_{kV} (A/MVA)	

Uneven, not multiple of 3				10 kV	20 kV	
$v = 5$	10 min	6.0 % of U_n	6.0 % of U_c	2.5	0.115	0.058
$v = 7$	10 min	5.0 %	5.0 %	2.0	0.082	0.041
$v = 11$	10 min	3.5 %	3.5 %	1.3	0.052	0.026

$v = 13$	10 min	3.0 %	3.0 %	1.0	0.038	0.019
$v = 17$	10 min	2.0 %	2.0 %	0.55	0.022	0.011
$v = 19$	10 min	1.5 %	1.5 %	0.45	0.018	0.009
$v = 23$	10 min	1.5 %	1.5 %	0.3	0.012	0.006
$v = 25$	10 min	1.5 %	1.5 %	0.25	0.010	0.005
Uneven, multiple of 3						
$v = 3$	10 min	5.0 %	5.0 %	4.0	—	—
$v = 9$	10 min	1.5 %	1.5 %	0.7	—	—
$v = 15, 21$	10 min	0.5 %	0.5 %	—	—	—
Uneven, $v > 25$	10 min	—	—	$0.25 \times 25/v$	$0.06/v$	$0.03/v$
Even harmonics						
$v = 2$	10 min	2.0 %	2.0 %	$1.5/v$	$0.06/v$	$0.03/v$
$v = 4$	10 min	1.0 %	1.0 %	$1.5/v$	$0.06/v$	$0.03/v$
$v = 6–24/6–38$	10 min	0.5 %	0.5 %	$1.5/v$	$0.06/v$	$0.03/v$
Subharmonics						
$\mu < 40$		—	—	$1.5/\mu$	$0.06/\mu$	$0.03/\mu$
$\mu > 40$		—	—	$4.5/\mu$	$0.18/\mu$	$0.09/\mu$
Voltage-related events (criterion is the number of limit violations)						
Rapid voltage changes (e.g. due to switching operations)	$U_n \pm 5\,\%$ (in general)		$U_c \pm 4\,\%$ (in general)		$\Delta u_{max} =$ 3 % (< 1/5 min)	$\Delta u_{max} = 2\,\%$ (< 1/1.5 min)
	$U_n \pm 10\,\%$ (several times per day)		$U_c \pm 6\,\%$ (several times per day)		—	—
Voltage dips ($t < s$; $U > 0.01\ U_n$ and $< 0.90\ U_n$)	$10/a–1000/a$		$10/a–1000/a$		—	—
Short-time interruptions ($t < 3$ min; $U < 0.01\ U_n$)	$10/a–300/a$		$10/a–300/a$		—	—
Long-time interruptions ($t < 3$ min; $U < 0.01\ U_n$)	$10/a–50/a$		$10/a–50/a$		—	—
Grid frequency overvoltages	$U_{eff} < 1.5$ kV		$U_{eff} < 1.7–2\ U_c$		—	—
Transient overvoltages	$U_{eff} < 6$ kV				—	—

Table 4.8 National and international standards on EMC including from electricity supply systems. From reference [4.75]

IEC standard	DIN number	VDE reference	Title/content
IEC 61 000-2-2 (2002-03)	DIN EN 61 000-2-2	VDE 0 839 Part 2-2 (2003-02)	Electromagnetic compatibility (EMC): environmental conditions – compatibility level for low-frequency conducted disturbances and signal transmission in public low-voltage networks
IEC 61 000-2-4 (2002-06)	DIN EN 61 000-2-4	VDE 0 839 Part 2-4 (2003-05)	Electromagnetic compatibility (EMC): environmental conditions – compatibility level for low-frequency conducted disturbances in industrial systems
IEC 61 000-2-12 (2003-04)			Electromagnetic compatibility (EMC), Part 2-12: environmental conditions – compatibility level for low-frequency conducted disturbances and signal transmission in public medium-voltage grids
IEC 61 000-3-4 (1999-02)			Electromagnetic compatibility (EMC), Part 3-4: limit values – limiting the emission of harmonic currents in the low-voltage supply grids for devices and equipment with rated currents above 16 A
IEC 61 000-3-5 (1994-12)			Electromagnetic compatibility (EMC), Part 3: limits; Section 5 – limits for voltage fluctuations and flicker in low-voltage grids for devices with an input current > 16 A
IEC 61 000-3-6 (1996-10)			Electromagnetic compatibility (EMC), Part 3: Limits; Section 6 – assessment of emission limits for fluctuating loads in MV and HV power supply grids; EMC basic standard
IEC 61 000-3-7 (1996-11)			Electromagnetic compatibility (EMC), Part 3: Limits; Section 7 – assessment of emission limits for fluctuating loads in MV and HV power systems; basic EMC publication
IEC 61 000-3-11 (2000-08)	DIN EN 61 000-3-11	VDE 0838 Part 11 (2001-04)	Electromagnetic compatibility (EMC): limits – limiting of voltage changes, voltage fluctuations and flicker in public LV supply grids for devices and equipment with a rated current ≤ 75 A, which are subject to a special connection condition

Table 4.9 National and international standards for wind turbines. From reference [4.75]

IEC standard	DIN number	VDE reference	Title/content
IEC 61 400-1 (1999-02)			Wind turbine generator systems – Part 1: safety requirements
IEC 61 400-2 (1996-04)	DIN EN 61 400-2	VDE 0127 Part 2 (1998-01)	Wind turbines – safety of small wind turbines
IEC 61 400-11 (2002-12)	DIN EN 61 400-11	VDE 0127 Part 11 (2000-02)	Wind turbines – noise measurement procedure
IEC 61 400-12 (1998-02)	DIN EN 61 400-12	VDE 0127 Part 12 (1999-07)	Wind turbines – measuring procedure to determine power behaviour
IEC 61 400-13 (2001-06)			Wind turbine generator systems – Part 13: measurement of mechanical load
IEC 61 400-21 (2001-12)	DIN EN 61 400-21	VDE 0127 Part 21 (2002-11)	Wind turbines – measurement and assessment of the grid compatibility of grid-connected wind turbines
IEC/TS 61 400-23 (2001-04)			Wind turbine generator systems – Part 23: full-scale testing of rotor blades
IEC/TR 61 400-24 (2002-07)			Wind turbine generator systems – Part 24: lightning protection

Table 4.10 Overview of the current FGW regulations. From reference [4.75]

Part	Revision	Date	Title
0	14	01.12.2001	General requirements (now only applicable together with Parts 1 and 2)
1	15	01.01.2004	Determination of noise emission values
2	14	01.03.2004	Determination of the power curve and standardized energy yields
3	15	01.09.2002	Determination of the electrical properties
4	0	01.09.2003	Determination of the grid connection values
5	2	01.03.2004	Determination and application of the reference yields
6	0	01.03.2004	65 % reference yield demonstration based upon the determination of wind potential and energy yields

technical requirements for connections to the high and extremely high voltage grid. Changes relate particularly to wind turbines. The massive expansion of wind farms makes changes necessary so that the stability and availability of the transmission grid is assured when higher levels of wind power are supplied. Directives of other grid operators (RWE, Vattenfall, EnBW) are awaited.

Up until now it has been stipulated that wind turbines must be disconnected from the grid within 100 to 200 ms in the event of grid voltage and frequency values exceeding or falling below limit values. This immediate disconnection of the wind turbines causes problems for the transmission grid operators with regard to the adherence to UCTE criteria for the primary regulation of frequency. For example, in the event of a severe grid fault in North Germany, the grid voltage could fall so far that 3000 MW of wind power could fail suddenly. This power deficit, which makes up around 1 % of the UCTE peak load, would disrupt the grid frequency of the entire mid-European interconnected system in sympathy. In view of the politically-desired expansion of onshore and offshore wind energy exploitation in Germany and Europe, E.ON, together with the other transmission grid operators, has been asked to compile rules [4.76, 4.77] that draw upon wind turbines to support the grid. It has been specified that wind turbines have to continue operating in a considerably wider voltage and frequency range than previously. Furthermore, the reactive power output from wind turbines or wind farms must be adjustable on-line by the specification of a target value for power factor or voltage, as described in Section 4.6.

Furthermore, in the event of faults in the grid, very high requirements should be imposed upon the stationary and dynamic behaviour of wind turbines and wind farms. In the event of voltage dips, wind turbines may not be disconnected from the grid in a wide voltage–time graph (up to 3 seconds). In the event of voltage dips at 15 to 60 % of the nominal voltage, they should supply the greatest possible apparent power (in overexcited generator mode) to support the grid. Further specifications that are in line with the large-scale use of wind energy will be provided in the framework of these rules regarding switching on, active power transmission and the disconnection of the wind farm from the grid. These will not be described further here. In the coming years these will shift from the current land-based to the increasingly sea-based installations of the future. Grid connections for these wind farms, which are generally very large, will be in the high and extremely high voltage levels. Therefore, comparable directives are required to those that have already been described for low- and medium-voltage grids [4.21, 4.22].

5

Control and Supervision of Wind Turbines

Intervention into turbine driving power is possible at all times in thermal power stations [3.12, 4.25, 4.26]. In this connection we must differentiate between reactions to necessary long-term changes by energy feed and short-term output smoothing carried out to a limited degree by the steam circuit or a corresponding energy transport route. In diesel units, gas turbines and other systems (see Figure 5.1(a)), the supply of fuel can be adjusted to suit the grid state and the plant control system (or drooping characteristic), thus matching energy supply in the long and short term to changing consumer conditions within the given output framework.

Wind turbines, on the other hand, are dependent upon airflows, which are themselves subject to weather conditions and local effects. This results in corresponding variations in the primary energy supply over which the turbines have no influence. Output can only be changed by reducing power generation. The grid is therefore not only influenced by fluctuations on the power consumption side, but also – in the case of uncoordinated feed from wind turbines – by the effects of the weather on the energy supply (see Figure 5.1(b)).

On the power-supply side, short-term wind-speed changes such as gusts have a particularly pronounced effect on the behaviour of wind turbines. They can lead to high component loading and fluctuations in electrical output variables (voltage, frequency, power). These transient processes thus influence the control characteristics of the system. Moreover, the demands on component groups and their reaction characteristics can be determined to ensure the functionality and integration of a turbine.

Unlike the control unit, the management system must both provide up-to-date desired values and react to medium- and long-term variations in the ranges from minutes to years. High availability can be achieved by load adaptation or energy storage (not considered further here). Moreover, energy management can be used to create power reserves.

Grid Integration of Wind Energy Conversion Systems, Second Edition S. Heier

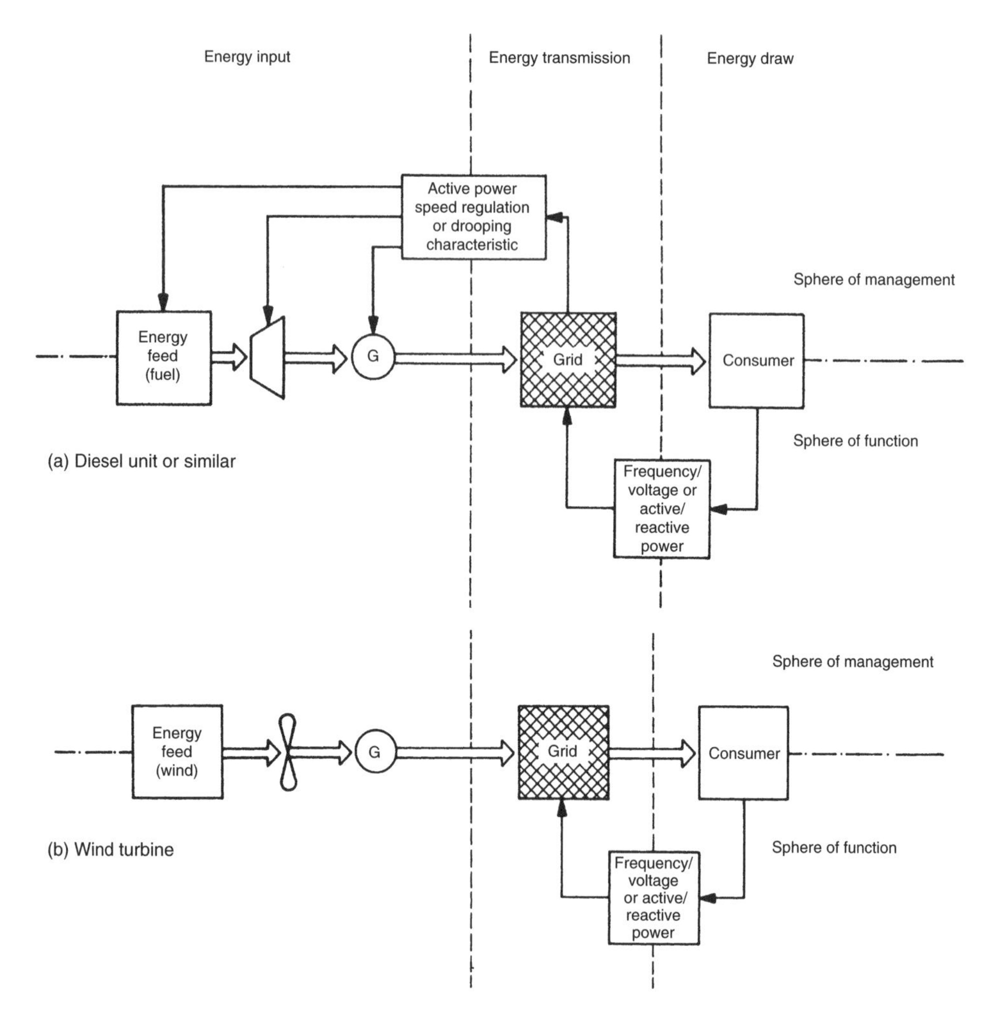

Figure 5.1 Energy flow and differences between the function and control of electricity supply systems based upon (a) diesel generators and (b) wind turbines in uncontrolled supply

Consumption values and wind conditions seem, at first glance, to be completely random variables. However, certain relationships exist, based on consumer habits in connection with environmental influences and physical conditions at the corresponding wind speeds. Structuring these precisely, converting them into layouts and using them in supply/load management, e.g. as shown in Figure 5.2, must take on particular importance if the value of wind power is to be increased.

To the wind turbine control system, only short-term wind-speed changes are of importance, whereas the management system must also take into account variations in the medium term. The management system, together with the control system, must take account of both consumer characteristics and conditions, and those of the turbine and its components. We shall briefly describe these demands below.

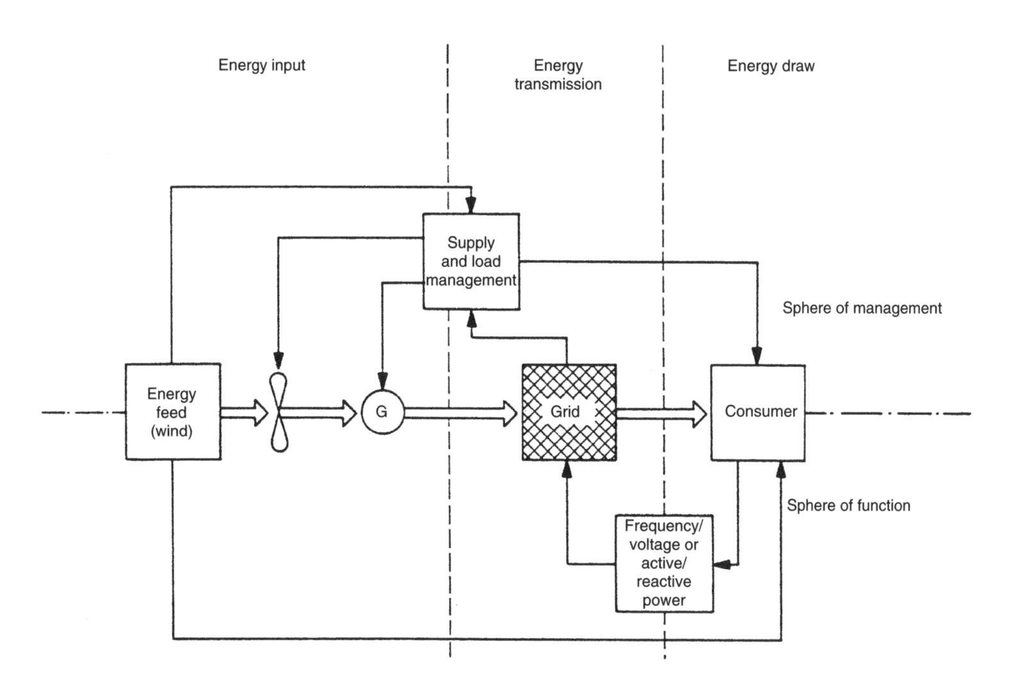

Figure 5.2 Energy flow in an electrical supply system with a controlled wind energy supply

5.1 System Requirements and Operating Modes

In the control and management of a wind turbine, both internal conditions (characteristics of component groups and their interaction) and external variables (consumer desires, conditions for grid-parallel operation) must be taken into account.

The management system takes decisions on logical connections. It monitors whether process plans are followed and limit values adhered to. The control system, on the other hand, must maintain the values specified for the turbine by the management system. Accordingly, it is necessary to ensure that management system decisions are not transferred directly to final control elements (e.g. the blade pitch adjustment device). Where this is compatible with the required reaction speed, the values determined by the management system should be implemented by the control system, which takes the dynamics of components and the turbine as a whole into account when intervention is called for. Exceptions should only be permitted for safety reasons (rapid shut-down processes in the event of faults, etc.).

Along with the normal characteristics for energy conversion plants, additional requirements for the operation of wind turbines are:

- automatic start-up and shut-down depending upon wind and turbine conditions;
- safety monitoring of turbine components by a management unit, with remote interrogation and fault indication available to the operator or maintenance service;
- the option of controlling turbine speed and electrical output;
- separate protection, independently of the control system, to prevent wind turbine output from increasing too fast in high winds;

- the characteristics of all electrical turbine components must be adapted to suit the electricity purchaser with regard to grid reactions, etc.;
- in future, the prerequisites for fault prediction should be met and the resulting advantages used.

Requirements differ for isolated and grid operation. In isolated operation the purely turbine-specific conditions must be taken into account. Moreover, the demands of the consumer must be considered, although these can only be precisely defined for individual cases. The relevant electrotechnical regulations must definitely be fulfilled, in particular those concerning earthing and protection against overvoltage, etc. (VDE 0100 and IEC 555). In addition to the above, for grid operation the local conditions for the parallel operation of electricity generating plants with the grid [3.1, 4.20–4.22] must be fulfilled.

The control system of a wind energy converter represents the link between the management unit and the actual turbine and its components. It must be oriented in particular towards the dynamic characteristics and load possibilities of the turbine, so that it can fulfil its adjustment task. Turbine-specific subsystem behaviour patterns, in particular, must be taken into account (e.g. hub and generator inertia, blade pitch adjustment moments, etc.).

The numerous different wind turbine configurations and applications require different types of control system [5.1–5.4]. These can be divided into isolated, grid and combined operation types to permit the allocation of different types of control system and thus allow application-specific differences to be discussed.

This is necessary in order to define the demands that will be placed on the system components and the operating options for the planned wind turbine. In addition, numerous criteria can be used, such as total turbine efficiency, costs, effects on the grid, operating reliability, use of proven mass-produced components, etc. [5.5].

The interaction between input systems and users strongly influences the behaviour of the entire configuration in weak grids, and in particular in isolated operation. This will be briefly described below.

5.2 Isolated Operation of Wind Turbines

In so-called isolated operation the wind turbine is not connected to an electrical supply grid, but feeds the connected consumers directly. The voltage-regulated synchronous generator is particularly suited for mechanical–electrical energy conversion in standalone operation, i.e. in a supply system with only one supply unit. To achieve the desired constant voltage with an asynchronous generator would require the availability of a regulated source of reactive power for excitation. The demands of the consumer regarding the maximum fluctuation of voltage and frequency at the generator and the maximum speeds – dictated by turbine components – limit the variation options and determine the design of the control system.

If sufficient wind is available, wind turbines in standalone operation can supply electrical consumers directly. The speed of the wind turbine is regulated by influencing the turbine output, thus maintaining the frequency of the generator. A prerequisite for this is that the required voltage can be maintained in the mechanical–electrical conversion system or at the consumer site. This is achieved by the adjustment of the required reactive power by varying the generator excitation, by the use of capacitors or by the use of static or rotating phase shifters (see Figure 5.3). The frequency value and the associated voltage are interdependent

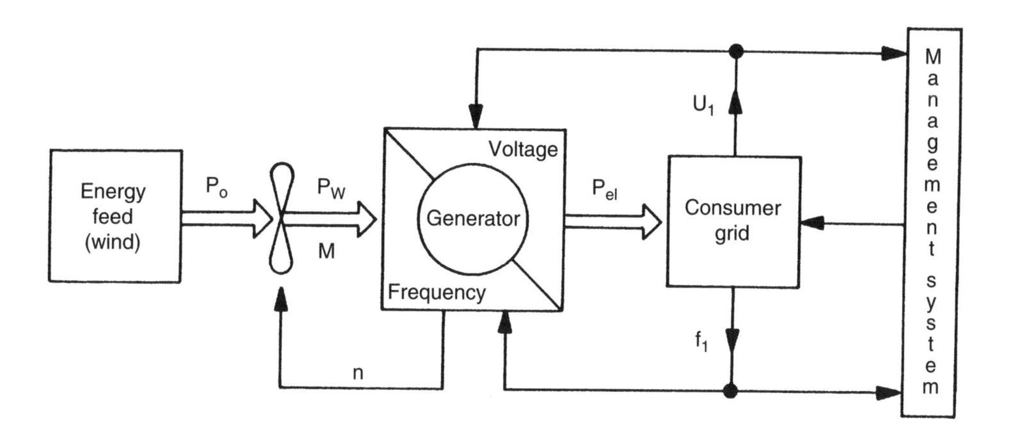

Figure 5.3 Spheres of influence in the separate operation of wind turbines

for synchronous generators. In a simplification, two separate control circuits can be formed, possibly using decoupling networks [5.6, 5.7].

According to the assignment of the spheres of influence within the generator of

- active power and frequency f_1 as well as
- reactive power and voltage U_1 (in connection with the electrical output power P_{el})

as shown above, both grid state variables (f, U) can be viewed, within certain limits, as largely decoupled effects. The power P_0 in the wind is passed on from the wind turbine as component P_w or as torque M according to the turbine state. The interaction between the drive and load torque, caused by the load, gives rise to the generator frequency or speed. If the consumer load exceeds the output capacity of the wind turbine, operation can only be maintained if

- the load is reduced and adjusted to the given output value of the prevailing wind conditions or
- an additional supply is connected.

With the help of the turbine management system

- users can be disconnected according to their supply priority,
- a battery with a bidirectional converter can be connected for short- or medium-term equalization or
- a diesel generator can be operated in the case of a long-term deficit.

The management systems, which prototypes are in operation and will become important, will be briefly mentioned here.

5.2.1 Turbines without a blade pitch adjustment mechanism

Due to the significant cost of a blade pitch adjustment, small turbines are usually designed with fixed blades. If the electrical output variables are not subject to any special demands, the simplest layout can be used for isolated operation, as shown in Figure 5.4(a). Self-excited synchronous generators are normally used for mechanical–electrical energy conversion. In this configuration the voltage and frequency of the generator vary with wind speed and load. Simple electricity consumers can be supplied. The generator must be designed such that the power in the wind can be utilized up to the shut-down speed v_{shut}. The turbine can be braked to low speeds via a mechanical cut-off device (e.g. an air brake mechanism on the blades) at speeds above v_{shut}.

Figure 5.4(b) shows a system in which the generator output current is rectified and direct current consumers are directly supplied. Sensitive alternating current consumers can be supplied with three-phase current of constant frequency and voltage via a self-commutated inverter.

The main disadvantage of all fixed-pitch turbines is that the maximum usable wind speed depends upon the size of the generator and the number of consumers connected at the time. As significantly lower wind speeds usually prevail, such turbines almost always operate in the lower part-load range due to their overdimensioned generators.

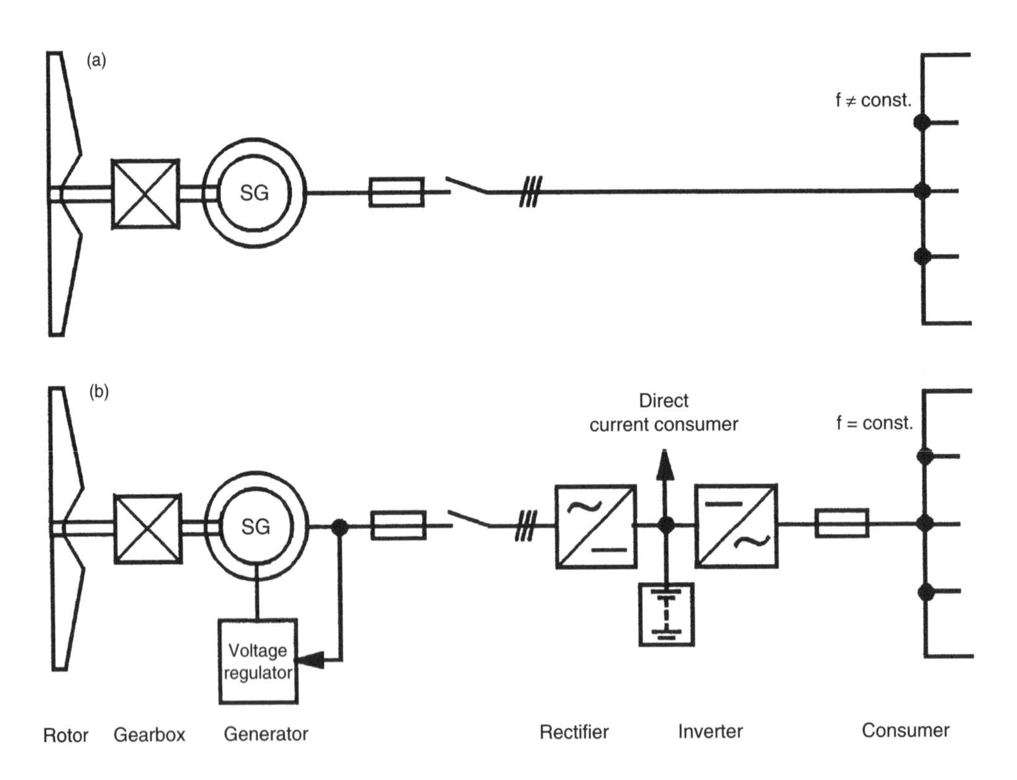

Figure 5.4 Control principles for wind turbines without a blade pitch adjustment mechanism for (a) a turbine without control intervention and (b) a turbine with variable speed and constant output frequency

5.2.2 Plants with a blade pitch adjustment mechanism

Depending upon the mechanical design, wind turbines with blade pitch adjustment can be operated up to very high wind speeds regardless of the power consumption at the time. Their start-up characteristics can also be influenced by the blade pitch.

At a sufficiently high level of wind, a blade pitch adjuster ensures that the turbine speed is kept roughly constant by altering the blade pitch angle. For reasons of stability and to reduce component loading, it is often advisable (in the case of larger turbines) to include a circuit for blade pitch adjustment and/or a circuit to control the rate and acceleration of blade pitch adjustment in the speed control circuit (see Figure 5.12).

A simple design for speed regulation by varying the blade pitch can be achieved using a hydraulic or mechanical centrifugal governor (Figure 5.5). Using such a device, the generator speed and thus the frequency can be controlled within a range of approximately $\pm 10\,\%$. This is adequate for the supply of numerous robust electrical consumers (e.g. simple motors, heat-exchanger units, etc.). Sensitive loads such as electronic devices, on the other hand, can only tolerate very small frequency fluctuations.

A clear improvement in control behaviour and frequency stability can be achieved by using an electrical or electrohydraulic blade pitch adjustment device. This can achieve high blade pitch adjustment speeds and (using electronic regulators) an exact matching of the control dynamics to the behaviour of the control system (Figure 5.6). Frequency fluctuations can, for example, be limited to a maximum of $\pm 1\,\%$.

5.2.3 Plants with load management

In order to account for the normal requirements of electrically sensitive consumers, generator voltage and frequency, and thus speed, should be kept almost constant, even in the part-load range of the wind turbine. To achieve this, the load must always be less than the power available from the wind. Thus the supplied consumers must be connected and disconnected according to frequency (Figure 5.6). However, power consumption need not be continuously

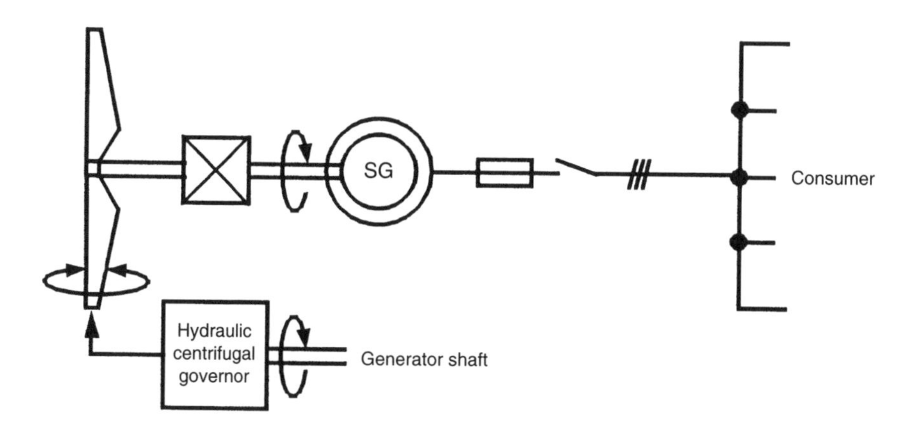

Figure 5.5 Control principle for wind turbines with blade pitch adjustment and a centrifugal governor for the direct connection of alternating current consumers

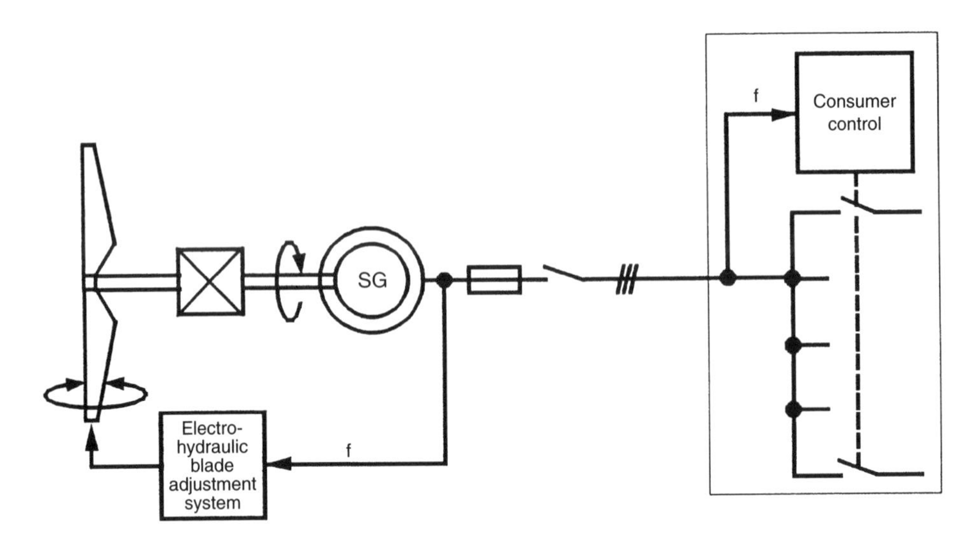

Figure 5.6 Control principle for wind turbines with blade pitch adjustment and electrohydraulic speed regulation for direct consumer connection via load management

variable, but can be altered in stages, i.e. by the connection and disconnection of individual consumers or groups of consumers. Excessively frequent switching processes, and the associated load shocks, are to be avoided.

5.2.4 Turbine control by means of a bypass

A reliable operation of 'stall-controlled' wind turbines is only possible if the turbine speed is maintained by the generator. This requires – as for grid operation – that the energy draw by the consumer is always ensured. If the consumer capacity is insufficient to maintain the drive train at its design speed and to operate the turbine in stall mode, additional consumers (so-called dump loads) must be connected via a bypass (Figure 5.7). This facilitates good frequency control at a reasonable cost for power electronics with the aid of three-phase a.c. power controllers. The generator is preferably an electronically regulated or permanently excited synchronous machine.

Control principles that provide further options for the integration of wind power into the grid can be derived from the designs for isolated operation.

5.3 Grid Operation of Wind Turbines

When operating wind turbines in the rigid combined grid, it is usually assumed that the output provided by the turbine, up to the generator capacity, can always be directly transferred to the grid. A high degree of energy utilization can be achieved in this manner. In order to protect the drive train, generator and grid connection from overload, it is sufficient to limit the output to the design-dependent nominal or maximum value. As shown earlier, this can be achieved in the short-term for control purposes by

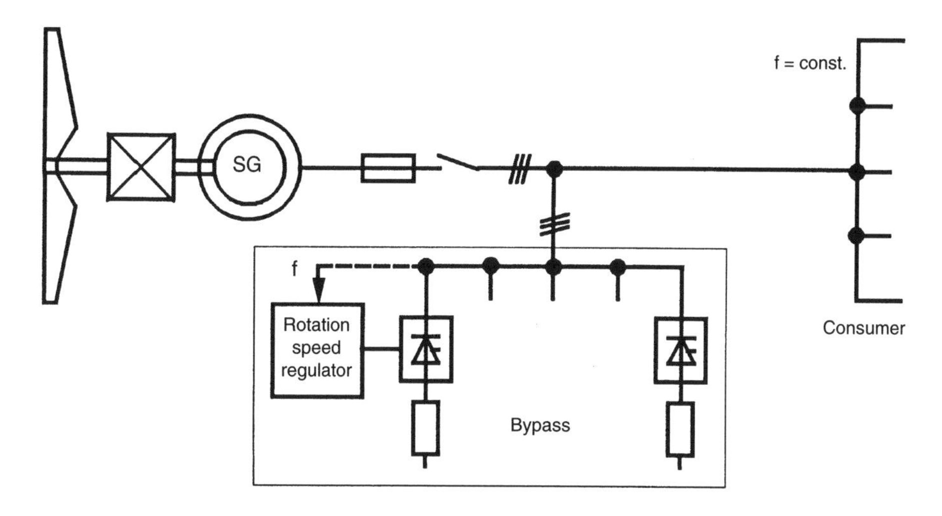

Figure 5.7 Control principle for stall-controlled wind turbines with bypass control

- active intervention at the turbine by changing the rotor blade position relative to the wind or flow direction or
- passive design measures by operating in stall.

For long-term regulation by the management system, it is possible to turn the rotational plane of the turbine into the wind (storm protection).

'Stall-controlled' wind turbines with asynchronous generators connected directly to the grid always exhibit inherent behaviour under different meteorological conditions (wind speed, wind direction, air density, etc.) and grid states (frequency, voltage), due to the constructional layout of the turbine and generator parameters. Moreover, they are subject to turbine-specific conditions caused by long-term changes such as dirty rotor blades, temporary influences such as blade icing or periodic fluctuations such as tower shadowing.

Site-specific harmonics or resonant frequencies in the grid can, if necessary, be reduced by a suitably designed generator winding. Operation-related intervention with regard to slight grid effects (e.g. power fluctuations) are, however, not possible.

Stall-controlled wind turbines of up to approximately 200 kW are encountered in large numbers in Californian and Danish wind farms, and stall-controlled turbines in the megawatt class are also used. The options for influencing this type of turbine by the control and management systems are largely limited to connection and disconnection processes, which can take place via a power converter (soft start) or at defined points in time. These processes can also be influenced by remote monitoring or control.

As described in Section 4.3.2, if this type of turbine is installed and operated in large numbers, greater smoothing effects are achieved in the case of power fluctuations as the number of turbines, and thus the area, increases. The operation of stall-controlled turbines with low individual output power connected to a common supply point is therefore entirely feasible.

If, however, grids are to be heavily loaded by (large) individual turbines or small groups (which for cost reasons is usually desirable in windy areas that previously had a low supply of power), the grid reactions of individual plants or of favourable combinations of systems that complement or compensate for each other (see Figure 4.45) must be assigned great importance.

In contrast to Figure 5.8, in Figure 5.9 the wind power or the drive torque can be brought into line with operating requirements by the management and control systems by adjusting the generator frequency f_G, i.e. the turbine speed n. The significance of remote monitoring systems could therefore increase greatly in the future for the presetting of maximum output values, etc. Such adjustments are most commonly achieved by frequency conversion by means of power electronic converter systems, with or without a constant current d.c. link. Alternatively, as shown by the dashed section in Figure 5.9, the supplied power can be altered by the turbine speed by varying the gear transmission [5.8]. In this manner, synchronous generators with a rigid connection to the grid can be used to support the grid, maintaining voltage and frequency and thereby achieving low grid reactions.

Blade pitch adjustment, as shown in Figure 5.10, permits the energy flow to be adjusted by means of the blade pitch angle β for the regulation or limitation of power output. State interrogation by the management system allows switching and regulation parameters to be stipulated by the remote monitoring system and suitably modified target values to be provided by the management system. External influences that affect the flow of power must be monitored at all times by the control system if enough wind energy is available. By contrast, deviations from the original turbine state, e.g. ice or dirty blades, can only be corrected by monitoring and correcting the turbine parameters.

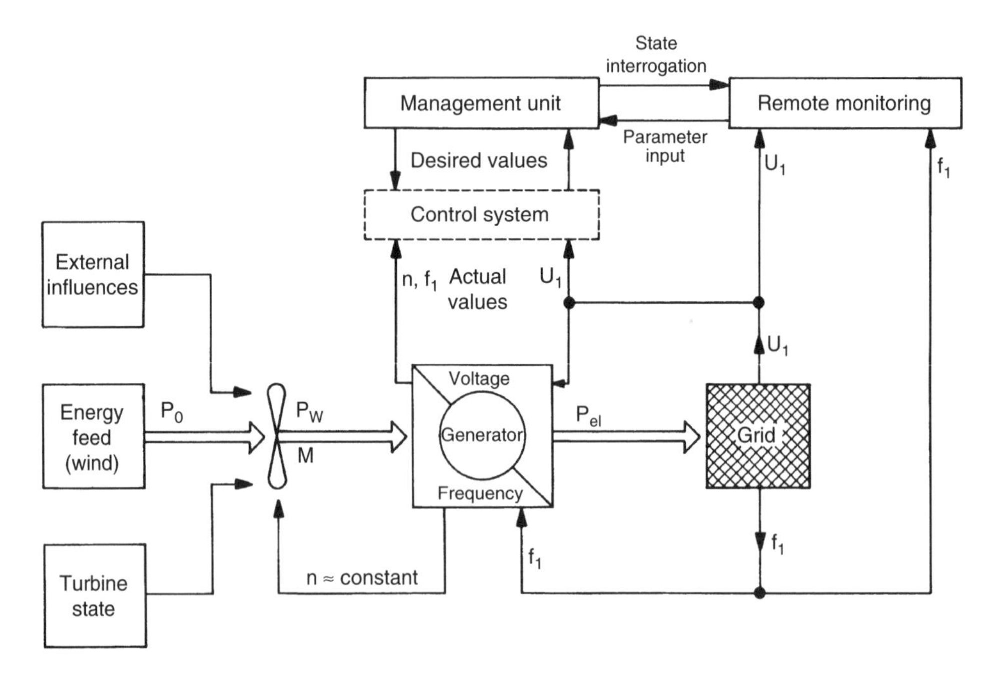

Figure 5.8 Control and management range of a wind turbine kept at a grid-specified fixed speed without blade pitch adjustment

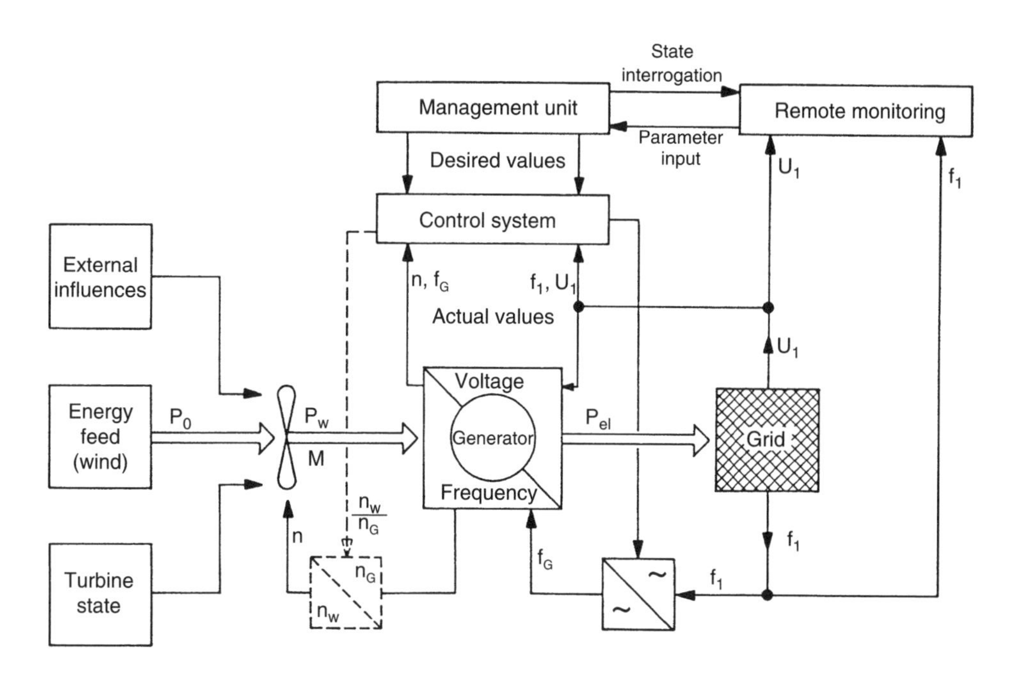

Figure 5.9 Control and management range for a variable-speed wind turbine without blade pitch adjustment

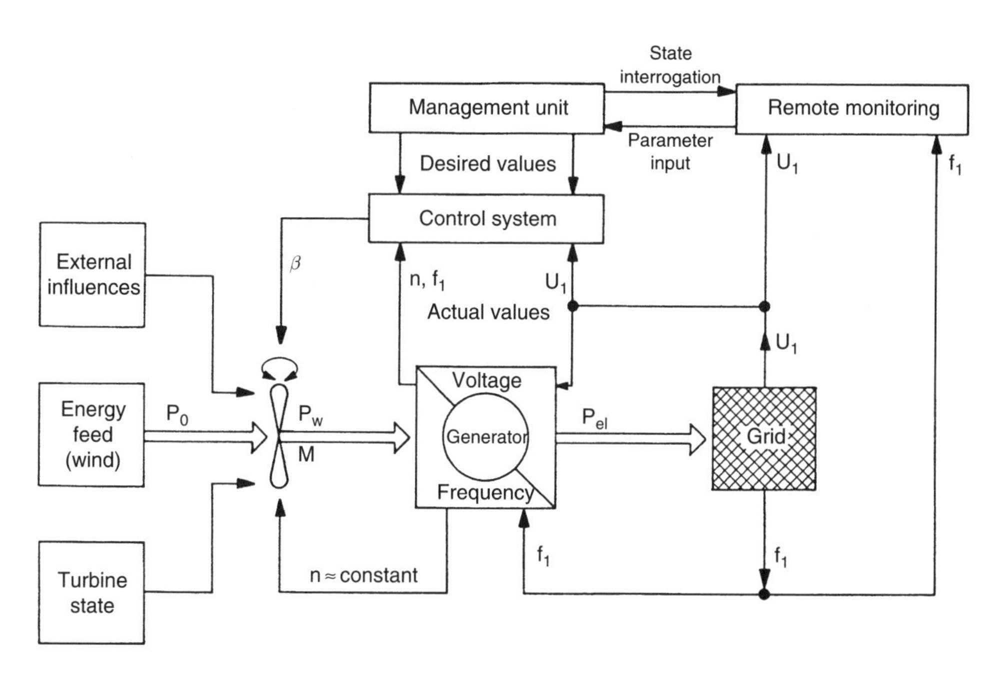

Figure 5.10 Control and management range of a wind turbine with a fixed-speed grid connection with blade pitch adjustment

Speed regulation by blade pitch adjustment, which is common in isolated operation, can also be used in grid operation to run up the wind turbine for synchronizing the generator. Unlike small turbines, which are only sometimes fitted with blade pitch control, wind turbines above approximately 500 kW nominal output are increasingly operated with this mechanism. The rotor blades are either adjusted along their entire length or only at the tip – the particularly active section. The wide range of aerodynamic effects on turbine loading and torque generation will not be considered in more detail here.

With regard to reducing grid reactions, the effects caused by output fluctuations can be reduced with particular effectiveness if, as is the case for this type of turbine, the energy uptake can be actively altered. According to Figure 4.29, the greatest output fluctuations can be expected from individual turbines in the part-load region. In these operating states, power limitation by blade pitch adjustment is rarely or never used. Figure 4.29, also shows (in the right-hand section) that in the operating range above the nominal wind speed, output variations can be significantly reduced. Blade pitch adjustment is fully utilized here.

Power variations, and the associated grid reactions in low-power or heavily loaded grids, can therefore be reduced if, for example, data supplied to the management unit by the remote monitoring system allow modified target values to be achieved by the control system. This process is, however, always associated with a reduced level of wind energy utilization. Moreover, the blade pitch adjustment mechanism of the type currently prevalent can be highly loaded by frequent interventions, and its service life thus significantly reduced. Wind turbines of the type shown in Figure 5.11, which are particularly common in medium- and

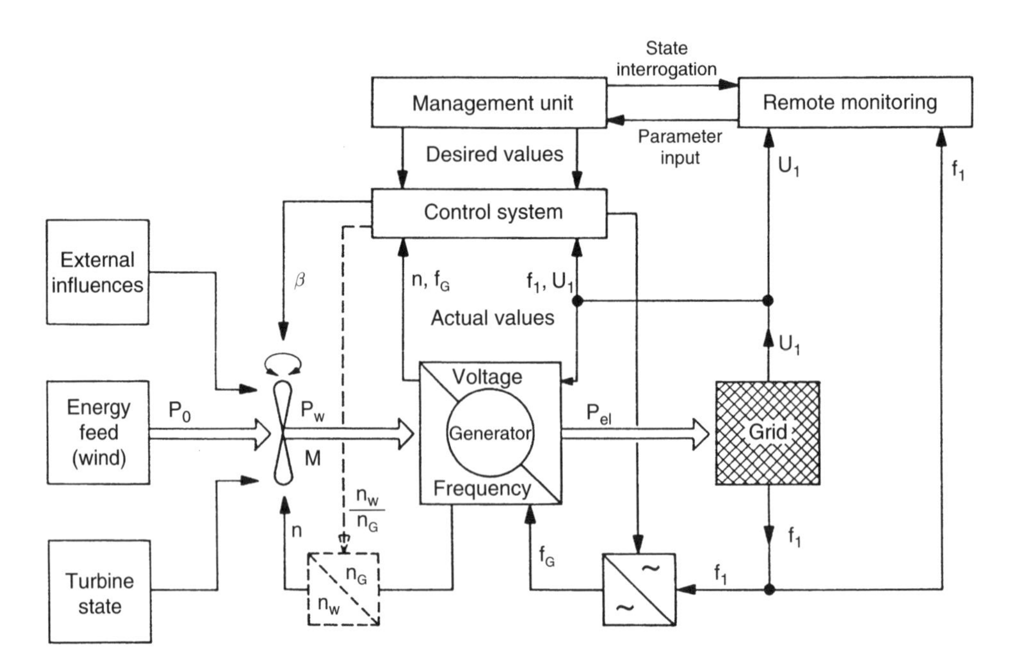

Figure 5.11 Control and management range of a variable-speed wind turbine with blade pitch adjustment

high-output units, provide the option of influencing the wind turbine output and the torque at the drive train, either by adjusting the blade pitch or by influencing generator torque and thus turbine speed. Such units therefore possess two independent intervention systems. This significantly increases operating reliability. Through splitting the influence options as targeted, mechanical regulation, procedures can be significantly reduced when the rotating mass is used as a fly wheel during speed variation. Furthermore, it reduces operational demands on the blade pitch adjustment system on the one hand and on the turbine, drive train and generator on the other hand. Wind turbines that can be operated in this manner can therefore reduce grid reactions caused by output fluctuations and minimize component loading. As a result, structural conditions can be set up much more favourably than is the case for other systems. The results of attempts to quantify this precisely are available. However, options for the definition of the construction and design of relevant elements call for extensive long-term investigations.

Particular attention must be paid to the safety of entire systems and subcomponents in connection with designs with built-in redundancy. In this respect, regulation systems of the type shown in Figure 5.11 with different intervention options offer great advantages. Knowledge of the faults and failures occurring in wind turbines (currently mainly caused by the control and electrical systems due to the as-yet low running times) can give rise to measures for improving reliability and increasing the service life of turbines. Particular attention should be paid to critical components with regard to service life in large-scale use. Owing to the low-cost contribution of, for example, electrical components, fundamental improvements to the entire system can be achieved with great effect at a relatively low financial cost. Repair and maintenance costs can thus be saved and plant availability increased.

Viewed as a whole, wind turbines can be set up using management and control systems such that they are suited for supplying electrical energy to isolated networks and particularly to the supply structures in the grid. This requires appropriate control concepts. These will be briefly described below.

5.4 Control Concepts

Wind turbines must provide reliable operation in all operating states. As well as appropriate dimensioning of the turbine, drive train and tower, the control and management systems, which are described in the next section, are of particular importance. For this reason, it is necessary to develop a local, turbine-specific profile of requirements, to expand this into a strategy that takes into account the desires of operators, manufacturers, power supply companies, etc., and to translate this into a control plan. This can be used for the dimensioning of the regulator, the simulation of the turbine and the design of the control mechanisms from a hardware point of view. Now that the requirements and possible strategies have been highlighted, the most important control concepts will be described below.

5.4.1 Control in isolated operation

Figure 5.12 shows the control diagram for a wind turbine fitted with a synchronous generator in isolated operation. A power regulator influences the power to be drawn from the wind by varying the blade pitch, allowing output to be held almost constant if the wind speed

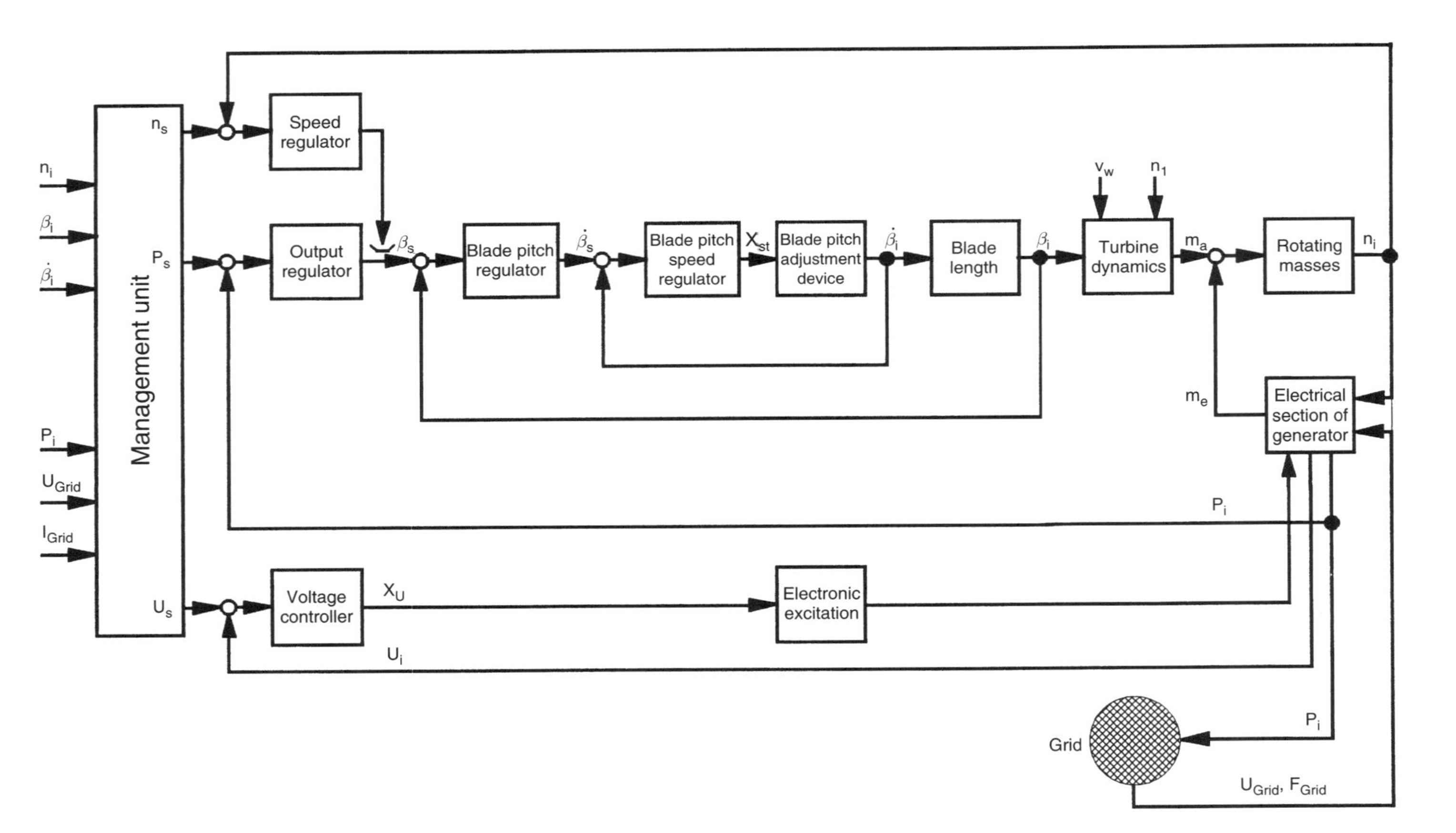

Figure 5.12 Structure for the control of a wind turbine with a synchronous generator in isolated operation

is sufficiently high. By limiting regulator output, this power control circuit can control a speed regulator. Controlled running-up and shut-down processes are thus achieved in the same manner as speed constancy within a tolerance range predetermined by the regulator and adjustment system.

For reasons of stability with regard to turbine characteristics and for the reduction of component loading, it is advisable, particularly for larger turbines, to add a blade pitch and/or blade pitch adjustment speed regulation circuit (internal cascade) and a blade pitch adjustment acceleration regulation circuit to the speed or output control circuit. The output voltage can be adjusted within narrow limits by using a separate voltage regulator if the generator excitation can be externally controlled. This regulation circuit is often integrated into the generator system. Voltage can thus be regulated to constant values or (as is usually the case) altered in accordance with the characteristic in Figure 5.13 with the aid of a reactive power drooping characteristic and matched to the reactive power conditions. A low inductive load leads to the predetermined desired value for the open-circuit grid voltage $U_{\mathrm{Ndes,O}}$. Highly inductive consumers, on the other hand, bring about voltage reductions.

An adjustment of voltage can take place such that the actual value of the grid voltage U_{Nact} arises synthetically from the grid voltage U_{N} by reactive power correction according to the quantity $\beta I_{\mathrm{B}}/I_{\mathrm{N0}}$ (Figure 5.14(a)) [3.12]. Alternatively, voltage regulation with a reactive power-dependent desired value is possible (Figure 5.14(b)). The grid voltage desired value U_{Ndes} is made up of the no-load desired value of grid voltage $U_{\mathrm{Ndes,L}}$.

Moreover, it is possible in a similar way to correct the turbine rotational speed or the grid frequency with the aid of an active power drooping characteristic according to the load state, as shown in the characteristic in Figure 5.15, in such a way that low loads lead to slightly increased rotation speeds and frequency values (e.g. 52 Hz at no-load). These fall to 48 Hz at high loads, for example, when operating at nominal ratings. Thus the actual values as well as the desired values (Figure 5.16) can be corrected, depending upon the load.

The characteristic of a supply system is altered by the active or reactive power drooping characteristic, such that the user is supplied at a high frequency and high voltage at low loads. At high loads these grid variables are reduced correspondingly. The control of the supply system is therefore supported by the stabilizing behaviour of the drooping characteristic.

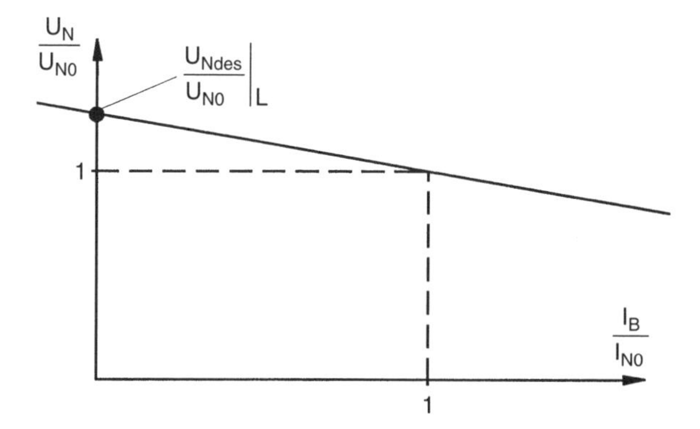

Figure 5.13 Voltage regulation by the reactive power drooping characteristic

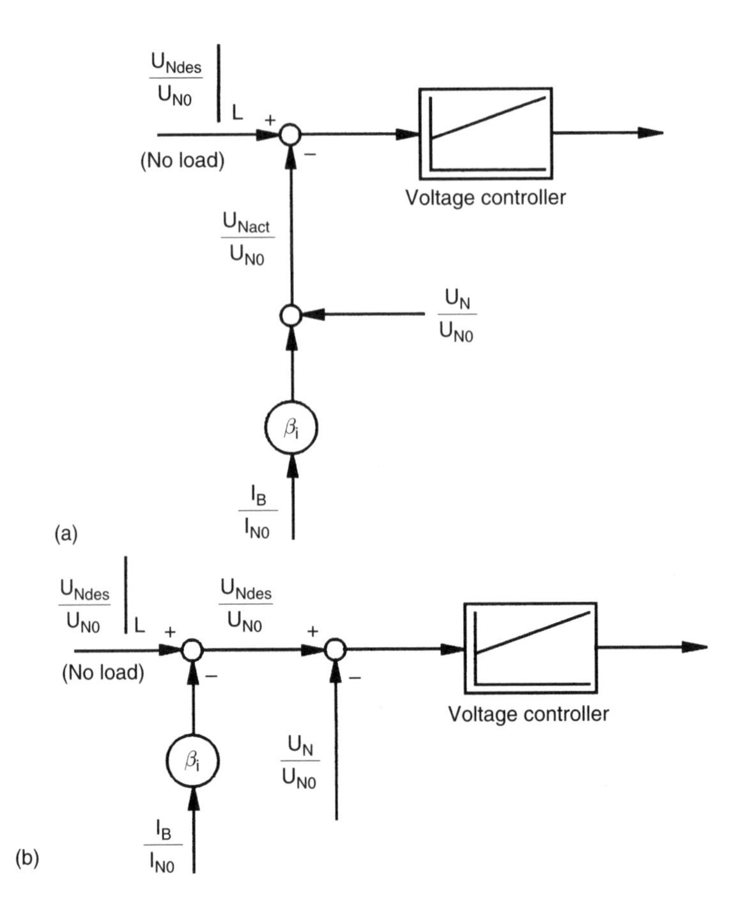

Figure 5.14 Voltage regulation with (a) reactive power-dependent actual value and (b) reactive power-dependent desired value

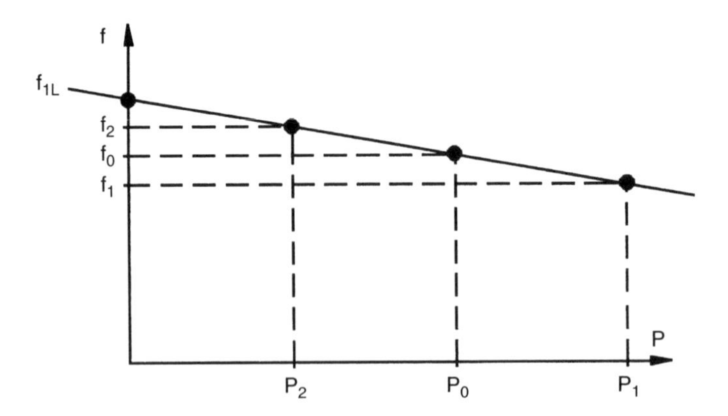

Figure 5.15 Load-dependent speed regulation of a wind turbine by the active power drooping characteristic (characteristic)

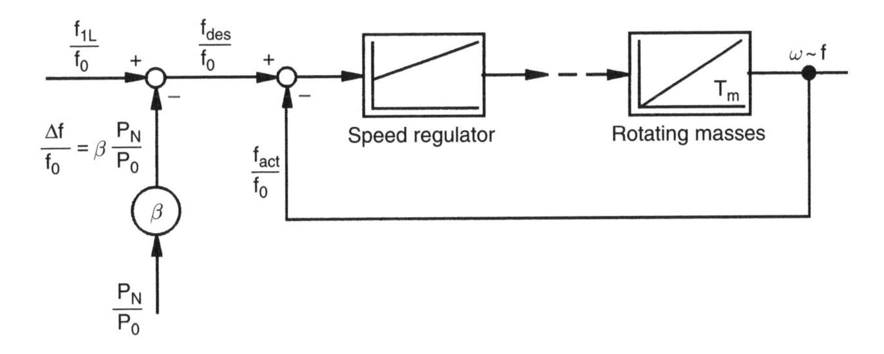

Figure 5.16 Load-dependent speed regulation of a wind turbine by the active power drooping characteristic (circuit with a desired value correction)

The structure shown in Figure 5.12 can also be used for the regulation of the blade pitch, output and speed of a wind turbine with an asynchronous generator in isolated operation. The voltage regulation shown in Figure 5.17(b) can take place within a large fluctuation range using capacitors that can be connected in stages, as shown in Figure 5.17(a). The voltage change can be reduced with an increasing number of capacitor stages. In practice $n = 2$ to 5 stages are used. Larger variation ranges, e.g. with 12 stages, are the exception. Voltage and frequency [5.10, 5.11] can also be kept within narrow limits using rapid switching systems (Figure 5.17(c) and (d)).

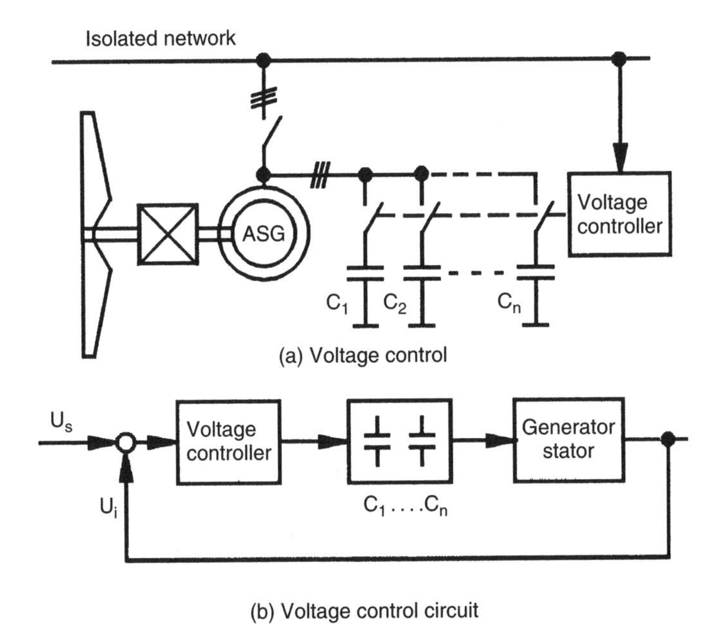

Figure 5.17 Voltage and frequency control of asynchronous generators

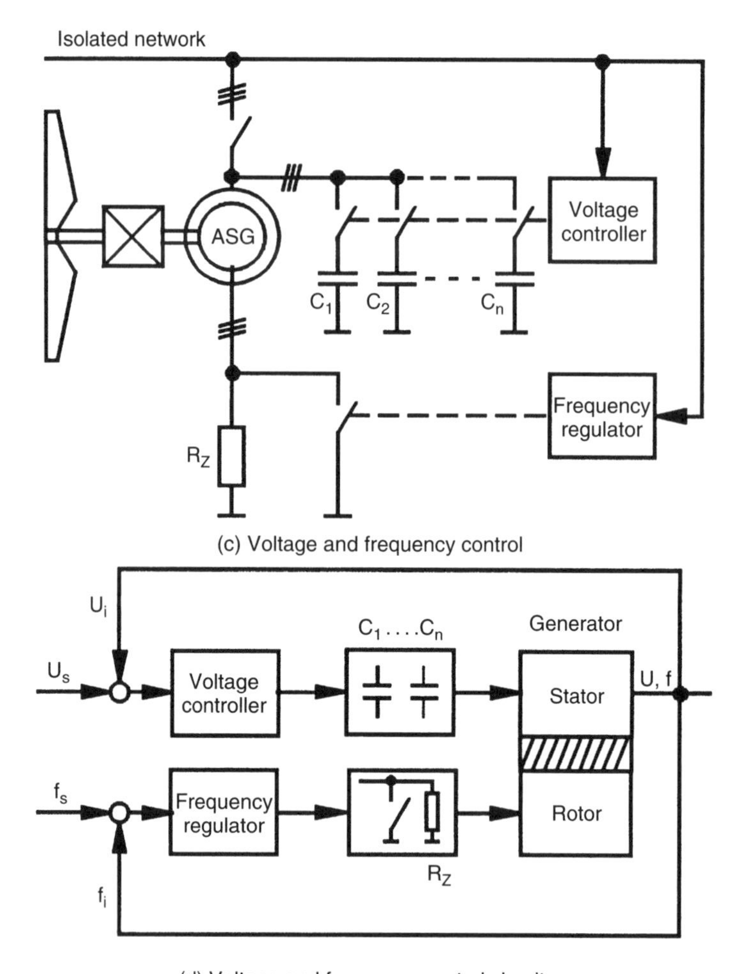

Figure 5.17 (Continued)

Just as in Figure 5.17(a), the capacitors provide the excitation reactive power for the asynchronous machine. The quasi-continuous bridging of the additional resistor R_z, e.g. by using IGBT switches, also permits the control of the stator frequency within narrow limits by slip variation.

For small wind turbines, sensing the blade pitch is relatively costly. To avoid this, the inclusion of pitch position and pitch speed regulation circuits can be dispensed with (Figure 5.18). The speed or power control circuit thus acts directly upon the blade pitch adjustment mechanism and regulates the blade pitch angle according to the prevailing speed and output values, without knowing the actual value of the angle.

However, a simple limiting effect can still be imposed upon the blade pitch adjustment speed in this case. This can be achieved by specifying the maximum value for the speed of the electric positioning motors or the flow in the hydraulic positioning cylinders. This

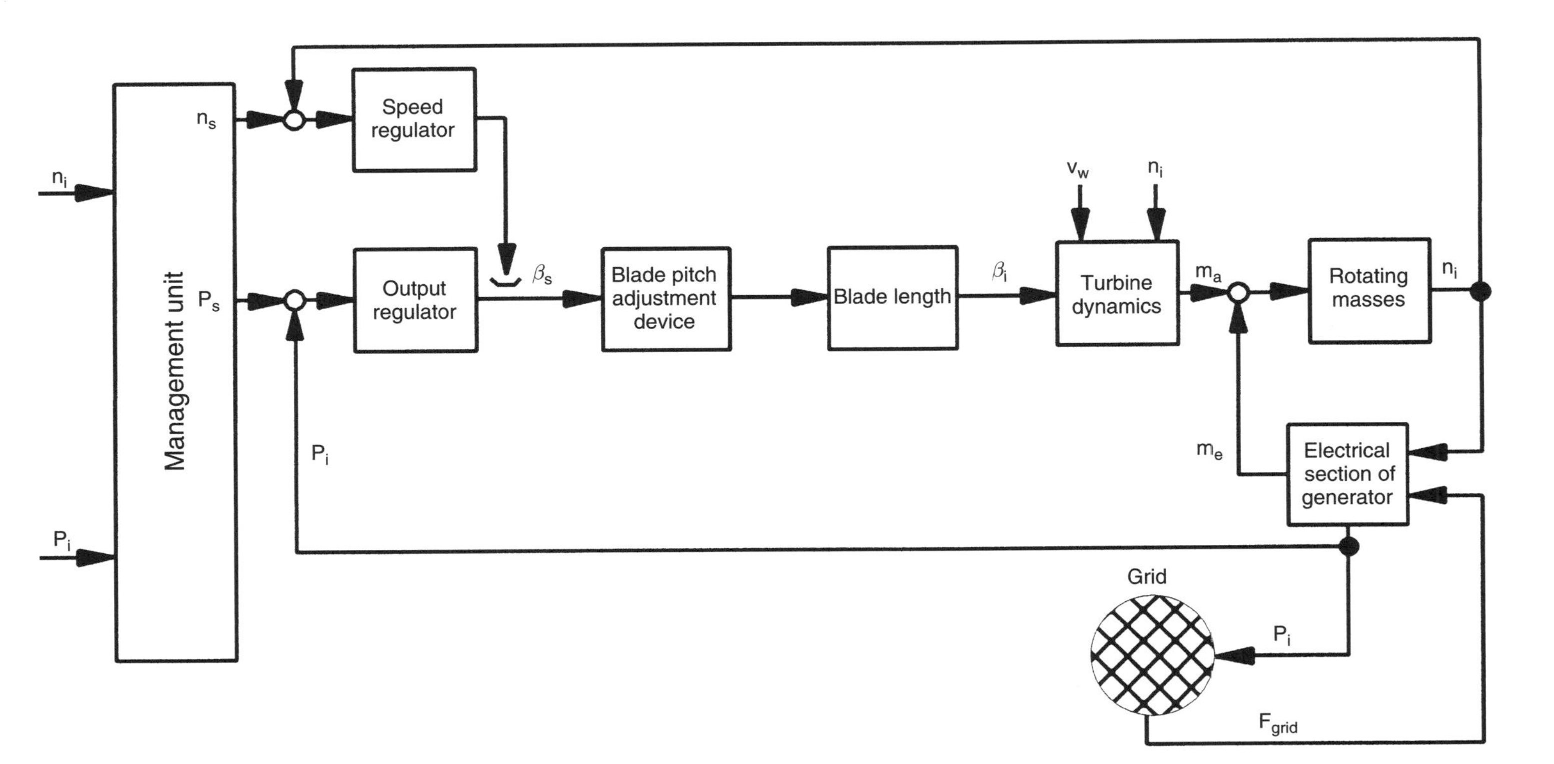

Figure 5.18 Structure for the regulation of a small wind turbine without blade pitch sensing in a separate operation

system simplification, however, does not permit the highly dynamic characteristics of the design shown in Figure 5.12.

5.4.2 Regulation of variable-speed turbines

Control structures for variable-speed turbines can easily be derived from the concepts developed for the control of wind turbines in separate operation. Instead of a fixed speed, or a speed predetermined by the power drooping characteristic, as shown in Figure 5.12, the speed producing the optimum output is determined (see Figures 2.58 and 3.35) and the turbine is run as close to this speed as possible. This type of system must therefore be designed with two different types of speed control circuit (Figure 5.19).

The regulation circuit in the upper section of the diagram has the task of limiting the input power and speed of the turbine at full-load operation to the nominal value, e.g. by adjusting the blade pitch. The second speed control circuit in the lower section of the diagram (e.g. as described in Section 4.1.5.6), on the other hand, must control the turbine speed by controlling the generator electric torque such that the turbine output takes on optimal values or achieves a reliable operating state and behaviour that protects components from excess loading. Using electric actuators, the output power can be adjusted ten times faster than the turbine output. Variable-speed turbines thus provide the option of reducing the load on drive-train components by rapid intervention. This can be achieved by targeted regulation or by limiting the generator torque and using the transient effects of all rotating masses in the drive train according to Equations (2.92) and (2.93).

The control structure as shown in Figure 5.19 is therefore suitable for both grid and isolated operation. The prerequisite for isolated operation is, however, that a self-commutated and self-clocked inverter, e.g. as shown in Figures 4.21 to 4.23, is used that is capable of forming a grid and covering the active and reactive power.

Different options exist for the desired output optimization. The energy conversion systems in Figure 3.4 (b) and (c) can be used and the systems in Figure 3.4 (i), (j), (k) and (l) are particularly suitable. Two well-tested methods are, firstly, the management of rotational speed according to a family of characteristics that depend upon the actual turbine values as shown in Figure 2.58 (c) or (d) and, secondly, approaching the optimum output by means of a search process based upon incremental changes to the rotational speed. The two methods can also be used in combination. Moreover, there are also optimization options that include wind speed and output values based upon the characteristic fields found.

These processes for the improvement of energy yields are not applicable in the case of systems with a rigid connection to the grid, as shown in Figure 3.4(a) and (g). Plants operated with variable slip, as shown in Figure 3.4(d) to (f), which will be considered below, only permit such optimization methods if the mechanical–electrical converter systems allow appropriately large turbine speed ranges, e.g. from 40 % of the nominal speed.

5.4.3 Regulation of variable-slip asynchronous generators

Asynchronous machines permit operation in the oversynchronous and undersynchronous range, i.e. above and below the synchronous speed, according to Equation (3.22). The speed of asynchronous machines can be varied if their generators have multiphase coils

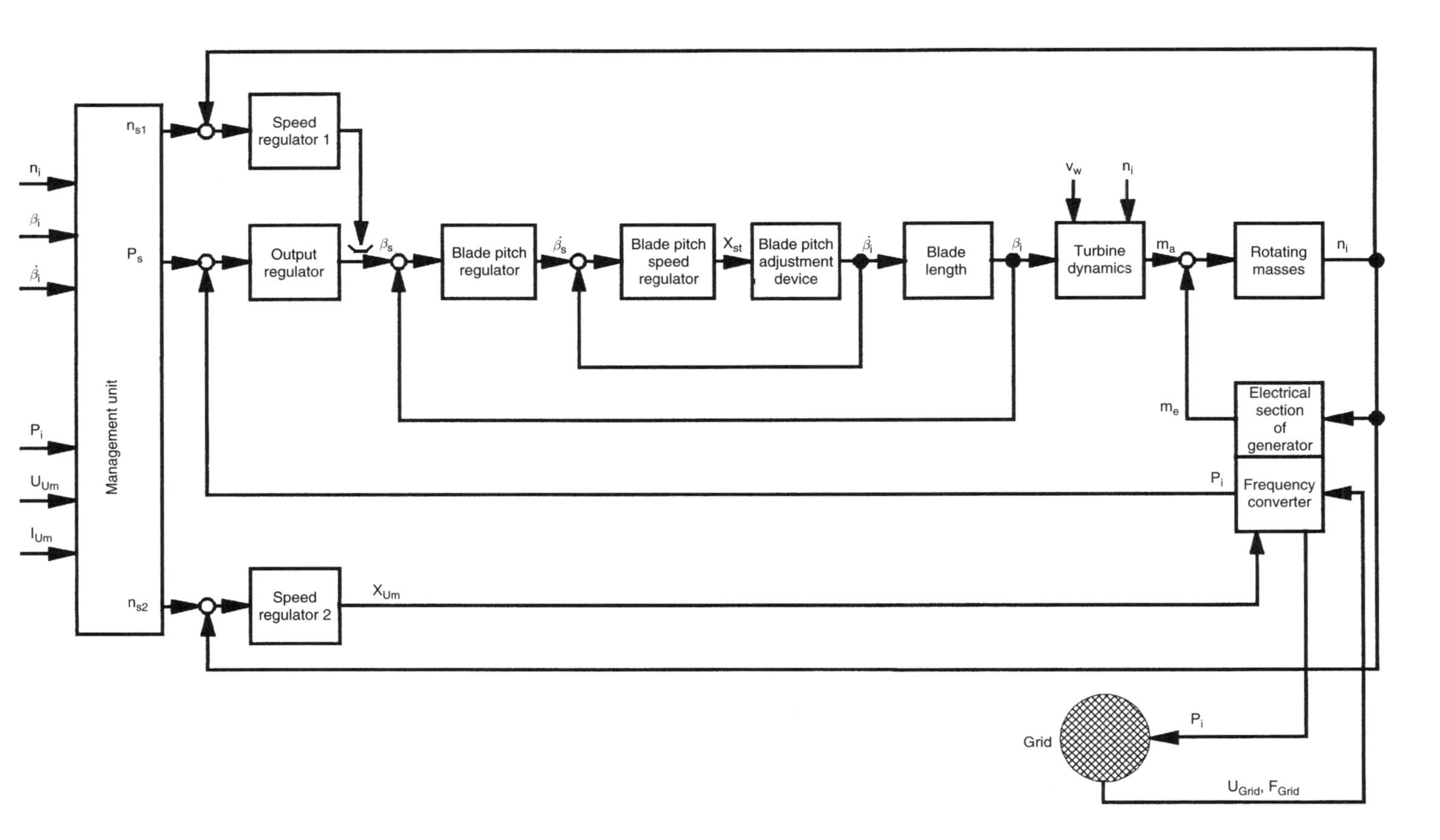

Figure 5.19 Structure for the control of a variable-speed wind turbine with a frequency converter for grid and separate operation

($m = 2, 3, \ldots$) and current transfer via slip rings. Speed can also be varied inductively, and thus without brushes, via additional coils, by

- up to around 10 % with dynamic slip control (Figure 3.4(d));
- approximately 30 % in oversynchronous operation using power converter cascades (Figure 3.4(e));
- around 40 % in under- and oversynchronous operation in double-fed systems (Figure 3.4(f)), i.e. a total of 0.6 to 1.4 times the synchronous speed.

So-called slip-ring rotor asynchronous machines give the option of output branching. In this system, the majority of the converted energy, e.g. 95 %, is passed directly from the stator to the grid. Rotor output, which is usually much lower and is proportional to slip, is drawn off into additional resistors in the form of loss by dynamic slip control. In the power converter cascade and double-fed asynchronous machines, on the other hand, this slip output is conditioned for transfer to the grid by frequency converters in oversynchronous operation. In double-fed machines, which will be considered below, the rotor power is taken from the grid via frequency converters in undersynchronous operation.

5.4.3.1 Double-fed asynchronous generators

Asynchronous machines with slip-ring rotors are generally designed with three-phase a.c. windings in the stator and rotor. Thus the windings in the stator and rotor can be supplied from three-phase systems of different frequency and voltage. Moreover, the magnetization (as in synchronous machines) can be wholly or partly supplied through either of the two electric circuits.

The stator is normally connected directly with the supply grid. The rotor is then supplied via a frequency converter. Thus the angular velocity of the stator rotary field

$$\frac{\omega_1}{p_1} = \omega_{\text{mech}} \pm \frac{\omega_2}{p_2} \tag{5.1}$$

is equal to the sum of angular velocity of mechanical rotation ω_{mech} and the rotor current frequency ω_2. Depending upon direction of the supply frequency, the machine can thus be operated in either the undersynchronous or oversynchronous mode. The pole numbers of the stator and rotor (p_1 and p_2 respectively) at the angular velocity of the rotary fields must be allowed for here.

The rotor active power

$$P_{\text{R}} = sP_{\delta} \tag{5.2}$$

is obtained from the product of the slip s and air-gap power P_{δ}. Thus the power that the rotor circuit transfers to or from the grid via a frequency converter is also proportional to slip. The slip-dependent speed-regulating range $\Delta n = s/n_0$ of the generator, e.g. $0.6 \leq n/n_0 \leq 1.4$, therefore determines maximum output (and thus to a large extent the cost) of the required frequency converter.

The magnetizing current of asynchronous machines is made up of the sum of stator and rotor currents as shown in Figure 3.16. Unlike, for example, short-circuit rotors, double-fed rotors allow the rotor current to be altered. Different phase positions can thus be achieved for the stator current. Similarly to synchronous machines, neutral and overexcited generator states as well as the normal underexcited operation can be achieved in this type of double-fed arrangement. This allows the machine speed and thus also the turbine speed to be varied. Figure 5.20 shows the structure for the regulation of a double-fed asynchronous generator and illustrates the complexity of such systems.

The transformation of the active and reactive components of the total output of large machines into the two-axis field coordinate system with d as the direct-axis and q as the quadrature-axis component (Figure 5.20, top centre) leads, if we disregard stator resistance, to a complete disconnection between the active power, or the largely corresponding machine torque variable, and reactive power [2.21, 2.22]. Thus the two state variables can be regulated separately without mutual interaction.

In order to make use of the advantages of variable-speed conversion systems, the total output of the machine P_G should be influenced by the regulation system, and the generator and turbine speed should be dictated in a controlled manner. To control the rotational speed (upper-left part of the block diagram) the speed–output characteristic (e.g. as shown in Figure 2.64(c)) can be used. Speed is influenced by torque, as is normal for rotating systems. This can be adjusted by means of the quadrature-axis component of the rotor current i_{Rq} by suitable control of the machine frequency converter via the rotor voltage according to the field [5.12, 5.13]. The reactive power can be predetermined in a similar manner, e.g. as a fixed desired value, or can be guided according to the requirements of the power generation company by means of the direct-axis component of rotor current i_{Rd} with the aid of the machine frequency converter (Figure 5.20, bottom left). Instead of reactive power Q_{Tdes}, the power factor can be predetermined and thus adjusted.

In both control circuits (i_{Rd} and i_{Rq}) the rotor current (as in the stator circuit) is detected in at least two or in all three phases by rotating pointer $\underline{i}_R$, converted into digital values and transformed into the field coordinates i_{Rd} and i_{Rq}. Thus the double-fed asynchronous generator can be adjusted or regulated via the pulse-controlled inverter on the machine side for the desired values of speed, torque and active power, as well as reactive power or the power factor.

In order to avoid disruptive connections between the grid-side pulse-controlled inverter and the overall regulation system via the machine-side pulse-controlled inverter, the grid frequency converters can be operated by the d.c. link voltage (see Figure 5.20, bottom right). This can be kept roughly constant by a preset hysteresis band [2.22]. In this manner, the grid-side frequency converter adapts to the value and phase of the required active power and works within the preset tolerance band in neutral operation. The grid-side reactive power in the rotor can therefore be kept close to zero and the total reactive power can thus be adjusted by means of the reactive power regulation system at the machine frequency converter.

High-speed current changes can occur in the stator circuit, particularly if the generator is suddenly disconnected from the grid for some reason, e.g. by the tripping of protective switches or by rapid auto-reclosure. These current changes induce high-voltage peaks in the rotor winding. To prevent this, overvoltage protection must be fitted in the rotor (see Figure 5.20, centre).

The double-fed asynchronous machine is particularly suitable for use as a generator in wind turbines. Due to the connection of the stator to the grid, these machines combine the beneficial electrical characteristics of synchronous machines with mechanical advantages

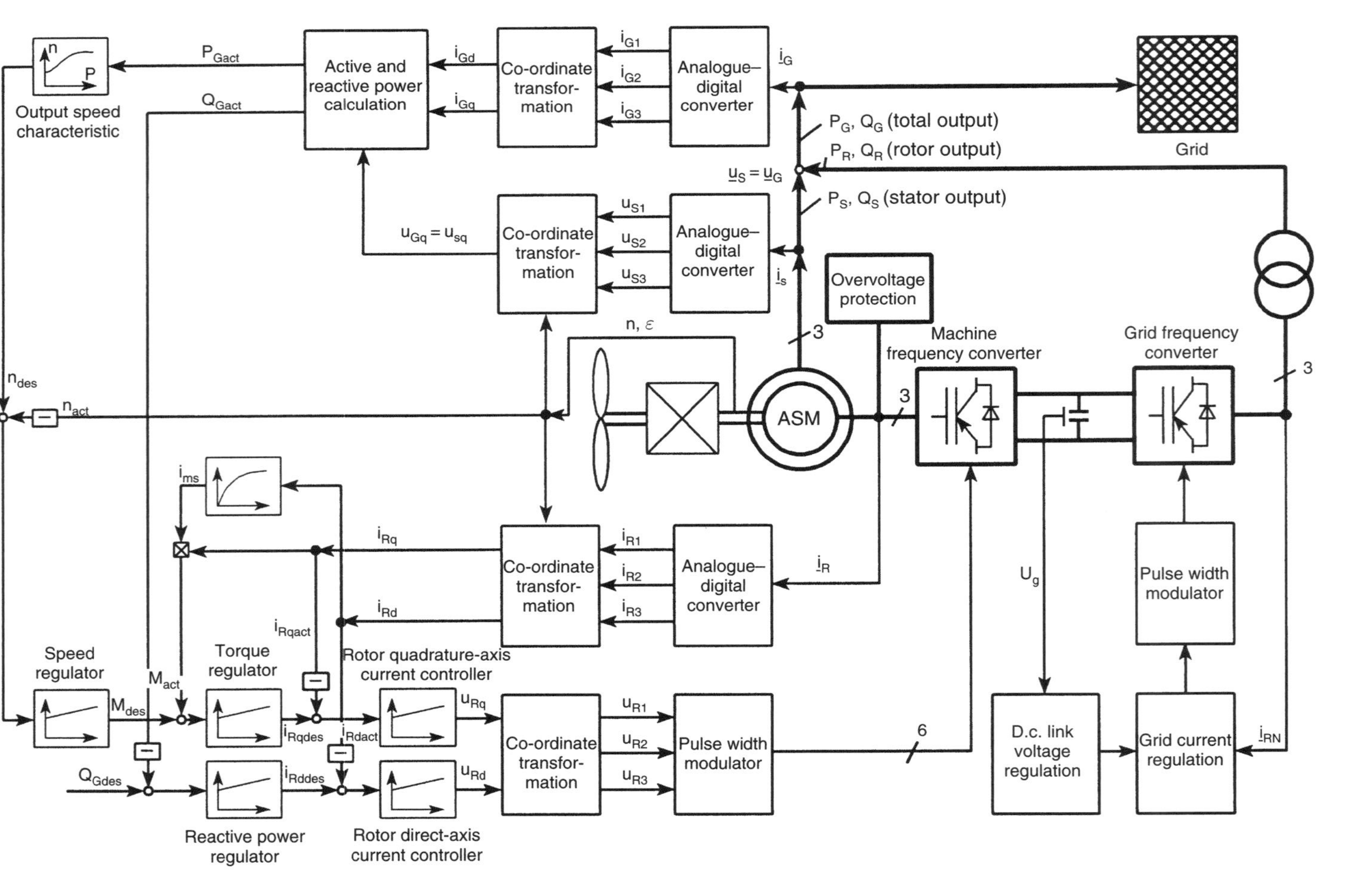

Figure 5.20 Block diagram for the control of double-fed asynchronous generators

associated with speed variability at very good efficiency. They thus allow the mechanical loading of the drive train to be significantly lessened and the fluctuations of electrical output power to be greatly reduced. By the use of IGBT pulse-controlled inverters in the rotor circuit, grid reactions caused by harmonics can be reduced to a noncritical level. Other options for capacitive and inductive operation can be used for grid support. Thus wind turbines with double-fed asynchronous generators can, particularly in weak grids, give access to significant connection capacities.

The double-fed asynchronous machine was first used with a direct frequency converter made up of thyristors, in the GROWIAN wind turbine, which was installed in Kaiser-Wilhelm-Koog on the North Sea coast at the beginning of the 1980s. The American MOD 5 B turbine in Hawaii was also fitted with an appropriate converter device.

During the operation of the 3 MW turbines it was possible to demonstrate the excellent regulating characteristics of these systems by the extremely well-smoothed output power [5.14]. The complete utilization of the so-called ceiling voltage proved its worth here, since it determines the rotation speed range of the generator and the reactive power requirement of the frequency converter [5.1]. By the capacitive operation of the stator, the phase control reactive power of the direct frequency converter can be generated to a large degree by the generator itself. However, the harmonic content of the generator current increases with rising capacitive stator reactive power, with the fifth, seventh and eleventh harmonics of grid frequency, in particular, dominating at high rotor frequencies and at high firing angle settings of the direct frequency converter.

Static reactive power compensation of the fundamental reactive power by filter circuits tuned to dominant harmonics provides a favourable compromise. However, this requires precise knowledge of grid characteristics in order to avoid possible grid resonances [4.53].

Both oversynchronous and undersynchronous operations are possible if the rotor of a slip-ring rotor machine, for example, supplies into a four-quadrant indirect converter. The load-side power converter can be designed with self-commutated thyristor valves or with a self-commutated or machine-clocked pulse-controlled a.c. converter. The regulation of the machine can be based on the block diagram in Figure 5.21. Thyristor valves, however, create high harmonic currents in the rotor of such systems, which, in combination with the high cost of the power converter (compared with oversynchronous cascades), makes this concept appear largely unsuitable for use in wind turbines [5.1].

Double-fed asynchronous machines with IGBT frequency converter systems allow particularly favourable operating characteristics to be achieved with regard to grid reactions. At the beginning of this section the advantages were described of how this type of converter device can be fully exploited in this type of system. Such systems are becoming increasingly important in turbines of the 0.5 to 5 MW class (e.g. DeWind 4 shown in Figure 1.28 to N80/90 shown in Figure 1.24(d) or V80/90 shown in Figure 1.25(f)) due to the possibilities that currently exist in power electronics and computing. The wear associated with slip-ring transmission is prevented if field-oriented currents are transmitted to the rotor via additional windings. Such machines are already available on the market. The additional delays that occur, however, play a decisive role in the field control.

If undersynchronous generator operation and reactive power regulation of the machine is avoided, the system based upon an oversynchronous power converter cascade, which operates in a narrow speed range, can be used for wind turbines. This system is described below.

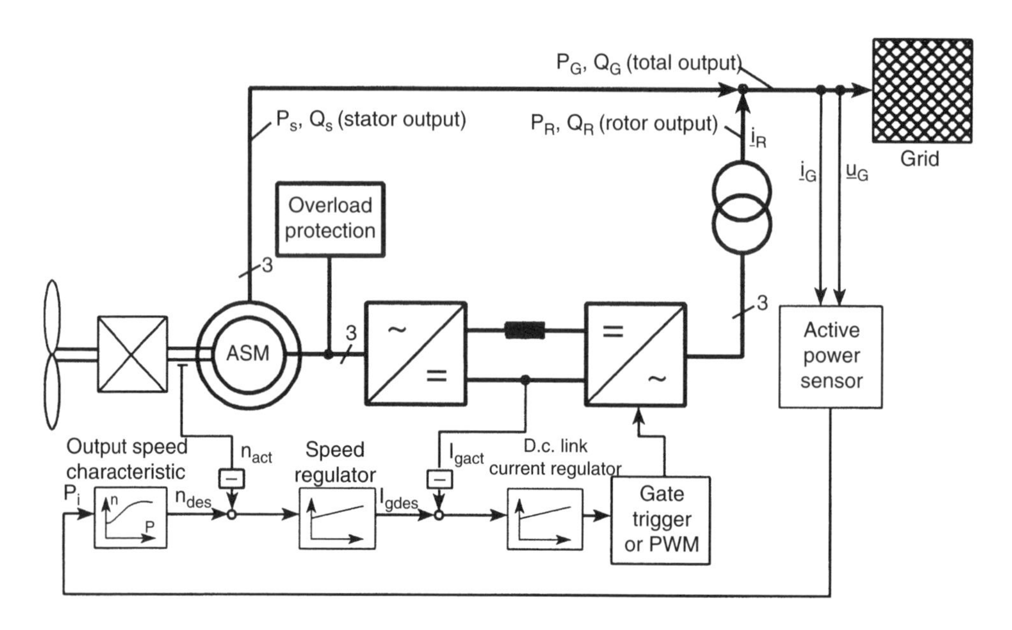

Figure 5.21 Block diagram for the control of an asynchronous machine with a slip-ring rotor and oversynchronous power converter cascade

5.4.3.2 Asynchronous generators with oversynchronous power converter cascades

Compared with double-fed asynchronous generators, much simpler conversion systems are achieved if slip-ring rotor machines are only operated in the oversynchronous speed range. In this system, the slip power of the rotor can be drawn from the rotor and supplied to the three-phase grid via an uncontrolled or controlled rectifier and via a line-commutated or self-commutated inverter (Scherbius principle). This type of converter system is characterized by its robust and low-maintenance construction and, moreover, represents a very reasonably priced variant.

Figure 5.21 shows the relatively simple structure for the control of the converter system. The machine-side rectifier converts the variable-frequency alternating variables of the rotor (voltage $\underline{u}_R$ and current $\underline{i}_R$) into direct variables (u_g, i_g). The direct current of the d.c. link is inverted to grid frequency by an inverter and matched to the grid voltage by a transformer. This system therefore brings about a complete decoupling of rotor speed and grid frequency. Rectifiers and inverters can take the form of diode rectifiers or thyristor-controlled or pulse-controlled a.c. converters.

Active power speed regulation is similar to that in a double-fed machine. However, the speed variations are limited to a much narrower range. Reactive power regulation is no simple matter. Overload protection, e.g. in the form of a switchable resistor, protects the d.c. link from overload and impedes uncontrolled operation of the generator.

The simple design of the system generally ensures unproblematic commissioning and operation. Moreover, if the d.c. link is favourably designed excellent dynamic converter characteristics can be achieved.

This type of conversion system was installed in the Spanish–German AWEC 60 on the north-west tip of Spain in Cabo Vilano at the end of the 1980s. The mechanical–electrical converter was able to demonstrate fully its good regulation potential and excellent dynamic characteristics in the 1.2 MW wind turbine with a 60 m rotor diameter, even under extreme operating conditions. This turbine was of almost identical design to the two German WKA 60 I and II wind turbines.

However, as is the case for cage rotor machines (see Figures 3.9 and 3.10), the reactive power for the excitation of the generator must be provided by the grid or by an additional compensation unit (see Figures 3.11 and 3.12). In an oversynchronous power converter cascade the commutation reactive power of the rotor-side rectifier, the phase-control reactive power of the inverter and the distortion reactive power of the entire turbine must also be provided if, for example, thyristor valves are used.

The phase-control reactive power of the inverter can be reduced to low values by specifying a narrow oversynchronous speed range. To achieve this, the inverter must be adjusted to the inverter stability limit when the generator is running at its maximum speed. Thus a reserve voltage must be dispensed with in order to fully utilize the d.c. link.

The slip-dependent stator current harmonics caused by the rotor currents represent a further disadvantage. These give rise to considerable grid reactions and mechanical loading of the machine, particularly due to oscillating torque. Moreover, the harmonics give rise to additional losses in the generator and reduce the efficiency of the conversion system [5.1]. It must also be borne in mind that the rising and falling edges of the rotor current are smoothed by the commutation of the rotor-side rectifier (see the sketch in Figure 5.22). This causes a reduction of harmonics in the generator current. However, the commutation causes a phase shift between the fundamental component of voltage and current in the rotor. This results in a weakening of the electrical moment and an increase in stator reactive power around the commutation component of the rectifier. The utilization of the slip-ring rotor machine is thus reduced.

Significant disadvantages of this system can be prevented by the use of pulse-controlled a.c. converters [5.15, 5.16]. These give rise to greater frequency converter losses. Because of

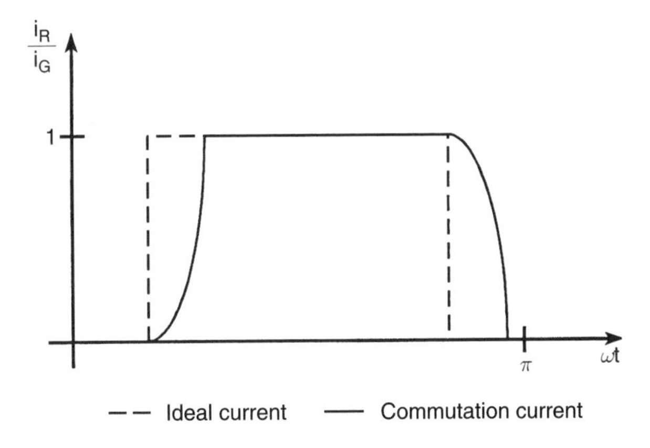

Figure 5.22 Sketch of the rotor current path

the small contribution of the rotor conversion system to total output (e.g. 0 to 15 %) these are hardly noticeable, since the greater part of the increase in losses can be largely balanced out by more favourable operating conditions (near-sinusoidal currents, reduction in oscillating torque, etc.).

The system is further simplified by dispensing with the return of a slip power predetermined by the slip regulation system from the rotor to the grid. The desired slip can be adjusted to the values necessary for dynamic adjustment (as low as possible). These will be considered below.

5.4.3.3 Asynchronous generators with dynamic slip control

Asynchronous generators with wire or bar windings in the rotor usually permit intervention via the slip rings. Moreover, the energy induced in the rotor can be converted in the rotor system or inductively transferred to the stator. As well as the methods that have already been described for converting slip energy and feeding it into the grid, there is also the option of completely dispensing with the use of this energy component in order to keep the cost of the conversion system low. Wind turbines are, however, subject to random and periodic output fluctuations due to wind-speed fluctuations, tower-shadowing effects, natural resonances of components, etc. In a fixed-speed connection of the generator to the grid, these are passed on to the grid at almost their full level via the drive train and the generator coil (see Figure 3.15(a)). Large slip values significantly reduce the load on the drive train and greatly reduce output fluctuations (see Figure 3.15(b)). However, if the slip energy is not recovered high losses in the rotor circuit are the result according to Equation (5.2). To achieve a high total efficiency, the rotor slip should be kept as low as possible in this relatively simple system. A favourable conversion system should therefore combine the two seemingly contradictory options of

- low slip energy losses and
- high dynamic speed and slip flexibility (see Figure 3.16).

Slip-ring rotor asynchronous machines and systems based upon the same operating principle permit three operating methods, as shown in Figure 5.23:

- In the case of a short-circuited rotor winding, the generator operates at the lowest slip and thus achieves the best machine efficiency. However, this results in high drive-train loading and large output fluctuations.
- By connecting resistors in the three phases of the rotor circuit, high slip values and thus high elasticity of speed can be achieved if the resistors are correctly dimensioned, with increased slip bringing about better dynamic characteristics but poorer efficiency.
- With a continuous adjustment of slip to the prevailing wind and grid conditions, a rapid change of rotor circuit resistance aided by alternating switching processes between short-circuited rotor winding and full resistance in the rotor circuit can bring about output-smoothed or efficient operating ranges [5.17]. Thus, in the part-load range, low slip values can be set and adjusted slightly (see Figure 3.16) in order to achieve a high level of efficiency.

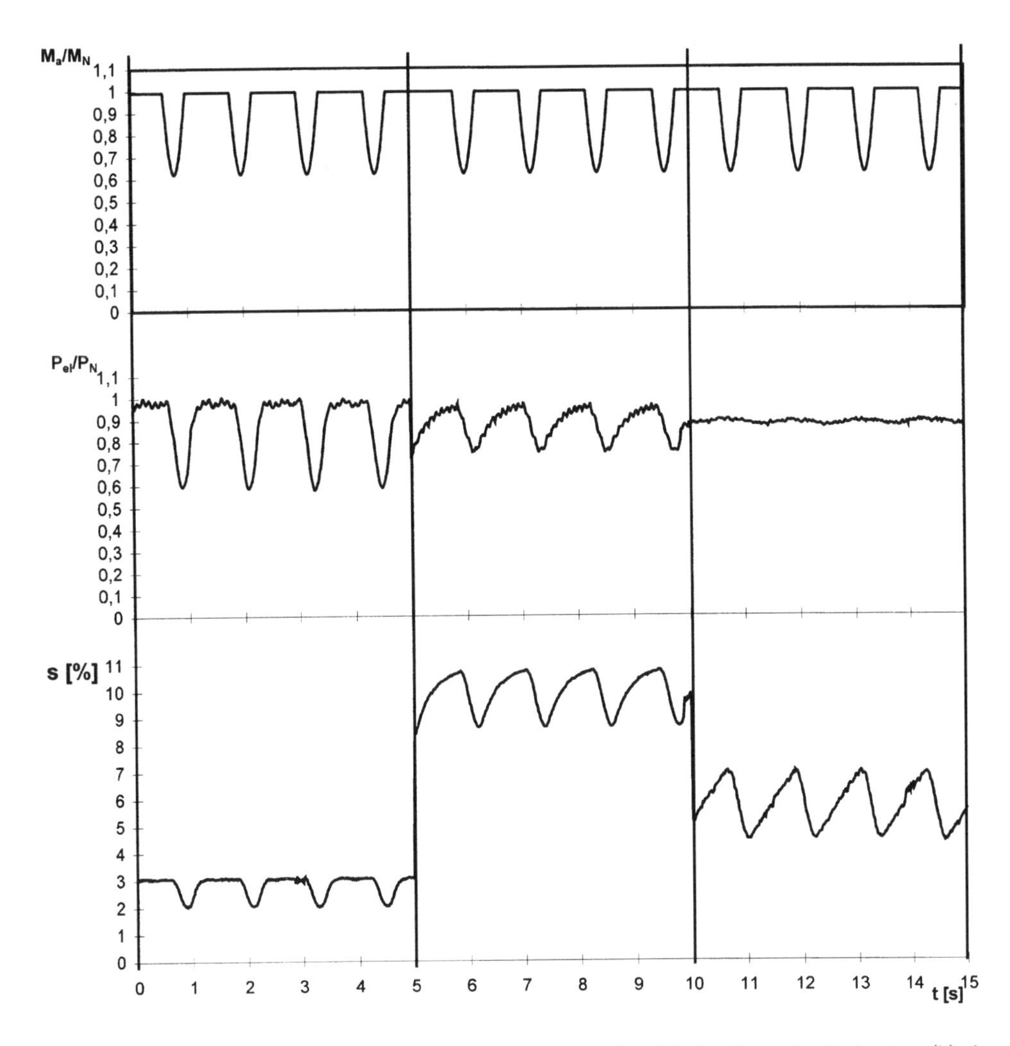

Figure 5.23 System behaviour of asynchronous generators for (a) the short-circuited rotor, (b) the rotor with fixed additional resistors and (c) slip control

Torque fluctuations arising from the drive train can be largely smoothed out in this manner [3.3]. If the wind supply is above the level required to produce nominal output, large losses due to increased slip can be accepted without sacrificing energy yield. Thus the advantages of torque- and output-smoothing characteristics can be combined with adequate speed-regulation reserves [5.18, 5.19].

Figure 5.23 illustrates the excellent characteristics of a system with slip control in comparison with the unregulated layout both with short-circuited rotor and with additional resistors connected in the rotor circuit. This shows that a slip-controlled system can smooth output fluctuations – caused, for example, by tower shadowing or similar effects – excellently. At the same time, average slip values, and thus losses, can be kept low.

Unlike the systems described previously, power electronic components do not affect the components that are electrically connected to the grid. They only act in weakened form via the air gap and the coupling inductances in the machine (see Figure 3.7(a)). This significantly reduces grid harmonics [5.17, 5.18]. The measurements in Figure 5.24 provide clear proof of this. The system configuration later in Figure 5.26 was used in these investigations. A comparison of the frequency spectra of the rotor and stator currents and voltages shows the 'filtering characteristics' of the generator.

Figure 5.25 shows the conventional concept for wind turbines with slip control, based upon the industry standard robust asynchronous generator with a slip-ring rotor. Using this system it is possible to design the rotor circuit with a rectifier bridge and single-phase pulsing (e.g. using IGBT) at the additional resistor (Figure 5.25(a)) or to design the rotor system with additional resistors and pulsing in all three phases (Figure 5.25(b)).

The additional rotor resistors and the slip regulation and its drive are generally fitted as a separate unit outside the generator. This makes it easy to perform maintenance and any necessary repairs to components. Moreover, the subsystem can be designed as a modular unit.

The investigations in references [3.3] and [5.17] have shown that a simple single-phase design, as shown in Figure 5.25(a), tends to cause mechanical vibrations in the drive train.

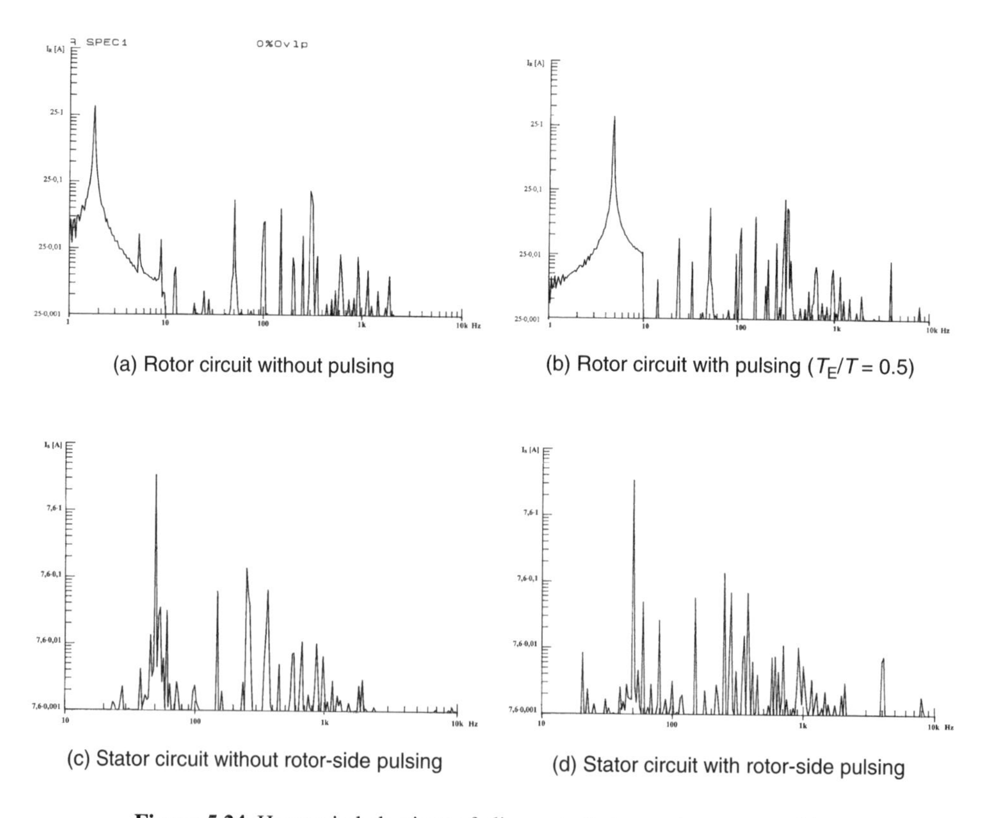

(a) Rotor circuit without pulsing

(b) Rotor circuit with pulsing ($T_E/T = 0.5$)

(c) Stator circuit without rotor-side pulsing

(d) Stator circuit with rotor-side pulsing

Figure 5.24 Harmonic behaviour of slip-controlled asynchronous machines

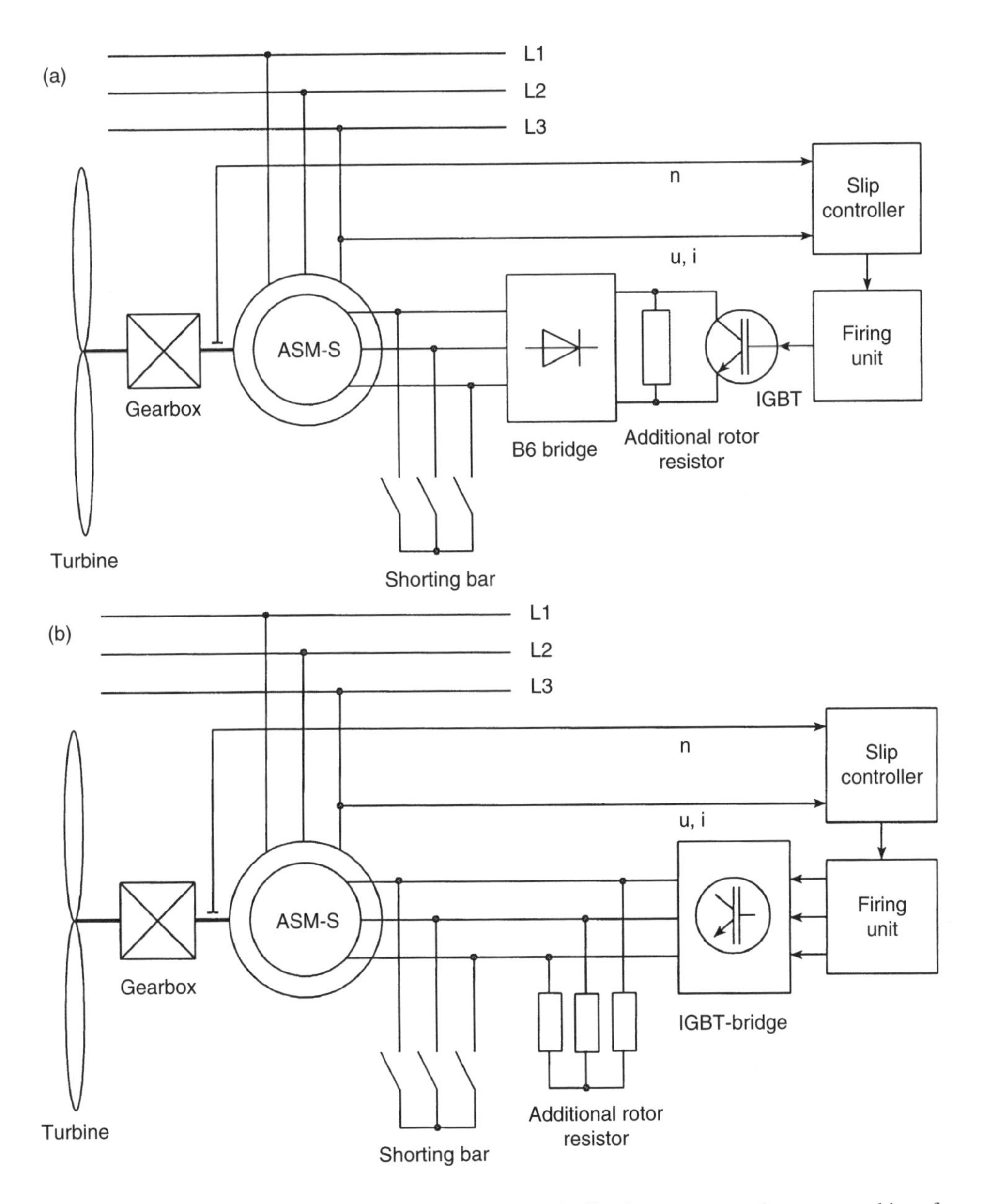

Figure 5.25 Dynamic slip regulation of wind turbines with slip-ring rotor asynchronous machines for (a) a rotor circuit with a rectifier bridge and single-phase pulsing at the additional resistor and (b) a rotor circuit with three-phase pulsing of the additional resistor

These result from the rectification of rotor currents and, due to the square-wave current in the case of a B6 rectifier bridge, for example, take on values six times the slip frequency. The rotor current harmonics this causes are also transferred to the stator and give rise to increased losses in the generator. Moreover, remedial measures are necessary to remove the vibrations. These disadvantages can be avoided by a symmetrical system with a three-phase rotor circuit design, as shown in Figure 5.25(b). However, due to the three-phase synchronized antiparallel design of the frequency converter that this calls for, the cost is increased to such a degree that, at only a little extra cost (rectifier, transformer), power recovery – and thus the utilization of the slip energy – is also possible.

By a combination of the two variants, the design shown in Figure 5.26 with a three-phase resistor and single-phase pulsing can be used to build a relatively simple system which can achieve good operating results at a reasonable cost. Unlike the single-phase system in Figure 5.25(a), this configuration does not cause mechanical vibrations in the drive train. However, due to the lack of a resistor in parallel with the 'IGBT switch' an additional protective circuit is required to protect against the overvoltage that is created in the direct current circuit due to current pulsing.

An annual running time of approximately 8000 hours must be assumed for generators in wind turbines. Components, particularly the slip-ring transmission system, are subject to wear. Thus, when these are used, regular maintenance and easy replacement of wearing elements must be ensured.

Brushless designs – normal in synchronous machines (see Figure 3.6) – are also desirable in asynchronous machines. It is necessary to make a fundamental differentiation here between brushless inductive transmission of the rotor-slip energy by secondary windings (Figure 5.27(a)) and the complete redesign of the power section with additional resistors, power electronics and regulation in the rotating part of the generator (Figure 5.27(b)), as in the OptiSlip® regulation system in the Vestas V 44 (600 kW) to V 66 (1.65 MW) and the V80/90 (2 or 3 MW), which is the US export variant.

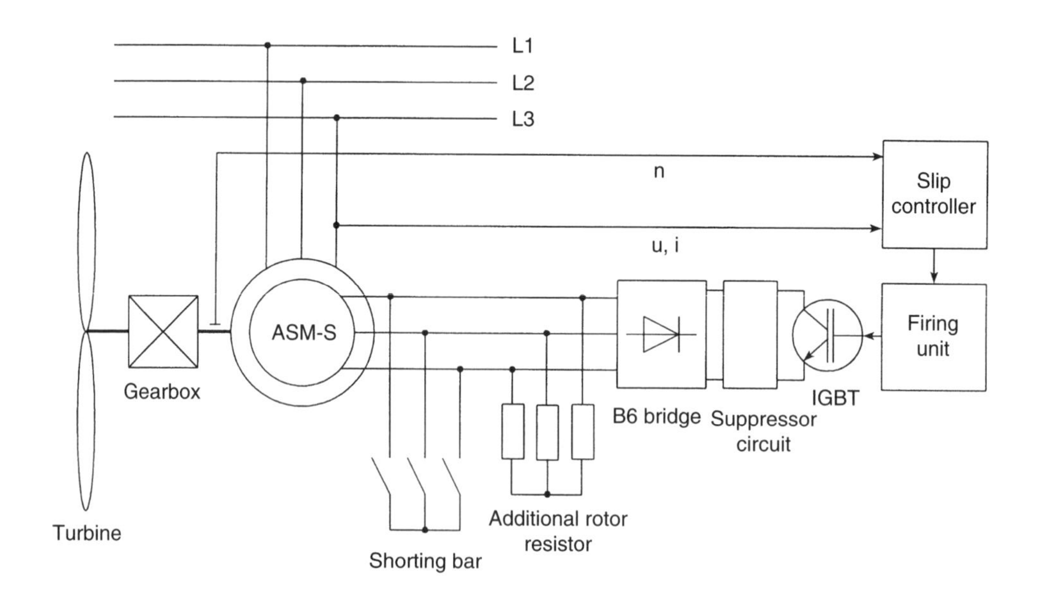

Figure 5.26 Dynamic slip regulation of wind turbines by three-phase additional resistors with direct current pulsing

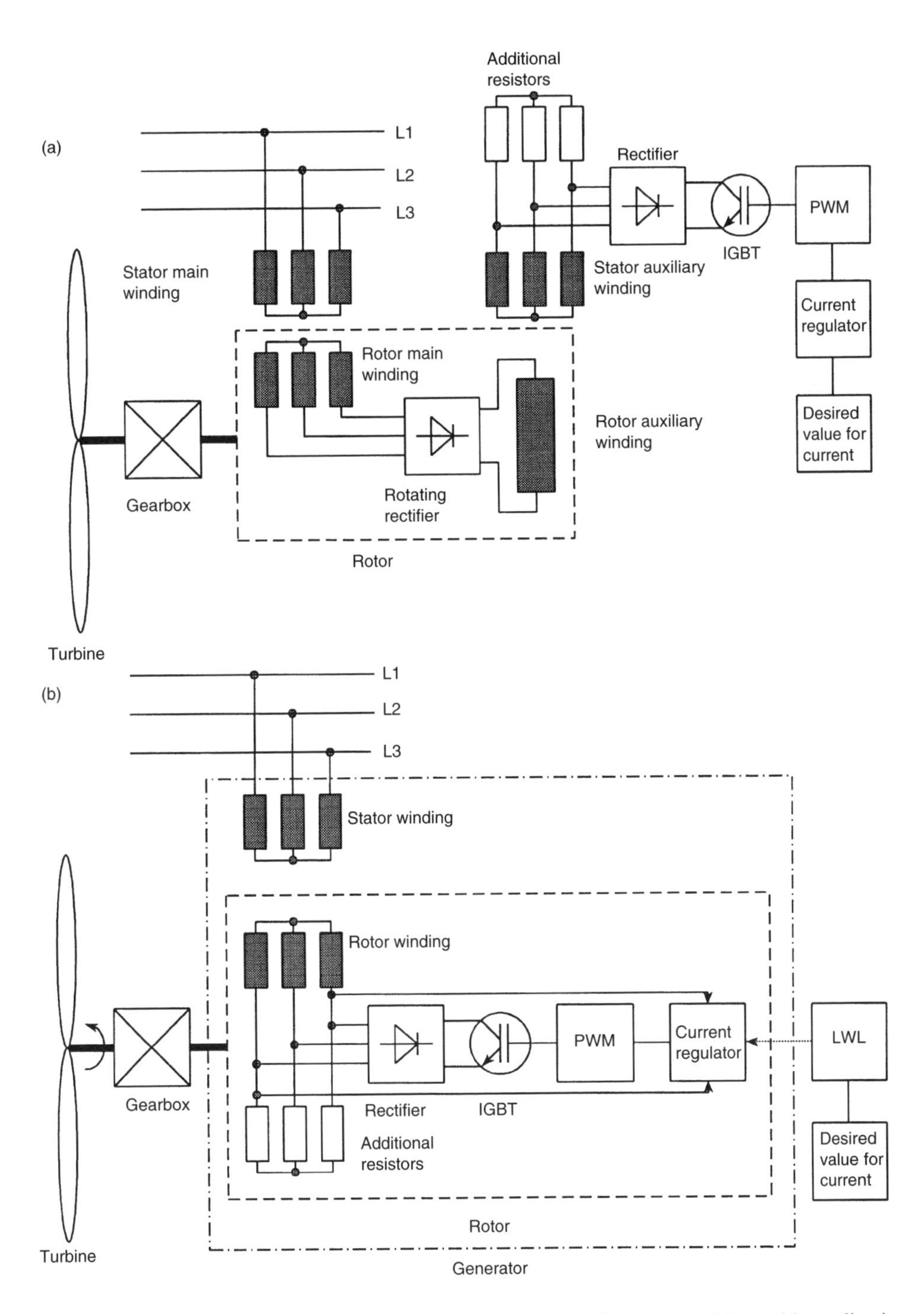

Figure 5.27 Dynamic slip regulation of wind turbines using asynchronous machines without slip-ring systems with (a) slip energy transmission by secondary windings and (b) power section and regulation in the rotor (Vestas)

The generators specifically designed for the OptiSlip® system are fitted with a wound rotor and an integrated current regulation system in the rotor (RCC = rotor current controller). This is installed at the rear of the generator on the end of the shaft, and consists of additional resistors, power electronics, current sensors and a microprocessor controller. The communications signals between the management system (VMP = Vestas multi-processor) and the current-regulation device are carried via a maintenance-free fibre-optic cable.

Figure 5.28 shows the block diagram for the dynamic slip regulation system of an asynchronous machine. During the running-up and shut-down processes the generator is disconnected from the grid. The speed of the turbine can be controlled and, if necessary, limited by the speed regulator using blade pitch adjustment.

Once the generator has been connected to the grid, the wind turbine can be operated at part load (i.e. below nominal output) at a constant blade pitch or, as in the Vestas design, brought into the optimal range according to the prevailing wind speed. Thus, with the aid of this so-called OptiSlip® function, the greatest possible energy yield can be achieved at part load.

Generator output is adapted to the momentary operating state (part or full load) with the aid of slip by means of current regulation, pulse-width modulation and the power controller in the rotor. Thus, low slip values (e.g. 2 %) can be set in the part-load range and high slip values (e.g. 5 %) can be set at nominal operation, so that the generator speed can be varied to smooth output power and drive-train torque.

If the above regulation options and their benefits with regard to reducing the loads on components, yield optimization, etc., are dispensed with, a very simple and robust system can be designed using wind turbines in rigid connection to the grid. This system is most widespread for turbines up to 600 kW nominal output. In wind turbines in the MW class, on the other hand, there is a clear trend in favour of flexible-speed units.

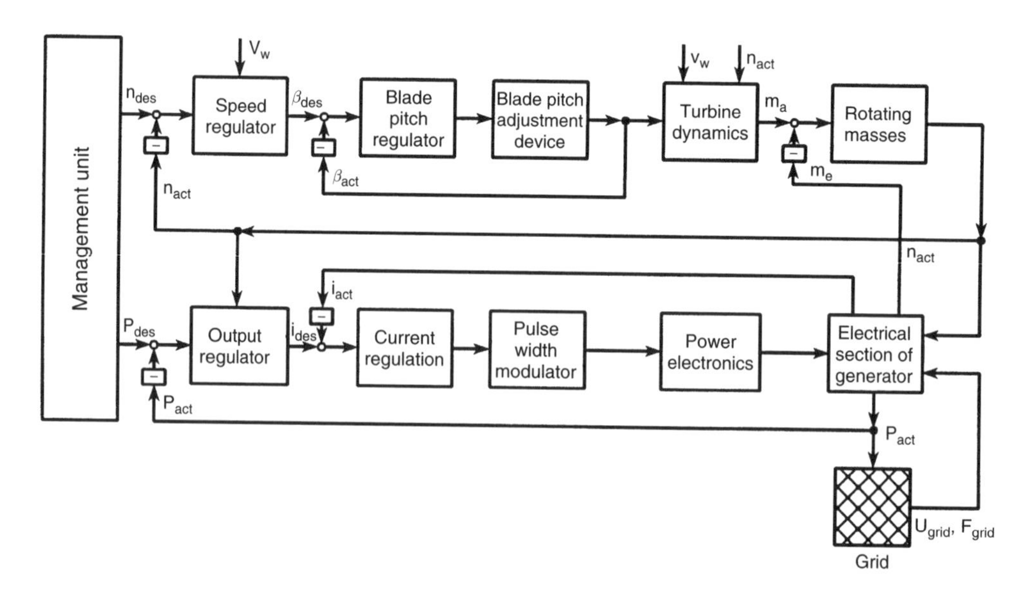

Figure 5.28 Block diagram for dynamic slip control of an asynchronous machine in the grid

5.4.4 Regulation of turbines with a rigid connection to the grid

For plants with a largely fixed-speed connection to the grid neither power optimization nor the option of power smoothing are available. Turbine speed is predetermined by the grid via the generator (Figure 5.29). The speed regulation circuit therefore only has the task of running the turbine up and limiting its speed. All further functions of the circuit for the regulation of power, blade pitch and blade pitch adjustment speed are similar to those in wind turbines in isolated operation, as shown in Figure 5.12. For small turbines, the system structure can be further simplified by not including a blade pitch sensing system, as shown in Figure 5.18, if the regulation dynamics are suitably limited. This must be particularly taken into consideration in the dimensioning of the system components and the regulator.

5.5 Controller Design

The reliable operation of wind turbines is possible only if physical and technical limit values are not exceeded. Turbine control and adjustment mechanisms must therefore be capable of maintaining specified speed and output limits at all times, both in normal operation and in extreme situations. Therefore particular attention must be paid to design of these systems.

The options for the dimensioning of actuators have already been described in Section 2.3.2.5. The following considerations therefore relate to the representation of control circuits, the determination of their parameters and the design of the required controller. The reader is referred to the description in Section 2.1 with regard to the forces acting upon the rotor blade and the driving torque and power at the turbine. Reference will also be made to blade pitch adjustment, described in Sections 2.3.2.1 to 2.3.2.3, drive-train influences, described in Section 2.4, and the behaviour of the generator and supply systems, as described in Chapters 3 and 4. The regulation structures in Sections 5.4.1 to 5.4.4 form the basis for further calculations.

It is evident from Figure 2.4 and Equations (2.24) and (2.25) that the tangentially acting driving force and the axially acting thrust value can be approximated by

$$\begin{bmatrix} \mathrm{d}F_{\mathrm{t}} \\ \mathrm{d}F_{\mathrm{ax}} \end{bmatrix} = z\frac{\rho}{2}v_{\mathrm{r}}^{2}t_{\mathrm{B}}\,\mathrm{d}r \begin{bmatrix} \sin\delta & -\cos\delta \\ \cos\delta & \sin\delta \end{bmatrix} \begin{bmatrix} c_{\mathrm{a}}(\alpha) \\ c_{\mathrm{w}}(\alpha) \end{bmatrix} \tag{5.3}$$

for a turbine circumferential element. The following equation describes the relationship between the blade pitch angle β, the resulting direction of flow δ and the angle of flow over the profile α:

$$\beta = \frac{\pi}{2} - \vartheta = \frac{\pi}{2} - \delta + \alpha \approx \frac{\pi}{2} - \arctan\left(\frac{v_2}{v_{\mathrm{u}}}\right) + \alpha. \tag{5.4}$$

Therefore, with the aid of a system for altering the blade pitch angle, the tangential component of the force acting on the rotor blades, and therefore the torque, can be influenced and the output and rotational speed of the turbine can be controlled. Furthermore, by deliberately influencing the thrust acting in the axial direction, the rotor blade and tower bending can also be limited.

By the superposition of the processes, it is also possible to reduce or damp the tower vibrations. However, this requires that system-specific changes and the resulting influences

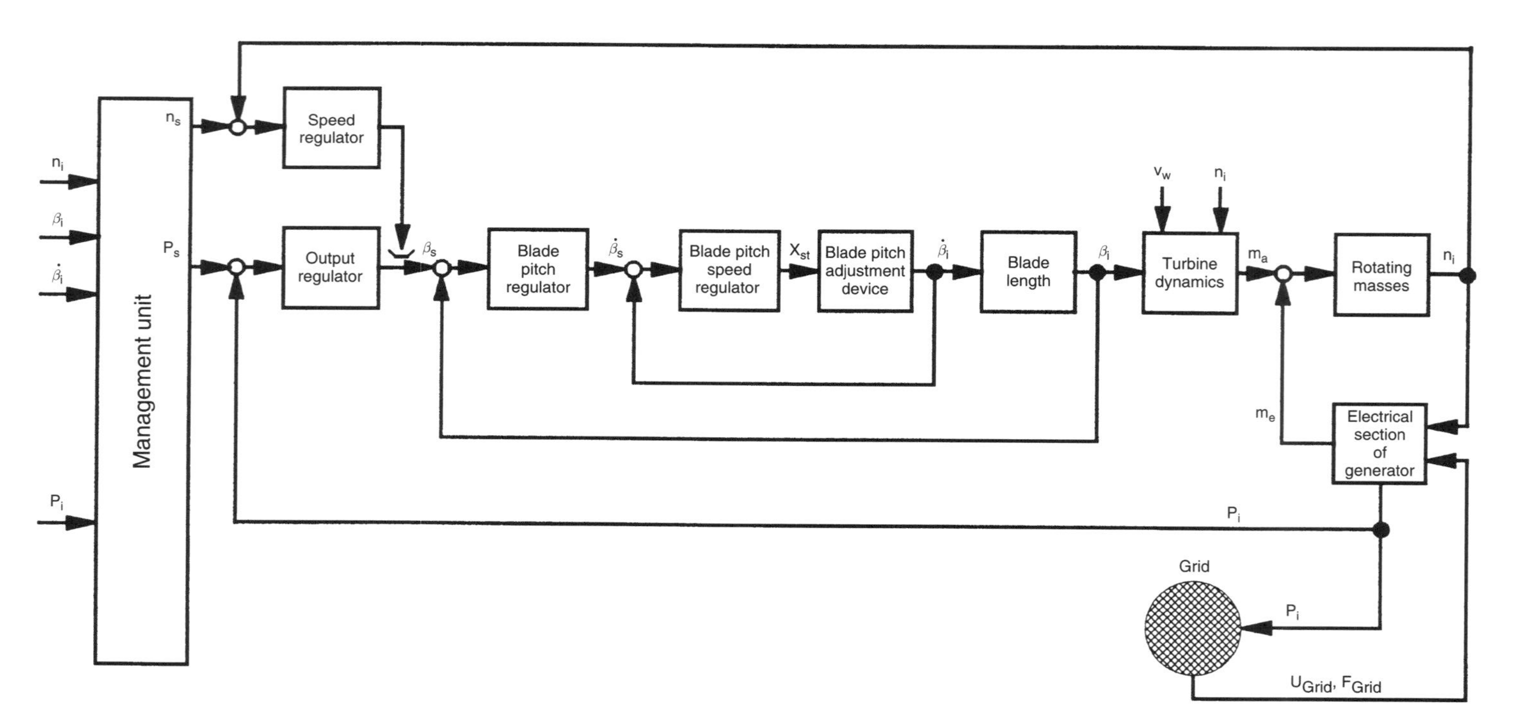

Figure 5.29 Structure for the control of a wind turbine with a fixed-speed grid connection (grid operation)

on the rotor blades during rotation (e.g. due to airstream height gradients, flow disturbance at the tower and partial gust effects on the surface of the turbine) be taken into account. Local conditions and the time-variant position of the turbine system thus have a decisive effect upon measures to limit or damp bending and vibrations in the rotor blades and the tower. For such configurations, forces and moments on one blade can be represented according to

$$\begin{bmatrix} F_{t1} \\ F_{ax1} \end{bmatrix} = \int_{R_i}^{R_a} \frac{\rho}{2} v_r^2 (r, \varepsilon, t)\, t_B(r) \begin{bmatrix} \sin\delta & -\cos\delta \\ \cos\delta & \sin\delta \end{bmatrix} \begin{bmatrix} c_a(\alpha, \varepsilon) \\ c_w(\alpha, \varepsilon) \end{bmatrix} \mathrm{d}r \tag{5.5}$$

and

$$\begin{bmatrix} M_{t1} \\ M_{ax1} \end{bmatrix} = \int r \begin{bmatrix} \mathrm{d}F_t \\ \mathrm{d}F_{ax} \end{bmatrix} = \int_{R_i}^{R_a} r \frac{\rho}{2} v_r^2 (r, \varepsilon, t)\, t_B(r) \begin{bmatrix} \sin\delta & -\cos\delta \\ \cos\delta & \sin\delta \end{bmatrix} \begin{bmatrix} c_a(\alpha, \varepsilon) \\ c_w(\alpha, \varepsilon) \end{bmatrix} \mathrm{d}r. \tag{5.6}$$

The vast majority of all installed wind turbines are designed with three-blade rotors. For this design, as shown in Figure 5.30, the total values at the turbine, which are dependent upon the rotor position, are found to be

$$\begin{bmatrix} F_t \\ F_{ax} \end{bmatrix} = \begin{bmatrix} F_{t1}(\varepsilon, \alpha, t) + F_{t2}(\varepsilon + 2\pi/3, \alpha, t) + F_{t3}(\varepsilon + 4\pi/3, \alpha, t) \\ F_{ax1}(\varepsilon, \alpha, t) + F_{ax2}(\varepsilon + 2\pi/3, \alpha, t) + F_{ax3}(\varepsilon + 4\pi/3, \alpha, t) \end{bmatrix} \tag{5.7}$$

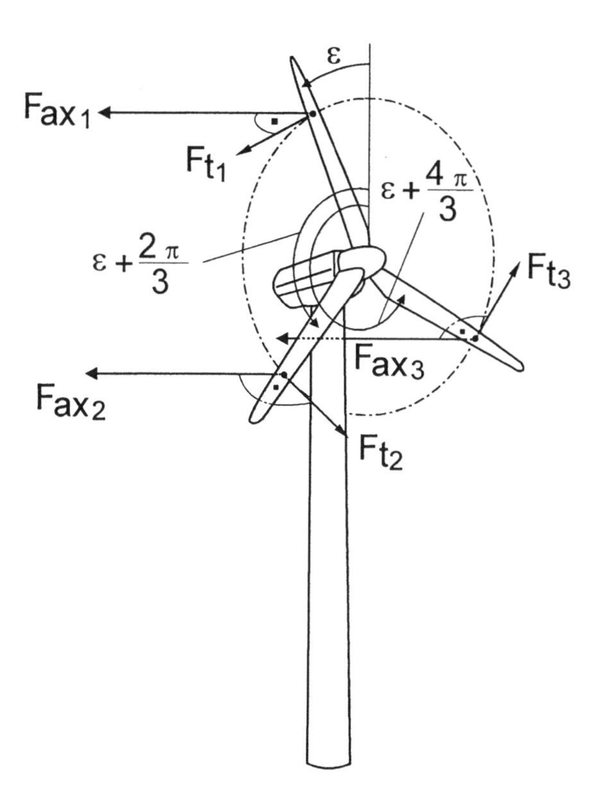

Figure 5.30 Dynamic effects and angular relationships at the turbine

and

$$\begin{bmatrix} M_{t} \\ M_{ax} \end{bmatrix} = \begin{bmatrix} M_{t1}(\varepsilon, \alpha, t) + M_{t2}(\varepsilon + 2\pi/3, \alpha, t) + M_{t3}(\varepsilon + 4\pi/3, \alpha, t) \\ M_{ax1}(\varepsilon, \alpha, t) + M_{ax2}(\varepsilon + 2\pi/3, \alpha, t) + M_{ax3}(\varepsilon + 4\pi/3, \alpha, t) \end{bmatrix} \tag{5.8}$$

and these values have a decisive effect upon measures and control strategies to limit bending or minimize vibrations. The short-term interventions necessary to achieve these goals can usually only be carried out by means of the blade pitch adjustment system. Although changes to flow can be initiated very quickly (at speeds below the millisecond range) on the generator side by rotational speed changes, such interventions are only conditionally suitable due to the high rotor system time constants that they require.

From the point of view of control engineering, wind turbines with blade pitch adjustment systems thus provide the option of actively altering the flow of energy in response to both changes in wind speed and changes in conditions in the turbine and supply system. Furthermore, the mathematical treatment of the control circuits in such designs covers a very comprehensive field of required dimensioning. The following discussions therefore relate mainly to turbines with blade pitch adjustment systems.

Very different procedures for the determination of the controller parameters are possible, depending upon the control philosophy. We can assume that a controller presetting is used that ensures stable system behaviour in relation to the design state or the operating ranges under consideration. This initial dimensioning can be carried out using the normal measuring procedures such as optimum amount, symmetrical optimum, etc., in accordance with references [5.20], [5.21], [5.22] and [5.23], and the stability of the control circuit can be investigated using the familiar methods of frequency response analysis, root locus curve procedure, Hurwitz criterion of stability, etc. However, this requires that considerable simplifications (see Section 2.3) and linearizations are made in the control systems, and partial restrictions of the operating range must be accepted. This default setting of the controller, which will be described in more detail in the following part, can be checked by means of simulation calculations and system trials, and can be empirically fine-tuned with regard to the strategies and goals of the control system. The following section considers the control circuits and preliminary dimensioning of controllers, based on the adjustment processes at the rotor blades.

5.5.1 Adjustment processes and torsional moments at the rotor blades

The necessary definitions and the effect of moments during the adjustment of turbine rotor blades have already been described in Section 2.3.2. The following representations are based upon this section.

If the inertia caused by accelerating air masses, which plays only a very minor role in comparison with the rotor blade component, is disregarded then the following relationship is found for nonteetering hub turbines, based on Equation (2.69):

$$J_{Bl}\frac{d^2\beta}{dt} + \left(\frac{dJ_{Bl}}{dt} + k_{DB} + k_{RL}\right)\frac{d\beta}{dt} + \left(\frac{dk_{DB}}{dt} + \frac{dk_{RL}}{dt}\right)\beta$$
$$+ M_{Pr} + M_{lift} + M_{T} + M_{bend} = M_{st}. \tag{5.9}$$

For deflection-resistant blades, $\mathrm{d}J_{\mathrm{Bl}}/\mathrm{d}t$ and M_{bend} can also be disregarded. Furthermore, for predimensioning, the time derivatives of the damping and frictional components ($\mathrm{d}k_{\mathrm{DB}}/\mathrm{d}t$ and $\mathrm{d}k_{\mathrm{RL}}/\mathrm{d}t$) can also be disregarded. This gives the greatly simplified differential equation

$$J_{\mathrm{Bl}}\frac{\mathrm{d}^2\beta}{\mathrm{d}t^2}+(k_{\mathrm{DB}}+k_{\mathrm{RL}})\frac{\mathrm{d}\beta}{\mathrm{d}t}+M_{\mathrm{Pr}}+M_{\mathrm{lift}}+M_{\mathrm{T}}=M_{\mathrm{St}}. \tag{5.10}$$

This includes moments arising as a result of inertia, damping, friction, propeller effects, lift and torsion due to aerodynamic resetting. These will be described in more detail in the following section.

5.5.1.1 Propeller moments

For turbine systems with non-bending rotor blades that rotate with their blade axis in the plane of rotation with no cone angle, the propeller moment as shown in Figures 2.37 and 5.31 can be determined for each blade by the equation

$$M_{\mathrm{Pr}}=-\int_{R_{\mathrm{i}}}^{R_{\mathrm{a}}}\omega_{\mathrm{R}}^2 a_{\mathrm{p}}^2 \sin(90^\circ-\beta)\cos(90^\circ-\beta)\,\mathrm{d}m. \tag{5.11}$$

The geometrical positions of the axis of rotation and the centre of gravity on the blade profile are of particular significance here. As shown in Section 2.1.4, calculations can again be based upon 20 subelements.

Figure 5.32 illustrates the different magnitudes and directions of propeller moments for a turbine of 40 m rotor diameter as a function of the blade pitch angle and speed for three different blade axes of rotation. Very different maximum moment values are found for the above turbine at different blade rotational axis positions, i.e.

$$\text{at } t_{\mathrm{B}}/8, \quad M_{\mathrm{Pr\,max}\,t/8}=-1700\,\mathrm{N\,m},$$

$$\text{at } t_{\mathrm{B}}/4, \quad M_{\mathrm{Pr\,max}\,t/4}=-400\,\mathrm{N\,m},$$

and near the blade centre of gravity

$$\text{at } 3t_{\mathrm{B}}/8, \quad M_{\mathrm{Pr\,max}\,3t/8}=-1.2\,\mathrm{N\,m}.$$

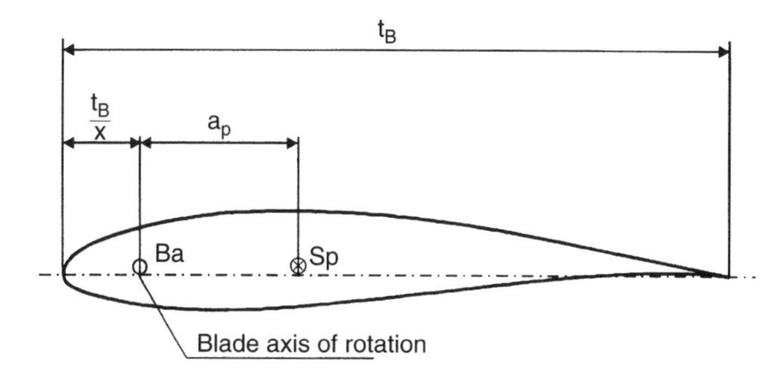

Figure 5.31 Geometrical position of the axis of rotation and the centre of mass on the blade profile (Ba = blade axis, Sp = centre of gravity)

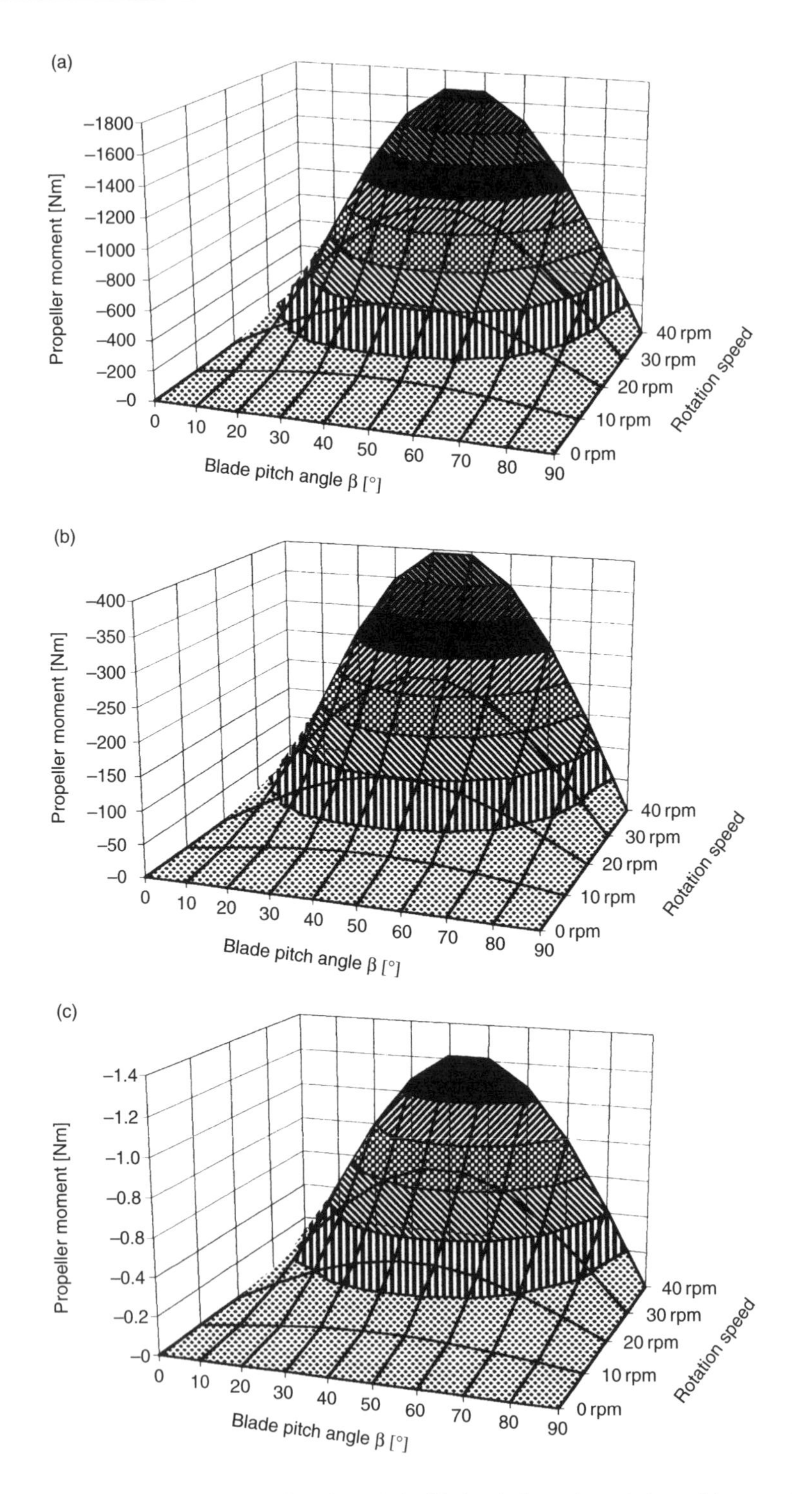

Figure 5.32 Propeller moments as a function of the blade pitch angle and the turbine rotation speed, with the blade axis of rotation at (a) $t_B/8$, (b) $t_B/4$ and (c) $3t_B/8$

In this connection, negative torques characterise a twisting of the blades in the direction of increasing blade pitch angle β according to the definition in Figure 2.34.

5.5.1.2 Torsional moments due to lifting forces

According to Figure 2.36, the torsional moment caused by lift, M_{lift}, is proportional to the lifting force and the distance a_{a} between the blade axis of rotation and the point of action of the lifting force or the active component $a_{\mathrm{a}} \cos \alpha$. Equation (2.22) therefore yields

$$M_{\mathrm{lift}} = \int_{R_{\mathrm{i}}}^{R_{\mathrm{a}}} a_{\mathrm{a}} \cos \alpha \frac{\rho}{2} t_{\mathrm{B}}(r)\, c_{\mathrm{a}}(\alpha)\, v_{\mathrm{r}}^2 \, \mathrm{d}r. \tag{5.12}$$

According to Section 2.1.4, the calculation can be based upon 20 blade elements in accordance with Equation (2.60). Figure 5.33 shows the torsional moments caused by lift on the profile of a turbine with a 40 m rotor diameter, as a function of wind speed and rotational speed, with the blade pitch angle as a parameter. The different effects caused by the positioning of the axis of rotation in front of or behind the point of action of the lifting force (at $t_{\mathrm{B}}/8$ or $3t_{\mathrm{B}}/8$) are again evident here. Furthermore, the above equation illustrates that, with an axis of rotation at $t_{\mathrm{B}}/4$, no moments caused by lift are present. This state only exists, however, with completely deflection-resistant blades that have a centre-of-pressure resistant profile.

5.5.1.3 Torsional resetting moments

Similarly to Equation (5.12), using Equation (2.61), the torsional moment caused by the resetting effect of the profile in the airstream based on $t_{\mathrm{B}}/4$ can be determined by

$$M_{\mathrm{T}} = \int_{R_{\mathrm{i}}}^{R_{\mathrm{a}}} c_{\mathrm{t}}(\alpha) v_{\mathrm{r}}^2 \frac{\rho}{2} t_{\mathrm{B}}^2 \, \mathrm{d}r \tag{5.13}$$

Here, c_{t} represents the angle-of-flow-dependent coefficient of the torsional moment for a profile. Figure 5.34 shows the result of a calculation with 20 blade elements for a 40 m turbine. For blade axes of rotation deviating from $t_{\mathrm{B}}/4$, appropriate recalculations of the torsional moments are necessary.

5.5.1.4 Total moments

Figure 5.35 illustrates the strong dependence of the blade pitch adjusting moment on the position of the blade axis of rotation for three characteristic operating states. These were selected close to the starting point of 6 m/s wind speed and a rotation speed of 14 rpm, in the nominal range (12 m/s, 28 rpm) and shortly before the shut-down of the turbine (24 m/s, 28 rpm). It is clear from this that the torque can take on high positive or negative values, or can be largely cancelled out at an axis position of $t_{\mathrm{B}}/4$.

As well as the torsional moments mentioned here, frictional moments caused by the blade bearings, aerodynamic damping components and any moments caused by pull-back springs, which are mainly dependent upon the blade pitch angle, should be taken into account as described in Section 2.3.

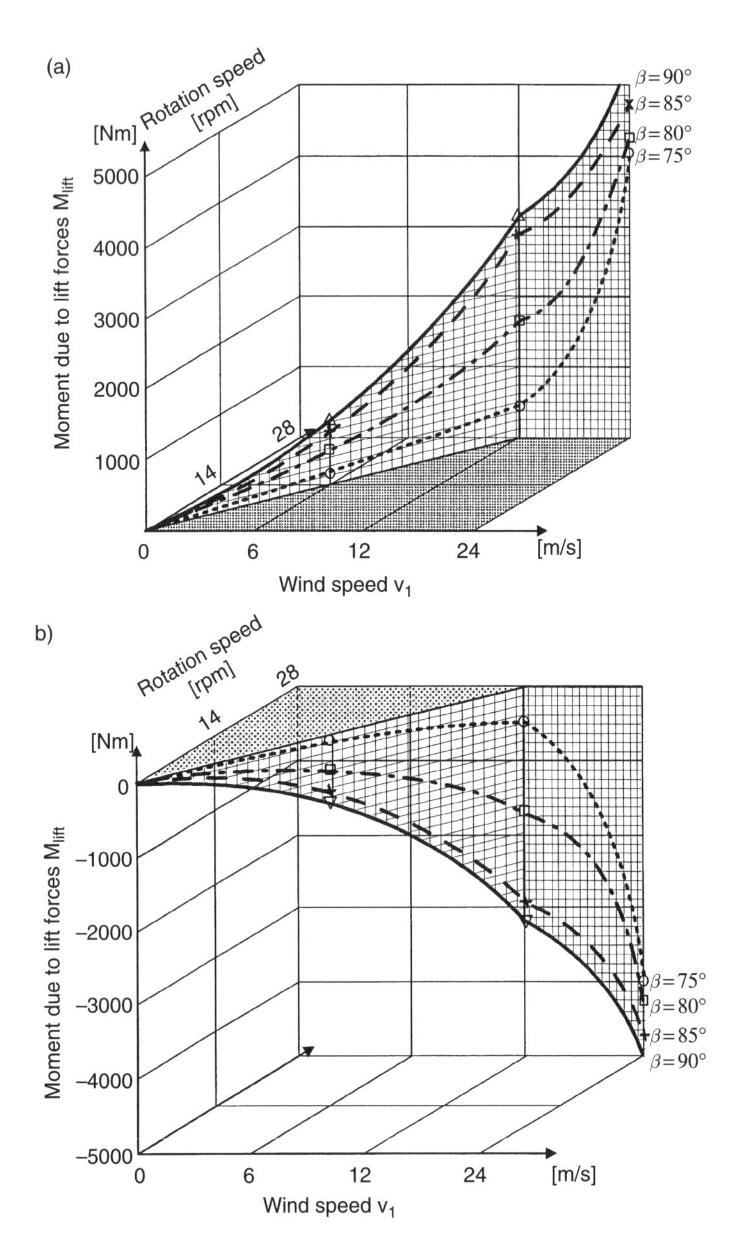

Figure 5.33 Torsional moments caused by lift with the blade axis of rotation at (a) $t_B/8$ and (b) $3t_B/8$

5.5.2 Standardizing and linearizing the variables

The variables of torsional moment, wind speed, blade pitch angle, rotation speed, output, etc., have different types of dimensions. It is therefore advisable to standardize all variables, e.g. to nominal or maximum values, thereby obtaining dimensionless units in similar ranges, such as between zero and one.

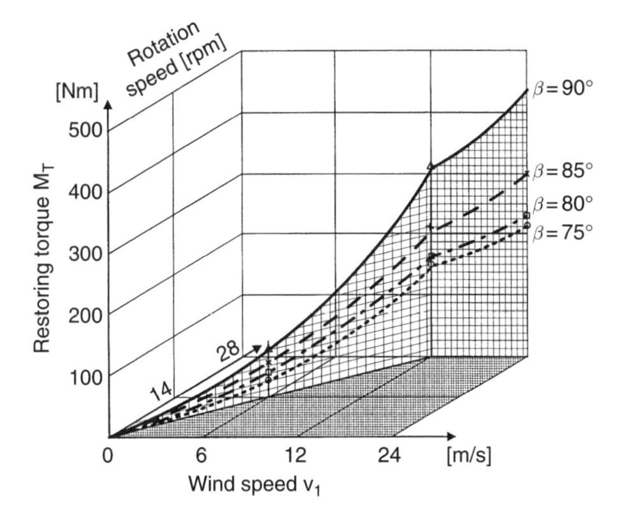

Figure 5.34 Torsional resetting moment as a function of wind speed and rotational speed, with the blade pitch angle as a parameter

Due to nonlinear relationships in the transmission, a quantitative analysis for the determination of control parameters is no simple matter. However, close to its operating point continuous families of characteristics can be approximated by the linear standardized elements, i.e. the start of a Taylor expansion in standard form, according to the equation

$$\Delta\left(\frac{M}{M_{\tau N}}\right) \approx (v_1 - v_{1N}) \frac{\partial f}{\partial\left(\frac{v_1}{v_{1N}}\right)\Big|_{v_{1N},\beta_N,n_N}} + (\beta - \beta_N) \frac{\partial f}{\partial\left(\frac{\beta}{\beta_N}\right)\Big|_{\beta_N,v_{1N},n_N}} + (n - n_N) \frac{\partial f}{\partial\left(\frac{n}{n_N}\right)\Big|_{n_N,v_{1N},\beta_N}} \tag{5.14}$$

Nonlinear characteristics around the operating or nominal operating point are hereby linearized [5.20]. We can thus obtain the linearized form of the standardized torsional moments due to:

- lifting force

$$\frac{M_{\text{lift}}}{M_{\tau N}} \approx \Delta\frac{M_{\text{lift}}}{M_{\tau N}} = k_{11}\frac{\beta}{\beta_N} + k_{12} + k_{13}\frac{v_1}{v_{1N}} + k_{14}\frac{n}{n_N}, \tag{5.15}$$

- resetting effects

$$\frac{M_T}{M_{\tau N}} \approx \Delta\frac{M_T}{M_{\tau N}} = k_{21}\frac{\beta}{\beta_N} + k_{22} + k_{23}\frac{v_1}{v_{1N}} + k_{24}\frac{n}{n_N}, \tag{5.16}$$

- propeller moments

$$\frac{M_{\text{Pr}}}{M_{\tau N}} \approx \Delta\frac{M_{\text{Pr}}}{M_{\tau N}} = k_{31}\frac{\beta}{\beta_N} + k_{32} + k_{33}\frac{v_1}{v_{1N}} + k_{34}\frac{n}{n_N} \tag{5.17}$$

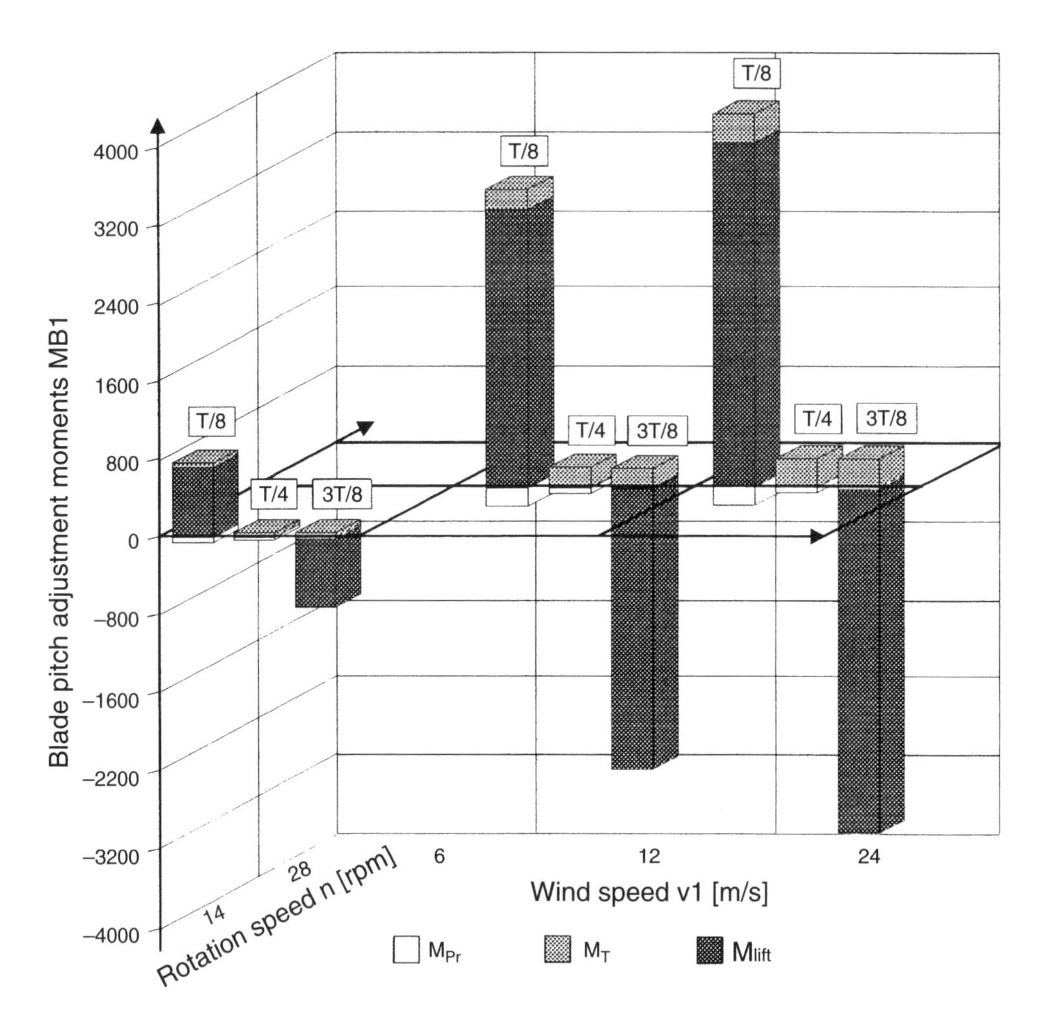

Figure 5.35 Blade pitch adjustment moments in relation to the wind speed and speed of rotation

- and resetting moments by springs (if these are used)

$$\frac{M_{\mathrm{F}}}{M_{\tau \mathrm{N}}} \approx \Delta \frac{M_{\mathrm{F}}}{M_{\tau \mathrm{N}}} = k_{41} \frac{\beta}{\beta_{\mathrm{N}}} + k_{42} + k_{43} \frac{v_1}{v_{1\mathrm{N}}} + k_{44} \frac{n}{n_{\mathrm{N}}}. \tag{5.18}$$

Furthermore, standardizing and linearizing the motion equation (5.10) with the moments acting on the rotor blades yields the following equation:

$$k_{\mathrm{b}} \frac{\mathrm{d}^2 \left(\frac{\beta}{\beta_{\mathrm{N}}} \right)}{\mathrm{d}t^2} + k_{\mathrm{d}} \frac{\mathrm{d} \left(\frac{\beta}{\beta_{\mathrm{N}}} \right)}{\mathrm{d}t} + k_{\beta} \frac{\beta}{\beta_{\mathrm{N}}} + k_1 + k_v \frac{v_1}{v_{1\mathrm{N}}} + k_{\mathrm{n}} \frac{n}{n_{\mathrm{N}}} = \frac{M_{\mathrm{St}}}{M_{\tau \mathrm{N}}} \tag{5.19}$$

where the acceleration-dependent component is

$$k_{\mathrm{b}} = \frac{J_{\mathrm{Bl}}(v_1)\, \beta_{\mathrm{N}}}{M_{\tau \mathrm{N}}},$$

the damping value, which depends upon the speed of blade pitch angle adjustment, is

$$k_{\mathrm{d}} = \frac{(k_{\mathrm{DB}} + k_{\mathrm{RL}})\,\beta_{\mathrm{N}}}{M_{\tau\mathrm{N}}},$$

the angle-dependent variable is

$$k_{\beta} = \sum k_{\nu 1},$$

the constant component is

$$k_{1} = \sum k_{\nu 2},$$

the wind-speed equivalent is

$$k_{v} = \sum k_{\nu 3},$$

and the value proportional to rotation speed is

$$k_{\mathrm{n}} = \sum k_{\nu 4}.$$

The dynamic behaviour of the transmission elements close to the selected steady-state operating point is therefore described by a linear differential equation with constant coefficients. This can be solved in closed form, and general statements can be made, for example, concerning stability [5.20].

Linearization is, however, not meaningful if marked nonlinearities exist. For noncontinuous characteristics, encountered, for example, in switching amplifiers, linearization is completely inappropriate and must be avoided.

The torque matrix of the control system for rotor blade pitch adjustment is found from Equations (5.15) to (5.19):

$$\frac{M_{\mathrm{St}}}{M_{\tau\mathrm{N}}} = \frac{\sum M}{M_{\tau\mathrm{N}}} = \frac{1}{M_{\tau\mathrm{N}}}\begin{bmatrix} M_{\mathrm{b}} \\ M_{\mathrm{d}} \\ M_{\mathrm{lift}} \\ M_{\mathrm{T}} \\ M_{\mathrm{Pr}} \\ M_{\mathrm{F}} \end{bmatrix}$$

$$= \begin{bmatrix} k_{\mathrm{b}} & 0 & 0 & 0 & 0 & 0 \\ 0 & k_{\mathrm{d}} & 0 & 0 & 0 & 0 \\ 0 & 0 & k_{11} & k_{12} & k_{13} & k_{14} \\ 0 & 0 & k_{21} & k_{22} & k_{23} & k_{24} \\ 0 & 0 & k_{31} & k_{32} & k_{33} & k_{34} \\ 0 & 0 & k_{41} & k_{42} & k_{43} & k_{44} \end{bmatrix} \begin{bmatrix} \ddot{\beta}/\beta_{\mathrm{N}} \\ \dot{\beta}/\beta_{\mathrm{N}} \\ \beta/\beta_{\mathrm{N}} \\ 1 \\ v_{1}/v_{1\mathrm{N}} \\ n/n_{\mathrm{N}} \end{bmatrix} \tag{5.20}$$

Depending upon the rotor configuration, the coefficients of the matrix take on characteristic values, which can be zero. For example, the lift-dependent components disappear in the case of a blade rotational axis position of $T/4$. Then, for nonbending blades,

$$k_{11} = k_{12} = k_{13} = k_{14} = 0.$$

Moments brought about by springs are usually independent of the prevailing wind speeds and rotational speed values. Therefore, in addition,

$$k_{43} = k_{44} = 0.$$

Moreover, in systems without return springs

$$k_{41} = k_{42} = 0.$$

If the blade axis of rotation is selected at the centre of mass of the section then some of the coefficients vanish, i.e.

$$k_{31} = k_{32} = k_{33} = k_{34} = 0.$$

Furthermore, torsional moments caused by blade bending can be considered in the components through lift M_{lift} with coefficients k_{11} to k_{14} and propeller effects M_{Pr} with coefficients k_{31} to k_{34} as through an increase in the moment of inertia in k_{a}.

5.5.3 *Control circuits and simplified dimensioning*

The linear transmission element for rotor blade pitch adjustment can be described using the second-order differential equation according to Equation (5.19) or in the form

$$\frac{k_{\mathrm{b}}}{k_{\beta}} \frac{\mathrm{d}^2\left(\frac{\beta}{\beta_{\mathrm{N}}}\right)}{\mathrm{d}t^2} + \frac{k_{\mathrm{d}}}{k_{\beta}} \frac{\mathrm{d}\left(\frac{\beta}{\beta_{\mathrm{N}}}\right)}{\mathrm{d}t} + \frac{\beta}{\beta_{\mathrm{N}}} = \frac{1}{k_{\beta}}\left(\frac{M_{\mathrm{St}}}{M_{\tau\mathrm{N}}} - \frac{M_{\tau}}{M_{\tau\mathrm{N}}}\right) \tag{5.21}$$

where $M_{\tau}/M_{\tau\mathrm{N}} = k_1 + k_{\mathrm{v}} v_1/v_{1\mathrm{N}} + k_{\mathrm{n}} n/n_{\mathrm{N}}$. Using the Laplace transform, this relationship can also be represented as a complex frequency response, by the transmission function of the so-called control system, where $s = \delta + \mathrm{j}\omega$, or, for the special case $s = p = \mathrm{j}\omega$,

$$F_{\mathrm{S}}(p) = \frac{\frac{\beta}{\beta_{\mathrm{N}}}}{\frac{M_{\mathrm{St}} - M_{\tau}}{M_{\tau\mathrm{N}}}} = \frac{\frac{1}{k_{\beta}}}{T_1 T_2 p^2 + (T_1 + T_2)\,p + 1} = \frac{V_{\beta}}{(T_1 p + 1)(T_2 p + 1)} \tag{5.22}$$

i.e. it is possible to substitute

$$\frac{k_{\mathrm{b}}}{k_{\beta}} = T_1 T_2, \qquad \frac{k_{\mathrm{d}}}{k_{\beta}} = T_1 + T_2 \qquad \text{and} \qquad \frac{1}{k_{\beta}} = V_{\beta}$$

Figure 5.36 shows the block diagram for the blade pitch control system of a wind turbine. Here the Laplace transforms characterize the following quantities:

$\beta_{\mathrm{s}}/\beta_{\mathrm{N}}$	the command variable,
$\beta/\beta_{\mathrm{N}} = \beta_{\mathrm{i}}/\beta_{\mathrm{N}}$	the regulating variable or its actual value,
$M_{\tau}/M_{\tau\mathrm{N}}$	the disturbance variable of the control circuit.

When designing the system, the question of whether the actuator is centrally controlled and acts upon all blades simultaneously or each blade is separately adjusted must be taken into account.

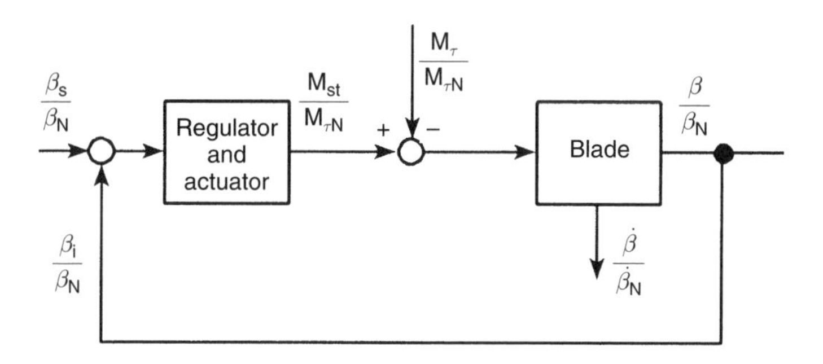

Figure 5.36 Control system with the control circuit for blade pitch adjustment

The transfer function of the open control circuit F_K is found to be the product of the transfer function of the controller F_R and the controlled section F_S:

$$F_K(p) = F_R(p)F_S(p). \tag{5.23}$$

By superimposing the two stimuli of command and disturbance variables, the regulation variable is found to be

$$\frac{\beta}{\beta_N}(p) = \frac{F_K}{1+F_K}\frac{\beta_S}{\beta_N}(p) + \frac{F_S}{1+F_K}\frac{M_\tau}{M_{\tau N}}(p). \tag{5.24}$$

This represents the sum of two products: the product of the so-called command transfer function $F_g = F_K/(1+F_K)$ and the command variable and the product of the disturbance transfer function $F_{gz} = F_g/F_R$ and the disturbance variable of the control circuit, or

$$\frac{\beta}{\beta_N}(p) = F_g(p)\frac{\beta_S}{\beta_N}(p) + F_{gz}(p)\frac{M_\tau}{M_{\tau N}}(p) \tag{5.25}$$

The purpose of the control system is to bring the regulated variable into line with the command variable over as wide a frequency range as possible, and to suppress the influence of the disturbance variable [5.20]. This means that the following should hold:

$$|F_g| \approx 1 \tag{5.26}$$

and

$$|F_{gz}| \approx 0. \tag{5.27}$$

Furthermore, if damping is sufficient, there must be stable control-circuit behaviour and adequate control speed. Therefore, controller characteristics are carefully selected in an attempt to compensate completely or partially for large delays in control sections, which are particularly detrimental to the control-circuit dynamics.

The conditions according to Equations (5.26) and (5.27) are fulfilled if $|F_K| \gg 1$ in the fundamental frequency range. This condition can be fulfilled by high amplification in the circuit or controller or with integrating controllers, the transfer function of which has one pole at zero frequency, so that $F_R(0) \to \infty$ if there are no differentiating sections and $F_S(0) \neq 0$.

Thus the dynamic characteristics of second-order control sections as shown in Figure 5.35 can be significantly improved by 'compensating' for large delay components with the aid of the derivative action of a controller (PI (proportional-integral), PID (proportional-integral-derivative)). This means that, in the case of dimensioning according to the so-called optimum amount system, the delay with the greater time constant T_1 is eliminated. The resulting transmission link would then only exhibit a parasitic delay of around $T_v' \approx 0.1T_1$ and the remaining smaller time constant T_2.

The 'system time constants' T_1 and T_2 can, however, vary greatly – due in particular to blade deformation and the increased moments of inertia that this causes – compared with nondeformed blades (with T_0), e.g. $T_1T_2 = 1\text{–}5(T_0)^2$. It should also be considered that measuring the blade angle is very expensive and is avoided in small systems in particular for reasons of cost. Blade pitch speeds or their equivalents such as the speed of actuators, the mass flow of hydraulic mechanisms and so on, on the other hand, are relatively simple to sense. It is therefore completely feasible to install blade pitch speed or even blade pitch acceleration control circuits in order to utilize fully the dynamic characteristics of this control system for speed or output control.

5.5.3.1 Blade pitch speed control circuit

The output or rotation speed of a turbine can be influenced, within the range of the available flow provided by the wind, by adjusting the blade pitch. Desired power or rotational speed gradients, on the other hand, can be maintained by appropriate blade pitch speeds. Figure 5.37 shows the simple structure for controlling the blade pitch speed of a wind turbine. The control circuit consists of the controller and the 'actuator' and 'blade' system sections. Therefore the open-loop transfer function is

$$F_{K\dot{\beta}} = F_{R\dot{\beta}} F_{s\ St} F_{S\ Bl}. \tag{5.28}$$

For a system controlling purely output and rotational speed respectively it is sufficient to use the dynamic characteristics of the blade pitch speed control circuit without knowing the precise blade pitch and its speed and accurately adjusting these values. In the simplest case this can be achieved using a P (proportional) controller, and the control processes can be designed using a converter-fed motor. The amplification of the actuating system and

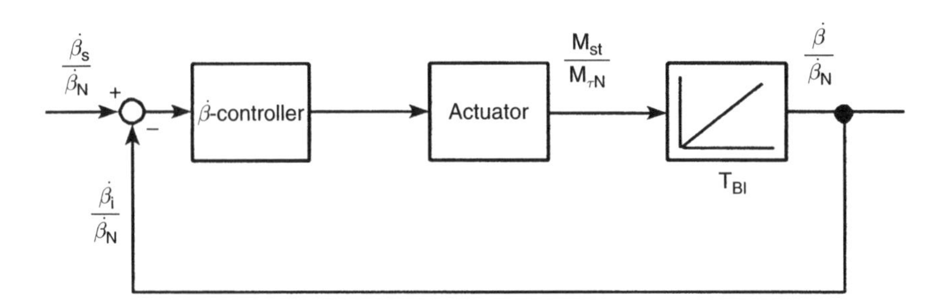

Figure 5.37 Structure of a blade pitch speed control circuit

the delay time for the torque generation of the actuator motor (approximately 10 to 50 ms) should be taken into account here. The integration time constant of the blade is

$$T_{\mathrm{Bl}} = J\dot{\beta}_{\mathrm{N}}/M_{\tau\mathrm{N}} \tag{5.29}$$

Therefore, by the selection of the controller circuit damping (e.g. $D_{\dot{\beta}} = 1,\ 1/\sqrt{2}, \ldots$), the controller amplification can be determined, or, at a given amplification, the damping behaviour and therefore the stability of the circuit ascertained. As described in Section 5.4, the blade pitch control circuit can also be dispensed with. In order to avoid control deviations the $\dot{\beta}$ controller should take over the integral effect. On the other hand (as is often the case) if output or speed tolerances are permissible then the immunity ranges can be adjusted by the selection of controller amplification. In order to maintain the predetermined turbine output and speed gradients can also be maintained on the drive side, for example, as well as controlling the speed the acceleration of the blade pitch should also be controlled using a cascade control circuit. However, this control system will not be described in further detail in what follows. A further option for the control of output and speed gradients is to control these variables on the frequency converter side via the drive train. The higher-level control circuit is described briefly in what follows.

5.5.3.2 Blade pitch control circuit

The so-called pitch speed integrator, which is characterized by the time constant $T_{\beta} = \beta_{\mathrm{N}}/\dot{\beta}_{\mathrm{N}}$, is connected after the closed blade pitch speed control circuit as an additional controlled system. A blade pitch controller is added in front of the control system to create a higher-level cascade as shown in Figure 5.38. This should exhibit integrating behaviour and derivation action (e.g. PI, PID controller) in order to maintain the output or speed of the turbine as precisely as possible and to adjust these quickly. The transfer factor of the open circuit is

$$F_{\mathrm{K}\beta} = F_{\mathrm{R}\beta}F_{\mathrm{g}\dot{\beta}}F_{\mathrm{S}\beta}. \tag{5.30}$$

A controller is connected in series with the $\dot{\beta}$ control loop described above, which can, for example, be represented as a second-order delay element. The controlled system therefore represents a delayed integrator. However, there is a delay element of the second order or higher. The integral term of the blade pitch controller, which is necessary for precise control

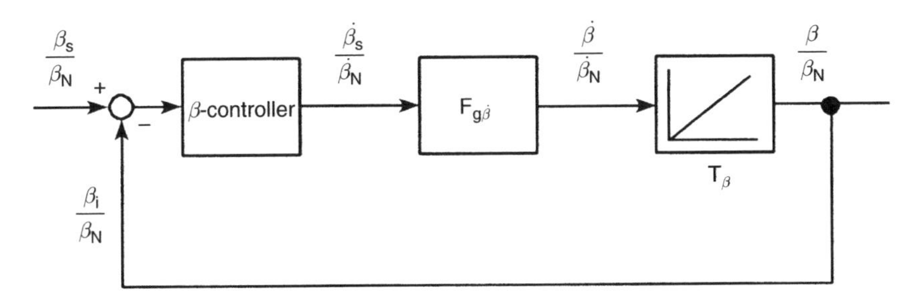

Figure 5.38 Structure of a blade pitch control circuit

of disturbance variables, leads, in connection with the system integrator, to a double pole at $p = 0$ in the transfer function of the circuit. The derivation action of the controller is necessary here, in order to be able to stabilize the system at all.

In order to dimension the control circuit in a simple manner, the blade pitch speed circuit with its higher-order delay can be approximated by a first-order substitution function. A common method of approximating the measured or calculated step response of a higher-order controlled system, for example, by means of a first-order delay element is the creation of a substitution function. The relationship

$$f_{\mathrm{sub}}(t) = V_{\mathrm{sub}} \left(1 - \mathrm{e}^{-t/T_{\mathrm{sub}}}\right) \tag{5.31}$$

is approximated by the step response for the same final value and equivalent control area. If the time constants of the system components differ greatly, e.g. between the actuator and the blade, with values of 20 ms compared with 500 ms, there is the option of replacing the blade pitch speed control circuit by the partial first-order function with the greatest time constant as an approximation. Stabilization of the system is only possible if the time constants in the integral section of the controller are greater than the delay time of the controlled system [5.20]. For the control circuit, a PI controller can be simply dimensioned according to the symmetrical optimum system. Strong overshooting of the step response in the control circuit, which is to be expected if the control variable is altered, can be eliminated by the connection of a desired value delay.

The circuits for the control of the turbine speed and output can be dimensioned in a similar manner. These will, however, not be described further here. However, the designs shown here – for which linearizations have been carried out and simplifying assumptions made – can, when dimensioned for safe damping behaviour, lead to critical operating states in certain operating ranges due to the sluggish control system. On the other hand, critical designs would lead to the risk of instabilities. In order to safely control all the operating ranges of the turbine, the control system therefore requires refinement.

5.5.4 Improving the control characteristics

Control procedures that can react to different operating states are designed to protect components, are orientated towards effectiveness criteria and can automatically perform adjustments in response to altered system and environmental conditions as well as improve the control characteristics and therefore increase the operating reliability and service life of wind turbines.

Figure 5.39 shows the structure for the control of turbine rotational speed. This includes the speed controller, the closed blade pitch adjustment circuit, the turbine driving torque generation and the inertia of the rotating masses of the drive train. Their integration and running-up time constant T_{R} can be determined by Equation (2.93). With regard to controller design, particular attention must be paid to the highly nonlinear characteristics of control sections. This particularly affects the generation of driving torque and driving power. The nonlinear c_{p}–λ family of characteristics (see Figure 2.6 or 2.8) causes a highly variable amplification of this system. This is the result of output changes in relation to blade pitch adjustment. Furthermore, it is necessary to differentiate here between idling and nominal operation of the turbine. The range of values covers around two powers of 10

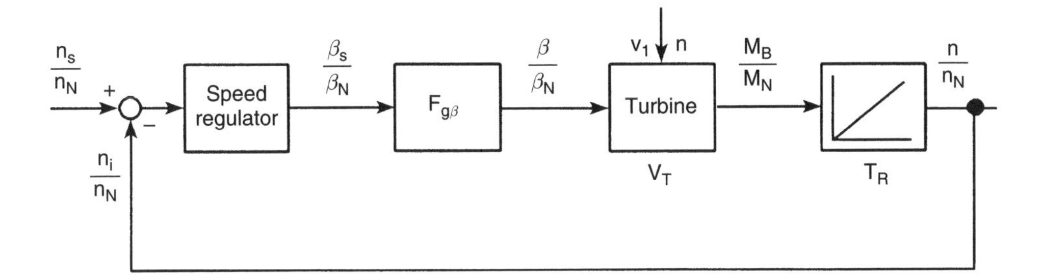

Figure 5.39 Structure of the rotation speed control circuit of the turbine

(e.g. amplification factors of approximately 1 to 100), with low values being encountered in nominal operation.

If fixed regulation parameters were to be used, dimensioning would have to be based on the highest amplification – i.e. on the highest permissible wind speed – in order to ensure stable regulation behaviour of the wind turbine in all operating ranges. This would mean that only very inadequate regulation characteristics could be achieved between idling and nominal operation. In particular, in the operating range in which the turbine is operated for most of the time, unfavourable types of behaviour and the resulting negative effects would be unavoidable.

Good regulation behaviour can be achieved in all operating ranges if the regulation parameters are adapted to the current operating state. This can be achieved in different ways. The simplest option for achieving good results is a fixed characteristic of the adaptation factor, e.g. in relation to the blade pitch adjustment angle (Figure 5.40).

5.5.4.1 Fuzzy controllers

The application of fuzzy logic [5.24] and the use of fuzzy controllers offers an innovative method for the control and management of wind turbines [5.25]. Unlike digital systems, which can only differentiate between and evaluate 'true' and 'false' values or equivalent

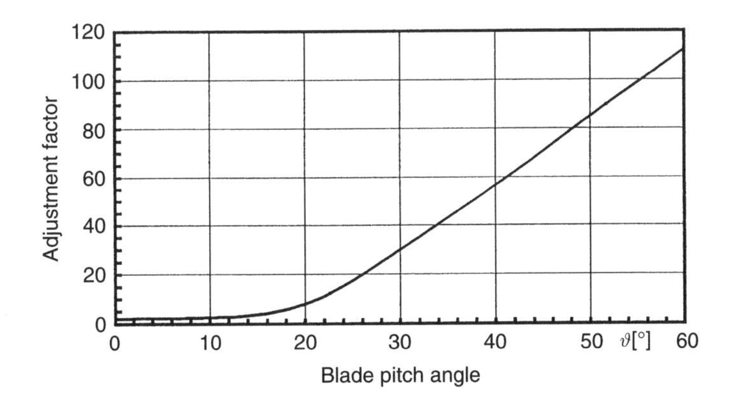

Figure 5.40 Adjustment factor for the regulation of turbine output as a function of blade pitch angle

contrasts, this technology, being based on imprecise technology, can also take into account intermediate levels. Such systems are therefore also attributed human characteristics. So-called membership functions represent the connection between linguistic statements and numerical values, and basic predictions can be linked to complex values by logical connections. Therefore linguistic statements on the input side can lead to a 'control engineering' operating instruction on the output side. The sum of all linguistic rules therefore represents the basis of the controller. In the design of a fuzzy controller the controller structure forms the framework within which the knowledge base can be defined in substages.

Linguistic variables and terms must be selected such that the complexity of the controller is kept low and the selected formulations must be expressively structured. Furthermore, the membership functions of the input variables must be defined and the controller base made up in matrix form. Performance characteristic controllers are the result of the design process.

The simulation results shown in Figure 5.41 for one such system [5.25] illustrate the different output characteristics of turbines with P controllers compared to those with fuzzy controllers. The comparison shows markedly smoother reactions for fuzzy controllers. Even at more rapid wind speed changes the fuzzy controller variant does not tend to oscillate. Similar results can, however, also be achieved with conventional control devices, e.g. with controllers that also have integral parameters and are designed for smooth system behaviour.

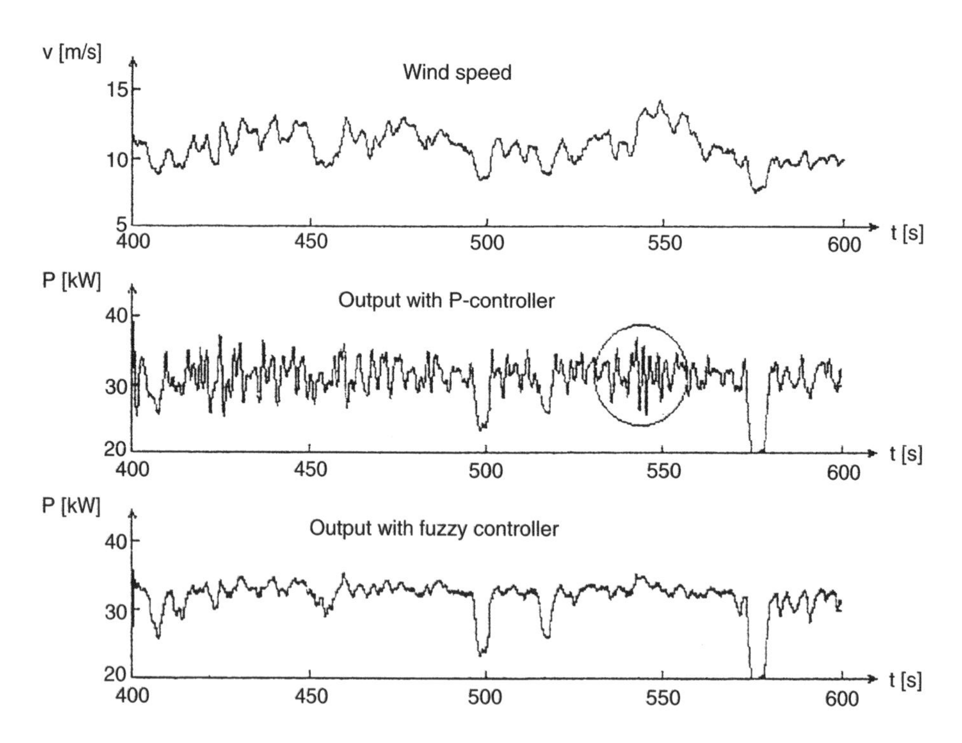

Figure 5.41 Output behaviour of a wind turbine with a P controller and a fuzzy controller as a function of the associated wind speed (simulation results)

5.5.4.2 Self-tuning control systems

Surface effects on the rotor blades, e.g. icing, roughening, dirt and moisture, should also be taken into account, as well as physical changes to the operating-state-dependent section amplification of wind turbines. Designing a controller set-up that takes all operating ranges and states of the turbine into account to the same high level is therefore no simple matter. Good controller behaviour can, however, be achieved at all turbine states if the linearized transfer behaviour of the turbine is identified at different operating points. Therefore, with the aid of the system parameters that have been acquired, control systems can be built for the controller that are dimensioned for the controller and are capable of learning.

The creation of a self-tuning turbine controller can, as described in reference [5.26], be based upon the identification of the transmission link between the wind speed and electrical power output. This process has the disadvantage that a prediction of the wind speed is required and the dynamics of the turbine must be known. In fact, it is almost impossible to fulfil both conditions in practical operation.

The process described in reference [5.12], on the other hand, requires only the measured output power and the desired value for the blade pitch adjustment speed. If computers are used for the control and supervision of turbines, these variables are already available within the system. No additional measuring devices are required for their identification. This method takes as its basis the general discrete-time process representation based upon the controlled and manipulated variables, a white noise signal, and the wanted and interference signals of the controlled variable. Identification in a closed control circuit takes place with an initially approximately adjusted controller, so that even extreme gusts cannot damage the turbine. System natural frequencies of interest are stimulated by a test signal at the controller output and identified with the aid of correlation analysis.

The estimated impulse response of the model exhibits a high level of scattering due to wind-speed fluctuations. The values are smoothed by regression. This process produces the coefficients of the discrete impulse transfer function. For this, the scanning time and the test signal must be determined in a suitable manner, since these parameters influence the effectiveness of the identification. After the identification of the impulse transfer function of the controlled system model, a model of the control circuit can be simulated on the computer in parallel to the actual control system. Thus predetermined responses by the controlled system model output variables to interference can be achieved by controller optimization.

By identifying the controlled system model transfer function at discrete operating points, optimal controller parameters can be determined and expanded to a control system with managed adaptation, i.e. to an adjustment of parameters. Between the detected values, the controller parameters can be found by interpolation at any operating point. However, particularly during turbulent wind conditions, the adaptation of controller parameters to each sampling step must be avoided. Good operating results can be achieved using quasi-continuous controller adjustment by averaging the values over a time period, e.g. in the seconds range [5.27].

5.5.4.3 System-oriented controller design

In addition to the classical methods for controller design given in references [5.1] and [5.28] to [5.33], etc., which are particularly oriented towards the function and stability of the control system, procedures can be employed that use control mechanisms to maintain maximum and minimum values of component or system parameters. Thus, for example, the use of 'linear–quadratic' (LQ) optimization [5.34] aims to reduce periodically induced blade noises, which are particularly associated with influences due to tower shadowing and the wind height profile. Furthermore, significant inherent values are taken into account in reference [5.35], while the use of Kalman filters to estimate the turbine state is considered in references [5.36] and [5.37], and references [5.38] and [5.39] take account of multivariable controllers. The use of the linear–quadratic output-feedback method (LQOFB) in reference [5.40] was aimed at reducing the torques in the rotor shaft of a fixed-speed turbine. In references [5.41] to [5.43] blade loading is reduced in variable-speed turbines by the use of linear–quadratic methods.

In a numerical design process for the optimization of quality vectors developed by Kreißelmeier and Steinhauser [5.44] the controllers are optimized on the basis of simulation results. This requires no special (e.g. linear) controller structure. Design criteria can be defined as turbine- or operation-specific. For this, control aims must be quantified in the form of quality criteria and summarized as quality vectors. These control aims may include the limiting of electrical output and rotor speed at full load, the maximization of electrical output at part load, the reduction of output fluctuations, the minimization of actuating processes and the reduction of load alternations in the drive train and rotor, at individual blades or at the tower [5.45–5.49].

Therefore particular requirements are imposed on the control system that can lead to conflicting goals. Moreover, since the limits of realization work are generally not known at the design stage, goals can only be defined within limits. Besides, the different results on the controller adjustment, which arise through many essential optimizations, demand many objective selection criteria.

A control structure must be determined before the use of the procedure. A quality criterion is linked to each required system characteristic, with lower values indicating better fulfilment of the criterion. The standard deviation, the control area and the displacement of a dominant pole of the transfer function from the desired pole position are examples of variables that can be used for evaluation.

The individual quality criteria and the resulting quality vector are determined by simulation and compared with the desired values or the corresponding desired vector. The aim of the optimization of controller parameters is to minimize all components of the quality vector. However, a performance function defined for this purpose exhibits step changes at points where the criteria change. For practical applications, therefore, if there are 10 to 20 criteria it is advisable to minimize the natural logarithm of the sum of all exponential functions of quotients of quality and preset values.

From general principles in references [5.50] and [5.51] and in representations, for example, according to references [5.52] and [5.53], so-called 'state monitors' can be introduced for the recovery of unacceptable state variables. Thus the 'estimation' of complete state vectors can take place, from which the best possible approximation can be expected.

An estimator is used in references [5.47] to [5.49] to take into account the mechanical loading variables in the control system. In reference [5.49] this derives, with the aid of a simulation model for the wind turbine, the required estimated values for the aerodynamic loading of the turbine and the effective wind speed in the rotor circuit from the electrical output, generator speed and blade pitch, these variables being available by measurement. In this manner, significant mechanical loading variables in the turbine, drive train and tower can be combined with the available measured values, to reconstruct and be included in the control process.

In reference [5.49] control is performed by two independent subcontrollers. A drive-train controller influences the generator torque and a turbine controller acts exclusively upon the blade pitch adjustment system. Both subcontrollers can be active. In this manner, the loading of the turbine and drive train can be kept within preset values independently of the operating state of the generator by limiting wind turbine output. Insignificant energy sacrifices thus lead to significantly reduced peak values of the rotor blade shock moment and tower deflection moments. Furthermore, with these methods it is possible to achieve better types of behaviour during the transitions between the part-load and nominal-load range than is possible when conventional controllers are used.

A one-year test run of the control process on an experimental system belonging to the Institut für Solare Energieversorgungstechnik (ISET, Institute for Solar Energy Supply Technology) in Kassel demonstrated an enormous alleviation of the load on the rotor. A comparison of the measured load collective between conventional and modified controllers at fixed-speed operation indicated a statistical extension of the service life of rotor blades on the order of magnitude of 30 %. For variable-speed turbines, the expected service life of rotor blades can be increased by around 60 % and the service life of the rotor shaft can be increased by around 40 % [5.54].

5.5.4.4 Neural networks

Innovative control procedures for wind turbines, e.g. with self-adjusting, component-oriented or system-oriented load-limiting controllers, usually require the processing of large amounts of measured data and consideration of as much additional information as possible. However, favourable modes of behaviour for components and systems remain largely limited to the selected design goals. A cumulation of as many advantages as possible is only attainable to a limited degree with a technology-orientated procedure.

Biological systems cope with the processing and compression of very different types and almost unlimited amounts of information by evolution. The human skin can be cited as an example, with approximately 500 000 touch receptors. It can be viewed as a comprehensive interface with our environment. The numerous very different sensors initiate influence-specific body reactions.

Neuroanatomical investigations in references [5.55] and [5.56] yielded topological relationships between neighbouring receptors and cerebral cortex areas. Spatial information was therefore retained in both sensing areas. It was also found that the size of the cerebral cortex areas was related to the density of the receptors and not the spatial extent of the skin zone, indicating that the sensing signal areas are represented according to their importance.

Neural connections, which are genetically influenced only in their basic structure, are largely created by self-organization processes between the cells [5.57]. Their activity depends upon a variety of environmental stimuli.

Biologically accurate models for self-organizing processes in systems involving topology have been developed for the simulation of neural processes. A generally valid principle in accordance with references [5.58] to [5.61] is based upon a two-dimensional neuron layer in the form of a grid. Equidistant nodes (so-called ‘neurons’) symbolize the nerve cells.

Excitations are transmitted by the comparison between defined pattern vectors and the input signal vector. The excitation response therefore decreases as the distance from the excitation centre increases. Limited learning steps ensure that the 'map' containing topology achieves quasi-stationary states, and slow changes in the input signal area can take place. Ritter and Schulten carried out investigations into the dynamics of the map generation processes and highlighted various possible applications [5.62–5.69].

The use of neural networks is particularly beneficial in systems where functional relationships cannot easily be sensed analytically or using measuring technology. Their application for storing characteristic lines for sensing the state of lead storage batteries is a classic example [5.70]. With a network of 40×40 neurons, the state of discharge of batteries in a typical application in hybrid systems can be illustrated with sufficient precision.

Neural networks can be characterized by the following valuable aspects: they can learn relationships between inputs and outputs (e.g. the c_p–λ family of characteristics of a wind turbine) without knowing the physics of the processes involved or the associated mathematical equations. Furthermore, they are capable of generalizing characteristics and thus reacting in an appropriate manner to unpredicted, i.e. nonlearned, events by interpolation or extrapolation. Moreover, if the task or problem formulation is altered, only the pattern files need be updated. The developed programs can, unlike in other procedures, be retained.

Investigations into the use of neural networks for the learning of power or power coefficient characteristic diagrams [5.71] have delivered satisfactory results in terms of precision, if a suitable network structure is selected. Owing to the excellent interpolation characteristics, even with a coarse interpolation-point density the characteristic diagrams can easily be determined. Furthermore, these networks have excellent generalization capabilities. For wind turbines, therefore, correspondingly good operating characteristics can be expected from a system-orientated controller. Progressive adaptation in the future for taking into account external and internal parameter changes could lead to the importance of neural networks increasing greatly.

5.6 Management System

The management system must ensure the reliable and automatic operation of wind turbines. To achieve this, the relevant components and system variables must be monitored continuously. By maintaining permissible values and value ranges for system variables, the management system can bring about predetermined operating states and recognition in emergency or fault situation.

To achieve this, the turbine management system must influence the operating behaviour of the wind turbine based on the preset control signals and desired values, and react to changes in system variables or to malfunctions. Along with reliable operation, another goal is to achieve the optimal compromise between the output and low mechanical and electrical loading of the turbine and its components [5.1, 5.29, 5.30].

Figure 5.42 shows the structure of the management system for a variable-speed wind turbine with a frequency converter supply (see Figure 5.19) and gives an overview of the most important operating states and changeovers, which will be outlined in what follows as an example based upon the main procedures.

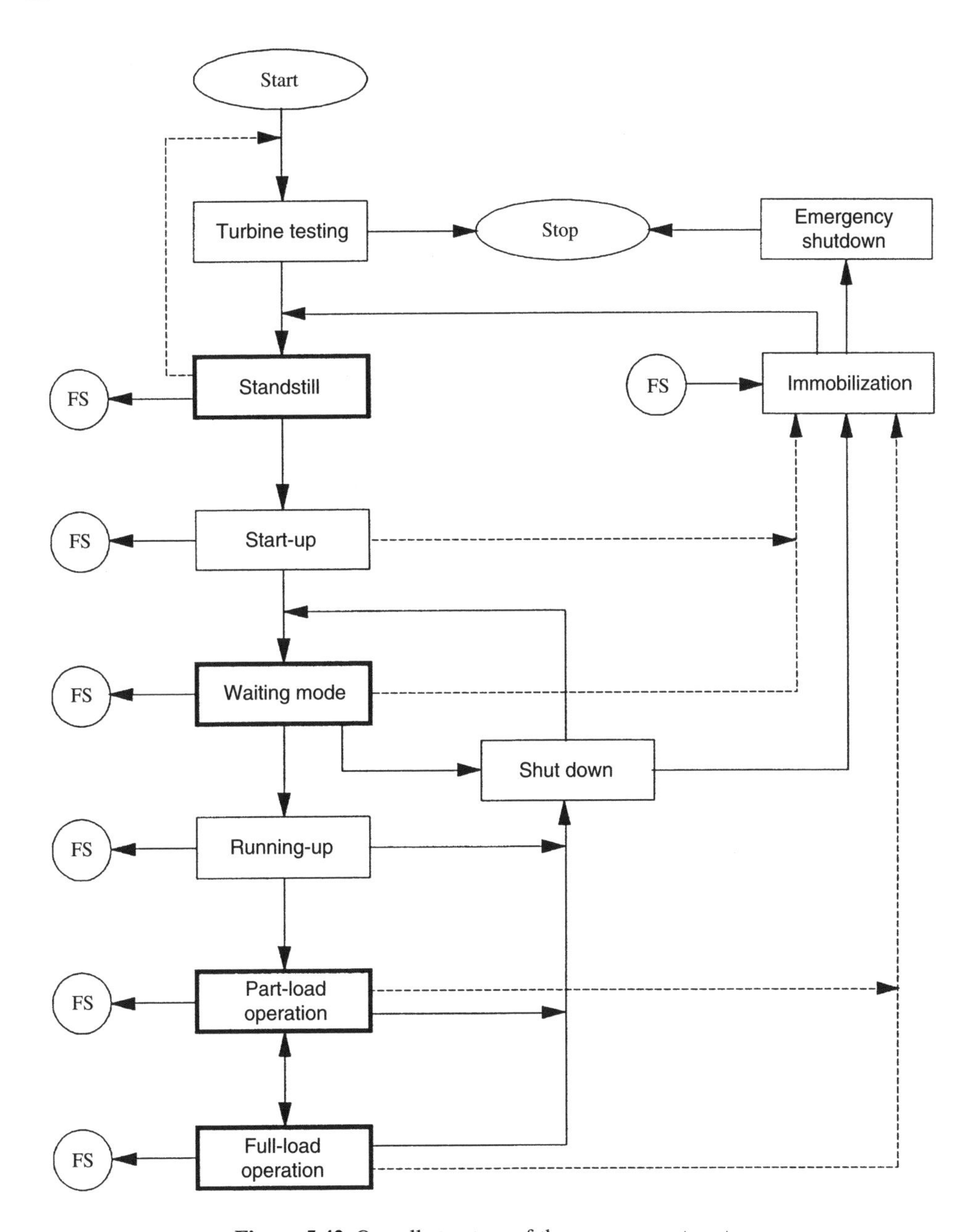

Figure 5.42 Overall structure of the management system

5.6.1 Operating states

The turbine normally runs in automatic mode. However, manual and semi-automatic operating modes with the manual input of desired values are necessary during commissioning and maintenance.

Transient operating states may last for a limited period only. Their duration is therefore monitored. After the predetermined maximum periods have been exceeded a fault shut-down is initiated, since it must be assumed that there is a malfunction.

The duration of steady-state operating states is not monitored by the management system. The turbine remains in these states as long as all normal operating conditions are fulfilled.

In all operating states the conditions for normal operation must be continuously interrogated. Only one condition is required for the transition to the immobilization, shut-down, fault shut-down or emergency shut-down operating states. In contrast, all conditions must be fulfilled for the initiation of the start-up or running-up operating states.

5.6.1.1 Turbine testing (transient)

After commissioning the management system, the monitored components, influencing variables and command variables must be checked and recorded. Figure 5.43 shows the turbine testing structure and the associated Table 5.1 lists the most important messages. The outputs of all subsystems must be interrogated for standstill values and all mechanical actuators driven for test purposes. The correct reactions of configurations can be checked by sensors. If faults occur, these must be recorded. Faults lead to the suspension of further operation until the faults have been rectified and the turbine manually released.

All turbine components and their limit values must be checked in all operating states. This system check-tests whether all systems are functioning properly, whether temperatures are within the operating range and whether the message 'system OK' is universally present. After successful testing, the turbine goes over into the next operating states; otherwise the testing of the turbine operating state is repeated until all release conditions are fulfilled, such as operator commands, unlocking after emergency shut-down, grid available and OK, component functionality, temperatures and limit values.

5.6.1.2 Standstill (steady-state)

The standstill turbine state is characterized by the stationary rotor. Moreover, in this operating state the rotor brakes are activated, the rotor blades are in their feathered position and the nacelle of the wind turbine is yawed out of the wind. If cable twist in the tower must be rectified, this can be carried out at low wind speeds. The electric generator is switched off and disconnected from the supply grid. First of all, those conditions that have prevented the turbine from going over into the start-up operating state are checked. Then a system check is carried out. If all conditions are now fulfilled the start-up conditions are interrogated. If these are also fulfilled, the turbine goes over into the start-up operating state. As in the turbine testing operating state, the messages 'supply grid disconnected', 'cable twist', etc., are identified by the appropriate operating state number.

5.6.1.3 Start-up (transient)

When starting the turbine at no load, i.e. with the rotor brakes released, the turbine is driven, by the wind alone with no power being drawn via the frequency converter, from standstill to the speed that has been predetermined by the control system. At this point, the rotor blades are driven by the blade pitch regulation mechanism from the feathered position to a defined

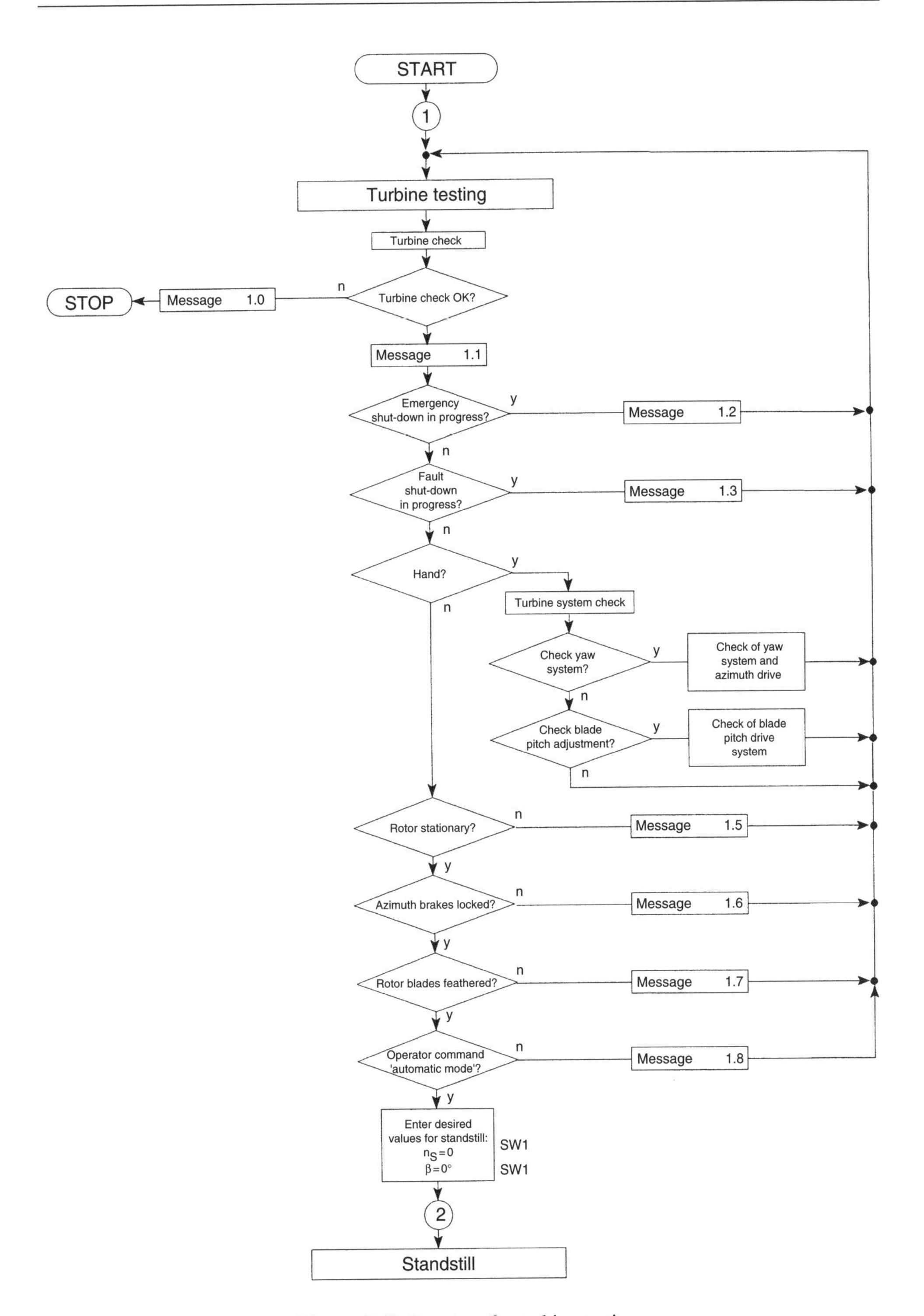

Figure 5.43 Structure for turbine testing

Table 5.1 Messages during turbine testing

Output	Meaning
1.0	STOP: turbine check negative
1.1	Turbine check positive
1.2	Emergency shut-down in progress
1.3	Fault shut-down in progress
1.4	STOP: maximum azimuth angle exceeded
1.5	STOP: rotor not stationary
1.6	STOP: azimuth brakes not locked
1.7	Rotor blades not in the feathered position
1.8	No operator command 'automatic mode'

angle. In a repeating sequence, the conditions for fault shut-down and immobilization are then checked and the appropriate routines initiated if required. Speed is also checked during this sequence. As soon as the minimum waiting speed is achieved, the turbine goes over into the steady-state waiting mode. During the start-up operating state the nacelle is again yawed out of the wind.

5.6.1.4 Waiting mode (steady-state)

In the waiting mode all the components of the wind turbine are ready for operation. The rotor speed lies within a range determined by the management system and is influenced by the blade pitch control system. The generator system is not yet connected to the supply system. The fault shut-down, immobilization and running-up conditions are checked one after the other. If the appropriate conditions are fulfilled the relevant operating states are initiated. The speed is maintained within a defined permissible range by the adjustment of the rotor blade pitch according to desired values. If the waiting mode is maintained for a long period then the operator is notified and, after a certain period, e.g. one day, a further turbine test is carried out. Moreover, in this operating state the conditions to be fulfilled are continuously checked and the nacelle is yawed in the direction of the wind.

5.6.1.5 Running-up (transient)

If the wind speed is high enough, the rotor speed of the wind turbine can be run up to a value at which it is possible to connect the generator system to the grid (see Figure 5.44). The frequency converter is first checked for its readiness for power so that the grid protection system can be connected. Then, with the aid of the blade pitch adjustment system, the rotor speed is adjusted to the speed determined by the management system. During running-up, the fault shut-down and immobilization conditions are continuously checked and the nacelle yawed according to the wind direction.

When the required desired speed is attained, the generator and frequency converter system is connected to the supply grid and electrical power can be supplied. The turbine is now in part-load operation. The necessary messages and limit values during this process are listed in Table 5.2.

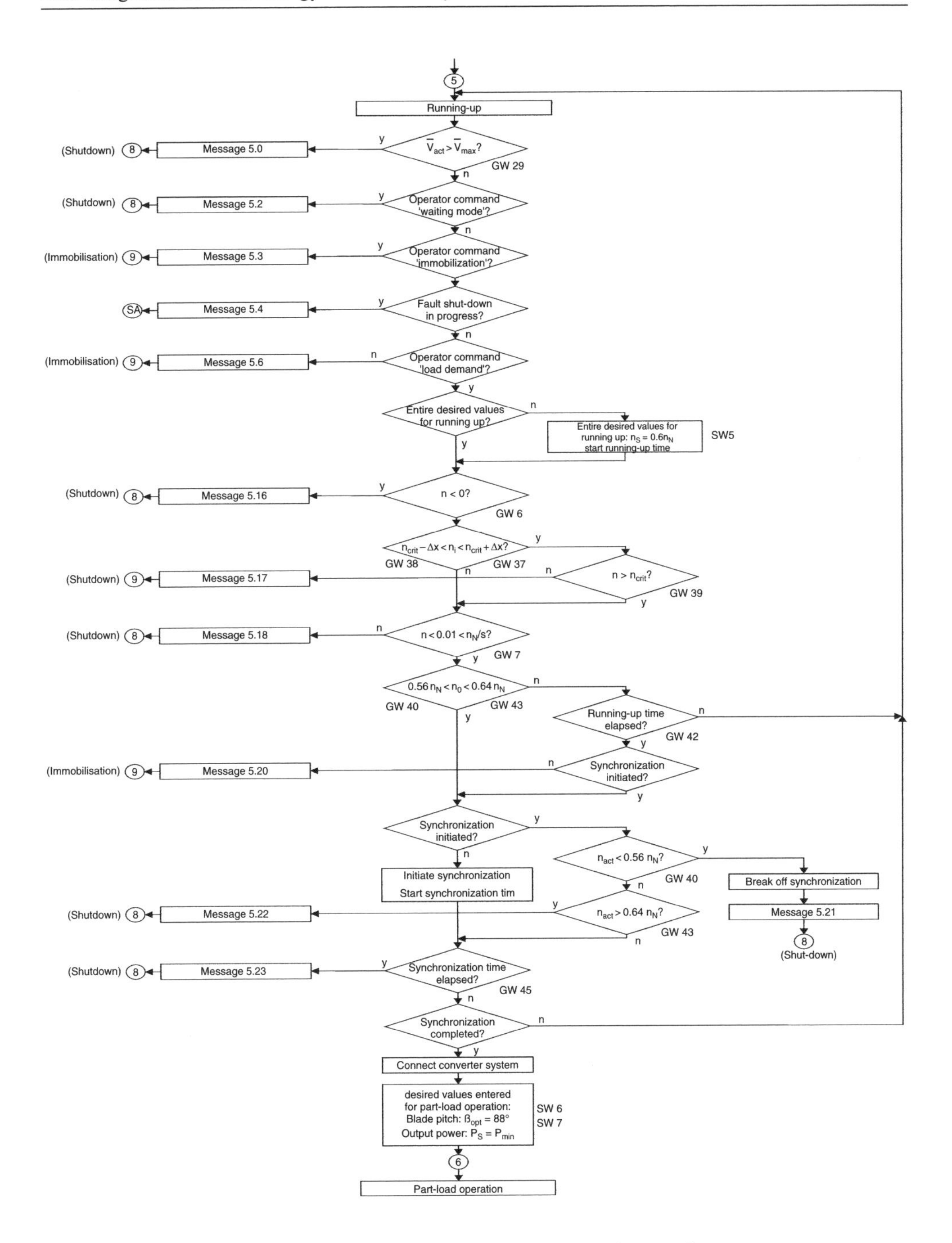

Figure 5.44 Management system structure during running-up

Table 5.2 Messages during running-up

Output	Meaning
5.0	RUNNING UP
5.1	SHUT-DOWN: average wind speed too high $v_{\mathrm{act\,av}} > v_{\mathrm{ru\,av\,max}}$
5.2	IMMOBILIZATION: supply grid disconnected
5.3	SHUT-DOWN: operator command 'waiting mode'
5.4	IMMOBILIZATION: operator command 'immobilization'
5.5	FAULT SHUT-DOWN: fault shutdown in progress
5.6	IMMOBILIZATION: temperature limit value exceeded
5.7	SHUT-DOWN: no operator command 'load demand'
5.8	IMMOBILIZATION: external temperature $T_{\mathrm{ext}} > T_{\mathrm{ext\,max}}$
5.9	IMMOBILIZATION: external temperature $T_{\mathrm{ext}} < T_{\mathrm{ext\,min}}$
5.10	IMMOBILIZATION: monitoring period exceeded; azimuth angle outside wind direction tolerance range
5.11	FAULT SHUT-DOWN: $\Delta\gamma_{\mathrm{Win/Azim}}$ exceeded 10 times /d
5.12	FAULT SHUT-DOWN: maximum azimuth angle γ_{max} exceeded
5.13	IMMOBILIZATION: monitoring period for maximum azimuth angle range exceeded ($\gamma > \gamma_{\mathrm{max}} - \Delta\gamma$)
5.14	IMMOBILIZATION: azimuth angle γ_2 exceeded
5.15	IMMOBILIZATION: monitoring period for azimuth angle range γ_2 exceeded ($\gamma > \gamma_2 - \Delta\gamma$)
5.16	IMMOBILIZATION: speed gradient $\mathrm{d}n_{\mathrm{ru}}/\mathrm{d}t < 0$
5.17	IMMOBILIZATION: speed gradient $\mathrm{d}n_{\mathrm{ru}}/\mathrm{d}t < \mathrm{d}n_{\mathrm{ru\,crit}}/\mathrm{d}t$ (for $n_{\mathrm{crit}} - \Delta x < n_{\mathrm{act}} < n_{\mathrm{crit}} + \Delta x$)
5.18	IMMOBILIZATION: speed gradient $\mathrm{d}n_{\mathrm{ru}}/\mathrm{d}t > \mathrm{d}n_{\mathrm{ru\,max}}/\mathrm{d}t$
5.19	FAULT SHUT-DOWN: $\mathrm{d}n_{\mathrm{ru}}/\mathrm{d}t > \mathrm{d}n_{\mathrm{ru\,max}}/\mathrm{d}t$ exceeded 10 times /d
5.20	IMMOBILIZATION: running-up time elapsed and synchronization not initiated
5.21	SHUT-DOWN: synchronization initiated, rotor speed $n_{\mathrm{act}} < n_{\mathrm{syn\,min}}$
5.22	SHUT-DOWN: rotor speed $n_{\mathrm{act}} > n_{\mathrm{syn\,max}}$
5.23	IMMOBILIZATION: synchronization time elapsed $t_{\mathrm{syn}} > t_{\mathrm{syn\,max}}$
5.24	FAULT SHUT-DOWN: five synchronization attempts reached /d

5.6.1.6 Part-load operation (steady-state)

In part-load operation (see Figure 5.45) the generator system supplies electrical energy into the supply grid. The blade pitch is set or adjusted to an optimal value, so that the maximum power output or minimum component loads are possible. The management system sets a value for output power in relation to speed (see Figure 2.64(c)). In part-load operation, the generator system's frequency converter regulates the speed and power output. In Table 5.3, no changes are made to the desired values within the adjustment range.

When the control reserve value is reached, the desired value for speed is altered according to the power–speed characteristic line. The nacelle is also continuously yawed into the wind. The blade pitch control system functions as part of the safety system, braking the rotor in the event of an emergency. Given a high enough wind speed, the turbine automatically goes over into the steady-state full-load operating state. Again, all conditions for normal operation are checked in part-load operation and, if necessary, the appropriate procedures are initiated.

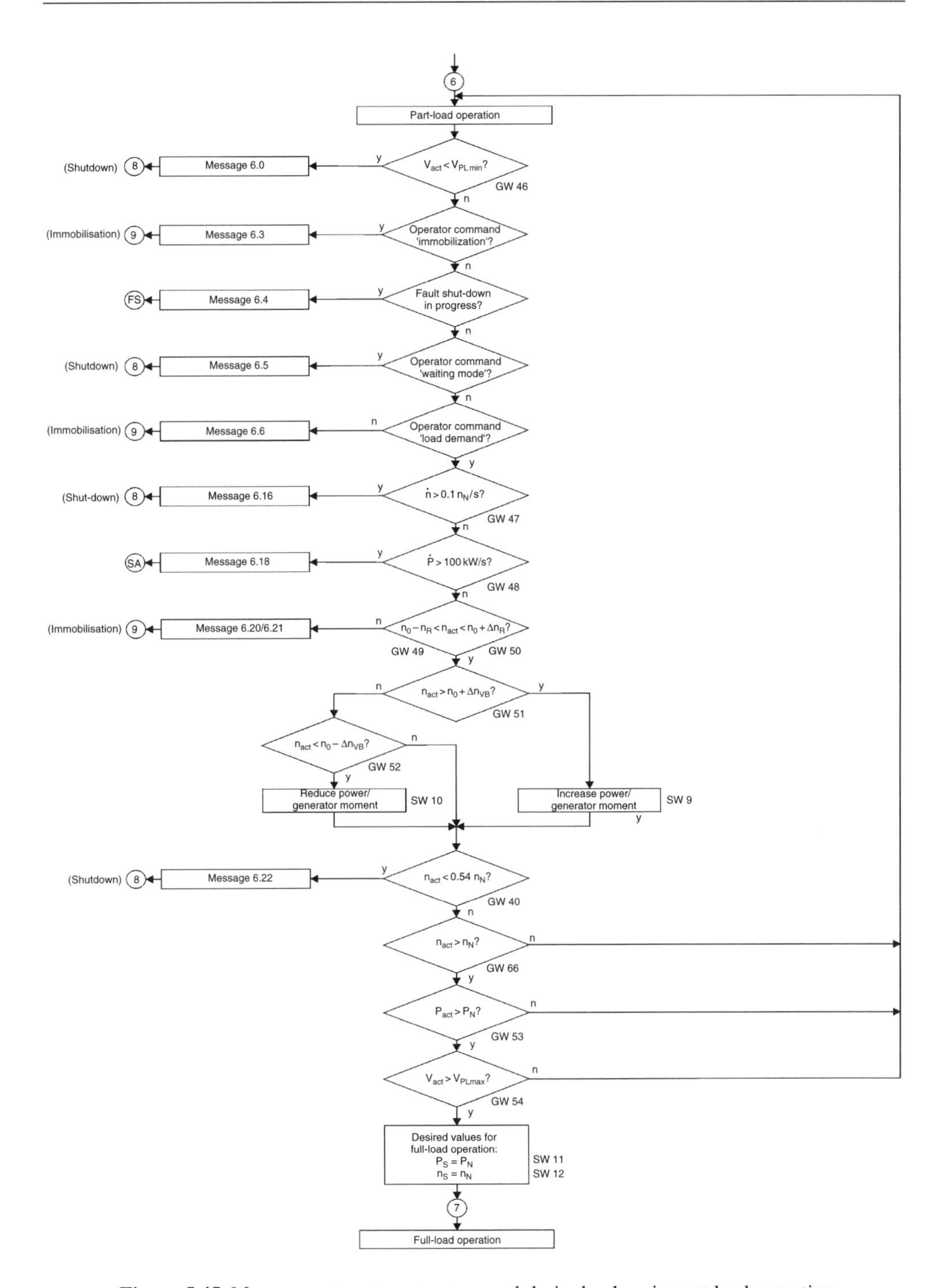

Figure 5.45 Management system structure and desired values in part-load operation

Table 5.3 Messages and desired values in part-load operation

Output	Meaning
6.0	PART-LOAD OPERATION
6.1	SHUT-DOWN: wind speed too low $v_{act} < v_{PL\,min}$
6.2	IMMOBILIZATION: supply grid disconnected
6.3	IMMOBILIZATION: temperature limit value exceeded
6.4	IMMOBILIZATION: operator command 'immobilization'
6.5	FAULT SHUT-DOWN: fault shutdown in progress
6.6	SHUT-DOWN: operator command 'waiting mode'
6.7	IMMOBILIZATION: no operator command 'load demand'
6.8	IMMOBILIZATION: external temperature $T_{ext} > T_{ext\,max}$
6.9	IMMOBILIZATION: external temperature $T_{ext} < T_{ext\,min}$
6.10	IMMOBILIZATION: monitoring period exceeded; azimuth angle outside wind direction tolerance range
6.11	FAULT SHUT-DOWN: $\Delta\gamma_{Win/Azim}$ exceeded 10 times /d
6.12	FAULT SHUT-DOWN: Maximum azimuth angle γ_{max} exceeded
6.13	IMMOBILIZATION: Monitoring period for maximum azimuth angle range exceeded $(\gamma > \gamma_{max} - \Delta\gamma)$
6.14	IMMOBILIZATION: azimuth angle γ_2 exceeded
6.15	IMMOBILIZATION: Monitoring period for azimuth angle range γ_2 exceeded $(\gamma > \gamma_2 - \Delta\gamma)$
6.16	IMMOBILIZATION: speed gradient $dn_{PL}/dt > dn_{PL\,max}/dt$
6.17	FAULT SHUT-DOWN: $dn_{PL}/dt > dn_{PL\,max}/dt$ exceeded 10 times /d
6.18	FAULT SHUT-DOWN: power gradient $dP/dt > 100\,kW/s$
6.19	IMMOBILIZATION: rotor speed below acceptable range $n_{act} < n_0 - 10\,\%$
6.20	IMMOBILIZATION: rotor speed above acceptable range $n_{act} > n_0 + 10\,\%$
6.21	IMMOBILIZATION: rotor speed too low $n_{act} < n_{PL\,min}$
6.22	FAULT SHUT-DOWN: rotor speed $n_{act} > n_{max}$

5.6.1.7 Full-load operation (steady-state)

If the wind speed is high enough, the turbine will go over from part-load operation to full-load operation (see Figure 5.46). In this operating state the management system sets desired values for nominal speed, its fluctuation range and the nominal output of the system. Speed and power output are regulated by blade pitch adjustment.

In full-load operation the frequency converter can maintain both the power output and the generator moment at a constant level or change them in relation to a specified function. Output fluctuations at the turbine therefore give rise to slight speed changes. The speed is maintained within the regulation reserve range by blade pitch adjustment. A small overload range is permissible in the case of gusts, so that the blades do not have to be adjusted as fast or as often. The overload range must, however, be of limited duration, depending upon the thermal behaviour of the entire system. The termination conditions of this operating state are checked continuously and the necessary messages given out (Table 5.4). The tower nacelle is yawed in the direction of the wind.

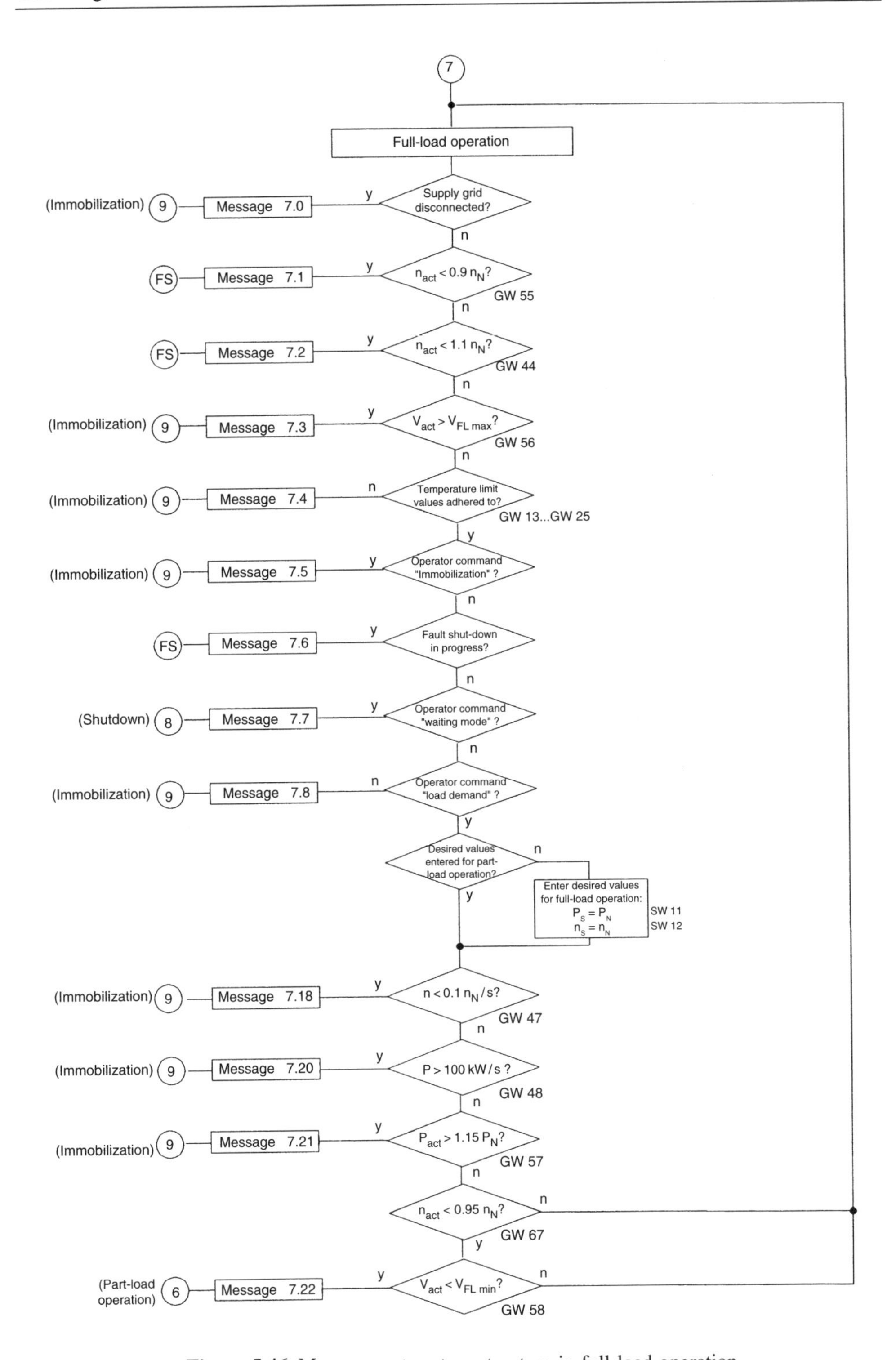

Figure 5.46 Management system structure in full-load operation

Table 5.4 Messages in full-load operation

Output	Meaning
7.0	FULL-LOAD OPERATION
7.1	IMMOBILIZATION: supply grid disconnected
7.2	IMMOBILIZATION: value below rotor-speed range $n_{act} < n_0 - 10\,\%$
7.3	IMMOBILIZATION: value above rotor-speed range $n_{act} > n_0 + 10\,\%$
7.4	IMMOBILIZATION: wind speed too high $v_{act} > v_{FL\,max}$
7.5	IMMOBILIZATION: temperature limit value exceeded
7.6	IMMOBILIZATION: operator command 'immobilization'
7.7	FAULT SHUT-DOWN: fault shutdown in progress
7.8	SHUT-DOWN: operator command 'waiting mode'
7.9	IMMOBILIZATION: no operator command 'load demand'
7.10	IMMOBILIZATION: external temperature $T_{ext} > T_{ext\,max}$
7.11	IMMOBILIZATION: External temperature $T_{ext} < T_{ext\,min}$
7.12	IMMOBILIZATION: Monitoring period exceeded; azimuth angle outside wind direction tolerance range
7.13	FAULT SHUT-DOWN: $\Delta\gamma_{Win/Azim}$ exceeded 10 times /d
7.14	FAULT SHUT-DOWN: maximum azimuth angle γ_{max} exceeded
7.15	IMMOBILIZATION: monitoring period for maximum azimuth exceeded $(\gamma > \gamma_{max} - \Delta\gamma)$
7.16	IMMOBILIZATION: azimuth angle γ_2 exceeded
7.17	IMMOBILIZATION: monitoring period for azimuth angle range γ_2 exceeded $(\gamma > \gamma_2 - \Delta\gamma)$
7.18	IMMOBILIZATION: speed gradient $dn_{FL}/dt > dn_{FL\,max}/dt$
7.19	FAULT SHUT-DOWN: $dn_{FL}/dt > dn_{FL\,max}/dt$ exceeded 10 times /d
7.20	IMMOBILIZATION: power gradient $dP/dt > 100\,kW/s$
7.21	IMMOBILIZATION: maximum permissible power exceeded $P_{act} > 1.15P_N$
7.22	Wind speed too low for full-load operation $v_{act} < v_{FL\,min}$

5.6.1.8 Shut-down (transient)

From part-load operation, full-load operation and running-up, it must be possible at all times to shut the turbine down, bring it into the waiting mode operating state and report the appropriate states. To achieve this, after desired values have been set by the management system, the power output is reduced by the frequency converter and the turbine is decelerated by adjusting the blade pitch towards the feathered position, so that values are reached that permit the generator system to be disconnected from the supply grid. The fault shut-down and braking conditions are checked in a recurring sequence. After a successful separation process, the turbine returns to the waiting mode operating state.

5.6.1.9 Immobilization (transient)

It must be possible to stop the turbine from any operating state. Immobilization is similar to shut-down. If the speed has fallen below the minimum value predetermined by the management system then the rotor and nacelle are braked and the turbine goes into the immobilization state. Again, the fault shut-down and braking conditions must be checked repeatedly and state messages displayed during the immobilization operating state.

5.6.1.10 Fault shut-down (transient)

Fault shut-down takes place in a similar manner to immobilization. The turbine management system can, however, impose steeper desired value ranges than are normally used when stopping the turbine. This operating state can also be initiated from higher speeds, in which case the rotor brake may be applied in a controlled manner if the rotor is not decelerating quickly enough. If the rotor speed falls below a minimum value predetermined by the management system then the rotor and nacelle are braked and the turbine goes over into the immobilization state. During fault shut-down, the emergency shut-down and braking conditions must be checked continuously.

5.6.1.11 Emergency shut-down (transient)

An emergency shut-down is triggered if a normal immobilization procedure is not possible. This state lasts until the turbine is stationary. It can be initiated either by the turbine management system or by a higher safety system. As an emergency shut-down can take place from full speed, all braking systems should be used to bring the turbine to a standstill as safely as possible. The safest, but mechanically least favourable, method is the abrupt operation of brakes and blade-adjusting hydraulics. The rotor is locked as soon as it is stationary. Further operation of the turbine is prevented by the management system. Recommissioning is only possible after manual release.

5.6.2 Faults

To ensure the reliable operation of the turbine and its components, disturbances to normal operation must be recognized by the management system. The management system should cut in before the safety system, so that the latter is used as seldom as possible. The safety system can be made up of a redundant monitoring computer or a speed sensor connected either directly to the hydraulic system or to an electric adjustment mechanism.

5.6.2.1 Rapid auto-reclosure of grid

In the event of grid failures – even those of only short duration – it is necessary to prevent an excessive increase in speed. In the designs selected here, only the frequency converter can recognize grid failures. It must therefore shut down immediately and send a message to the management system. As the turbine's generator is no longer opposed by a load moment, speed increases. Using blade pitch adjustment and, if necessary, the brake (in the upper speed range), the turbine is decelerated into the waiting mode. As soon as all conditions (usually grid OK) are again fulfilled, running-up can be automatically initiated once again. If the grid is not available after a certain time, a fault shut-down (i.e. immobilization) must take place.

5.6.2.2 Short-circuits

Short-circuits bring about high currents that can damage or even destroy turbine components, circuitry and protective devices. To prevent damage, short-circuits must be recognized quickly and protective measures initiated. This process is concluded by the tripping of the main switch. At the same time, the frequency converter reports the short-circuit to the management system, which initiates a fault shut-down.

A generator short-circuit can result in a sudden reduction in voltage in one or more phases of the generator-side input to the frequency converter, despite adequate speed. As soon as the management system recognizes this error, a fault shut-down is initiated.

A short-circuit within the frequency converter must be recognized independently. The internal frequency converter electronics then switch off the power section and report a fault. The management system then initiates a fault shut-down.

5.6.2.3 Overspeed

When the turbine is at full load, i.e. at wind speeds above the nominal range, the speed is held within the control range by adjustments to the blade pitch angle. A regulation reserve allows delayed reaction to increases in speed. If the speed nevertheless climbs above the maximum permissible operating speed (e.g. 10 % above the nominal range), a fault shut-down is initiated. If the rotor continues to run too fast despite the intervention of the management system and reaches the tripping speed, the safety system must work to limit the speed. In this case, the safety system immediately initiates the emergency shut-down procedure.

5.6.2.4 Overtemperature

All turbine components are designed such that in normal operation no impermissably high temperatures occur. If these temperature limits are exceeded, it can be assumed that there is a fault or overload in the system. Therefore the fault shut-down must be initiated.

5.6.3 Determining the state of system components

This description relates to a gearless wind turbine with a blade pitch adjustment system, fitted with a permanent-magnet synchronous machine and connected to the grid via a pulse-controlled a.c. converter. The turbine components will be described below and the options for ensuring the sensing of relevant states investigated.

For the blade adjustment device, the adjustment angle and adjustment speed must be monitored and limited. The pitch adjustment angle and direction, as well as the adjustment speed and locking of the blades, are predetermined. Moreover, the hydraulic or power supply to the device must be ensured.

Braking intervention and drive-train locking are predetermined for the generator. Output values are speed, rotational direction, temperature and electrical variables such as current, voltage, power, power factor and frequency.

The frequency converter takes on the functions of grid and generator monitoring, temperature limitation and grid synchronization. The power output or electrical moment of the generator are specified. As well as checking that the system is ready to be powered up and connected to the grid, the current and voltage in the d.c. link, and the output power are sensed.

For the azimuth drive, the adjustment of the nacelle and the adjustment speed and locking of the machine housing must be predetermined, and the nacelle angle and any cable twist reported. The drive and braking function and the brake lining thickness must be monitored. As well as the speed and direction of the wind, vibrations in the nacelle, tower and foundations should be sensed. On the grid side, the energy must be sensed, fuses monitored and an uninterruptible power supply provided for the management system, safety system, emergency lighting, etc.

For the control and monitoring of all operating states, the management system must receive not only operator commands but also all measured and monitored variables, so that it can stipulate the desired values for the frequency converter, blade pitch adjustment system and azimuth adjustment system, and display the status of all turbine components. Moreover, fault messages, remote interrogation and monitoring should be possible, and error diagnosis and fault prediction systems integrated to ensure reliable operation of wind turbines.

5.7 Monitoring and Safety Systems

Besides the normal turbine and operational management components, further monitoring and safety systems should be taken into account when considering the management and safety of the turbine. These may depend upon requirements relating to the turbine, grid or location. Such systems cover measuring and monitoring systems for temperature, pressure, moisture, acceleration, oscillations, voltage, etc. Furthermore, illumination systems for the tower, nacelle and grid station, a system for the automatic rectification of cable twisting and navigation lights should also be considered. Measures to protect against lightning and other extreme effects such as earthquakes, tornadoes, etc., should also be taken into account. Aerodynamic, mechanical and electrical braking systems (see Sections 2.3.2.4 and 3.6.2.2 to 3.6.2.4) protect against overspeed and serve to bring the rotor to a standstill. Requirements and design notes for safety systems are listed in reference [5.74].

In addition to air density and humidity, the wind conditions at turbine sites are particularly important for determining the drive power of a wind turbine. These will be briefly described in what follows.

5.7.1 Wind measuring devices

Relevant flow conditions for wind turbines are determined by the air speed and its direction in relation to the horizontal. To determine this, individual or combined wind gauges can be fitted on the nacelle, comprising an anemometer (usually a cup anemometer) and a vane. A lightning rod can also be fitted on the wind gauge to protect against a direct lightning strike.

The measuring range of the device must cover the cut-in ($v_{\text{cut-in}}$), nominal (v_N) and shut-down (v_{shut}) wind speeds of the turbine. To determine the average values over 1, 3, 5, 10 or 15 minute periods, the minimum and maximum occurring values must also be determined, which means that the measuring range must cover at least

$$v_{\text{meas}} = (0\text{–}1.5)v_{\text{shut}}$$

Therefore, for example, at a shut-down wind speed of 25 m/s, a measuring range up to approximately 40 m/s is required.

5.7.2 Oscillation monitoring

In order to protect the turbine from severe jarring and high-amplitude movements in the nacelle, imbalance in the rotor system and similar effects, vibrations are monitored. If limit values are exceeded, the turbine is brought to a standstill.

Vibrations in the longitudinal and transverse (and, if required, vertical) directions can be determined as a vector variable with frequency and amplitude dependences by an acceleration sensor in the bottom of the nacelle. A reliable and robust design option for the acceleration sensor is offered by piezo elements. The measured values can be processed with the aid of charge amplifiers. Critical operating conditions, e.g. caused by natural resonances of the tower, rotor blade deflection, etc., must be terminated as quickly as possible, e.g. by ensuring that the turbine only passes through the speed range in question for a brief period. All vibrations are monitored, and if the limit value is reached a message must be sent to the turbine management system. At amplitudes of 50 to 60 % of the applicable limit value a (delayed) fault shut-down should be initiated and at a maximum of 90 % an immediate emergency shut-down should be initiated.

If the option of selecting different amplitudes and acceleration values for the initiation of shut-down procedures is rejected then much cheaper designs can be used. Mechanical systems offer very simple but effective options for the monitoring of vibrations. These are often used in small (and medium)-sized wind turbines. In most cases a ring-and-ball system is used, fitted in the nacelle or the top of the tower. In this case, the relative diameters of the ball and ring should be selected such that the free-lying ball falls from the ring if the acceleration limit value is reached. This trips an emergency stop switch that immediately brings the turbine to a standstill. Another design for the detection of vibrations in the nacelle is the fitting of a pendulum rod. This pendulum is made of an electrically conductive material and passes through a metal ring so that in the event of oscillation the pendulum makes contact with the ring and an electrical signal initiates the shut-down of the turbine. The frequency and amplitude of the vibrations can be set to suit the turbine parameters by the selection of the length and mass of the pendulum length and the internal diameter of the ring.

5.7.3 Grid surveillance and lightning protection

In the case of voltage or frequency deviations exceeding, for example, 10 or 5 % of the nominal values, the turbine must be disconnected from the grid to prevent unwanted separate operation in grid branches. The turbine is protected from overvoltage damage caused by overvoltage at the generator or by direct or indirect lightning strikes by means of powerful coarse and fine protective devices in the measurement and control circuits, at the generator and at the supply mechanisms, etc.

Direct lightning strikes usually result in serious damage. Diverters in the rotor blades specifically designed to conduct current through connections to the shaft and tower and through an effective (low-resistance) foundation ground connection allow damage to be limited. For this purpose, metal caps are fitted on the blade tips and coarse copper mesh is fitted on to the blade surfaces to conduct lightning currents away without causing much damage.

5.7.4 Surveillance computer

Wind turbines are usually built at some distance from towns and the operator. Visual monitoring is therefore not usually possible. To keep the down-time of turbines low, remote diagnosis systems are necessary. These require suitable measuring, transmission and monitoring units for individual turbines and wind farms.

The analogue and digital data collected can include turbine states plus grid and meteorological conditions such as output, rotation speed, turbine position, temperature, etc. These data are generally processed to ensure fault-free transmission. To this end, the information to be transmitted is divided into blocks of information, provided with error protection, error checking and error correction, and synchronized in blocks. Physical signal preparation takes place by means of encoding and modulation [5.75].

The data collected can be used for control and management, for error checking and for statistical evaluation by the operator, maintenance company and manufacturer. Data transmission is therefore necessary for the transfer of statistical data or the immediate reporting of faults.

As shown in Figure 5.47, individual turbines can be connected to the monitoring computer by copper or fibre-optic cables, modem and telephone or radio connections at the turbine and computer. For data transmission, analogue equipment such as the telephone system or the C1 radio system, or digital equipment such as ISDN or the D1 or D2 network, can be used.

Analogue transmissions are sometimes subject to serious interference. Digital data, on the other hand, can be checked and corrected using codes. Transmission errors can be significantly reduced in this manner and monitoring systems made relatively reliable.

Wind farm monitoring can take place in different ways. Depending upon factors such as the distance between turbines and the system configuration, the data from individual turbines may be transmitted via integral modems as shown in Figure 5.47, brought together as a group

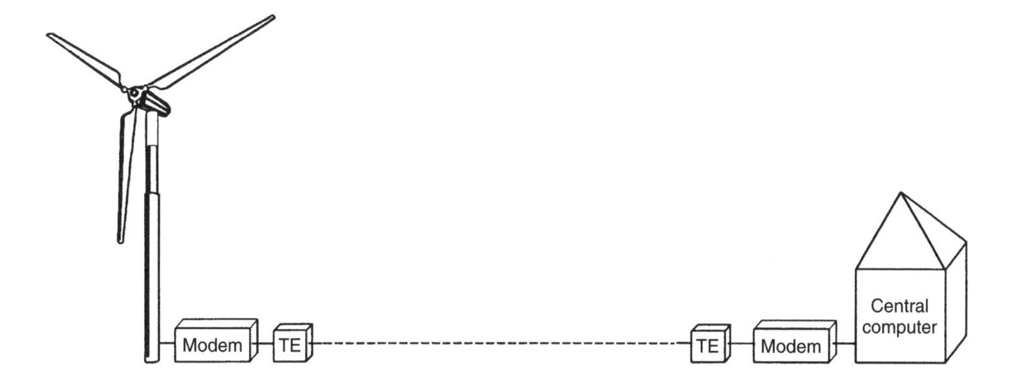

Figure 5.47 Individual wind turbine monitoring

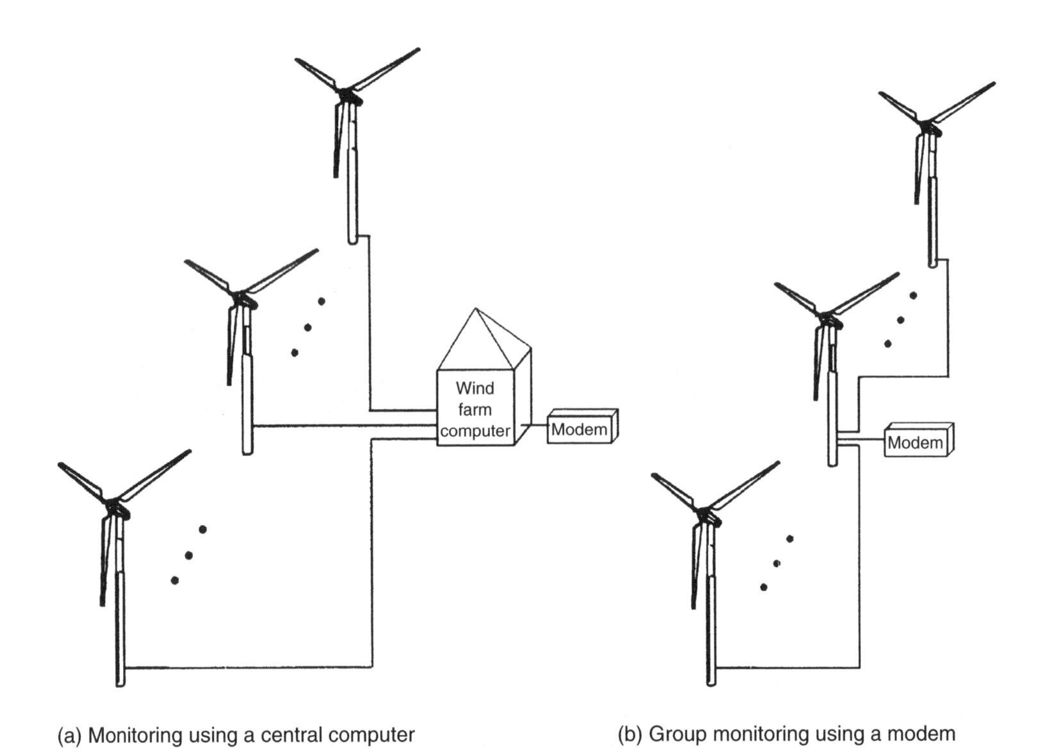

Figure 5.48 Different methods of wind farm monitoring

(Figure 5.48(b)) or transmitted to the central computer (Figure 5.48(a)). Moreover, with the use of a wind farm computer, it is possible to process, evaluate and compress data on site and send it to the central computer. In this case, the connection between the turbines can be made via cables or the local radio network. As well as cost and safety aspects, upgradability, e.g. with regard to fault prediction, should play a major role in system selection. This aspect will be described briefly in the following section.

5.7.5 Fault prediction

Fault prediction is taking on increasing importance in the field of quality assurance and in the monitoring of technical plant and equipment. By monitoring and evaluating relevant measured signals of a wind turbine, fault indications can be determined before visual, vibration or acoustic changes become apparent and serious damage is done to subcomponents or the system as a whole. In this manner, secondary damage can be avoided, subsequent costs reduced, maintenance intervals adjusted to the state of the turbine and necessary repair work planned in advance and carried out in periods of low wind for safety reasons. Such a system also permits remote monitoring and remote diagnoses to be carried out. Therefore the down-time of the turbine can be reduced, reliability and economic viability improved and the service life of the turbine increased. The most common causes of faults, which are listed in detail in reference [5.76], are defective components and the turbine control system.

External effects due to storms, lightning strikes and grid faults, and turbine-specific effects caused by the loosening of components are also of significance. Significant causes of faults in the mechanical components of a wind turbine are the fatiguing of materials and wear and loosening of components. Changes observed in the event of such defects, e.g. in relation to vibration behaviour, can generally be recognized before they become critical. It is thus possible to rectify expected faults in advance.

In fault-prediction systems, relevant measuring signals are continually captured and evaluated in relation to fault-related characteristics [4.40]. The most expressive measuring variables are mainly used, which are always available in running operation. State-related information that is relevant to faults can be determined from electrical power (see Figure 4.34), generator currents, turbine rotation speed and the acceleration of vibration-monitoring systems. Furthermore, body noise and possibly also air noise measurements can be used for fault prediction.

Spectral analysis processes are particularly suitable for the evaluation due to the permanent random and periodic excitations to the turbine caused primarily by the wind and the rotation of the rotor. At this point, measuring signals are divided into deterministic and random components, broken down into sections of equal length and weighted using a window function. This allows direct components with occurring trends to be filtered out. Using a fast Fourier transform (FTT), the spectra of the filtered components are calculated and the mean taken. By comparison of the measurements with known spectra of fault-free turbines and turbines with faults, changes and the development of faults can be recognized.

Precise knowledge of the turbine behaviour in normal operation and in fault states permits a detailed diagnosis to be made of the current turbine state, and allows necessary measures for fault diagnosis to be introduced [5.77–5.85]. In modern high-output turbines, fault predictions are expected to form a fixed component of the turbine monitoring system in the near future.

6

Using Wind Energy

The use of wind power for the supply of electricity broadens the energy base and reduces environmental pollution. It is particularly practical if it can be made to be economically competitive with conventional energy sources [6.1, 6.2]. In countries such as Denmark or regions such as Schleswig-Holstein in Germany wind power already makes a significant contribution to the electricity supply.

Knowledge of system costs and the expected energy yield are of fundamental importance in this context [6.3, 6.4]. Good wind conditions at the planned site of a turbine or wind farm must be viewed as being the most important prerequisite for the economical exploitation of wind energy [6.5]. Moreover, in densely populated, coastal and offshore areas planning permission issues take on a critical role.

6.1 Wind Conditions and Energy Yields

The Earth is surrounded by an atmosphere in which various physical processes influence the weather, which includes the winds. The atmosphere is kept in motion primarily by differences between heating processes. The contours of the Earth's surface have a decisive influence on local wind speeds. Good conditions for the exploitation of wind energy can be expected near water and in smooth areas of land. Trees, buildings and hills in the immediate vicinity, on the other hand, impair the flow of air.

6.1.1 Global wind conditions

The rate at which the speed of the wind increases with increasing height above ground depends upon the roughness of the landscape (e.g. water, meadow, grassland with bushes, trees, buildings) (see Figure 2.9). Figure 6.1 illustrates the concentration of favourable wind conditions – with regard to their exploitation – in coastal areas. High inland areas can also offer similar conditions.

Grid Integration of Wind Energy Conversion Systems, Second Edition S. Heier

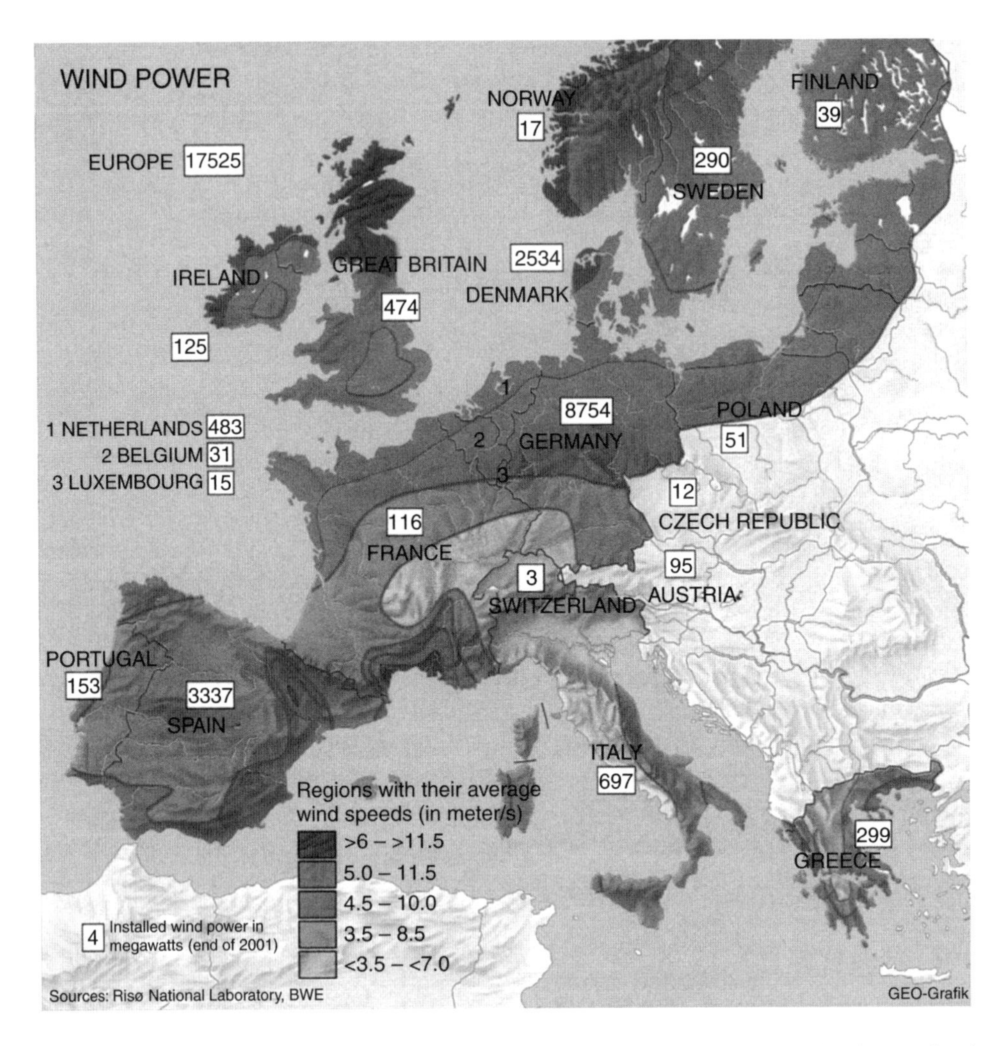

Figure 6.1 Wind speeds in Europe and installed wind turbine outputs in the countries in question in 2001

The economic operation of wind turbines is completely dependent upon the local wind conditions. These can often deviate significantly from the values stated on wind maps. Statistically determined wind velocity and their wind velocity distributions are decisive for the expected energy yield. These indicate the percentage of the time or the number of hours per year for which every relevant wind speed occurs. Moreover, the daily and yearly wind-speed graphs, the height dependence of wind speeds, the contours and roughness of the landscape and the influence of obstacles must also be taken into account. The gustiness of the area, the degree of turbulence and the maximum wind speeds all impose requirements with regard to the stability of the structure and the control of the turbine. Large inland turbines in particular are often subject to dynamic loads that should not be underestimated.

Before wind turbines are erected, the expected energy yields should be predicted as precisely as possible to determine the economics of the project for the operator and to minimize the investment risk. Site surveys and energy yield predictions based on measurements and calculations are thus required. For cost reasons, measurements are generally only carried out for wind farm projects or in the case of sites for which insufficient reliable data are available.

6.1.2 Local wind conditions and annual available power from the wind

Relatively precise forecasts can currently be obtained by model calculations to determine the local wind potential and turbine-specific energy yields. The limitations of these must, however, be kept in mind.

Precise knowledge of local wind conditions is of fundamental importance for the assessment of a site since wind turbine output and energy yields are proportional to the cube of wind speed. As well as climatological factors such as the shape of the land (orography), surface roughness (topography) and obstacles near the location (mechanical turbulence) influence air density, temperature and sunshine (thermal turbulence) and the direction and strength of the wind [6.5].

Energy predictions based on local wind conditions measured at the hub height of a planned turbine give the most precise results. However, this involves an expensive and time-consuming process. For today's turbine sizes, measurement at the hub height (50 to 100 m) is barely feasible for cost reasons and due to the cumbersome nature of large measuring masts. Therefore, wind speed and direction are measured at lesser heights (10, 20, 30 and 40 m) and the measurements arithmetically extrapolated to the hub height (see Figure 6.2). A measured

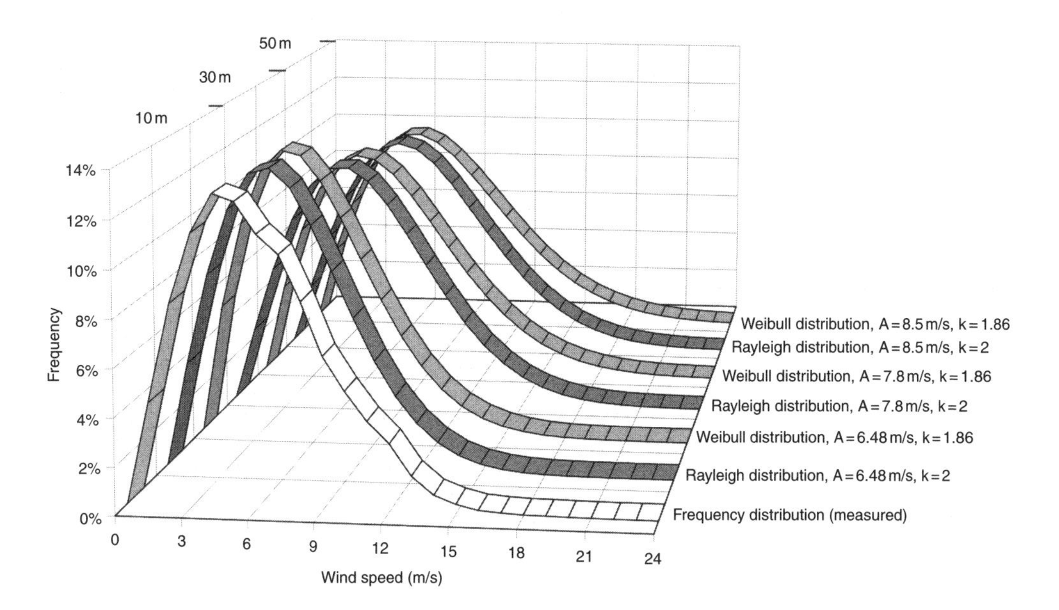

Figure 6.2 Measurement of the frequency distribution at 10 m and calculation of the Rayleigh and Weibull distributions at heights of 30 and 50 m (ISET)

or numerically determined frequency distribution of the wind speed is approximated by an analytical function. The Weibull frequency distribution of wind speeds is usually used for this.

The frequency density

$$h_{\text{Weibull}}(v) = \frac{k}{A}\left(\frac{v}{A}\right)^{k-1} \mathrm{e}^{-(v/A)^k} \tag{6.1}$$

is completely determined, in addition to wind speed, by the dimensionless form parameter k with values from 1 to 3 and the scaling factor A with units of m/s. The mathematically simpler special case $k = 2$, known as the Rayleigh distribution function, is used to describe wind conditions in the event that more precise site data are not available, and is generally sufficiently precise

$$h_{\text{Rayleigh}}(v) = \frac{2v}{A^2}\mathrm{e}^{-(v/A)^2}. \tag{6.2}$$

In the Rayleigh distribution the form factor A is calculated directly from the average wind speed

$$A = v_m \frac{2}{\sqrt{\pi}}. \tag{6.3}$$

A further special case, $k = 3.5$, represents the approximation of the Gaussian distribution.

To determine the relative frequency of a certain wind class the wind speed at the centre of the class is established and the calculated frequency density is multiplied by the breadth of the class (e.g. 1 m/s). The summed frequency of the wind speed

$$F(v) = 1 - \mathrm{e}^{(-v/A)^k} \tag{6.4}$$

is determined by the same parameters. For its calculation below a certain wind speed the upper limit of the last class to be included is used.

The main components of modern wind measurement systems are the anemometer, anemometer mast and measuring computer. Such systems permit fully automatic and maintenance-free operation if they are weatherproof and have internal lightning protection and an effective power supply. Primarily cup anemometers are used for the measurement of wind speed. Ultrasound, hydrometric vane, hot-wire anemometers and venturi nozzles are also occasionally encountered. To record wind speeds and directions over longer periods of time, automatic recording devices (so-called 'data loggers') are required, which facilitate a computerized evaluation of the data. The system often also incorporates a radio modem for the remote interrogation of the measurements.

In areas for which no data are available a measuring period of at least a year is necessary to enable seasonal differences to be taken into account. Moreover, deviations from the long-standing average, the so-called normal wind year, must also be taken into account by means of correlation. This is achieved by drawing upon statistically treated data from numerous measuring stations in the wider area. Figure 6.3 shows the clear differences in the measured annual available wind energy between the individual years, and also between the four site categories listed, throughout the 10 years shown. At the coast and on islands (average power 178 W/m^2 at a height of 10 m) the annual values vary by around $\pm$15 % (at most) between approximately 210 W/m^2 (1994) and 150 W/m^2 (1996). In the North German planes and in wooded areas, the annual values deviate from the average far more, by 88 and 67 W/m^2 respectively. In low mountain ranges, where the average wind power is 106 W/m^2, annual variations reach 20 % and above.

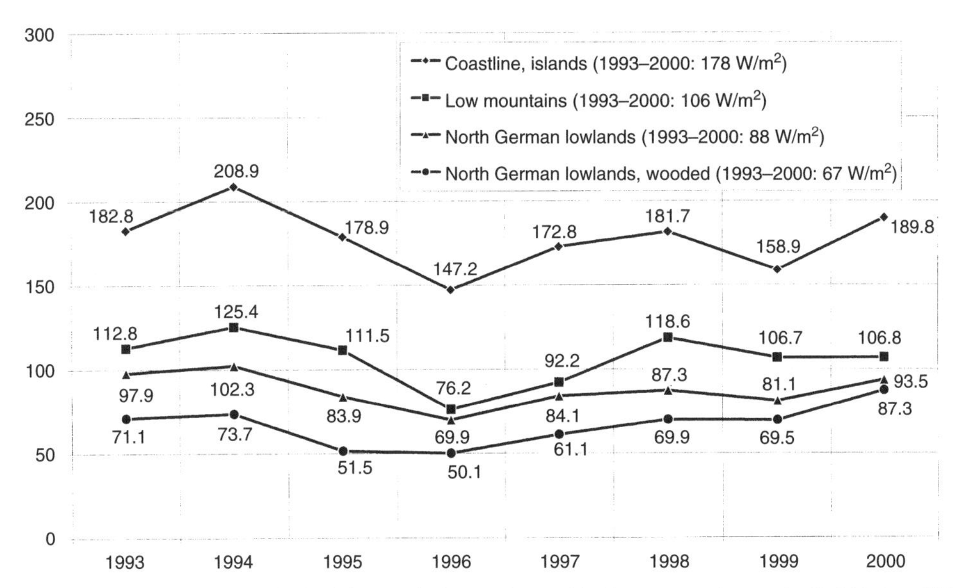

Figure 6.3 Gross available wind energy in the period 1993 to 2001. *Source*: WMEP-Messungen in 10m Höhe – Windenergie-Report, Germany 2001, ISET

6.1.3 Calculation of site-specific and regional turbine yields

Using various calculation procedures, relatively precise energy yield forecasts are drawn up for certain wind turbines based upon measured or calculated average wind speeds and frequency distributions for wind speed and wind direction. Figure 6.4 illustrates this. As an example, a Weibull distribution at the hub height is extrapolated from the frequency distribution of wind speeds measured at a height of 10 m (bottom left in Figure 6.4). Based upon the relative frequencies, the duration of time of each individual wind speed, which will prevail each year, can be determined. With the multiplication of this figure by the output of the wind turbine at the wind speed in question we find the so-called class yields. Summing these gives the annual energy yield in the form of the cumulative curve (top right in Figure 6.4).

In order to determine the power values for each wind speed from the wind velocity distributions of the wind value in question, aerodynamic turbine behaviour, system design (generator power, rotor transmission), the operational characteristics of the turbine and the influences of control and management must be taken into consideration. Since energy yields can be derived from the power characteristic (third graph from the top in Figure 6.4) in connection with the distribution of the wind speed in question, these procedures permit the determination of the available power or energy during a period of one year. In newer large turbines in particular, turbulence, gustiness and the unevenness of the wind speed in relation to the entire swept area of the rotor also play an important role. The quality and details of turbine control can also influence the economics. Practical experience in the operation of wind turbines is therefore indispensable both when making preliminary calculations and when drawing up estimates of economic viability [6.6]. In the framework of large-scale

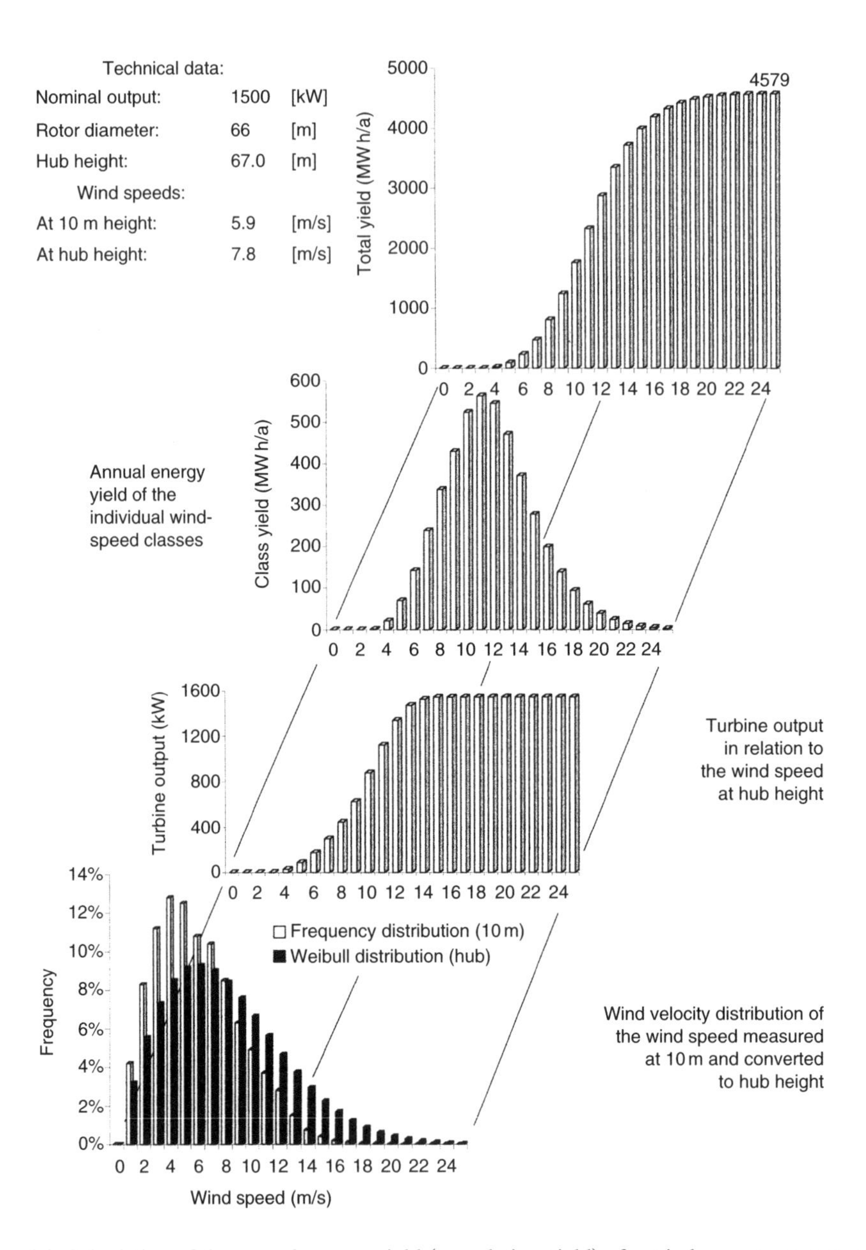

Figure 6.4 Calculation of the annual energy yield (cumulative yield) of a wind energy converter from the measured (or calculated) frequency distribution of the wind speed at a height of 10 m

investigations [6.7], measurements were carried out on wind turbines of different design and size throughout the whole of Germany. The results of site-specific evaluations for the duration of one year for a 1500 kW turbine in northern Germany are shown as an example in Figure 6.5. In addition to the monthly average wind speeds, the figure also shows the wind

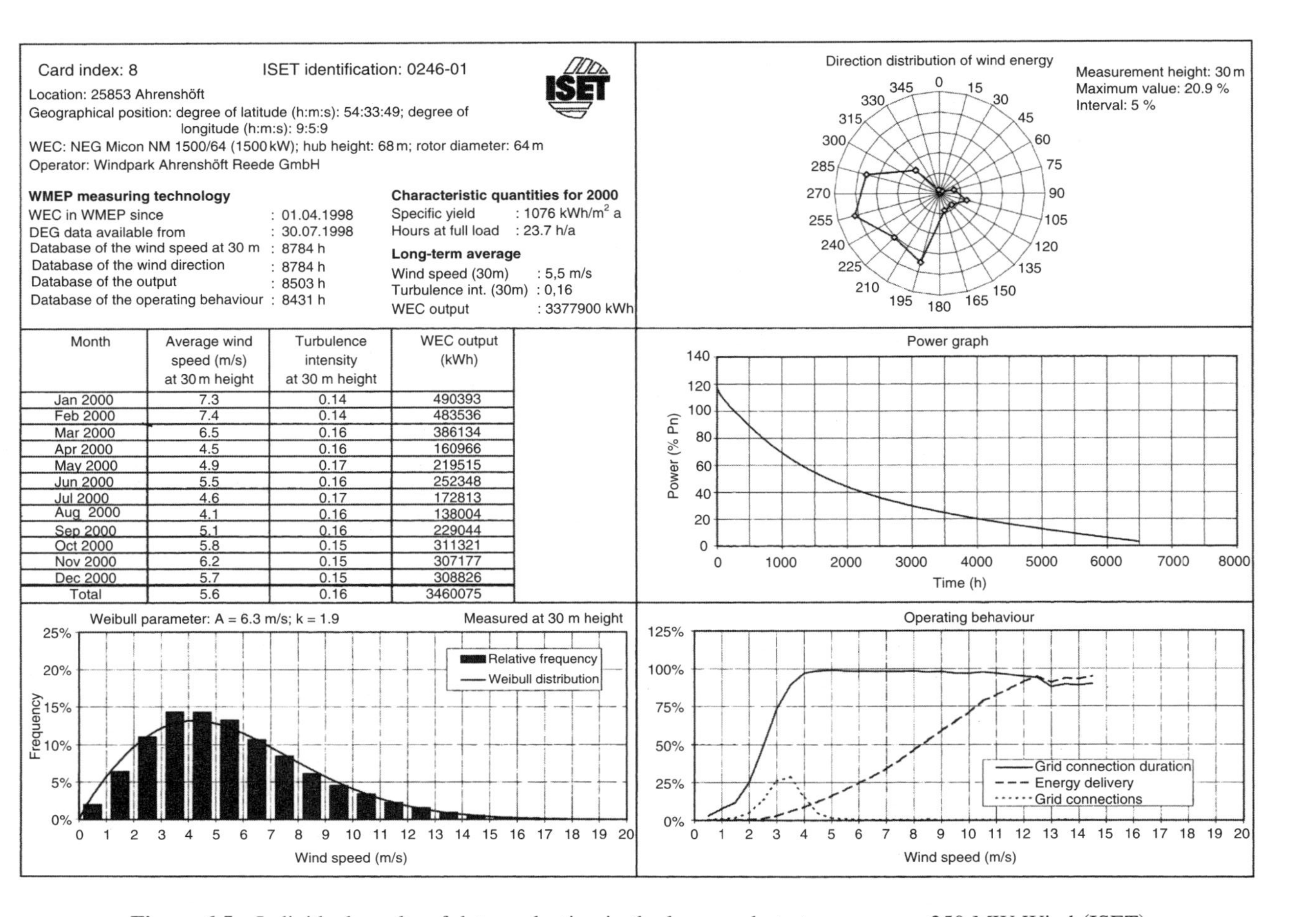

Card index: 8 ISET identification: 0246-01

Location: 25853 Ahrenshöft

Geographical position: degree of latitude (h:m:s): 54:33:49; degree of longitude (h:m:s): 9:5:9

WEC: NEG Micon NM 1500/64 (1500 kW); hub height: 68 m; rotor diameter: 64 m

Operator: Windpark Ahrenshöft Reede GmbH

WMEP measuring technology

WEC in WMEP since : 01.04.1998
DEG data available from : 30.07.1998
Database of the wind speed at 30 m : 8784 h
Database of the wind direction : 8784 h
Database of the output : 8503 h
Database of the operating behaviour : 8431 h

Characteristic quantities for 2000

Specific yield : 1076 kWh/m^2 a
Hours at full load : 23.7 h/a

Long-term average

Wind speed (30m) : 5,5 m/s
Turbulence int. (30m) : 0,16
WEC output : 3377900 kWh

Month	Average wind speed (m/s) at 30 m height	Turbulence intensity at 30 m height	WEC output (kWh)
Jan 2000	7.3	0.14	490393
Feb 2000	7.4	0.14	483536
Mar 2000	6.5	0.16	386134
Apr 2000	4.5	0.16	160966
May 2000	4.9	0.17	219515
Jun 2000	5.5	0.16	252348
Jul 2000	4.6	0.17	172813
Aug 2000	4.1	0.16	138004
Sep 2000	5.1	0.16	229044
Oct 2000	5.8	0.15	311321
Nov 2000	6.2	0.15	307177
Dec 2000	5.7	0.15	308826
Total	5.6	0.16	3460075

Figure 6.5 Individual results of data evaluation in the large-scale test programme 250 MW Wind (ISET)

velocity distribution of wind speeds and arithmetic values such as the turbulence intensity and Weibull parameter. Monthly energy yields, the energy-weighted wind direction distribution and the graph showing the number of hours per year for which the various power levels are achieved are also shown. According to this, the turbine may, for example, deliver nominal output or above for 300 hours per year and around half this value for 1700 hours, with the turbine operating for around 7000 hours.

With the aid of the large-scale tests important findings could be obtained about the wind conditions and energy yields at various sites and in various regions. Figure 6.6 shows the wind velocity distribution of wind speed in the site categories of coastal and islands, North German lowlands and low mountain ranges. This clearly shows that in relation to lowlands and low mountain ranges, windy coastal sites exhibit lower frequencies for low wind speeds and greater distribution for higher wind speeds.

The specific annual energy yield in kilowatt hours per square metre turbine rotor swept area per year ($kWh/m^2 a$) is drawn upon in order to be able to compare energy yields from different wind turbines. Investigations in the large-scale testing program have shown that at coastal sites wind turbines of the 50 kW class achieve specific annual energy yields of around $500\,kWh/m^2 a$, turbines of the 100 kW class achieve $600\,kWh/m^2 a$, turbines of the 200 kW class achieve $900\,kWh/m^2 a$, turbines of the 300 kW class achieve $1000\,kWh/m^2 a$ and turbines of the MW class achieve approximately $1200\,kWh/m^2 a$. Inland turbines achieve values of around half these, whereas turbines in low mountain ranges are significantly better, currently achieving values of a good 60 % of those quoted above. Since roughness and orography have a particularly great influence upon the wind conditions nearer to the ground, greater hub heights should be chosen for inland sites in comparison to coastal areas.

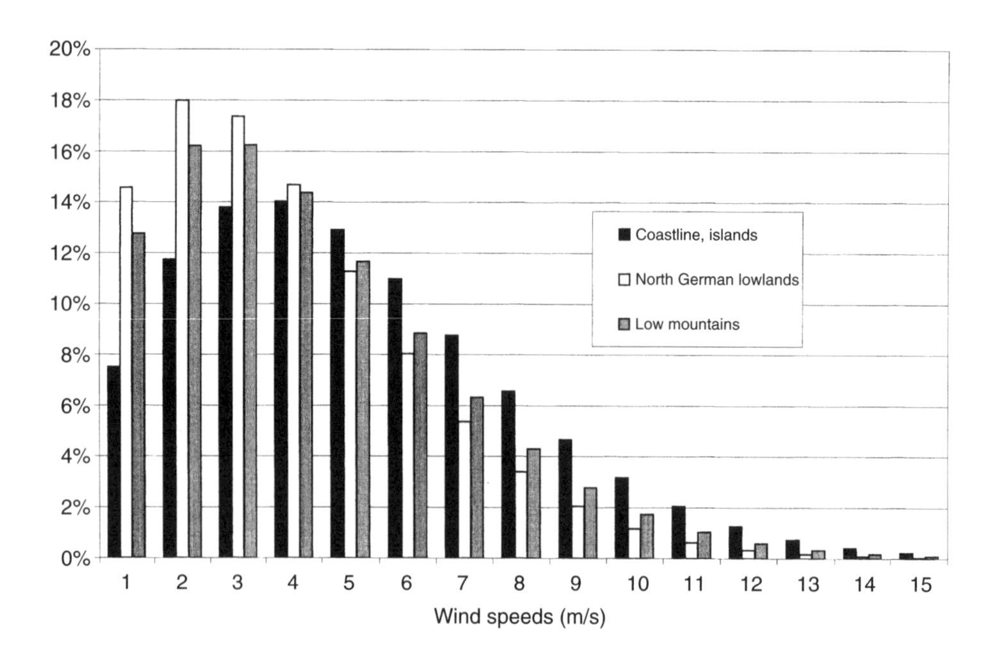

Figure 6.6 Wind-speed wind velocity distribution at various site categories (ISET)

In site-related economic considerations, the annual energy yield calculations can firstly be oriented towards the considerations of different turbine configurations and models (fixed or variable-speed systems, geared or gearless designs, etc.), regulation systems used (stall/pitch), tower heights, etc. Secondly, calculations for various turbine sizes can also be of great importance, e.g. in order to achieve the most economically favourable construction density for a site. Figure 6.7 aims to highlight the basic procedure on the basis of four relatively diverse turbine sizes, since this helps to illustrate the large differences between the individual power and yield values. As in Figure 6.4, based upon the wind-speed distribution at a height of 10 m (bottom left in Figure 6.7) the Weibull distribution of wind speed at hub height is determined for each turbine (not shown in the figure). Based upon the wind velocity distribution in question, we find the class and cumulative yields per year associated with the turbine outputs in question (top right in the figure).

Wind conditions at the site have a decisive influence upon energy yields. Figure 6.8 shows how wind conditions affect the annual yield of a 2 MW plant. Based upon the three frequency distributions of wind speed at hub height for coast, low mountain ranges and lowlands (bottom left in the figure) the annual class and cumulative yields (top right in the figure) are determined with the aid of the associated power values. These illustrate the great differences in the annual energy yields to be expected, with values varying from almost 1300 MW h in the lowlands to around 1700 MW h in low mountain ranges and approximately 2700 MW h in coastal areas.

6.1.4 Wind atlas methods

For locations for which no measurements are available, model calculation procedures have been developed that allow the potential of the wind to be estimated with a reasonable degree of precision. Such calculations can be performed using commercial programs on standard PCs. This process is based on the 'European Wind Atlas' compiled at the Danish Risø Research Centre on behalf of the European Union and the so-called 'Wind Atlas Analysis Application Programme (WASP)' [6.8], which uses the wind Atlas method (Figure 6.9). In this approach, specific measurements recorded over many years [6.9] are incrementally standardized – taking into account local conditions such as obstacles, surface roughness and orography – to standard environments (flat land, no obstacles, etc.) (left-hand side in Figure 6.9 with the arrow pointing upwards). Taken together, these data represent the European Wind Atlas, and reflects the regional wind conditions disregarding landscape influences. One hundred and seven sites have as yet been recorded throughout the whole of Germany.

In order to calculate the conditions at the site under consideration, we now move in the opposite direction (see the right-hand side of Figure 6.9 with the arrow pointing downwards). Regional statistics are included in the WASP programme using local parameters. Wind climatological factors such as the structure of the landscape (see Figure 6.10), surface texture or roughness and obstacles at the site are particularly important. Shadowing losses in wind farms can also be taken into account.

The WASP programme was developed for use in areas without complex orography. It therefore provides reliable information about local wind conditions for the analysis of areas in coastal regions. However, in highly structured landscapes inland and in low mountain ranges, such calculation procedures are of limited use. However, more complicated methods, e.g. the so-called Mesoskala model which takes into account factors like jet effects in the

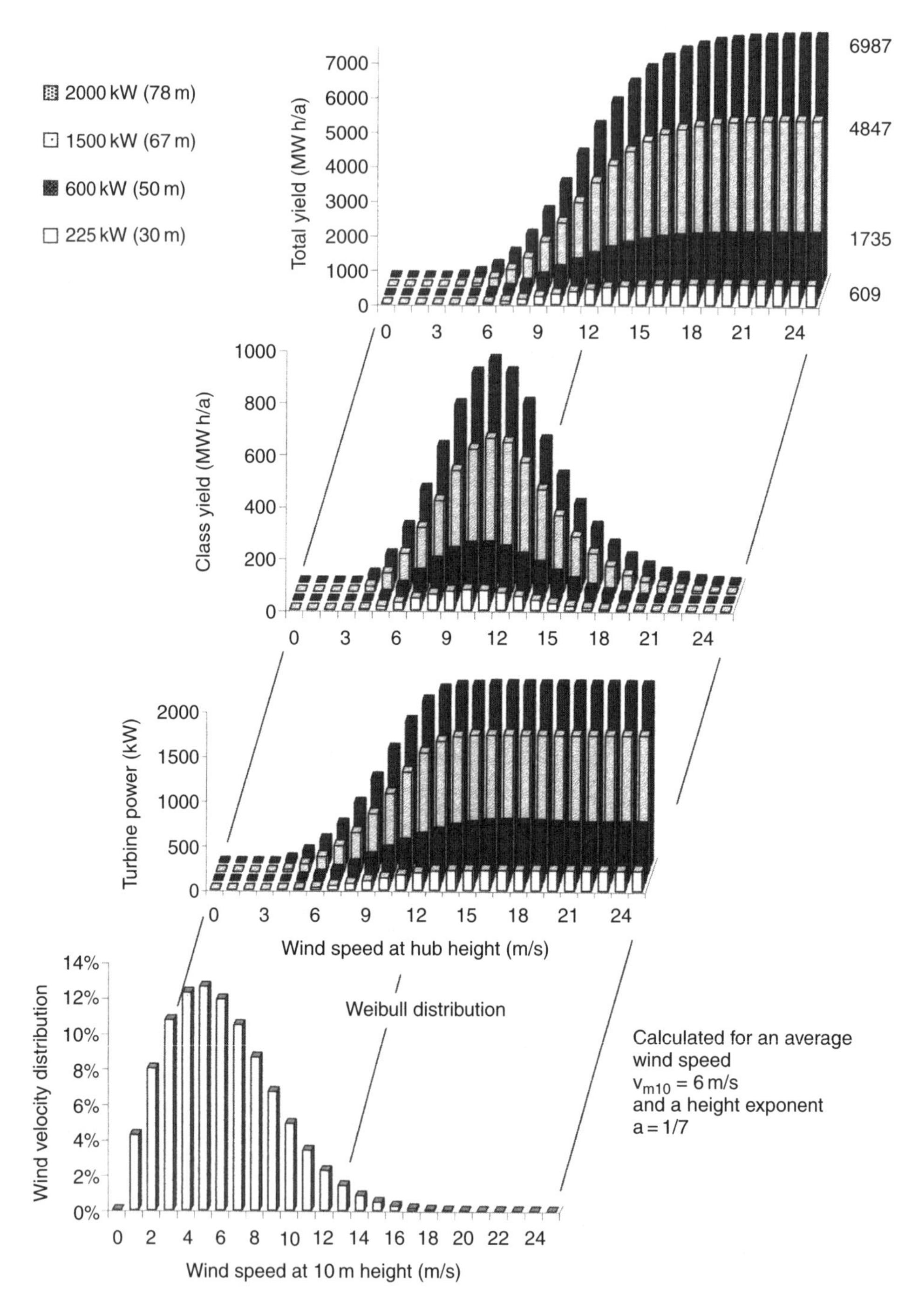

Figure 6.7 Calculation of the annual energy yield on the basis of wind measurements at a height of 10 m for various turbine sizes

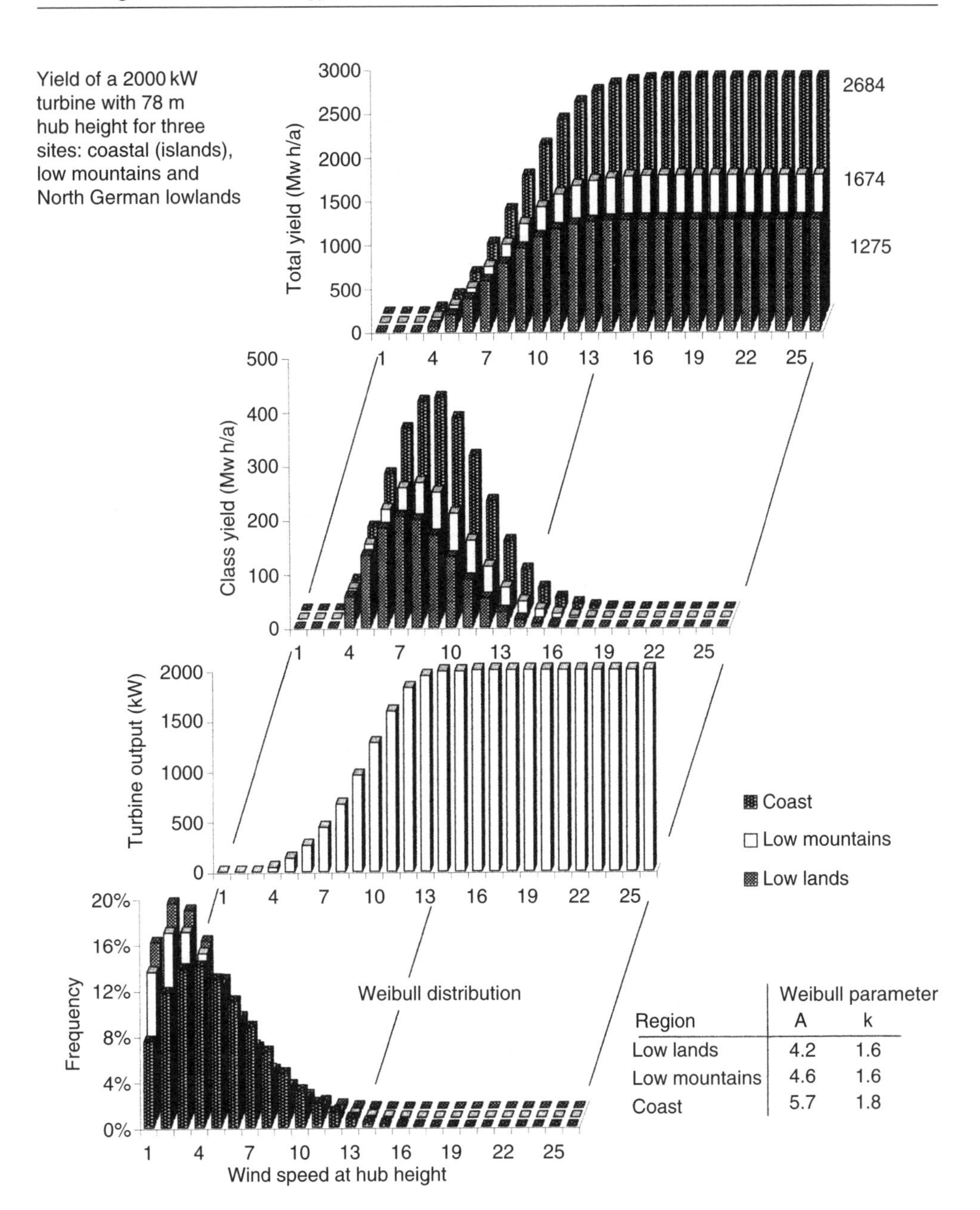

Figure 6.8 Calculation of annual energy yields of a 2 MW turbine at different sites in Germany

landscape, do supply relatively good predictions even for complex landscape structures, but they are much more expensive.

The suitability of the landscape along with wind speed plays a decisive role for wind energy utilization. In coastal areas, locations directly next to the water are preferable. Surface roughness is lowest in such areas. Turbines installed at a distance of 5 km from the coastline

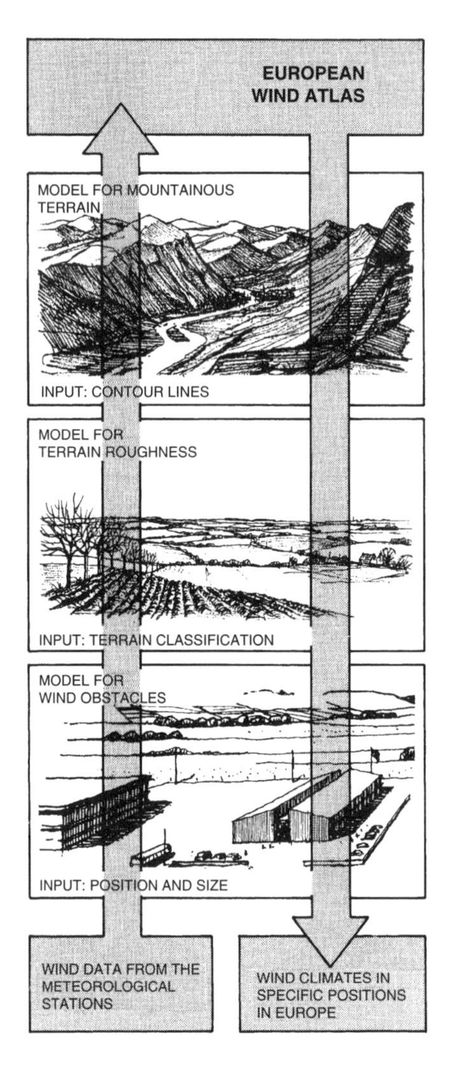

Figure 6.9 Procedure for resource analysis according to the wind atlas method. Reproduced by permission of National Laboratory Risø

achieve significantly lower energy yields than turbines located directly on the coast. Inland, exposed locations are of particular interest. Elevated plains and mountain ranges, with as little woodland as possible, and in which the wind can flow freely in the most common wind direction (usually south-west in Germany), are the preferred locations. There should be no other hills or obstacles in the immediate vicinity. It is difficult to obtain planning permission for sites close to nature reserves or national parks and buildings or towns. The distance from the supply grid should be kept as short as possible for cost reasons. Other factors that should be particularly taken into consideration in the selection of a location – again primarily for cost reasons – are land ownership, roads that are available or under construction and the solidity of the building land.

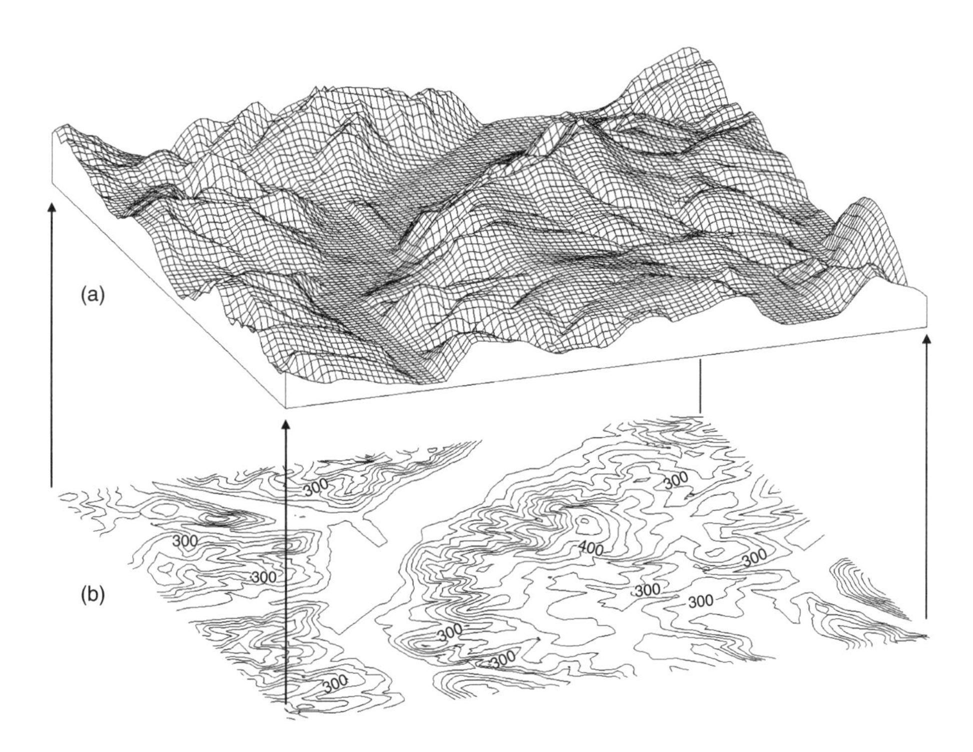

Figure 6.10 Surface structure (orography) of (a) a location based upon (b) map contours

6.2 Potential and Expansion

In the context of wind energy exploitation, we must differentiate between the site and economic potential. The former covers the conditions for the erection of a turbine, which depend upon meteorology, topography, buildings and conditions relating to the permissible operating methods. Economic potential, on the other hand, relates to those locations that offer the possibility of profitable operation under the prevailing economic conditions for energy. In Germany, the coastal locations dominate this category.

Some capacity estimates have been made since as early as the mid-1960s for the generation of electricity from wind energy [6.10–6.16], in particular for mainland Germany. These yielded very different results. Designs for Lower Saxony [6.17] and Schleswig-Holstein [6.18] based on site analyses from approximately 10 years ago also led to different expected capacities. Elimination criteria played an important role here. Today, conservative estimates are sometimes significantly exceeded [6.2, 6.19–6.21].

Figure 6.11 shows the growth in installed wind turbine power over the past few years in those countries that are the most advanced worldwide. It is noteworthy here that large countries with excellent wind conditions, which thus have the greatest technical and economic potential, are not among the front-runners. The energy policy of the country plays a decisive role.

Figure 6.11 shows that there was no noteworthy increase in the United States for the 10 years after the first great wind power boom in California in the mid-1980s. Only in recent years did larger new installations get underway again. Among the worldwide wind turbine

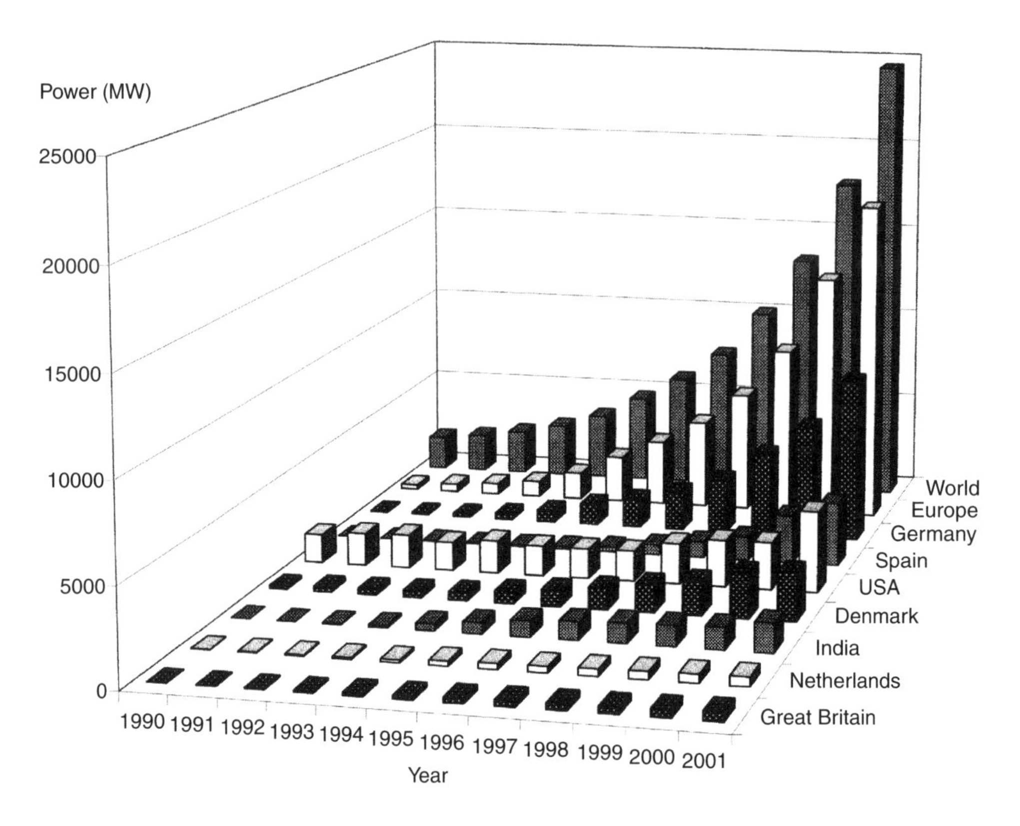

Figure 6.11 Development of the installed wind turbine power worldwide, Europe-wide, and in the most important countries (BWE, EWEA data, etc.)

output of around 25 GW, more than two-thirds is generated in Europe and, of this, approximately half is generated in Germany. A further large sales market is currently developing in Spain.

A significant expansion of energy capacity on the mainland can be achieved by the development and installation of wind turbines in coastal waters (offshore). Investigations into this field were being carried out as early as the beginning of the 1980s [6.22, 6.23] and both worldwide [6.24] and country-specific (e.g. reference [6.25]) studies on potential are still being continued. Further studies have been conducted throughout Europe and Germany [6.26, 6.27] and indicate technical feasibility, but place economic considerations in the foreground.

Large areas that encroach upon traffic routes, military areas, pipelines and sea cables, as well as national parks, have to be ruled out. Only those areas that are at least 30 km from the coast are termed 'offshore'. However, at these distances the water depth in the North Sea is already 15 to 30 m. The associated high costs for the building of foundations and the much higher costs for grid connection and maintenance in comparison to coastal or inland operation must be balanced by higher yields and larger farms. However, as yet insufficient experience has been gained regarding flow [6.28], waves, ice, etc., and there is a lack of wind data for heights of approximately 60 to 100 m over the surface of the water. Even countries such as Denmark [6.29], the Netherlands [6.30] or Sweden [6.31–6.33] (Figures 6.12, 6.13 and 6.14) only have experience regarding coastal waters. Due to their visibility from the shore, in the future such locations will only come into consideration to a limited degree for large wind farms. Since offshore plants will reach the order of magnitude of 5 MW, there is still (as previously inland) significant potential for cost-reduction. The first projects are expected to be realized in 2003 to 2004. Optimistic predictions [6.34] assume offshore wind power exploitation of up to 25 GW by 2030 in Germany.

All estimates show that despite the currently developed areas wind power still has a significant potential for expansion. Current levels of growth will be maintained, particularly as a result of the increasing size of individual turbines, the replacement of old, smaller turbines by new larger ones (so-called repowering) and the offshore construction that will start in a few years time.

Schleswig-Holstein in Germany has, since the year 2000, covered 20 % of its net power consumption by wind power. This value was the target set for 2005 in the mid-1990s. According to the Bundesverband Windenergie (Federal Association for Wind Power) 3.5 % of the net electricity requirement throughout Germany was generated by wind power in 2001.

Wind power can thus contribute significantly to the goal declared in the 1997 EC White Book [6.35] of doubling the proportion of the gross primary energy requirement generated by renewable sources from 6 to 12 % by 2010.

Figure 6.12 Turbine rotor transport (1.5 MW, GE/Tacke). Reproduced by permission of GE Wind Energy

Figure 6.13 Turbines in the Utgrunden wind farm. Reproduced by permission of GE Wind Energy

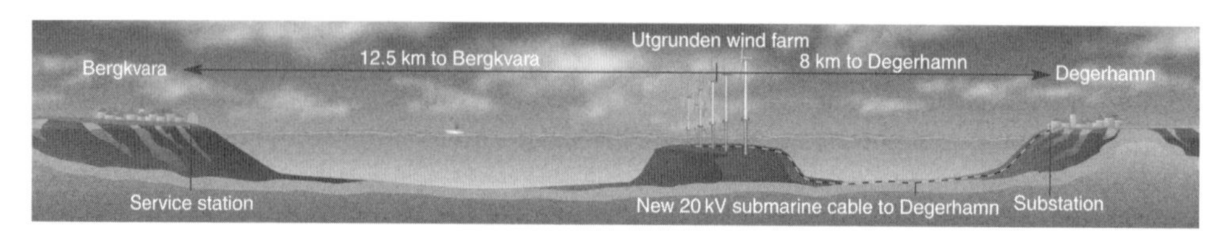

Figure 6.14 Height profile of the Utgrunden installation site in Sweden (GE/Tacke). Reproduced by permission of GE Wind Energy

6.3 Economic Considerations

Renewable energy systems can only compete with conventional installations in the long term if their use brings about significant advantages in the

- technical,
- political,
- economic,
- labour-market and
- ecological

fields. Such wide-ranging improvements cannot, however, be achieved in all these aspects compared with new and established systems. Partial advantages, on the other hand, are certainly possible. Particular importance must therefore be assigned to the weighting of individual aspects. In the future, in addition to operational economic considerations, social prospects for the future and ecological effects will increase in importance in this weighting. Special consideration should be given to the fact that the wind energy industry is relatively labour-intensive in comparison to the conventional power sector, as shown in Figure 6.15. This shows that for every direct job approximately another two indirect jobs are created in supplying industries. Danish manufacturers also contribute to this by importing a significant

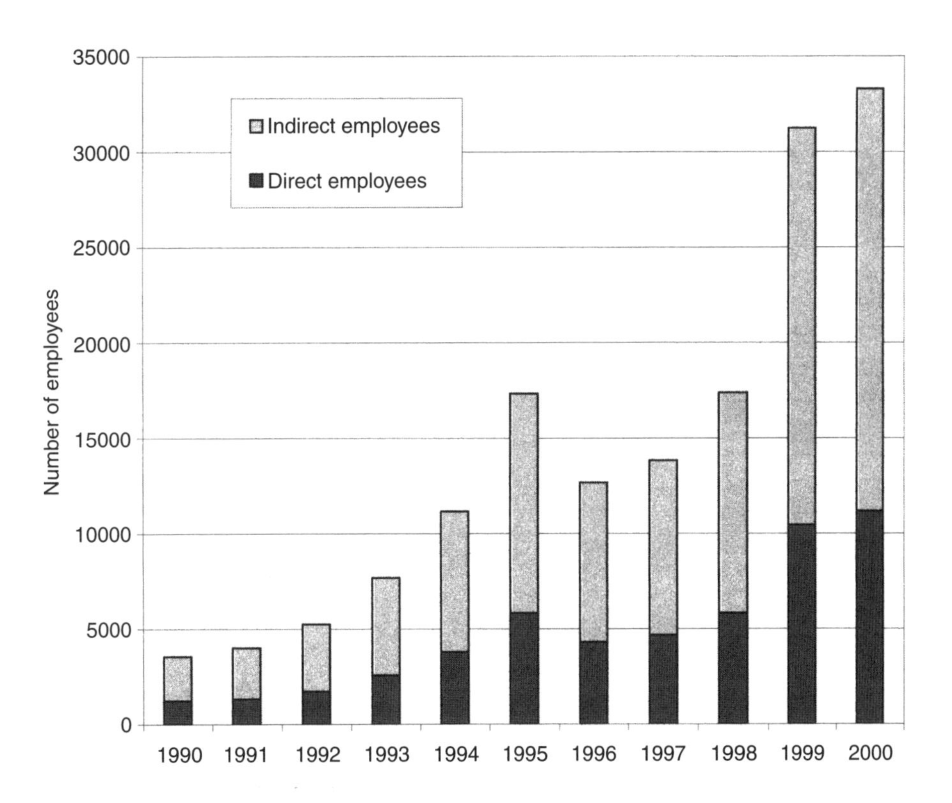

Figure 6.15 Growth in direct and indirect jobs as a result of the wind power industry (DEWI)

quantity of German components (gearboxes, generators, etc.). Moreover, the disturbance of the landscape will have to be given more consideration with regard to its ecological effects, such as the effects associated with brown-coal power stations with their mines and open-cast workings compared with wind and solar farms. From this point of view, energy sources that are still considered to be uneconomic today could bring many social and ecological advantages in the short, medium and long term.

These aspects are evaluated very differently in different countries. Accordingly, there are also large differences in the evaluation and payment systems for electrical energy from conventional installations with fossil or nuclear conversion units and renewable systems. A unified tariff system within Europe or internationally cannot be expected in the foreseeable future.

6.3.1 Purchase and maintenance costs

Turbine costs (reference values are listed in Figures 2.78 to 2.80) can be determined after the compilation of a comprehensive profile of requirements and the definition of usage details, including technical turbine data and predicted costs for maintenance and repairs by the manufacturer or supplier. It is relevant to the cost calculation whether the converter will be used in isolated operation or combined with other power generation units (in grids, with diesel generators, etc.) and what demands will be made of the safety-related component groups (e.g. for installation in water protection areas). Costs for local transport, foundation, cabling and connection should also be taken into account. The total secondary costs lie somewhere between 15–30 % of the cost of the turbine itself. If investment cost subsidies are granted in the framework of promotional measures, these can be drawn into the equation as a reduction of the acquisition costs.

Figure 6.16 shows the average annual operating costs for a turbine, the guarantee period of which has already expired. In addition to insurance, rental, remote monitoring, etc., these costs also include the average values for maintenance and servicing, which are shown in Figure 6.17 with an appropriate breakdown.

Figures 6.16 and 6.17 show that turbines of around 1 MW exhibit the lowest operating costs at approximately 10 euro per kW installed power per year and that maintenace, servicing and package contracts are also the cheapest for this class at 4 euro per kW per year. In turbines of less than 300 kW these figures are more than twice as high.

6.3.2 Power supply and financial yields

In Germany, the ‘Energiewirtschaftsgesetz’ (Act for the Promotion of the Fuel and Electricity Industries) obliges the operator of an electricity-generating wind turbine to inform the responsible power supply utility. In addition to technical requirements to preclude the possibility of damage to the public supply grid, the provisions of the VDE (Verband Deutscher Elektrotechniker, or Association of German Electrical Engineers) must be adhered to. The technical connection conditions can differ from one supply area to the next. The precise matching of the protective devices and instrumentation should be performed in consultation with the power supply utility.

Financial yield – found by taking the product of the energy supplied and the tariff at which it is sold – is of fundamental importance for economic operation and thus also for

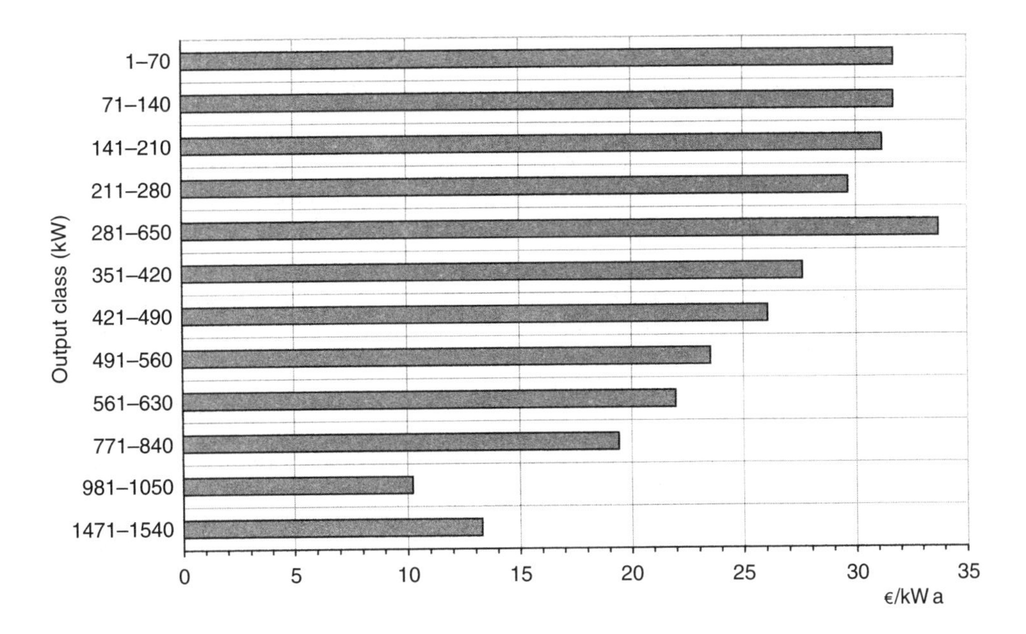

Figure 6.16 Average annual operating costs for wind turbines (ISET)

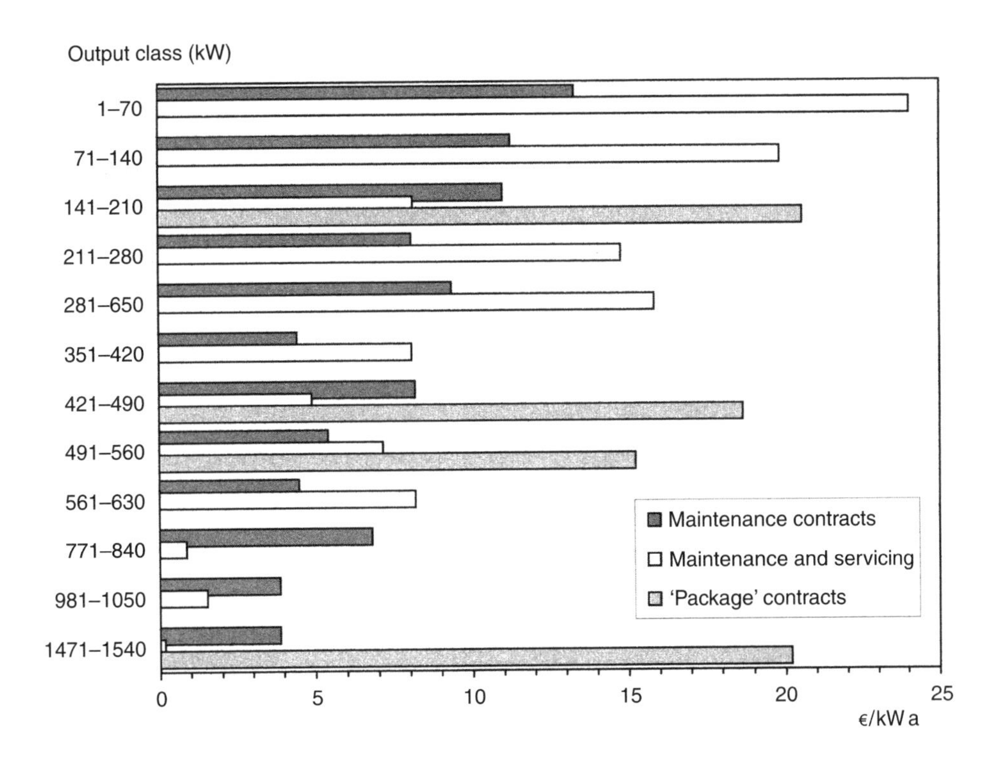

Figure 6.17 Maintenance and servicing costs of wind turbines as a yearly average (ISET)

the long-term use of power supply plants. This aspect will be considered further in what follows.

The 'Erneuerbare Energien Gesetz (EEG)' (Renewable Energy Act) was introduced in Germany in the year 2000. This replaced the 'Stromeinspeisungsgesetz' (Electricity Supply Act) that had been in force since 1991. The EEG sets the minimum prices that the power supply utility has to pay for the supply of the power from renewable energy. The tariff for electricity from wind turbines that were put into service before the end of 2001 was at least 9.1 eurocent per kW h, for a period of five years calculated from the time of commissioning onwards. Thereafter the tariff for turbines that in this time have achieved 150 % of the yield calculated for the reference turbine according to the annex to the law is 6.2 eurocents per kW h. At sites that do not achieve the reference yields, the period of higher remuneration is extended by two months for every 0.75 % that the yield is below the reference yield.

Spain is also currently using a minimum price arrangement. Other countries are following different routes, but despite the fact that these countries often have a greater exploitable potential than Germany these approaches have not led to such large increases in wind power generation as the German minimum price system. Great Britain has carried out tendering rounds in connection with its 'Non-Fossil Fuel Obligation' (NFFO). Denmark has recently gone over from a minimum price system to a certificate trade in green power, which has led to a sharp decrease in growth (see Figure 6.11).

These descriptions show that the systems of remuneration for wind-generated electricity are very differently managed, even across Western Europe. Investigations regarding the achievable yields were carried out for differently sized turbines at selected sites or the corresponding wind conditions for a few Western European countries under the applicable conditions [6.36]. These studies used yield data for different wind turbines at different times and locations determined by the '250 MW Wind Scientific Measurement and Evaluation Programme', so that the labour and service components of the payment in the countries under consideration could be accounted for according to the currently applicable conditions. Considerations thus relate to wind conditions at German sites. Under the conditions mentioned, the investigations showed that Switzerland, Luxembourg and Germany pay the highest tariffs, while Sweden, France and Ireland pay the lowest. Results from the large-scale testing programme (WMEP) show the dependency of the monetary annual yields upon the turbine size and in particular upon the site category in Germany. In order to be able to undertake a comparison between different sizes, etc., the monetary yields for the year 2000 were related to the installed turbine output. Figure 6.18 shows, for example, that for turbines of the 50 kW class the yields amount to around 180 euros per year and per kW installed power in the low mountain ranges and in the North German lowlands or 320 euros per kW a at the coast. Turbines of 1.5 kW, on the other hand, achieve annual yields from approximately 250 to 500 euros/kW a.

The annual financial yields from wind turbines and wind farms can be estimated from the specific values or calculated from the energy yields and tariffs. They are balanced by the electricity generation costs, which will be considered in what follows.

6.3.3 *Electricity generation costs*

The specific electricity generation costs in eurocents per kilowatt hour are shown in Figure 6.19 for the market-dominating 0.5 to 1.5 MW turbines as a function of the annual

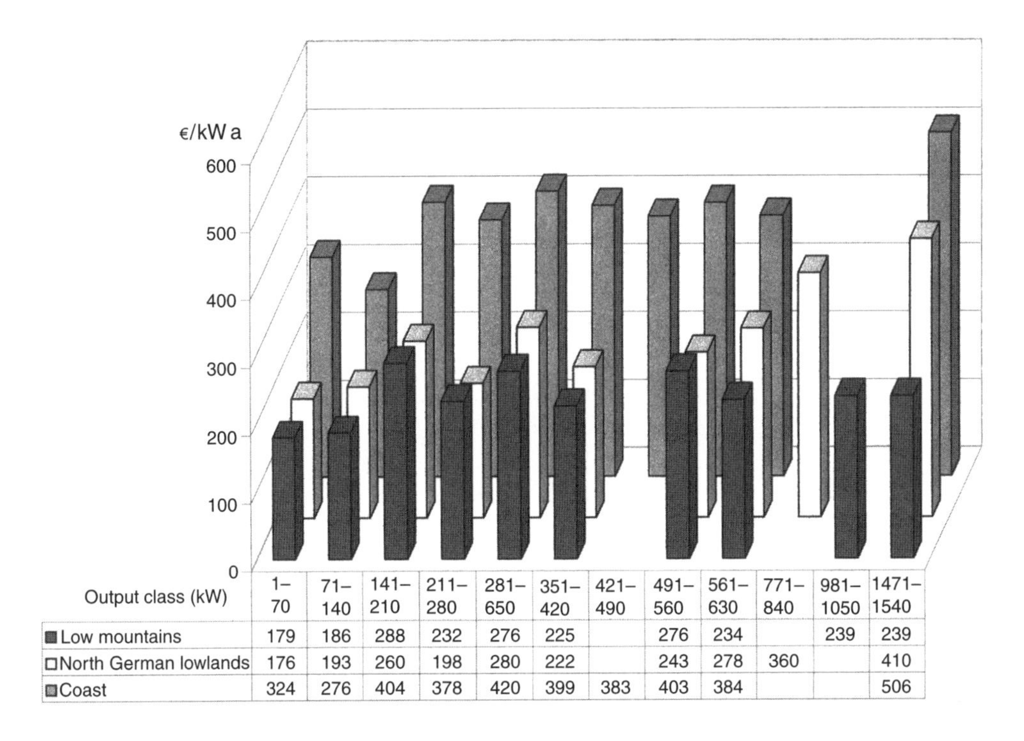

Output class (kW)	1–70	71–140	141–210	211–280	281–650	351–420	421–490	491–560	561–630	771–840	981–1050	1471–1540
Low mountains	179	186	288	232	276	225		276	234		239	239
North German lowlands	176	193	260	198	280	222		243	278	360		410
Coast	324	276	404	378	420	399	383	403	384			506

Figure 6.18 Annual monetary yields per kW installed power (ISET)

energy yields. The preconditions upon which the turbine costs of between 800 and 900 euros per kilowatt are based, i.e. additional investment costs, operating costs, financing and timeframe or interest rate, are individually listed in Figure 6.19 and the supply tariff framework is also identified. The representation shows that 1 MW turbines currently achieve the lowest specific costs and even at 1800 full-load hours these are less than the supply tariff. Turbines of 0.5 and 1.5 MW do not achieve this until they have been in use for around 2000 full-load hours.

The specific electricity generation costs for an extended spectrum of turbine sizes from 150 to 1500 kW nominal power are shown in Figure 6.20. These are based upon the framework conditions in accordance with Figure 6.19. Figure 6.20 illustrates that 150 kW turbines require around 2300 full-load hours, 300 and 1500 kW turbines require approximately 2000 full-load hours, 500 kW turbines require more than 1800 full-load hours and 1000 kW systems require less than 1800 full-load hours of use before their generation costs are less than the supply tariff, thus achieving economical use.

6.3.4 *Commercial calculation methods*

Static and dynamic calculation methods can be used to assess the economic viability of wind turbines. In the annuity method, returns and costs are assumed to remain constant (static) over the entire depreciation period. In contrast to this, the capital value method

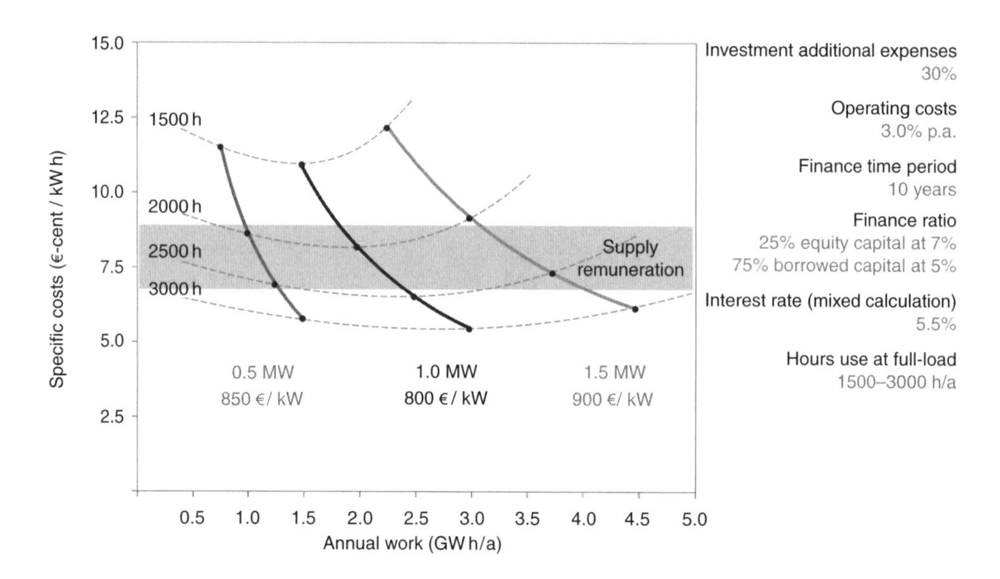

Figure 6.19 Specific electricity generation costs for wind turbines as a function of work done per year with full-load hours as a parameter

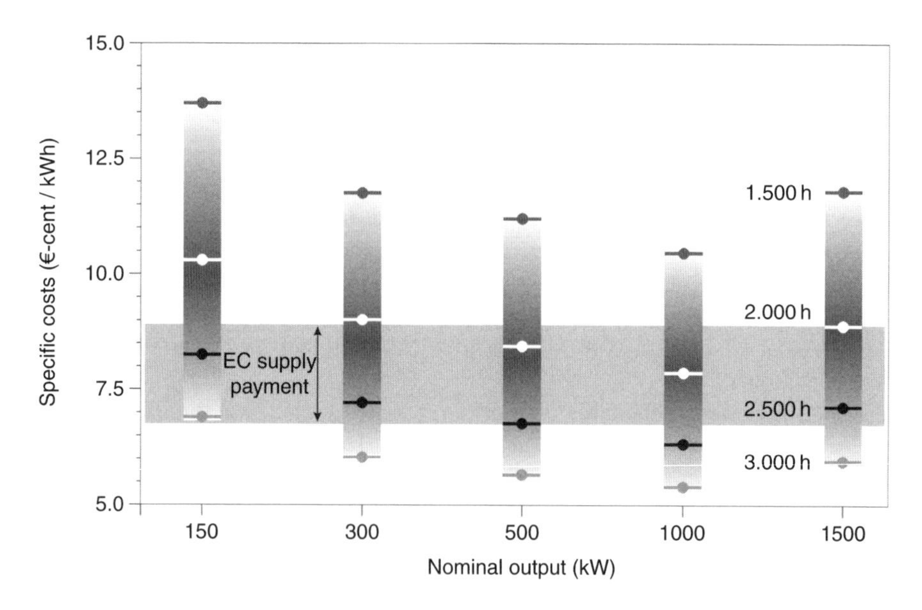

Figure 6.20 Specific electricity generation costs as a function of the turbine nominal power with number of full-load hours as a parameter (ISET)

takes into account the loss of value of the loan due to inflation and increasing returns due to increasing remuneration for supply. Moreover, promotion schemes in the European Community, Germany and individual regions play an important role in questions of economic viability.

6.3.4.1 The annuity method

The annuity reflects the annual percentage of interest and repayments for externally financed loans. Capital costs can be easily determined from the relationship

$$K = p + \frac{p}{\left(1 + \frac{p}{100}\right)^z - 1}$$

where

p is the interest rate in %
z is the repayment term in years and
K is the annuity in %

Annual operating costs, capital costs and taxes must be included in the calculation. The operating and additional costs incorporated should be approximately 1.25 % for maintenance, approximately 0.9 % for insurance and around 0.4 to 0.6 % for percentage excess and other costs. For a 10 year duration and an interest rate between 6 and 8 %, an annuity of 13.5 to 15 % can be expected.

6.3.4.2 The capital value method

Dynamic calculations using the capital value method can provide a long-term point of view in the financial evaluation of wind turbines. The starting point is the equation

$$C_0 = \sum_{i=1}^{n} \left(\frac{1+r}{1+p}\right)^i (E_i - K_i) - I_0$$

where

C_0 is the capital value
p is the rate of interest
n is the duration
r is the rate of inflation
r_f is the real increase in energy cost
i is the year
k_f is the energy costs
K_i $f(r_b)$ costs in year i
I_0 is the invested capital
E_i $= E_0 \times k_f(1 + r_f)^i \times \gamma$ is the return for energy generated (in year i)
r_b is the percentage of I_0 for maintenance and servicing
γ is the technical availability

Iterative solution methods permit, for example, the determination of the redemption period A_z, i.e. the year i in which the capital value $C_0 = 0$. The required energy costs k_f in the base year can be calculated for the redemption period. Figure 6.21 shows a graphical representation of the relationships described by the equation. In order to be able to work with the outer areas of the nomogram, a scaling factor m can be introduced, which must have the same value both for turbine costs and for the supply of energy. The nomogram is based on a constant proportion for maintenance and service, $r_b = 2.5$ % of investment costs. Different results can be read off from the nomogram in the selected parameter ranges.

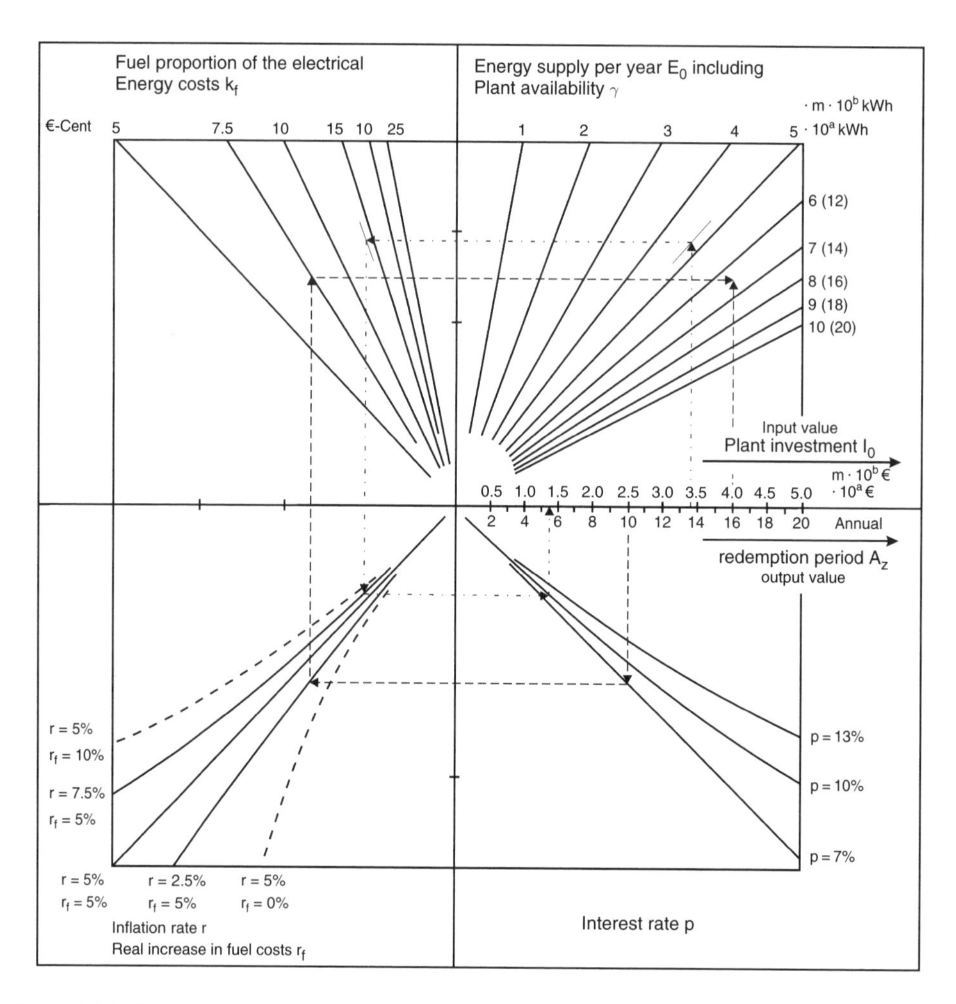

Figure 6.21 Nomogram for the determination of the economic viability of wind with a and b as exponents and m as the scaling factor [6.36]

The redemption period A_z can be determined for the use of a wind turbine in isolated operation, as a fuel saver or in parallel with the grid. Based on a turbine investment of 34000euro $= 3.4 \times 10^4$ euro for a 20 kW turbine, the energy supply of 48000 kW h $=$ 4.8×10^4 kW h per year and the electrical energy costs or the feed tariff $k_f = 14$ eurocents per kW h is drawn orthogonally to the axes and extended into a polygon via the inflation rate (7.5 %) in connection with the real increase in fuel costs (5 %) to the applicable rate of interest (10 %). Thus the redemption period of $5\frac{1}{2}$ years can be read off the initial axis. In Figure 6.21 this procedure is shown in the form of a dotted line.

The energy costs or the payment for supply k_f necessary for economical operation can be determined, firstly, by starting from the turbine investment and energy supply. Secondly, the redemption period A_z can be taken as a basis, and the straight lines are connected via the applicable rate of interest and the rate of inflation or fuel cost increase at the intersection point. The intersection point of the two polygon lines permits an interpolation, which gives a minimum tariff for supplied electricity at which economical turbine operation is ensured.

There are further options for determining the

- maximum investment,
- minimum energy returns,
- highest interest rate and
- the associated rate of inflation and fuel cost increase.

We will now show, as an example, a further estimation option for 1.5 MW turbines, which are currently the class with the highest sales (the dotted line in Figure 6.21). The total investment is assumed to be around 1.6 million euro – $4 \times 4 \times 10^5$ euro. Thus the scaling factor m is 4 and the exponent b is 5. Assuming a 10 year redemption period, an interest rate of 7 %, a rate of inflation of 2.5 % and a real increase in the cost of fuel of 5 %, the turbine must supply at least $4 \times 6.5 \times 10^5$ kW h, i.e. 2600 megawatt-hours of energy, from the wind at a tariff of 7.5 cents per kilowatt-hour. These yields could be achieved at an average annual wind speed of around 5 m/s.

Economic operation of wind turbines is only possible if a redemption period can be achieved that is less than the service life of the turbine. Manufacturers cite values of 20 to 30 years for service life. If possible, redemption periods of 10 years should not be exceeded.

Serious deviations in energy costs in Danish and US discussions on the economic viability of wind energy, where values of, for example, 6 eurocent/kW h and 5 US cents/kW h are assumed, can be mainly attributed to the redemption periods set there of 20 years.

6.4 Legal Aspects and the Installation of Turbines

The densely populated nature of Germany and the enormous growth in the use of wind power over recent years mean that it is crucial to harmonize the interests and objectives of turbine planners and those living in the vicinity of the planned site. The main considerations here are the casting of shadows, the propagation of noise and the long-distance visibility of the now usually very large turbines. Beneficial effects of development have been that rotor speeds have fallen continuously as turbine diameters have risen and that large turbines give a much more peaceful impression than the older, smaller types in the 100 kW class.

The realization of the EU directive on environmental testing [6.37] should lead to the above-mentioned problem areas being countered and the interests of the environment being taken into account. Formal environmental compatibility tests are required for groups of more than 20 turbines. For groups of up to 19 turbines there must be a general preliminary test to determine the necessity for an environmental compatibility test. For three to six turbines only a site-specific preliminary test is required. The precise implementation of the environmental compatibility test for wind turbines has not yet been set down in the statutes in Germany. Therefore the regulations vary between different regions and often between individual communities. Schleswig-Holstein, however, is taking the lead in harmonizing the regulations [6.38]. Even without unified regulations, many planners have calculations of the noise emissions, shadow casting and the visibility of their turbines performed in order to rule out from the outset conflicts with those living nearby.

6.4.1 Immission protection

In Germany, the Bundesemissionsgesetz (Federal Emission Act) forms the legal basis for emission testing [6.39]. Where wind turbines are planned close to residential areas, the main considerations are noise emissions and the casting of shadows by the turbines. The previously much-discussed disco effect, caused in particular by reflective coatings on the rotor blades, has now been countered by the use of further-developed matt coatings.

6.4.1.1 Noise propagation

The propagation and effects of noise are broken down into the following categories:

- emission [6.40],
- transmission [6.41] and
- immission [6.42].

The DIN and ISO standards mentioned and the VDI directives deal with the category in question. Further standards and directives are concerned with the minimization of noise emissions [6.43].

The effects of noise are ultimately judged on the basis of immissions. Values for immissions are given in the Technischen Anleitung Lärm (TA Lärm) (Technical Directive on Noise Control) [6.44] and in the Baunutzungsverordnung (Ordinance on the Use of Buildings) [6.45]. Furthermore, an immission protection ranking is specified in this, adherence to which is evaluated by the immission protection authorities as part of the Gewerbeaufsichtsamtes (Trade Supervisory Agency) or the Umweltamtes (Environment Agency).

For night-time hours, when stricter measures are specified, the following maximum noise levels apply:

- 35 dB(A) for purely residential, leisure or health resorts;
- 40 dB(A) for general residential areas and small estates (primarily homes);
- 45 dB(A) for central, mixed and village areas where no usage type predominates;
- 50 dB(A) for industrial areas (primarily industrial plant).

The logarithmic dB(A) scale reflects to some degree the specific noise-sensitivity of the human ear. If individual values are exceeded, but not the limit values for the day, an order may be imposed for the shutting down of the turbine overnight. However, this would result in a severe sacrifice in terms of returns.

In order to ascertain the level of noise emissions with a reasonable level of certainty in the planning phase, commercially available programmes are now used that permit a relatively precise prediction of the risk areas. To this end, the orography of the surrounding area, as described in Section 6.1.2, is first recorded and relevant immission areas, such as estates, mapped out together with their limit values. The wind turbine is entered at the planned site and the noise immission calculated according to the standards [6.40–6.44]. The emission of the wind turbine is determined by the manufacturer according to a standardized procedure [6.46] and calculated at the wind speeds available at the site. Tones and pulses subjectively perceived as annoying, such as those caused by gearboxes, are evaluated by adding a supplement to the noise level. Existing noise levels due to roads, etc., are also taken into account, as are reflections and absorption, for example, due to the ground, the air or obstacles.

The result of such a calculation is a map as shown in Figure 6.22, in which along with the wind turbine and the relevant immission areas the ISO lines of the above-mentioned noise limit values are drawn in. Vibrations in the ground give rise to so-called 'infrasound', which can now be classified as harmless [6.48, 6.49].

6.4.1.2 Shadow casting

Wind turbines are usually installed at exposed sites, e.g. on a hill. As a result they cast long shadows. This effect can be precisely determined by means of standard programs [6.50].

To determine the shadows cast, the orography data that are available from energy and noise predictions are drawn into the calculation along with the path of the sun (which can be determined by the geographic coordinates) and the turbine dimensions. This information is then represented in the form of a map (Figure 6.23). The characteristic path of the areas in shadow usually has a butterfly-like shape, which is caused in particular by the low positions of the sun in the morning and evening.

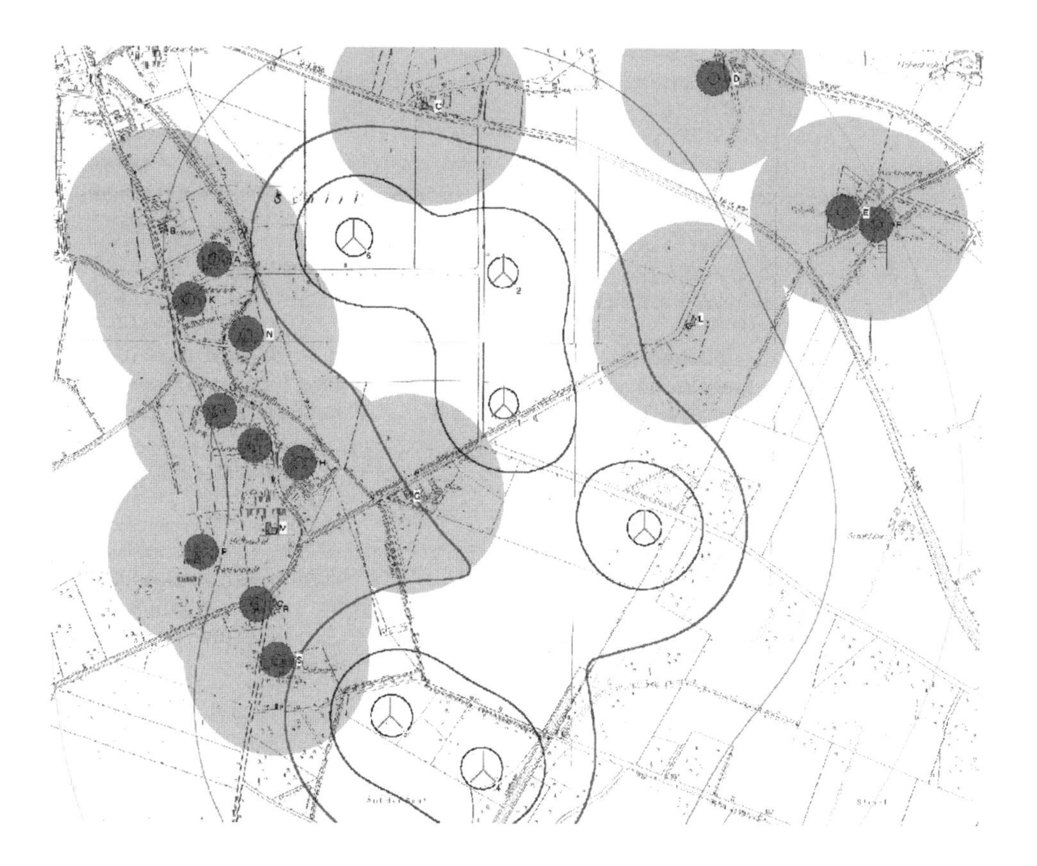

Figure 6.22 Noise immission map for wind turbines (drawn up using the forecasting program [6.47])

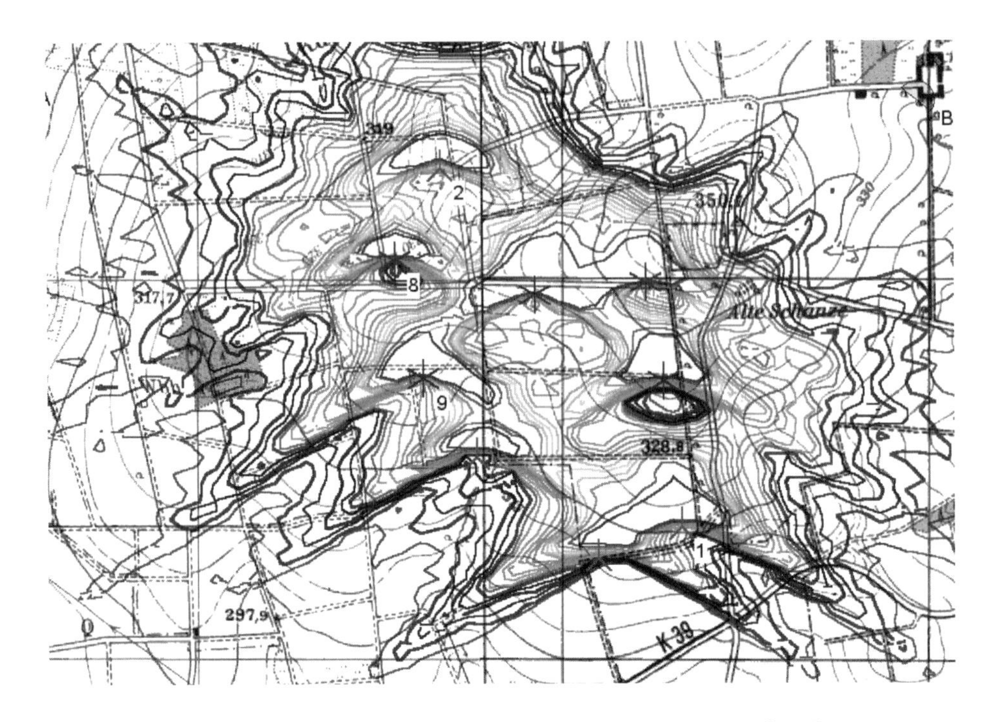

Figure 6.23 Shadowed area of a wind farm of eight turbines [6.50]

The calculations can be based upon the worst case for the time of day using the assumption of permanent sunshine. This yields the maximum possible shadow casting periods. For a more realistic estimation, meteorological data on the statistical sunshine hours at the site in question is drawn upon. The prediction can be refined still further by including turbine stationary times. Intelligent turbine control systems can position the rotor with one blade downwards during the stationary period so that the total height of the turbine including rotor, and thus the shadow casting, is minimized. If predetermined limit or guide values are exceeded, the turbine can be switched off at critical times of the day.

Since rotating turbines of this order of magnitude, which can exceed 150 m blade tip height, are new to a field of building regulations that usually relates to buildings, statutory guidelines and limit values have yet to be revised. Up until now, limit values of 30 hours per year or 30 minutes per day have normally been used in worst-case calculations. Furthermore, the criterion developed at the Technical University of Kiel, according to which a brightness change with less than 20 % coverage of the visible disc of the sun by the rotor blade is not subjectively perceived, is also used. Thus the distance of the immission point from the turbine takes on increasing importance.

6.4.2 Nature and landscape conservation

In Germany, in contrast to many other countries, nature and landscape conservation is assigned particular importance. In this connection, disruption of the landscape due to visibility of turbines and of the migration and breeding of birds must be considered as well as effects upon the ecology of land and water animals, plants and the water balance (pollution caused by hydraulic and gear oil, etc.). Often compensatory landscaping measures are required. The following brief descriptions should, however, be limited to the two first and last areas mentioned.

Wind turbines, due to their size, represent a disruption to the landscape, although in terms of time this is limited due to the service life of approximately 20 years.

6.4.2.1 Visibility

First of all, the question of where the planned wind turbines are actually visible from must be answered during the planning phase. In addition, computers are used to check whether the view to the topmost point of a turbine, the rotor tip, is interrupted by landscape elements (mountains) and obstacles (buildings). A visibility map drawn up in this manner shows both visibility and shadowing areas around the turbine. In the case of wind farms, the number of turbines visible at the location in question is determined. The evaluation of this map is, however, subject to subjective points of view. Figure 6.24 shows an example of such a visibility map in which the shadowing of individual obstacles is indicated by means of dark specks.

The visual impression of a planned wind turbine can be illustrated with the aid of commercial photo-processing programs [6.52], as shown in Figure 6.25. At this point three-dimensional models of the planned turbines are incorporated into a photo and visually displayed. Furthermore, the simulated rotor movement can also be considered on the computer. Visualizations are particularly common in public presentations of projects.

6.4.2.2 Compensatory measures

In Germany, if the landscape is disrupted compensatory or replacement measures must be put in place. To this end, a survey is drawn up in which the environment surrounding

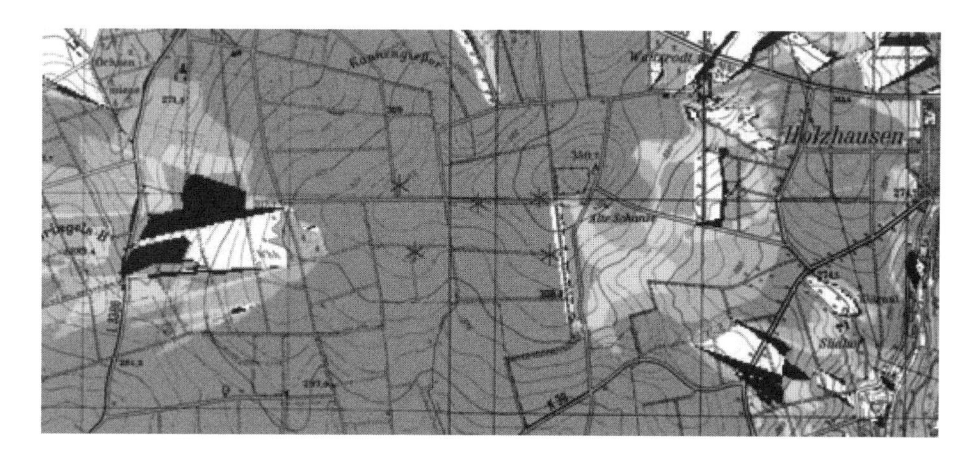

Figure 6.24 Results map of a visibility analysis [6.51] (visibility shadows of obstacles are shown as dark specks)

Figure 6.25 Visualization of a wind farm showing existing turbines with planned turbines superimposed with the aid of a computer [6.52]

the wind turbines are first of all divided into so-called landscape-aesthetic units, which are then classified according to their value. This is then used to evaluate the intensity and importance of the disruption for each unit. Additional factors, such as perception coefficients and compensation area values ultimately determine the size of the compensation area that must be upgraded in aesthetic terms by special measures such as tree planting. Several procedures have been developed for this calculation, e.g. the landscape compensation method by Dr W. Nohl.

6.4.3 Building laws

With regard to building laws and the planning of wind turbines, the Baugesetzbuch (BauGB) (Building Code) [6.53] and Baunuzungsverordnung (BauNVO) (Ordinance on the Use of Buildings) [6.54] must be observed within the jurisdiction of Germany, and the relevant Landesbauverordnungen (LBO) (Regional Building Ordinances) must be observed within the jurisdiction of the German regions.

In Germany, basic requirements concerning planning permission for turbine sites are regulated on a national level. According to the German Baugesetzbuch:

1. A general development plan must be submitted, with a construction and area utilization plan covering constructional and other uses.
2. The constructional use must be specified within the general development plan together with the BauNVO, taking into account areas that can be built upon and those that cannot be built upon and the permissible constructions on the construction areas.

According to the Baugesetzbuch, planning competence is transferred to the local planning authority (usually the municipal authority). In built-up areas (local turbines) the turbine must blend in with existing buildings and may not detract from the view of the area.

According to section 35 of the German Baugesetzbuch, projects in outlying areas are only permissible if the project is not detrimental to the public interest, if the public is adequately informed and if the project supplies the public with electricity, heat, water, etc. Due to the particular environmental requirements of wind turbines, e.g. favourable wind conditions, the possibility of connection to the grid and the like, the prerequisites for planning permission may be granted. In the updating of the Baugesetzbuch (BauGB) 1997 [6.55] wind turbines were categorized as so-called privileged plant, which makes it much easier to obtain planning permission. Permission can only be refused if it is not in the public interest. Encroachment alone is not sufficient. Due to this privileged position the allocation of prime sites for wind turbines should be promoted by local authorities as part of a space utilization plan.

6.4.4 Planning and planning permission

The rules of procedure for plant construction planning permission stipulate that it is compulsory to obtain planning permission for the erection of wind turbines. There are no guidelines that specify a maximum tower height and rotor diameter at which wind turbines can be erected without permission. For smaller turbines, however, the rules of procedure and inspection procedures are normally relaxed and turbine construction is simplified.

Therefore, in principle, permission must be requested for all projects. It is advisable to submit an application for outline planning permission to the responsible inspector of works through the municipal authorities. This application should include a description of the entire site, site plan, ground plan and views, possibly with a drawing or photograph of the wind turbine. The authorities will reply, explaining the basic erection options and giving notes on the further procedure with regard to changes to the construction plan or to the turbine. They will also indicate any additional licensing procedures that are required, together with the responsible authorities (e.g. nature conservation or countryside preservation authorities) or the requirement for a regional planning order (if not already initiated).

In the application for planning permission, differentiation must be made between

- private or commercial users, who must direct their application to the responsible building supervisory board (local authority, District Office, president of the administrative district) and
- authorities, who must initiate a consent procedure via the president of the administrative district to the responsible state building supervisory board (e.g. the University Building Office).

The quantity and scope of documentation to be submitted are not regulated in a unified manner. They are, however, precisely specified by the local authorities. The following must normally be submitted in triplicate:

- building specifications;
- layout plan or copy of the cadastral map (1:1000 or 1:5000) showing the location of the turbine;
- construction drawings with view, ground plan and at least one sectional drawing (1:100), which will normally be provided by the manufacturer;
- static calculations for the tower and foundations to prove its stability, and a certificate of operating safety;

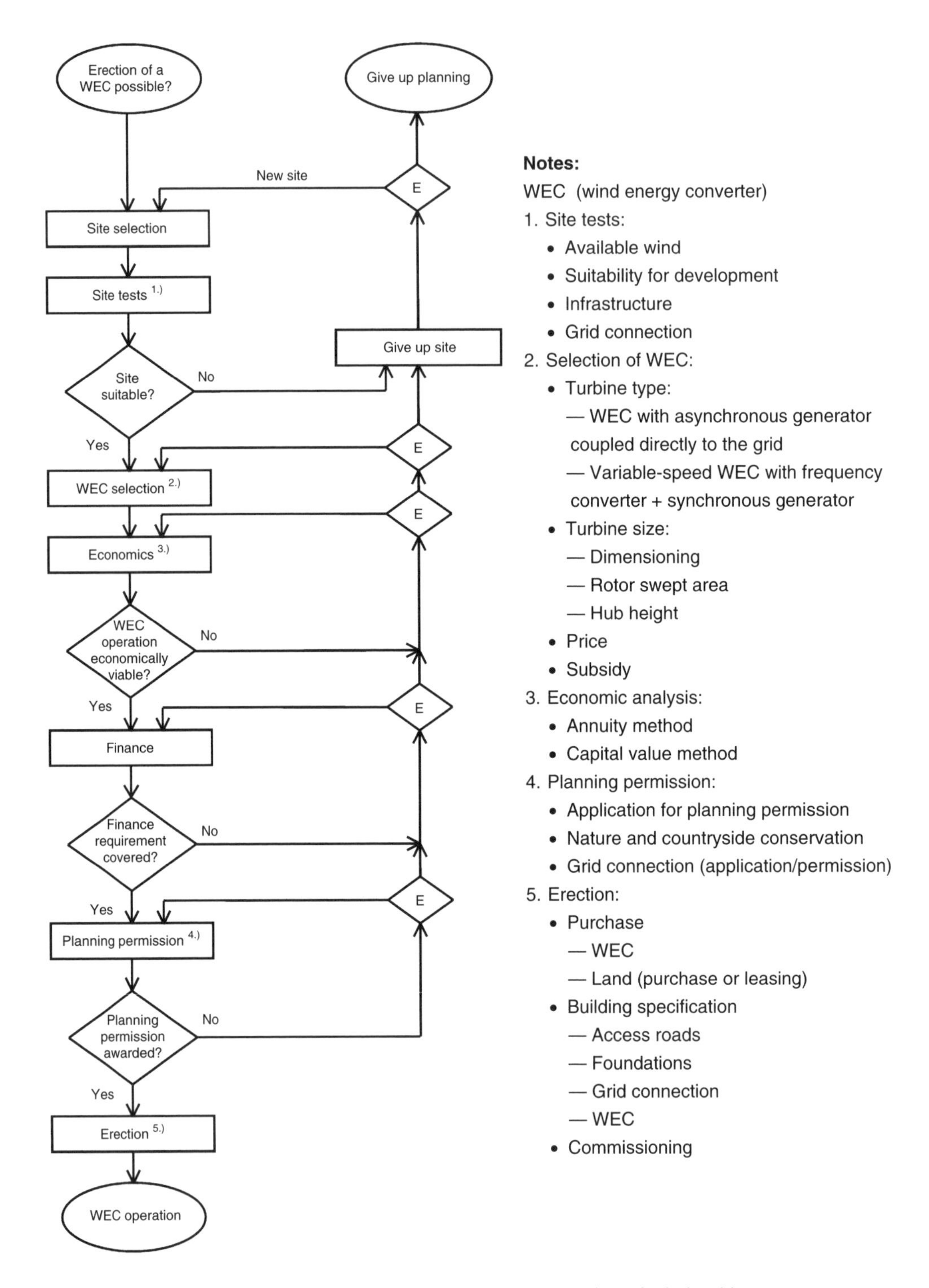

Figure 6.26 Procedure for the planning and erection of wind turbines

- type approval is normally necessary for mass-produced plants – if this is not available then the following individual licences are required:
  - — appraisal report, certificate, etc., for construction and components, according to the applicable guidelines;
  - — certificate covering the turbine's safety equipment;
  - — technical certification of the nacelle and rotor;
  - — test results concerning noise measurements and possibly turbine vibration behaviour;
  - — instructions for the operator;
- maps for
  - — noise immission,
  - — shadow impact,
  - — visibility and the
  - — visualization of the turbine or wind farm.

This completes the documentation. If the application is rejected, an appeal can be lodged, which leads to an investigation of the process.

Procedures and guidelines for checking the compatibility of wind turbines with nature and the countryside are not currently regulated in a unified manner. This can significantly extend the licensing procedure. Therefore, a time schedule incorporating all stages up to the operation of the turbine must incorporate certain planning uncertainties.

6.4.5 Procedure for the erection of wind turbines

Figure 6.26 shows the procedure for the planning, licensing and erection of wind turbines. The purpose of this flowchart is to illustrate important stages and decisions on the route to turbine operation in a coarsely structured and greatly simplified manner.

References

Chapter 1

[1] *Meyers Enzyklopädisches Lexikon*, Vol. 25.9, Completely Revised Edn, Bibliographisches Institut AG, Mannheim, 1979.
[2] Kleinhenz, F., Projekt eines Großwindkraftwerkes, *Der Bauingenieur*, 1942, 23/24.
[3] Honnef, H., *Windkraftwerke*, Vieweg, Braunschweig, 1932.
[4] Zephyros brochure, *Permanent Performance*.

Chapter 2

[1] Schlichting, H. and Truckenbrodt, E., *Aerodynamik des Flugzeuges*, Vol. 2, Springer, Berlin, 1960.
[2] Betz, A., *Wind-Energie und ihre Ausnutzung durch Windmühlen*, Vandenhoeck und Ruprecht, Göttingen, 1926.
[3] Gasch, R., *Windkraftanlagen. Grundlagen und Entwurf*, 3rd Edn, Teubner, Stuttgart, 1996.
[4] Molly, J.P., *Windenergie. Theorie, Anwendung, Messung*, 2nd Edn, Müller, Karlsruhe, 1990.
[5] Rijs, R.P.P. and Smulders, P.T., Blade Element Theory for Performance Analysis of Slow Running Wind Turbines, *Wind Engineering*, 1990, 14(2), 62–79.
[6] Wilson, R. and Lissaman, P., *Applied Aerodynamics of Wind Power Machines*, Oregon State University, 1974.
[7] Hernandez, J. and Crespo, A., Aerodynamic Calculation of the Performance of Horizontal Axis Wind Turbines and Comparison with Experimental Results, *Wind Engineering*, 1987, 11(4), 177ff.
[8] Kraft, M., *Verfahren zur Leistungsbestimung für Windturbinen*, Thesis, Kassel University, 1990.
[9] Man, D. T., Sullivan, J. P. and Wasynczuk, O., Dynamic Behavior of a Class of Wind Turbine Generators During Random Wind Fluctuations, *IEEE Transactions on Power Apparatus and Systems*, June 1981, PAS-100(6).
[10] Anderson, P. M. and Bose, A., Stability Simulation of Wind Turbine Systems, *IEEE Transactions on Power Apparatus and Systems*, December 1983, PAS-102(12).
[11] Amlang, B., Arsurdis, D., Leonhard, W., Vollstedt, W. and Wefelmeier, K., *Elektrische Energieversorgung mit Windkraftanlagen*, Abschlußbericht BMFT-Forschungsvorhaben 032-8265-B, Braunschweig, 1992.
[12] Moretti, P. and Divone, L., Moderne Windkraftanlagen, *Spektrum der Wissenschaft*, August 1986.
[13] Hau, E., *Windkraftanlagen*, 2nd Edn, Springer, Berlin, 1996.
[14] Divalentin, E., The Application of Broad Range Variable Speed for Wind Turbine Entrancement, in EWEA Conference, Rome, Italy, 1986.

Grid Integration of Wind Energy Conversion Systems, Second Edition S. Heier

[15] Eppler, R., *Vorlesungsmanuskript Technische Mechanik*, Stuttgart University, 1968.
[16] Försching, H.W., *Grundlagen der Aeroelastik*, Springer, Berlin, 1974.
[17] Heier, S. and Kleinkauf, W., *Regelung einer großen Windenergie-anlage (GROWIAN)*, Final Report, Part III: *Wirkleistungs-Drehzahlregelung*, Kassel University, Kassel, 1978.
[18] Cramer, G., Drews, P., Heier, S., Kleinkauf, W. and Wettlaufer, R., *Mitarbeit bei der Entwicklung der Windenergieanlage Aeroman. Regelung, Betriebsarten und Betriebsverhalten. Abschlußbericht*, Kassel University, Kassel, 1979.
[19] Cramer, G., Drews, P., Heier, S. and Kleinkauf, W., *Regelung und Simulation des Betriebsverhaltens der Windenergieanlage Näsudden/Schweden. Abschlußbericht*, Kassel University, Kassel, 1980.
[20] Albrecht, P., Cramer, G., Drews, P., Grawunder, M., Heier, S., Kleinkauf, W., Leonhard, W., Speckheuer, W., Thür, J., Vollstedt, W. and Wettlaufer, R. *Betriebsverhalten von Windenergieanlagen. Abschlußbericht zum BMFT-Forschungsvorhaben*, O3E-4362-A, BMFT-FB-T 84-154, Part I, Department of Electrical Energy Supply Systems, Kassel University, and Institute for Control Technology Braunschweig Technical University, Karlsruhe, 1984.
[21] Vollstedt, W., Variable Speed Turbine Generator with Low Line Interactions, in International Energy Agency (IEA) Expert Meeting, Göteborg, Sweden, October 1991.
[22] Vollstedt, W., Braussemann, H., Hanne, R. and Horstmann, M., *Kostengünstiges Generatorsystem für drehzahlvariable Windturbinen mit guter Netzeinspeisequalität. BMFT-Abschlußbericht F 0812.00*, 1995.
[23] Kleinkauf, W., *et al.*, *Windenergieanlagen im Verbundbetrieb, Abschlußbericht zum BMFT-Forschungsvorhaben 03E8160-A*, Part 1, Department of Electrical Energy supply, Kassel University, Kassel, 1986.
[24] Hackenberg, G., *Interner Bericht zur Meßkampagne Aeroman 12,5/33*, Department of Electrical Energy Supply Systems, Kassel University, 1985.

Chapter 3

[1] Vereinigung Deutscher Elektrizitätswerke (VDEW), *Grundsätze für die Beurteilung von Netzrückwirkungen*, 3rd Revised Edn, VDEW, Frankfurt a.M., 1992; corrected reprint 1997.
[2] Heier, S. and Kleinkauf, W., Windpark Kythnos, in International Energy Agency (IEA) Conference, Palo Alto, California, October 1983.
[3] König, V., *Simulation der dynamischen Betriebseigenschaften einer Asynchronmaschine mit Schleifringläufer beim Einsatz in Windkraftanlagen*, Thesis, Kassel University, 1992.
[4] Heier, S., *Windenergiekonverter und mechanische Energiewandler. Anpassung und Regelung*, Tagung: *Energie vom Wind*, Deutsche Gesellschaft für Sonnenenergie, Bremen, 1977.
[5] Kovacs, K.P. and Racz, J., *Transiente Vorgänge in Wechselstrommaschinen*, Vols 1 and 2, Hungarian Academy of Science, Budapest, 1959.
[6] Schuisky, W., *Induktionsmaschinen*, Springer, Vienna, 1957.
[7] Wüterich, W., Übersicht über die Einschaltmomente bei Asynchronmaschinen im Still-stand, *Elektrotechnische Zeitschrift-A*, 1967, 88, 555–9.
[8] Jordan, H., Lorenzen, H.W. and Taegen, F., Über den asynchronen Anlauf von Synchron-maschinen, *Elektrotechnische Zeitschrift-A*, 1964, 85, 296–305.
[9] Keil, F., Zur Berechnung des asynchronen Anlaufs von Schenkelpol-Synchronmaschinen mit massiven Polen, *Elektrotechnische Zeitschrift-A*, 1969, 90, 396–9.
[10] Siemens AG, *Formel-und Tabellenbuch für Starkstrom-Ingenieure*, 3rd Edn, Girardet, Essen, 1965.
[11] Leonhard, W., Vollstedt, W., *et al.*, *Windenergieanlagen im Verbundbetrieb. Verbundbetrieb von großen Windenergieanlagen mit Gleichstrom- und Drehstromsammelschiene*, Bundesministerium für Forschung und Technologie, Statusreport Windenergie, Lübeck, 1988, pp. 1–20.
[12] Caselitz, P., Heier, S. *et al.*, *PV Simplorer*, Zwischenbericht, 2001.
[13] Leonhard, W., *Regelung in der elektrischen Energieversorgung*, Teubner, Stuttgart, 1980.
[14] Kleinrath, H., *Stromrichtergespeiste Drehfeldmaschinen*, Springer, Vienna, 1980.
[15] Richter, R., *Elektrische Maschinen*, Vols I–IV, Birkhäuser, Basel/Stuttgart, 1967.
[16] Bödefeld, T. and Sequenz, H., *Elektrische Maschinen*, 8th Edn, Springer, Vienna, 1971.
[17] Klamt, J., *Berechnung und Bemessung elektrischer Maschinen*, Springer, Berlin/Göttingen/ Heidelberg, 1962.
[18] Weiher Electric, *Generatorsysteme für Windenergieanlagen Prospekt.*
[19] Weiher, Spezialentwicklung für den drehzahlvariablen WKA-Betrieb. Bürstenloser doppeltgespeister Asynchrongenerator für Pitchanlagen, *Wind Kraft Journal*, 1998, 2/98, 42.

[20] Engel, U. and Wickboldt, H., Explosionsgeschützte Drehstrommotoren und die neuen Normspannungen, *Elektrotechnische Zeitschrift-A*, 1991, 112, 1082–6.
[21] Weiher Electric GmbH, Weiher Vari Slip-Generator erfolgreich auf einer Megawatt-Stallanlage getestet, *Wind Kraft Journal*, 2000, 6/2000, 44.
[22] Seinsch, H.-O, *Oberfeldschwingungen in Drehfeldmaschinen*, Teubner, Stuttgart, 1992.
[23] Brach, K., *Wellenspannungen bei Drehstrominduktionsmaschinen mit Käfigläufer*, Dissertation, TH Hannover, 1990.
[24] DIN VDE 0530, Beiblatt 2, *Umrichtergespeiste Induktionsmotoren mit Käfigläufer*, Application Guidelines (IEC 60034-17), January 1999.
[25] Chen, S. and Lipo, T., Circulating Type Motor Bearing Current in Inverter Drives, in IEEE IAS Annual Meeting, 1996.
[26] Binder, A. and Schrepfer, A., Lagerströme bei umrichtergespeisten Drehstrommotoren, *Antriebstechnik*, 1999, 38.
[27] Herrmann, T. and Wimmer, J., Wellenspannungen und Lagerströme bei Drehstrommaschinen in Verbindung mit Umrichterbetrieb, *Wind Kraft Journal*, 2000, 6/2000, 30–3.
[28] Wimmer, J. and Stadler, H., Lagerströme bei umrichtergespeisten Drehstrommaschinen, *Antriebstechnik*, 2000, 39.
[29] Hansberg, V. and Seinsch, H.-O., Kapazitive Lagerspannungen und -ströme bei umrichter-gespeisten Induktionsmaschinen, *Electrical Engineering*, 2000, 82.
[30] Vensys brochure.
[31] Weh, H., Permanenterregte Synchronmaschinen hoher Kraftdichte nach dem Transversalflußkonzept, *Elektrotechnische Zeitschrift Archiv*, 1988, 10(5), 143–9.
[32] Weh, H., Hoffmann, H, Landrath, J., Mosebach, H. and Poschadel, J., Directly-Driven Permanent-Magnet Excited Synchronous Generator for Variable Speed Operation, in European Community Wind Energy Conference, Herning, Denmark, June 1988, pp. 566–72.
[33] Hill, W., *Konzeption und Bau eines getriebelosen Windgenerators*, Thesis, Kassel University, 1995.
[34] Heier, S., Hill, W. and Kleinkauf, W., Neuartige Konzeption eines permanenterregten Vielpolgenerators für getriebelose Windkraftanlagen, in 3 Deutsche Windenergie-Konferenz DEWEK 96, Deutsches Windenergie Institut, Wilhelmshaven, 23–24 October 1996, pp. 121–4.
[35] Jöckel, S., Gearless Wind Energy Converters with Permanent Magnet Generators – An Option for the Future?, in European Wind Energy Confererence, Göteborg, Sweden, pp. 414–7.
[36] Nürnberg, W., *Die Asynchronmaschine. Ihre Theorie unter besonderer Berücksichtigung der Keilstab- und Doppelkäfigläufer*, Springer, Berlin, Göttingen, Heidelberg, 1963.

Chapter 4

[1] Apel-Hillmann, F.M., *Windkraftanlagen im wissenschaftlichen Meß- und Evaluierungs-programm (WMEP) – Datenerfassung und Netzanschluß*, Thesis, Kassel University, 1994.
[2] Heier, S., Kleinkauf, W. and Sachau, J., *Power Conditioning – The Link between Solar Conversion and Consumer. Advances in Solar Energy*, Vol. 9: *An Annual Review of Research and Development*, American Solar Energy Society, Boulder, Colorado, 1994, pp. 161–244.
[3] Heumann, K., *Grundlagen der Leistungselektronik*, Teubner, Stuttgart, 1991.
[4] Bystron, K., *Leistungselektronik*, Hanser, Munich, Vienna, 1979.
[5] Wasserrab, T., *Schaltungslehre der Stromrichtertechnik*, Springer, Berlin, 1962.
[6] Meyer, M., *Leistungselektronik. Einführung, Grundlagen, Überblick*, Springer, Berlin, Heidelberg, New York, London, Paris, Tokyo, Hong Kong, Barcelona, 1990.
[7] Leonhard, W., *Regelung in der elektrischen Antriebstechnik*, Teubner, Stuttgart, 1974.
[8] Hienz, G., Neuer Direktumrichter für Industrieantriebe, *Antriebstechnik*, 1989, 28(3).
[9] Arlt, B., Der MOS Controlled Thyristor MCT, *Elektronik Industrie*, 1992, 11/92, 56–60.
[10] Bober, G. and Heumann, K., Vergleich der Eigenschaften von MCT und IGBT unter 'harten' Schaltbedingungen, Vol. 115, Books 13–14, ETZ, 1994.
[11] Heumann, K., Die Intelligenz hält Einzug – Umrichter in der Antriebstechnik, in *Elektronik*, Book 1, 1995, pp. 83–96.
[12] Salzar, L. and Jos, G., PSPICE Simulation of Three-Phase Inverters by Means of Switching Functions, *IEEE Transactions on Power Electronics*, 1994, 9(1), 35–42.

[13] Apel-Hillmann, F.M., *Konzeption eines Stromrichters zur Netzanbindung von permanterregten Synchrongeneratoren*, Thesis, Kassel University, 1995.

[14] Deisenroth, H. and Trabert, C., Vermeidung von Überspannungen bei Pulsumrichterantrieben, Vol. 114, Book 17, ETZ, 1993, pp. 1060–6.

[15] Schmid, W., Pulsumrichterantriebe mit langen Motorleitungen, *Antriebs- und Getriebetechnik*, Book 4, 1992, pp. 58–64.

[16] Aninger, H. and Nagel, G., *Vom transienten Betriebsverhalten herrührende Schwingungen bei einem über Gleichrichter belasteten Synchrongenerator*, Part 1: *Theoretische Untersuchungen*, Siemens Forschungs- und Entwicklungsberichte, Vol. 9, No. 1, Springer-Verlag, 1980, pp. 1–7.

[17] Ernst-Cathor, J., *Drehzahlvariable Windenergieanlage mit Gleichstromzwischenkreis-umrichter und Optimumsuchenden Regler*, Dissertation, TU Braunschweig.

[18] Schmeer, H.R., EMV 1994, in 4 Internationale Fachmesse und Kongress für EMV, February 1994.

[19] Bösterling, W., Jörke, R. and Tscharn, M., IGBT-Module in Stromrichtern: steuern, regeln, schützen, Vol. 110, Book 10, ETZ, 1989.

[20] Vereinigung Deutscher Elektrizitätswerke (VDEW), *Technische Anschlußbedingungen für den Anschluß an das Nie derspannungsnetz*, TAB 2000, VWEW, Frankfurt am Main (Hrsg), Edn 2000.

[21] Fördergesellschaft Windenergie e.V., *Technische Richtlinie für Windenergieanlagen*, revision 13, 1.1.2000, Part 3: *Bestimmung der Elektrischen Eigenschaften*, Fördergesellschaft Windenergie e.V. (FGW), Hamburg, 2000.

[22] Vereinigung Deutscher Elektrizitätswerke (VDEW), *Eigenerzeugungsanlagen am Mittelspannungsnetz. Richtlinie für Anschluß und Parallelbetrieb von Energieerzeugungsanlagen am Mittelspannungsnetz*, VWEW, Frankfurt am Main (Hrsg), 2nd Edn, 1998.

[23] Curtice, D.H. and Patton, J.B., Analysis of Utility Protection Problems Associated with Small Wind Turbine Interconnections, *IEEE Transactions on Power Apparatus and Systems*, 1982, PAS-101(10), 3957–66.

[24] Dugan, R.C. and Rizy, D.T., Electric Distribution Protection Problems Associated with the Interconnection of Small, Dispersed Generation Devices, *IEEE Transactions on Power Apparatus and Systems*, 1984, PAS-103(6), 1121–7.

[25] Flosdorff, R. and Hilgarth, G., *Elektrische Energieverteilung*, Teubner, Stuttgart, 1994.

[26] Happoldt, H. and Oeding, D., *Elektrische Kraftwerke und Netze*, Springer, Berlin, 1978.

[27] Brown, M.T. and Settebrini, R.C., Dispersed Generation Interconnections via Distribution Class Two-Cycle Circuit Breakers, *IEEE Transactions on Power Delivery*, 1990, 5(1), 481–5.

[28] Webs, A., *Einfluß von Asynchronmotoren auf die Kurzschlußstromstärken in Drehstromanlagen*, VDE Technical Reports Vol. 27, VDE, Offenbach, 1972.

[29] Wachenfeld, V., *Netzrückwirkungen durch Windkraftanlagen*, Thesis, Kassel University, 1994.

[30] Krämer, T., *Untersuchung von Netzstrukturen zur Bewertung von Netzrückwirkungen durch Windkraftanlagen*, Thesis, Kassel University, 1994.

[31] Westinghouse Electric Corporation, *Electrical Transmission and Distribution*, Reference Book, East Pittsburgh, Pennsylvania, 1964.

[32] Krause, P.C. and Man, D.T., Transient Behavior of Class of Wind Turbine Generators during Electrical Disturbance, *IEEE Transactions on Power Apparatus and Systems*, 1981, PAS-100(5), 2204–10.

[33] Durstewitz, M., Heier, S., Hoppe-Kilpper, M. and Kleinkauf, W., *Meßtechnische Untersuchungen am Windpark Westküste. Untersuchung der elektrischen Komponenten und ihrer Integration in schwache Netze. BMFT-Abschlußbereicht*, April 1992.

[34] Enßlin, C., Durstewitz, M., Heier, S. and Hoppe-Kilpper, M., Wind Farms in the German '250 MW Wind'-Programme, in European Wind Energy Association Special Topic Conference '92 on *The Potential of Wind Farms*, Herning, Denmark, 1992, pp. B4.1–B4.7.

[35] Enßlin, C., Hoppe-Kilpper, M., Kleinkauf, W., Koch, H. and Schott, T., The WMEP in the German '250 MW-Wind' Programme – Evaluations from the large scale WMEP measurement network, in ISES Solar World Congress, Budapest, August 1993.

[36] Heier, S., *Wind Power Generation. Decentralized Energy. Options and Technology*, Omega Scientific Publishers, New Delhi, 1993, pp. 65–76, ISBN 81-85399-26-3.

[37] Tande, J.O.G. and Landberg, L., A 10 Sec. Forecast of Wind Output with Neural Networks, in European Wind Energy Conference, 1993.

[38] Menze, M., *Leistungsprognose von Windkraftanlagen mit Neuronalen Netzen*, Thesis, Kassel University, 1996.

[39] Rohrig, K., Online Monitoring of 1700 MW Wind Capacity in a Utility Supply Area, in European Wind Energy Conference and Exhibition, Nice, France, March 1999.

[40] Beyer, H.G., Heinemann, D., Mellinghoff, H., Mönnich, K. and Waldl, H.-P., Forecast of Regional Power Output of Wind Turbines, in 1999 European Wind Energy Conference and Exhibition, Nice, France, March 1999.

[41] Rohrig K., Online Monitoring and Short Term Prediction of 2400 MW Wind Capacity in a Utility Supply Area, in *Wind Forecast Techniques*, 33 Meeting of Experts, Technical Report from the International Energy Agency, R&D Wind, Ed. S.-E. Thor, FFA, Sweden, July 2000, pp. 117–119.

[42] Rohrig K., Ernst, B., Ensslin, C. and Hoppe-Kilpper, M., Online Supervision and Prediction of 2500 MW Wind Power, in *Wind Power for the 21st Century*, Special Topic Conference and Exhibition, Convention Centre, Kassel, Germany, 25–27 September 2000.

[43] Landberg, L., Joensen, A., Giebel, G., Madsen, H. and Nielsen, T.S., Zephyr: The Short Term Prediction Models, in *Wind Power for the 21st Century*, Special Topic Conference and Exhibition, Convention Centre, Kassel, Germany, 25–27 September 2000.

[44] Moehrlen, C, Jorgensen, J.U., Sattler, K. and McKeogh, E.J., On the Accuracy of Land Cover Data in NWP Forecasts for High Resolution Wind Energy Prediction, in 2001 European Wind Energy Conference and Exhibition, Bella Center, Copenhagen, Denmark, 2–6 July 2001.

[45] Moehrlen, C., On the Benefits of and Approaches to Wind Energy Forecasting, invited speaker at the Irish Wind Energy Association Annual Conference on *Towards 500 MW*, Ennis, 25 May 2001.

[46] Focken, U., Lange, M. and Waldl, H.-P., Previento – A Wind Power Prediction System with an Innovative Upscaling Algorithm, in 2001 European Wind Energy Conference and Exhibition, Bella Center, Copenhagen, Denmark, 2–6 July 2001.

[47] Rohrig, K., Ernst, B., Schorn, P. and Regber, H., Managing 3000 MW Wind Power in a Transmission System Operation Center, in 2001 European Wind Energy Conference and Exhibition, Bella Center, Copenhagen, Denmark, 2–6 July 2001.

[48] Heier, S., *Grid Connected Wind Energy Converters. Commercialization of Solar and Wind Energy Technologies*, Amman, Jordan, April 1992, pp. 519–32.

[49] Heier, S., *Technical Aspects of Electrical Supply Systems for Villages. Decentralized Energy. Options and Technology*, New Delhi, 1993, pp. 107–27, ISBN 81-85399-26-3.

[50] Caselitz, P., Hackenberg, G., Kleinkauf, W. and Sachau, J., *Windenergieanlagen in elektrischen Energieversorgungssystemen kleiner Leistung*, BMFT-Final Report, Kassel University, 1985.

[51] Caselitz, P., Giebhardt, J. and Mevenkamp, M., On-line Fault Detection and Prediction in Wind Energy Converters, in European Wind Energy Association Conference – Macedonia EWEA '94, Thessaloniki, Greece, October 1994.

[52] Heier, S., *Elektrotechnische Konzeptionen von Windkraftanlagen im Vergleich. Windenergie Bremen '90*, DGS-Sonnenenergie Verlags-GmbH, Munich, 1990, pp. 91–106.

[53] Heier, S., Grid Influences by Wind Energy Converters, in International Energy Agency (IEA) Expert Meeting, Göteborg, October 1991, pp. 37–50.

[54] Heier, S., Windkraftanlagen im Netzbetrieb, in German Wind Energy Conference DEWEK 92, Wilhelmshaven, 28–29 October 1992, pp.141–5.

[55] Heier, S., Netzintegration von Windkraftanlagen, in Fördergesellschaft Windenergie (FGW)-Workshop on *Netzanbindung von Windkraftanlagen*, Hanover, 23 March 1993.

[56] Heier, S., Technical Aspects of Wind Energy Converters and Grid Connections, in British Wind Energy Association (BWEA) Workshop on *Wind Energy Penetration into Weak Electricity Networks*, 10–12 June 1993, Rutherford Appleton Laboratory, Chilton, Didcot, pp. 38–55, ISBN 1-870064-17-8.

[57] Heier, S., Netzeinwirkungen durch Windkraftanlagen und Maßnahmen zur Verminderung, in Husumer Wind Energy Conference, 22–26 September 1993, Husum, pp. 157–68.

[58] Heier, S., Grid Influences by Wind Energy Converters and Reduction Measures, in American Wind Energy Association 24th Annual Conference, Minneapolis, Minnesota, 1994.

[59] Dangrieß, G., Heier, S., König, V., Kuntsch, J. and Müller, J., Konzeptionen zur Auslastung der Netzkapazität, in German Wind Energy Conference DEWEK '94, Wilhelmshaven, 22–23 June 1994, pp. 163–70.

[60] Dangrieß, G., Heier, S., König, V., Kuntsch, J. and Müller, J., *Konzeption zur Ausnutzung der Netzkapazität*, Neue Energie Heft 11/94, Steinbacher Druck GmbH, Osnabrück, 1994, pp. 44–45.

[61] Krengel, U., *Untersuchung der Rückwirkungen eines 1 kW-Photovoltaik-Wechselrichters auf das Versorgungsnetz*, Thesis, Kassel University, 1990.

[62] Chun, S. and Damm, F., *Netzrückwirkung im Windpark Westküste – Oberschwingungen*, Thesis, Kassel University, 1990/1991.

[63] Götze, F., Schäfer, H. and Schulz, D., *Impedanz-Simulation beliebiger Netzkonfigurationen*, Project II, Kassel University, 1991.

[64] Heier, S. and Kleinkauf, W., Grid Connection of Wind Energy Converters, in European Community Wind Energy Conference, Lübeck–Travemünde, March 1993, pp. 790–793, ISBN 0-9521452-0-0.

[65] Cramer, G., Durstewitz, M., Heier, S. and Reinmöller-Kringel, M., *1,2 MW-Stromrichter am schwachen Netz. Filterauslegung zur Reduzierung von Stromoberschwingungen*, SMA info 10, April 1993, pp. 10–11.

[66] Oort, H.A. van, Numerical Model for Calculating the Power Output Fluctuations from Wind Farms, *Journal of Wind Engineering and Industrial Aerodynamics*, 1988, 27.

[67] Cretcher, C.K. and Simburger, E.J., Load Following Impacts of a Large Wind Farm on an Interconnected Electric Utility System, *IEEE Transactions on Power Apparatus and Systems*, 1983, PAS-102 (3), 687–92.

[68] Büchner, J., Beyer, H.-G., Eichelbrönner, M., Haubrich, H.-J., Steinberger-Willms, R., Stubbe, G. and Waldl, H.-P., *Modellierung der Netzbeeinflussung durch Windparks. Energiewirtschaftliche Tagesfragen 43. Jg.*, 1993, Book 5, pp. 332–5.

[69] Enßlin, C., Durstewitz, M., Heier, S. and Hoppe-Kilpper, M., Wind Farms in the German '250 MW-Wind'-Programme, in European Wind Energy Association Special Topic Conference '92 on *The Potential of Wind Farms*, Herning, Denmark, 1992, pp. B4.1–B4.7.

[70] Arnold, G. and Heier, S., Netzregelung mit regenerativen Energieversorgungssystemen, in Kasseler Symposium Energie-Systemtechnik '99, ISET, Kassel, 1999, pp. 166–77.

[71] Heier, S., Arnold, G., Durstewitz, M., Perez-Spiess, F., Meyer, R. and Juarez-Navarro, A., Grid Control with Renewable Energy Sources, in European Wind Energy Association (EWEA) Special Topic Conference on *Wind Power for the 21st Century – The Challenge of High Wind Power Penetration for the New Energy Markets*, International Conference, Kassel, Germany, 25–27 September 2000.

[72] Heier, S., Grid Integration of Wind Energy Converters and Field Applications, in Wind Energy Symposium, Alacati-Izmir, Turkey, 5–7 April 2001, pp. 151–164, ISBM 975-395-425-5.

[73] Arnold, G. and Heier, S., Grid Control with Wind Energy Converters, in Second International Workshop on *Transmission Networks for Offshore Wind Farms*, Royal Institute of Technology, Electric Power Systems, Stockholm, Sweden, 30–31 March 2001.

[74] Arnold, G., Heier, S., Perez-Spiess, F. and Lopez-Manzanares, L., Grid Control with Renewable Energy Sources – Results to the Field Tests, in European Wind Energy Conference, Copenhagen, Denmark, 2–6 July 2001.

[75] Arnold, G., *System zur stützung von Elektrizitätsnetzen mit windkraftan lagen und anderen Erneuerbaren Energien*, Dissertation, Universität Kassel, 2004.

[76] Verband der Netzbetreiber VDN e.V. beim VDEW: *Netzcodes*, Elektronische Dokumente (PDF), 2004; www.vdn-berlin.de/netzcodesl.asp.

[77] E.ON-Netz GmbH, *Netzanschlussregeln für Höch- und Höchstspannung*, Elektronisches Dokumente (PDF), 1 August 2004; www.con-netz.com/Ressources/downloads/ENE_NAR_HS_01082003.pdf.

Chapter 5

[1] Albrecht, P., Cramer, G., Drews, P., Grawunder, M., Heier, S., Kleinkauf, W., Leonhard, W., Speckheuer, W., Thür, J., Vollstedt, W. and Wettlaufer, R., *Betriebsverhalten von Windenergieanlagen, Abschlußbericht zum BMFT-Forschungsvorhaben O3E-4362-A, BMFT-FB-T 84-154*, Part II, Department of Electrical Energy Supply Systems, Kassel University, and Institute for Control Technology Braunschweig Technical University, Karlsruhe, 1984.

[2] Heier, S. and Kleinkauf, W., Regelungskonzept für GROWIAN (Große Windenergieanlage) in Seminar- und Statusbericht Windenergie, Kernforschungsanlage Jülich GmbH, Projektleitung Energieforschung (Hrsg), October 1978, pp. 407–18.

[3] Heier, S., Regelungskonzepte für Windenergieanlagen, in Wind Energy Conference of the German Association for Wind Energy (DGW) and KFA Jülich, Oldenburg, 27–28 March 1987, pp. 123–40.

[4] Heier, S., Regelungskonzepte für Windenergieanlagen, *Elektrotechnik/Schweiz*, 1988, 9, 51–6.

[5] Heier, S., Generatoren für kleine Windkraftanlagen im Netzbetrieb unter Berücksichtigung der regelungstechnischen Konzeption, in VI Symposium on *Micromachines and Servosystems*, Warszawa, Poland, May 1988.

[6] Heier, S., Kleinkauf, W. and Sachau, J., Wind Energy Converters at Weak Grids, in European Community Wind Energy Conference, Herning, Denmark, May 1988, pp. 429–33.

[7] Durstewitz, M., Heier, S., Hoppe-Kilpper, M., Kleinkauf, W. and Sachau, J., *Elektrische Energieversorgung mit Windenergieanlagen. Auslegung und Regelung von verbraucher-orientierten Versorgungseinheiten und Inselnetzen*, BMFT Final Report, March 1992.

[8] Caselitz, P. and Krüger, T., Drehzahlvariable Windkraftanlagen mit Überlagerungs-getriebe, *Windkraft-Journal*, 1993, 13.

[9] Leonhard, W., *Einführung in die Regelungstechnik. Lineare Regelvorgänge*, 2nd Improved Edn, Friedrich Vieweg & Sohn, Braunschweig, 1972.

[10] Nigim, K.A., *Static Exciter for Wound Rotor Induction Machine*, IEEE, 1990, pp. 933–7.

[11] Holmes, P.G. and Nigim, K.A., A Stand-Alone Induction Generator with Secondary Control to Give Constant Frequency and Voltage, in 22nd UPEC, Sunderland Polytechnic, UK, 14–16 April 1987.

[12] Arsudis, D., *Doppeltgespeister Drehstromgenerator mit Spannungszwischenkreis-Umrichter im Rotorkreis für Windkraftanlagen*, Dissertation, Braunschweig Technical University, 1989.

[13] Kiel, E. and Schumacher, W., Der Servocontroller in einem Chip, *Elektronik*, April 1994.

[14] Körber, F., Besel, G. and Reinhold, H., *Meßprogramm an der 3 MW-Windkraftanlage GROWIAN. BMFT-Forschungsbericht Förderkennzeichen 03E-4512A*, Hamburg, 1988.

[15] Arafa, O. and Heier, S., *Accurate Modelling of Parallel Chopper-Controlled Induction Motor Drives*, Report DAAD, University Gh Kassel, 1996.

[16] Arafa, O., *A Study of the Performance Characteristics of the Asynchronous Cascade in the Driving Mode*, Final Report DAAD, University Gh Kassel, 1996.

[17] Ritter, P. and Rotzsche, L., *Schlupfsteuerung von Asynchrongeneratoren für Windkraftanlagen zur Minderung ihrer Leistungsschwankungen*, Thesis, University Gh Kassel, 1995.

[18] Ritter, P., *Schlupfregelung einer Windkraftanlage mit Asynchrongenerator*, Thesis, University Gh Kassel, 1996.

[19] Hawranke, I., *Betriebsführung eines Teststandes zur Nachbildung von Windkraftanlagen*, Thesis, University Gh Kassel, 1996.

[20] Leonhard, W., *Einführung in die Regelungstechnik. Nichtlineare Regelvorgänge*, 2nd Revised Edn, Friedrich Vieweg & Sohn, Braunschweig, 1977.

[21] Buxbaum, A. and Schierau, K., *Berechnung von Regelkreisen der Antriebstechnik*, Elitera-Verlag, Berlin, 1974.

[22] Dörrscheidt, F. and Latzel, W., *Grundlagen der Regelungstechnik*, 2nd Revised Edn, B.G. Teubner, Stuttgart/Leipzig, 1993.

[23] Pfaff, G., *Regelung elektrischer Auftriebe*, 2nd Edn, R. Oldenburg Verlag, Munich, 1984.

[24] Bothe, H.-H., *Fuzzy Logic. Einführung in Theorie und Anwendung*, 2nd Enlarged Edn, Springer-Verlag, Berlin, Heidelberg, New York, London, Paris, Tokyo, Hong Kong, Barcelona, Budapest, 1995.

[25] Kähny, H., *Eigenschaften eines Fuzzy-Reglers zur Leistungsregelung einer Windkraftanlage mit Blattverstellung*, Dissertation, Universität Gesamthochschule Kassel, 1994.

[26] Danesi, A. *et al.*, A Self Adaptive Pitch Blade Control of a Large Wind Turbine with Predictive Wind Velocity Measurements, in European Wind Energy Conference, Rome, Italy, 1986, p. 641.

[27] Arsudis, D. and Bönisch, H., Self-Tuning Linear Controller for the Blade Pitch Control of a 100 kW WEC, in European Community Wind Energy Conference, Madrid, Spain, 10–14 September 1990, pp. 564–8.

[28] Barton, R.S., Bowler, C.E.J. and Piwko, R.J., *Control and Stabilization of the NASA / DOE MOD-1 Two Megawatt Wind Turbine Generator*, American Chemical Society, 1979, pp. 325–30.

[29] Rothmann, E.A., The Effects of Control Modes on Rotor Loads, in Second International Symposium on *Wind Energy Systems*, 1978, pp. 107–17.

[30] Kos, J.M., Online Control of a Large Horizontal Axis Wind Energy Conversion System and Its Performance in a Turbulent Wind Environment, in Proceedings of 13th Conversion Engineering Conference, San Diego, California, 1978.

[31] Hinrichsen, E.N. and Nolan, P.J., Dynamics and Stability of Wind Turbine Generators, *IEEE Transactions on Power Apparatus and Systems*, 1982, PAS 101, 2640–8.

[32] Svensson, J.E. and Ulen, E. The Control System of WTS-3 Instrumentation and Testing, in 4th Symposium on *Wind Energy Systems*, BHRA, Stockholm, 1982.

[33] Hinrichsen, E.N., Controls for Variable Pitch Wind Generators, *IEEE Transactions on Power Apparatus and Systems*, 1984, PAS 103, 866–92.

[34] Liebst, B.S., *Pitch Control Systems for Large Scale Wind Turbines*, American Institute of Aeronautical Engineering and Mechanics, Vol. 7, 1982, pp. 182–92.

[35] Murdoch A., Winkelman, J.R., Javid, S.H. and Barton, R.S., Control Design and Performance Analysis of a 6 MW Wind Turbine Generator, *IEEE Transactions on Power Apparatus and Systems*, 1983, PAS102, 1340–7.
[36] Mattson, S.E., *Modelling and Control of Large Horizontal Axis Wind Power Plants*, Lund, Sweden, 1984.
[37] Grimble, M.J., Two and a Half Degrees of Freedom LQG Controller Solution and Wind Turbine Control Applications, in Proceedings of American Control Conference, 1992, pp. 676–80.
[38] Steinbuch, M., *Dynamic Modelling and Robust Control of a Wind Energy Conversion System*, Dissertation, Delft University of Technology, 1989.
[39] Steinbuch, M. and Bosgra O.H., Optimal Output Feedback of a Wind Energy Conversion System, in Proceedings of 9th IFAC on *Power Systems: Modelling and Control Applications*, Brussels, Belgium, 1989, pp. 313–9.
[40] Makila, P.M. and Toivonen, H.T., Computational Methods for Parametric LQ Problems – a Survey, *IEEE Transactions on Automatic Control*, 1987, AC-32, 658–71.
[41] Bongers, P.M.M and Schrama, R.J.P., Application of LQ-Based Controllers to Flexible Wind Turbines, in Proceedings of 1st European Control Conference, Grenoble, France, 2–5 July 1991, pp. 2185–9.
[42] Bongers, P.M.M. and Dijkstra, S., Control of Wind Turbine Systems Aimed at Load Reduction, in Proceedings of American Control Conference, Chicago, Illinois, 24–26 June 1992, pp. 1710–4.
[43] Bongers, P.M.M., *Modeling and Identification of Flexible Wind Turbines and a Factorizational Approach to Robust Control*, Dissertation, Delft University of Technology, The Netherlands, 1994.
[44] Kreißelmeier, G. and Steinhauser, R., *Systematische Auslegung von Reglern durch Optimierung eines vektoriellen Gütekriteriums, Regelungstechnik*, 1979, Book 3, pp. 76–9.
[45] Reck, T., *Untersuchung zur aktiven Dämpfung von Turmschwingungen an einer Windkraftanlage mit Blattverstellung*, Dissertation, University Gh Kassel, 1996.
[46] Adam, H., *Entwurf eines Mehrgrößenreglers für eine drehzahlvariable Windkraftanlage durch Gütevektoroptimierung*, Dissertation, Universität Gesamthochschule Kassel, 1996.
[47] Caselitz, P., Krüger, T. and Petschenka, J., Load Reduction by Multivariable Control of Wind Energy Converters–Simulations and Experiments, in European Union Wind Energy Conference, Göteborg, Sweden, 1996.
[48] Caselitz, P., Giebhardt, J. Krüger, T., Mevenkamp, M., Petschenka, J. and Reichardt, M., Neue Verfahren zur Regelung von Windkraftanlagen, in Forschungsverbund Sonnenenergie, Annual Conference, 1996.
[49] Krüger, T., *Regelungsverfahren für Windkraftanlagen zur Reduktion der mechanischen Belastung*, Dissertation, Universität Gesamthochschule Kassel, 1997.
[50] Kalman, R.E. and Bucy, R.S., New Results in Linear Filtering and Prediction Theory, *Transactions of the ASME, Series D, Journal of Basic Engineering*, 1961, 83, 95–108.
[51] Luenberger, D.G., An Introduction to Observers, *IEEE Transactions on Automatic Control*, 1971, 16(6), 596–602.
[52] Weinmann, A., *Regelungen – Analyse und technischer Entwurf*, Vol. 2: *Nichtlineare, abtastende und komplexe Systeme; modale, optimale und stochastische Verfahren*, Springer-Verlag, Vienna, 1984.
[53] Isermann, R., *Digitale Regelsysteme*, 2nd Revised and Enlarged Edn, Vol. 1, *Grundlagen, Deterministische Regelungen*, corrected reprint, Springer-Verlag, Berlin, Heidelberg, New York, London, Paris, Tokyo, 1988.
[54] Störzel, K., *Untersuchungen zur mechanischen Beanspruchung drehzahlvariabler Windkraftanlagen unter Mittelgebirgsbedingungen im WKA-Testfeld Vogelsberg*, LBF Report No. 7545, Frauenhofer Institute für Betriebsfestigkeit, Darmstadt, 1996.
[55] Changeux, J.-P., *Der neuronale Mensch*, Rowohlt-Verlag, Reinbeck, 1984.
[56] Eccles, J.C and Popper, K.R., *Das Ich und sein Gehirn*, R. Piper Verlag, Munich, 1989.
[57] Shatz, C.J., Das sich entwickelnde Gehirn, *Spektrum der Wissenschaft*, 1992, 11, 44–52.
[58] Kohonen, T., Self-Organized Formation of Topologically Correct Feature Maps, *Biological Cybernetics*, 1982, 43, 59–69.
[59] Kohonen, T., Analysis of a Simple Self-Organizing Process, *Biological Cybernetics*, 1982, 44, 135–40.
[60] Kohonen, T., *Self-Organization and Associative Memory*, Springer-Verlag, Berlin, 1984.
[61] Kohonen, T., Adaptive, Associative and Self-Organizing Functions in Neural Computing, *Applied Optics*, 1987, 26(23), 4910–8.
[62] Ritter, H. and Schulten, K., Topology Conserving Mappings for Learning Motor Tasks, in Tagungsband zur AIP Conference, Snowbird, Utah, 1986, pp. 376–80.
[63] Ritter, H. and Schulten, K., Extending Kohonen's Self-Organizing Mapping Algorithm to Learn Ballistic Movements, in *Neural Computers*, Springer-Verlag, Heidelberg, 1988, pp. 393–406.

[64] Ritter, H. and Schulten, K., Convergence Properties of Kohonen's Topology Conserving Maps: Fluctuations, Stability, and Dimension Selection, *Biological Cybernetics*, 1988, 60, 59–71.
[65] Ritter, H., *Selbstorganisierende neuronale Karten*, Dissertation, Munich University, 1988.
[66] Ritter, H., *et al.*, *Ein Gehirn für Roboter*, mc, No. 2, 1989, pp. 48–61.
[67] Ritter, H., *et al.*, Topology-Preserving Maps for Learning Visuomotor-Coordination, *Neural Networks*, 1989, 2, 159–68.
[68] Ritter, H., *et al.*, 3D-Neural-Network for Learning Visuomotor-Coordination of a Robot Arm, in Tagungsband zur IJCNN-89 Conference, Washington, Vol. II, 1989, pp. 351–6.
[69] Ritter, H., *et al.*, *Neuronale Netze: Eine Einführung in die Neuroinformatik selbstorganisierender Netzwerke*, Addison-Wesley, Bonn, 1990.
[70] Stoll, M., *Ein Schätzverfahren über den inneren Zustand geschlossener Bleiakkumulatoren*, VDI-Verlag GmbH, Düsseldorf, 1994.
[71] Watschke, H., *Untersuchung zum Einsatz neuronaler Netze in der Regelung und Betriebsführung von Windkraftanlagen*, Dissertation, Universität Gesamthochschule Kassel, 1993.
[72] Gockel, M., *Betriebsführung für einen getriebelosen Windkraftgenerator,* Thesis, Universität Gh Kassel, 1994.
[73] Appel, D., *Entwurf einer Betriebsführung für eine getriebelose Windkraftanlage*, Dissertation, Universität Gh Kassel, 1995.
[74] Germanischer Lloyd, *Vorschriften und Richtlinien.* IV – *Nichtmaritime Technik.* Part 1 – *Richtlinie für die Zertifizierung von Windkraftanlagen*, Selbstverlag des Germanischen Lloyd, Hamburg, 1999.
[75] Adzic, L., *Fernüberwachung von Windkraftanlagen*, Thesis, Universität Gh Kassel, 1997.
[76] Institut für Solare Energieversorgungstechnik (ISET), *Windenergie Report Deutschland, Wissentschaftliches Meß- und Evalierungsprogramm zum Breitentest '250 MW-Wind'. Jahresauswertungen 1990 bis 2001*, Eigendruck Kassel.
[77] Caselitz, P., Giebhardt, J. and Mevenkamp, M., Fehlerfrüherkennung in Windkraftanlagen, in Kongreßband Husum Wind '95, Husum, 1995, pp. 143–51.
[78] Caselitz, P., Giebhardt, J. and Mevenkamp, M., *Verwendung von WMEP-Onlinemessungen bei der Entwicklung eines Fehlerfrüherkennungssystems für Windkraftanlagen, Jahresauswertung 1994, Wissenschaftliches Meß- und Evaluierungsprogramm zum Breitentest '250 MW Wind' im Auftrag des Bundesministeriums für Forschung und Technologie*, ISET, Kassel, 1995, pp. 155–61.
[79] Caselitz, P., Giebhardt, J. and Mevenkamp, M., Development of a Fault Detection System for Wind Energy Converters, in European Union Wind Energy Conference and Exhibition, Göteborg, Sweden, 1996.
[80] Morbitzer, D., *Simulation und meßtechnische Untersuchung der Triebstrangdynamik von Windkraftanlagen*, Dissertation I, ISET, Universität Gesamthochschule Kassel, 1995.
[81] Osbahr, S., *Untersuchung von Parameterschätzverfahren für die Fehlerfrüherkennung in Windkraftanlagen*, Dissertation, ISET, Hannover University, 1995.
[82] Eibach, T., *Untersuchung von Verfahren der Lager- und Getriebeüberwachung für die Fehlerfrüherkennung in Windkraftanlagen*, Dissertation I, ISET, Universität Gesamthochschule Kassel, 1995.
[83] Adam, H., *Implementation und Untersuchung Künstlicher Neuronaler Netze zur Fehlerfrüherkennung in Windkraftanlagen*, Thesis, ISET, Universität Gesamthochschule Kassel, 1995.
[84] Hobein, A., *Entwicklung eines Hardware-Moduls zur analogen Leistungsberechnung für ein PC-gestütztes Meßdatenerfassungssystem*, Thesis, ISET, Universität Gesamthochschule Kassel, 1995.
[85] Werner, U., *Entwicklung eines Hardware-Moduls zur Drehzahlmessung für ein PC-gestütztes Meßdatenerfassungssystem*, Thesis, ISET, Universität Gesamthochschule Kassel, 1995.

Chapter 6

[1] Haas, O., Heier, S., Kleinkauf, W. and Strauß, P., Zukunftsaspekte regenerativer Energien und die Rolle der Photovoltaik. Fortschrittliche Energiewandlung und -anwendung. Schwerpunkt: Dezentrale Energiesysteme, in VDI Conference, Bochum, 13–14 March 2001, VDI reports 1594, pp. 3–16, ISBN 3-18-091594-3.
[2] Heier, S., Situation und Perspektiven der Windenergie. Fortschrittliche Energiewandlung und -anwendung. Schwerpunkt: Dezentrale Energiesysteme, in VDI Conference, Bochum, 13–14 March 2001, VDI reports 1594, pp. 523–37, ISBN 3-18-091594-3.
[3] Heier, S., *Nutzung der Windenergie*, 4th Edn, TV Rheinland, Cologne, 2000.
[4] Berlipp, A., *Standort- und anlagenspezifische Energieerträge von Windkraftanlagen*, Thesis, Universität Gh Kasel, 1996.

[5] Döpfer, R. and Otto, K., Untersuchung eines Mittelgebirgsstandortes im Hinblick auf die Eignung zur Windenergienutzung, in *Abschlußarbeit Energie und Umwelt*, Kassel University, 1994.

[6] Durstewitz, M., Enlin, C., Hahn, B., Hoppe-Kilpper, M. and Rohrig, K., Ausgewählte Betriebserfahrungen mit Windkraftanlagen im Binnenland, in *Wind Energie Aktuell*, Book 10, January 1994, pp. 22–6.

[7] Institut für Solare Energieversorgungstechnik (ISET), *Wissenschaftliches Meß- und Evaluierungsprogramm zum Breitentest '250 MW-Wind', Jahresauswertungen 1990–2001*, Eigendruck, Kassel.

[8] Mortensen, N.G., Landberg, L., Troen, I. and Petersen, E.L., *Wind Atlas Analysis Programme (WASP)*, Risø National Laboratory, Roskilde, Denmark.

[9] Troen, I. and Petersen, E.L., *European Wind Atlas*, Risø National Laboratory, Roskilde, Denmark.

[10] Kleinkauf, W., Melíß, M., Molly, J.-P., *et al.*, *Energiequellen für Morgen?* Part III: *Nutzung der Windenergie*, BMFT-Study, Umschau, Frankfurt, 1976.

[11] Windheim, R., *Nutzung der Windenergie*, KFA Jülich, 1980.

[12] Selzer, H., *Solar Energy R & D in the European Community*, D. Reidel, 1986.

[13] Bierbrauer, H. von, *et al.*, *Darstellung realistischer Regionen für die Errichtung insbesondere großer Windenergieanlagen in der BRD*, BMFT-FB-T 85-053, Lahmeyer International, Frankfurt.

[14] Fichtner Development Engineering, *Abschätzung des wirtschaftlichen Potentials der Windenergienutzung in Deutschland und des bis 2000/2005 zu erwartenden Realisierungsgrades sowie der Auswirkung von Fördermaßnahmen*, BMFT Bonn/Forschungszentrum Jülich GmbH, Stuttgart, 1991.

[15] Consulectra, *Wind Power Penetration Study of the European Commission*, Federal Republic of Germany, 1991.

[16] European Wind Energy Association, *Time for Action/Wind Energy in Europe*, CEC DG XVII, October 1991.

[17] DEWI, *Feststellung geeigneter Flächen als Grundlage für die Standortsicherung von Windparks im nördlichen Niedersachsen*, German Wind Energy Institute Commissioned by the Lower Saxony Ministry for the Environment, Wilhelmshaven, January 1993.

[18] Glocker, S., Richter, B. and Schwabe, J., Methoden und Ergebnisse bei der Ermittlung von Windenergiepotentialen und Flächen in Mecklenburg-Vorpommern, Hamburg und Schleswig-Holstein, in German Wind Energy Conference 92, Wilhelmshaven, 1992, pp. 93–9.

[19] Durstewitz, M., Heier, S., Hoppe-Kilpper, M. and Kleinkauf, W., Entwicklung der Windenergietechik in Deutschland, in *Plenarvortrag Windenergie*, Internationales Sonnenforum, Cologne, 26–30 July 1998.

[20] Durstewitz, M., Heier, S. and Hoppe-Kilpper, M., Ausbaustrategien für die Windenergienutzung in Deutschland, in *Forschungsverbund Sonnenenergie-Jahrestagung*, Bonn, Forschungsverbund Sonnenenergie Themen 98/99, Cologne, 8–9 September 1998, pp. 40–5.

[21] Heier, S., Kleinkauf, W., Durstewitz, M. and Hoppe-Kilpper, M., Anwendung der Windenergie in Deutschland, in VDI-Tagung Regenerative Energien (Hungary), Budapest, Hungary, 14–16 September 2000.

[22] Östergrad, C., *Randbedingungen zur Seeaufstellung großer Windkraftanlagen*, Report STB No. 924, Germanischer Lloyd, Hamburg, 1982.

[23] Pernpeitner, R., Offshore Siting of Large Wind Energy Converter Systems in the North Sea and Baltic Sea Regions, *Wind Engineering*, 1985, 203–313.

[24] Leutz, R., Akisawa, A., Kashiwagi, T., Ackermann, T. and Suzuki, A., World-Wide Offshore Wind Energy Potential, in Second International Workshop on *Transmission Networks for Offshore Wind Farms*, Stockholm, Sweden, 29–30 March 2001.

[25] Smith, K. and Hagermann, G., The Potential for Offshore Wind Development in the United States, in Second International Workshop on *Transmission Networks for Offshore Wind Farms*, Stockholm, Sweden, 29–30 March 2001.

[26] Matthies, H. G. and Garrad, A. D., *Study of the Offshore Wind Energy in the EC, JOULE I (JOUR 0072)*, Verlag Natürliche Energie, Brekensdorf, 1985.

[27] *Windenergienutzung auf See*, Position paper by the Bundesministeriums für Umwelt, Naturschutz und Reaktorsicherheit on the use of wind power in the offshore area, 2001.

[28] Henderson, A.R. and Camp, T.R., Hydrodynamic Loading of Offshore Wind Turbines, in European Wind Energy Conference, Copenhagen, Denmark, 2–6 July 2001, pp. 561–6.

[29] Sorensen, H.C., Hansen, J. and Volund, P., Experience from the Establishment of Middelgrunden 40MW Offshore Wind Farm, in European Wind Energy Conference, Copenhagen, Denmark, 2–6 July 2001, pp. 541–7.

[30] van Bussel, G.J.W. and Zaaijier, M.B., DOWEC Concepts Study, Reliability, Availability and Maintenance Aspects, in European Wind Energy Conference, Copenhagen, Denmark, 2–6 July 2001, pp. 557–60.

[31] Stalin, T., Utgrunden Offshore Wind Energy Project, Sweden, in Second International Workshop on *Transmission Networks for Offshore Wind Farms*, Stockholm, Sweden, 29–30 March 2001.

[32] Kühn, M. and Sievers, T., Utgrunden Offshore Windfarm – First Results of Design Verification by Measurements, in Second International Workshop on *Transmission Networks for Offshore Wind Farms*, Stockholm, Sweden, 29–30 March 2001.
[33] Loman, G., The Offshore Wind Farm at Lillgrund, Application and Environmental Impact Assessment, in Second International Workshop on *Transmission Networks for Offshore Wind Farms*, Stockholm, Sweden, 29–30 March 2001.
[34] Rehfeldt, K., Gerdes, G.J. and Schreiber, M., *Weiterer Ausbau der Windenergienutzung im Hinblick auf den Klimaschutz – Teil 1*, Bundesumweltministerium/DEWI, 2001, p. 104.
[35] EU Commission, *Energy for the Future: Renewable Sources of Energy*, White Paper for a Community Strategy and Action Plan, 26 November 1997.
[36] Eischen, T., *Wirtschaftliche und technische Behandlung von Eigenerzeugungsanlagen in Europa*, Thesis, Kassel University, 1996.
[37] European Union, Directives 97/11/EC and 85/337/EEC on environmental compatibility testing.
[38] Amtsblatt für Schleswig-Holstein, 2001, No. 16/17, p. 216ff.
[39] Bundesimmissionsschutzgesetz BImSchG 1974, 1990.
[40] DIN 45635, VDI 2571.
[41] DIN ISO 9613-2, VDI 2720.
[42] DIN 45641, DIN 45645, TA Lärm, VDI 2058.
[43] VDI 2570, VDI 3720, DIN 4109, VDI 271.
[44] Technische Anleitung zum Schutz gegen Lärm, 1998.
[45] Baunutzungsverordnung BauNVO, 1990.
[46] Fördergesellschaft Windenergie e.V., *Technische Richtlinien zur Bestimmung der Leistungskurve, der Schallemissionswerte und der elektrischen Eigenschaften von Windenergieanlagen*, 1 April 1998.
[47] EMD Deutschland GmbH, *WindPRO, Modul Decibel*, Kassel, 2002.
[48] Ising, H., Markert, B., Shenoda, F. and Schwarze, C., Infraschallwirkungen auf den Menschen, in *Bundesministerium für Forschung und Technologie*, VDI Verlag, 1982.
[49] Buhmann, A., *Keine Gefahr durch Infraschall*, Neue Energie 1/98.
[50] EMD Deutschland GmbH, *WindPRO, Modul SHADOW*, Kassel, 2002.
[51] EMD Deutschland GmbH, *WindPRO, Modul ZKI*, Kassel, 2002.
[52] EMD Deutschland GmbH, *WindPRO, Modul VISUAL*, Kassel, 2002.
[53] Baugesetzbuch (BauGB), Bekanntmachung vom 8.12.86 (BGBl. I p. 2253), Änderung durch Artikel 24 Jahressteuergesetz 1997 vom 20.12.96 (BGBl. I p. 2049).
[54] Baunutzungsverordnung (BauNVO), Bekanntmachung vom 23.1.90 (BGBl. I p. 132), Änderung durch Artikel 3 Investitionserleichterungs- und Wohnbauland-Gesetz vom 22.4.94 (BGBl. I p. 466).
[55] Änderungsgesetz vom 30.07.1996 (BGBl. I p. 1189).

Index